Clinical Applications of Biomolecules in Disease Diagnosis

Ram Lakhan Singh • Pankaj Singh •
Neelam Pathak
Editors

Clinical Applications of Biomolecules in Disease Diagnosis

A Comprehensive Guide to Biochemistry and Metabolism

Springer

Editors
Ram Lakhan Singh
Department of Biochemistry
Dr. Rammanohar Lohia Avadh University
Ayodhya, Uttar Pradesh, India

Pankaj Singh
Department of Biotechnology
Dr. Rammanohar Lohia Avadh University
Ayodhya, Uttar Pradesh, India

Neelam Pathak
Department of Biochemistry
Dr. Rammanohar Lohia Avadh University
Ayodhya, Uttar Pradesh, India

ISBN 978-981-97-4722-1 ISBN 978-981-97-4723-8 (eBook)
https://doi.org/10.1007/978-981-97-4723-8

This Springer imprint is published by the registered company Springer Nature Singapore Pte Ltd.
The registered company address is: 152 Beach Road, #21-01/04 Gateway East, Singapore 189721, Singapore

To
BHAGWAN SHRI RAM JI
Whose divine inspiration guided us to this valuable contribution

Preface

Metabolism is the sum of biochemical processes of more than 8700 reactions in living organisms. Anomalies in these metabolic processes develop a number of metabolic disorders either related to degradative pathways of specific molecules, i.e., catabolism of molecules, or due to error in synthetic reactions of molecules, i.e., anabolism. Changes in the concentrations and fluxes of specific endogenous metabolites in biological system provide a sign of perturbations in specific cellular pathways, which leads to development of diseases. Recent data on prevalence of diseases due to metabolic disorders predict that one billion people may suffer from consequences of metabolic disorders on a global scale till 2030, and this will lead to an increased number of persons suffering from noncommunicable diseases. Several risk factors like lack of exercise, dietary intake, and other lifestyle behaviors play important role in the development of metabolism-related disorders. Rapid unfolding of clinical biochemistry resulted in addition of outstanding features in the qualitative and quantitative clinical analysis of disease during the past 20–30 years due to improvement of analytical methodology for individual biomolecules such as protein, carbohydrates, lipids, hormones, etc. in blood plasma, spinal fluid, urine, and other body fluids. The quantity of these substances in the blood or other fluids builds the basis for a diagnosis of the disease or treatment. Disease diagnosis is a complex, collaborative approach that involves clinical reasoning and information gathering to determine a patient's health status. Metabolomics, an outstanding emerging field, supports the clinical investigation of patients in which large numbers of metabolites from body fluids or tissues are detected quantitatively that can be used in early diagnosis of disease, treatment, and in understanding the pathogenesis. Metabolomics has now become a powerful tool in the detection of metabolic disorders. Recent advancement in technologies such as next-generation sequencing (NGS), nuclear magnetic resonance (NMR) spectroscopy, tandem mass spectrometry (MS/MS), etc. has expanded our knowledge of metabolic disorders, in general, and inborn error of metabolism, in particular. The ability of metabolomics to detect large numbers of metabolites at a time provides an efficient tool for study of altered chemistry of macromolecules for disease diagnosis, prognosis, and therapeutic development.

This book illustrates the importance and significance of the biomolecules such as carbohydrates, proteins, lipids, nucleic acids, vitamins, enzymes, and hormones and their role in disease onset due to error in metabolism. It is presented in a sharply focused, logical format and well-illustrated discussions to ensure a strong grasp about concepts of clinical biochemistry for disease development. This book also explains about metabolomics, which provide improved knowledge of genomics, proteomics, and interaction of these two fields in disease diagnosis, to understand the mechanism of disease onset, prognosis, and therapeutic interventions. This book also focuses on clinical aspects of nutritional and hormonal disorders, free radical diseases, oncogenesis, and inherited metabolic disorders. Each chapter begins with a short and clear explanation on the essential physiology and biochemistry of biomolecules, followed by in-depth descriptions of changes and appearance of diseases due to alterations in its metabolism. In summary, this book presents a balanced, comprehensive knowledge of all major biomolecules and its clinical roles in the diagnosis of diseases and fulfills the literature requirement of students studying medical biochemistry and biotechnology, clinical research, medical physiology, and nursing in healthcare sector.

Key Features

This book:

- Identifies and explores the role of biomolecules in diagnosing and managing specific metabolic diseases or disorders,
- Discusses use of modern automated high-throughput cost-effective technologies such as chromatography, western blotting, ELISA, 2D-electrophoresis, ion mobility mass spectrometry (IM-MS), and tandem mass spectrometry (MS/MS), nucleic acid sequencing methods, nuclear magnetic resonance, etc. for the detection of metabolic abnormalities, and
- Focuses on the mechanisms involved in diseases development due to change in metabolism of biomolecules.

Organization of the Book

This book *Clinical Applications of Biomolecules in Disease Diagnosis: A Comprehensive Guide to Biochemistry and Metabolism* consists of 16 chapters that focus on the role of biomolecules in specific metabolic diseases or disorders and modern approaches and strategies involved in their diagnosis.

Chapter 1: **Introduction to Clinical Applications of Biomolecules in Disease Diagnosis** covers the brief introduction, clinically important biomarkers and their classification, and applications of different biomolecules in the diagnosis of diseases or related disorders.

Chapter 2: **Carbohydrate Metabolism in Health and Diseases** elucidates carbohydrate metabolic pathways, impaired glucose metabolism in metabolic diseases, dysregulation of glucose uptake and utilization, consequences of altered carbohydrate metabolism on metabolic pathways, glycoxidation and lipo-oxidation pathways, and receptor-mediated metabolic diseases.

Chapter 3: **Protein Metabolism and Its Profiling for the Diagnosis of Metabolic Disorders** focuses on the introduction to amino acid metabolism, aromatic amino acid disorders such as phenylketonuria, alkaptonuria, and tyrosinemia, branched chain and sulfur-containing amino acid disorders, urea cycle disorders, protein deficiency disorders, and diagnostic techniques for amino acid metabolic disorders.

Chapter 4: **Disorders of Lipid Metabolism** describes normal lipid metabolism and primary lipid metabolic disorders such as hypertriglyceridemia, hypercholesterolemia, hyperalphalipoproteinemia, and hypolipoproteinemia and secondary lipid metabolic disorders such as cardiovascular diseases, obesity, diabetes, and other diseases.

Chapter 5: **Nucleic Acid Metabolism and Disorders** discusses the processes of purine and pyrimidine nucleotide biosynthesis and disorders and diagnosis of inborn errors of purine and pyrimidine metabolism.

Chapter 6: **Vitamins and Their Impact on Human Health** focuses on structure and physiochemical properties of fat and water soluble vitamins, dietary sources and its bioavailability, different forms of vitamins, recommended dietary intake and diseases due to their deficiency, and roles of vitamins in human health.

Chapter 7: **Electrolytes and Acid–Base Disorders** describes the electrolytes and buffers, role of HCO_3^-/CO_2 buffer in acid–base homeostasis, role of major ions in the human body, electrolytes as osmotic regulators, maintenance of plasma osmolarity, major organs involved in the acid–base balance regulation, renal regulation of acid–base balance, metabolic acidosis and alkalosis, electrolyte disorders, and factors causing electrolyte imbalance.

Chapter 8: **Metabolism of Macro-elements (Calcium, Magnesium, Sodium, Potassium, Chloride, and Phosphorus) and Associated Disorders** covers the metabolic regulation, functions, and disorders of calcium such as hypocalcemia, hypercalcemia, hyperparathyroidism, nutritional rickets, osteomalacia, and osteoporosis; metabolic regulation, functions, and disorders of phosphate metabolism like hypophosphatemia, hyper- and hypoparathyroidism, oncogenic osteomalacia, etc.; metabolic regulation, functions, and disorders of magnesium sodium, potassium, and chloride metabolism.

Chapter 9: **Metabolism of Essential Trace Elements and Associated Disorders** focuses on the metabolism of essential trace elements, i.e., iron, iodine, zinc, fluoride, copper, manganese, selenium, magnesium, cobalt, and chromium and related diseases due to its alteration and improves our understanding of the connection between trace elements and how diseases are affected by them.

Chapter 10: **Enzymes and Their Clinical Applications** highlights the properties of enzymes, enzymes for estimation of macromolecules, enzymes used in clinical diagnosis of bone diseases, autoimmune disorders, cancer, diabetes, Gaucher's and liver diseases, pancreatitis, myocardial infarction, and renal disorders; use of enzymes as therapeutics; factors affecting enzyme activities; and advantages and disadvantages of use of enzymes in clinical diagnosis.

Chapter 11: **Hormonal Manifestations of Endocrine Diseases** deals with the overview of endocrine organs, hormones, and receptors, hormonal manifestation of endocrine diseases related to hypothalamus, pituitary gland, thyroid gland, adrenal gland, pancreas, gonads, and placenta and pregnancy-associated endocrine diseases, and hormonal manifestation of secondary endocrine organs.

Chapter 12: **Biomolecular Components of Blood and Their Role in Health and Diseases** covers the blood and its characteristics, components of blood, functions of blood, blood and its related diseases such as development of anemia, anemia of chronic disease, thalassemia, multiple myeloma, thrombocytopenia, disseminated intravascular coagulation hemophilia, myeloproliferative neoplasms, leukemia, and lymphoma.

Chapter 13: **Free Radicals and Their Relation to Diseases and Protection Against Them** deals with free radical types, mechanism of free radical induced DNA damage, mechanism of lipid peroxidation, free radicals and lipid peroxidation, free radicals and protein damage and consequences of protein damage, impact of free radicals and its relation to various metabolic disorders, scavengers of free radicals, and strategies to mitigate free radical induced damage.

Chapter 14: **Clinical Biochemistry of Cancer** covers the current understanding of cancer development, characteristic-benign tumor or malignant tumor, types of cancer, adenocarcinoma, basal cell carcinoma, squamous cell carcinoma, sarcoma, causes of sarcoma, cancer biomarkers, oncogene and tumor suppressor genes, factors/environmental contaminants, mechanisms of action, statistics: international and national, multiple-omics approaches in cancer research, detection of cancer, and potential agents for cancer prevention and treatment.

Chapter 15: **Outbreak of Deadly Coronavirus (COVID-19) Pandemic: Diagnosis and Management** highlights the outbreak of COVID-19, structure of SARS-CoV-2, transmission modes of SARS-CoV-2 (COVID19), clinical manifestation and diagnosis, treatments and management of COVID-19, antiviral agents, vaccines, herbal remedies, and lessons from COVID-19 pandemic.

Chapter 16: **Techniques for the Identification and Characterization of Biomolecules** covers the extraction, purification, and detection methods for carbohydrate proteins, antigens, antibodies, enzymes, and lipids and identification and characterization of nucleic acids and low- and high-throughput methods to sequence nucleic acids.

Ayodhya, Uttar Pradesh, India — Ram Lakhan Singh
Ayodhya, Uttar Pradesh, India — Pankaj Singh
Ayodhya, Uttar Pradesh, India — Neelam Pathak

Acknowledgements

We are grateful to all the dedicated contributing authors who assisted materially in the preparation of this book with their tremendous efforts. We are extremely thankful to the editorial and production staffs of Springer Nature for their kind supports and excellent work. Special thanks are due to Ms. Aakanksha Tyagi, Senior Editor, Life Sciences, and Dr. Bhavik Sawhney, Senior Editor, Biomedicine, with whom we started the project at proposal level and got constant and invaluable supports throughout the project. We are also very thankful to Ms. Aishwarya Thyagarajan, Production Editor, and her entire team for bringing this book to its final shape. Last but not least, we wish to extend appreciation to our family and friends for providing strong moral support and encouragement.

Contents

Editors and Contributors

About the Editors

Ram Lakhan Singh is presently holding the position of Professor of Biochemistry at Dr. Rammanohar Lohia Avadh University, Ayodhya, UP. He has shaped Nilamber-Pitamber University, Medininagar, a state University of Jharkhand, as its Vice-Chancellor for a tenure of 3 years (June 2020–June 2023). He is having more than 36 years of teaching, research, and administrative experience of which 22 years as Professor of Biochemistry. His main areas of research are clinical/nutraceutical biochemistry, environmental biotechnology, and toxicology. He has published 93 research papers in national and international journals of repute and guided 26 students for their Ph.D. degree. He has edited 11 books and contributed 23 chapters in various books published by international publishers. He is on the panel of experts in various academic bodies and selection committees of various universities and funding agencies. He has been honored as Best Teacher by International Association of Lions Clubs in 1999. He has been awarded IUTOX Senior Fellowship by International Union of Toxicology during XI International Congress of Toxicology at Montreal, Canada, in 2007. He has been conferred with "Shikshak Shree Samman" by Government of Uttar Pradesh in 2012 and admitted to the Fellowships of the Society of Toxicology, India, in 2011 and Academy of Environmental Biology, India, in 2015. He has been conferred with "Lifetime Achievement Award" by Society of Life Sciences, India (2018), and Astha Foundation, India (2022).

Pankaj Singh is presently working as Assistant Professor in the Department of Biotechnology, Dr. Rammanohar Lohia Avadh University, Ayodhya, since 2019. He has more than 9 years teaching experiences at graduate and postgraduate levels. He completed his master's degree in biochemistry from Dr. Rammanohar Lohia Avadh University, Ayodhya, in 2007 and was awarded Research Fellowship under the UGC program of "Research Fellowship in Sciences for Meritorious Students" from March 2008 to March 2013. He joined the Department of Biochemistry at Dr. Rammanohar Lohia Avadh University, Ayodhya, as Ph.D. scholar in 2008 and worked extensively on antioxidant and hepatoprotective activities of medicinal plants and awarded Ph.D. degree in 2014. His research interest is in nutraceutical/clinical biochemistry, pharmacology, and toxicology. He has published 38 research papers in peer-reviewed journals and contributed 11 chapters in various books published by international publishers. He has participated in various national and international scientific conferences and symposia and presented 12 research papers. He has also participated in four national training programs related to molecular analysis, identification of phytochemicals, and faculty development program.

Neelam Pathak is currently working as a Professor at the Department of Biochemistry and Dean Students' Welfare at Dr. Rammanohar Lohia Avadh University, Ayodhya, India. Earlier she has worked as a Professor at the Department of Biosciences, Integral University, Lucknow, India. She has a master's in biochemistry from ANDUAT, Ayodhya India, and a Ph.D. in biochemistry from the University of Lucknow, Lucknow, India. After her doctoral degree, she worked as Research Associate at the Plant Gene Expression Lab, NBRI, Lucknow. Thereafter, she has also worked as a Faculty Research Associate at the University of Maryland, College Park, USA. She has received several fellowships (CSIR/UGC-JRF, SRF during her Ph.D., and CSIR-RA Award and Pool Officer Fellowship at post-Ph.D. level). She has also been the recipient of the Young Scientist Fast Track Award from the DST Govt. of India. She has ~20 years of teaching experience in various undergraduate and postgraduate life sciences courses. Moreover, she has ~30 years of research experience in plant biochemistry and molecular biology. She is a recipient of several research schemes sponsored by various funding agencies. Her research focuses on fruit ripening, medicinal plants, plant biodiversity, and plant growth-promoting microorganisms. She has co-authored a book on *Genetic Engineering* which is one of the bestsellers in the life sciences category of Oxford University Press. She has published 90 research papers in international journals of high repute. She has also authored 13 book chapters. She has presented papers at various national and international conferences and received several best and eminent scientist awards. She is a fellow of ISHRD and has been conferred with various prestigious awards.

Contributors

Mohammad Kaleem Ahmad Department of Biochemistry, King George Medical University, Lucknow, Uttar Pradesh, India

Parvej Ahmad IIRC-5, Clinical Biochemistry and Natural Product Research Lab, Department of Biosciences, Integral University, Lucknow, U.P., India

Saheem Ahmad Department of Medical Laboratory Sciences, College of Applied Medical Sciences, University of Hail, Hail, Saudi Arabia

Jalaluddin Mohammad Ashraf Department of Biochemistry, Faculty of Medicine, Samarkand State Medical University, Samarkand, Uzbekistan

Preeti Bajpai Department of Zoology, Mahatma Gandhi Central University, Motihari, Bihar, India

Monisha Banerjee Molecular and Human Genetics Laboratory, Department of Zoology, University of Lucknow, Lucknow, Uttar Pradesh, India

Institute of Advanced Molecular Genetics and Infectious Diseases (IAMGID), ONGC Centre for Advanced Studies (OCAS), University of Lucknow, Lucknow, Uttar Pradesh, India

Farhad Bano Special Centre for Molecular Medicine, Jawaharlal Nehru University, New Delhi, India

Vikas Chandra Department of Biotechnology, Guru Ghasidas Vishwavidyalaya, Bilaspur, Chhattisgarh, India

C. P. Chaturvedi Stem Cell Research Centre, Department of Hematology, Sanjay Gandhi Post Graduate Institute of Medical Sciences, Lucknow, Uttar Pradesh, India

Mohammad Farhan Department of Integrative Biology and Pharmacology, McGovern Medical School University of Texas Health Science Center at Houston, Houston, TX, USA

Minal Garg Department of Biochemistry, University of Lucknow, Lucknow, Uttar Pradesh, India

Prachi Garg Department of Biochemistry, Dr. Rammanohar Lohia Avadh University, Ayodhya, India

Sachin Gupta Department of ENT, SMGS Hospital Jammu, Jammu, Jammu & Kashmir, India

Safia Habib Department of Biochemistry, J.N. Medical College, Aligarh Muslim University, Aligarh, India

Harit Jha Department of Biotechnology, Guru Ghasidas Vishwavidyalaya, Bilaspur, C.G., India

Kirtanjot Kaur University Centre for Research and Development, Chandigarh University, Mohali, Punjab, India

Hafizurrahman Khan Department of Zoology, Maharani Janki Kunwar College, Bettiah, Bihar, India

Hamda Khan Department of Biochemistry, J.N. Medical College, Aligarh Muslim University, Aligarh, India

M. Salman Khan IIRC-5, Clinical Biochemistry and Natural Product Research Lab, Department of Biosciences, Integral University, Lucknow, U.P., India

Sarat Kumar Kottarath Department of Experimental Therapeutics, The University of Texas MD Anderson Cancer Center, Houston, TX, USA

Dharmendra Kumar Department of Biotechnology, Guru Ghasidas Vishwavidyalaya, Bilaspur, C.G., India

Sandeep Kumar Department of Medicine, University of Oklahoma Health Sciences Center, Oklahoma City, OK, USA

Osaid Masood Molecular and Human Genetics Laboratory, Department of Zoology, University of Lucknow, Lucknow, Uttar Pradesh, India

Shireen Masood Molecular and Human Genetics Laboratory, Department of Zoology, University of Lucknow, Lucknow, Uttar Pradesh, India

Neha Mehrotra FEST Division, CSIR-Indian Institute of Toxicology Research, Lucknow, Uttar Pradesh, India

Manish Kumar Mishra Centre for Molecular Biology, Central University of Jammu, Jammu, Jammu & Kashmir, India

Moinuddin Department of Biochemistry, J.N. Medical College, Aligarh Muslim University, Aligarh, India

Syed Faiz Mujtaba Department of Zoology, Faculty of Science, Shia Post Graduate College, Lucknow, Uttar Pradesh, India

Niharika Department of Biochemistry, University of Lucknow, Lucknow, U.P., India

Ramendra Pati Pandey School of Health Sciences and Technology (SOHST), UPES, Dehradun, Uttarakhand, India

Shivani Pandey Department of Biochemistry, King George's Medical University, Lucknow, U.P., India

Neelam Pathak Department of Biochemistry, Dr. Rammanohar Lohia Avadh University, Ayodhya, India

Zeeshan Rafi Department of Bioengineering, Integral University, Lucknow, Uttar Pradesh, India

Fareha Rayeen Department of Biochemistry, Dr. Rammanohar Lohia Avadh University, Ayodhya, India

Kehkashan Rehman IIRC-5, Clinical Biochemistry and Natural Product Research Lab, Department of Biosciences, Integral University, Lucknow, U.P., India

Shelly Sehgal Centre for Molecular Biology, Central University of Jammu, Jammu, Jammu & Kashmir, India

Sameen Shafi Department of Zoology, Mahatma Gandhi Central University, Motihari, Bihar, India

Uzma Shahab Department of Biochemistry, King George Medical University, Lucknow, U.P., India

Arunim Shah Stem Cell Research Centre, Department of Hematology, Sanjay Gandhi Post Graduate Institute of Medical Sciences, Lucknow, Uttar Pradesh, India

Shalini National Forensic Science University, Gandhinagar, Gujarat, India

Sapna Sharma Faculty of Biosciences, Institute of Biosciences and Technology, Shri Ramswaroop Memorial University, Lucknow, Uttar Pradesh, India

Uttar Pradesh State Institute of Forensic Science (UPSIFS), Lucknow, Uttar Pradesh, India

Vinod Pravin Sharma CSIR-Indian Institute of Toxicology Research, CSIR-IITR, Lucknow, U.P., India

Dhananjay Shukla Department of Biotechnology, Guru Ghasidas Vishwavidyalaya, Bilaspur, C.G., India

Shikha Shukla Department of Zoology, RRPG College, Amethi, India

Divyanshu Singh Department of Biotechnology, Guru Ghasidas Vishwavidyalaya, Bilaspur, C.G., India

Jyotsana Singh Department of Hematopoietic Biology and Malignancy, University of Texas, UT MD Anderson Cancer Center, Houston, TX, USA

Pankaj Singh Department of Biotechnology, Dr. Rammanohar Lohia Avadh University, Ayodhya, India

Pradeep Kumar Singh Department of Biochemistry, Dr. Rammanohar Lohia Avadh University, Ayodhya, India

Rajat Pratap Singh Department of Biotechnology, Guru Ghasidas Vishwavidyalaya, Bilaspur, C.G., India

Ram Lakhan Singh Department of Biochemistry, Dr. Rammanohar Lohia Avadh University, Ayodhya, India

Sarthak Srivastava Stem Cell Research Centre, Department of Hematology, Sanjay Gandhi Post Graduate Institute of Medical Sciences, Lucknow, Uttar Pradesh, India

Yogesh Srivastava Department of Genetics, The University of Texas MD Anderson Cancer Center, Houston, TX, USA

Ankit Tiwari Department of Biotechnology, Guru Ghasidas Vishwavidyalaya, Bilaspur, C.G., India

Neeraj Tiwari Faculty of Biosciences, Institute of Biosciences and Technology, Shri Ramswaroop Memorial University, Lucknow, Uttar Pradesh, India

Amit Kumar Tripathi Department of Microbiology, Immunology and Genetics, The University of North Texas Health Science Center, Fort Worth, TX, USA

Manikant Tripathi Department of Biotechnology, Dr. Rammanohar Lohia Avadh University, Ayodhya, India

Prabhanshu Tripathi Food Drug and Chemical Toxicology Division, Council of Scientific and Industrial Research (CSIR)-Indian Institute of Toxicology Research, Lucknow, Uttar Pradesh, India

Saba Ubaid Department of Biochemistry, King George's Medical University, Lucknow, U.P., India

Naveen Kumar Vishvakarma Department of Biotechnology, Guru Ghasidas Vishwavidyalaya, Bilaspur, C.G., India

Mohd. Waiz IIRC-5, Clinical Biochemistry and Natural Product Research Lab, Department of Biosciences, Integral University, Lucknow, U.P., India

Jyoti Yadav Department of Medicine, University of Oklahoma Health Sciences Center, Oklahoma City, OK, USA

Shivmurat Yadav Department of Medicine, University of Oklahoma Health Sciences Center, Oklahoma City, OK, USA

Introduction to Clinical Applications of Biomolecules in Disease Diagnosis

1

Pankaj Singh, Fareha Rayeen, Neelam Pathak, and Ram Lakhan Singh

Abstract

Early disease diagnosis has an important role in the prognosis and treatment of acute as well as chronic diseases. The identification and quantification of clinically important biomolecules such as enzymes, proteins, nucleic acids, antibodies, and hormones are developed as diagnostics tools and are necessary to find out the condition of the body's physiological and biological functions. Proteins have a large number of biological functions especially they act as biological catalysts, i.e., enzymes. Hormones and enzymes are responsible for many complex reactions and regulate the entire metabolism directly or indirectly. Thus, they are involved in the degradation and synthesis of biomolecules, nutrients, etc. Molecular diagnostics has transformed tremendously the clinical diagnosis and treatment of many diseases. It utilizes advanced molecular technologies which concentrate on the detection of pathogenic organisms at the genome level and led to the development of genomics. The structural and functional genomics approaches involved the exploration of disease-related genes and the identification of their structural alterations. Diagnostic applications of structural genomics are to find out genetic alterations and to prepare a database of disease-causing genes whereas functional genomics explores the expression genetics which utilizes PCR-based technologies, next-generation sequencing with bioinformatics, and computational biology. The quantification of these clinically important bio-

Authors Pankaj Singh and Fareha Rayeen have contributed equally to this work and share the first authorship.

P. Singh (✉)
Department of Biotechnology, Dr. Rammanohar Lohia Avadh University, Ayodhya, India
e-mail: pankaj@rmlau.ac.in

F. Rayeen · N. Pathak · R. L. Singh
Department of Biochemistry, Dr. Rammanohar Lohia Avadh University, Ayodhya, India

R. L. Singh et al. (eds.), *Clinical Applications of Biomolecules in Disease Diagnosis*, https://doi.org/10.1007/978-981-97-4723-8_1

molecules assists in knowing the type of damaged tissues and the extent of damage and finally helps to monitor the course of the disease which can also be used as a therapeutic.

Keywords

Biomolecules · Enzymes · Proteins · Molecular diagnostics · Genomics · Therapeutic

1.1 Introduction

Living cells exist in extraordinarily complex systems that are made up of a multitude of different macromolecules such as proteins, nucleic acids, lipids, and carbohydrates. They collaborate to sustain a variety of biological processes, such as DNA replication, transcription, translation, enzymatic reactions, cell signaling, etc. These interactions are referred to as biomolecular recognition. Recent research efforts in the utilization of biomolecules as clinical markers and their detection have mostly focused on two aspects: specificity and affinity. Specificity permits biological molecules to do their task without competing with other processes. Affinity confirms that molecular interactions are stable enough for biological activities to take place. Many factors, including the shape and size of the binding site, the charge state, and the type of chemical interactions that cause diseases, can influence specificity and affinity. These biomolecules are analytical agents that can be tracked and detected to give crucial information for disease diagnosis, and therapeutic efficacy (Huang et al. 2017; Prudent et al. 2020). Precision screening is aimed at modifying the treatment based on the characteristics of the disease and various reactions to the treatments they receive, because it is understood that all treatments are personalized (Johnson et al. 2021). Thousands of findings on the role of biomolecules in reactive oxygen species (ROS) activity that can cause cellular and tissue injury have been identified (Gupta et al. 2023; Singh et al. 2023). Oxidative stress is a major contributor to numerous diseases, including atherosclerosis, chronic obstructive pulmonary disease, Alzheimer's disease, and cancer (Singh et al. 2014; Forman and Zhang 2021). ROS are widely thought to play a role in a variety of pathologies, including inflammation, asthma, muscular dystrophy, dementia, anaphylaxis, rheumatoid arthritis, heart attack, carcinogenicity of various chemicals, and smoking (Yang and Lian 2019). It has not always been possible to accurately determine illness heterogeneity using only routinely collected clinical data (See 2023). Recently, Omics science has been used in healthcare facilities to personalize therapy in a variety of medical sectors, enhancing diagnosis, prognosis, and treatment. As a result, genomic, transcriptomic, metabolomic, and proteomic technologies have been developed to enable the identification and quantitative analysis of biomolecules such as proteins, nucleic acids, lipids, and carbohydrates from biological samples like blood, urine, respiratory secretions, and tissue (Table 1.1). These elements may aid in distinguishing between different types of disease-related damage and predicting clinical risk and outcomes.

Table 1.1 Important biomolecules used in clinical diagnosis

Nature of biomarker	Target	Samples taken from	Action/effect/function	Uses	References
Protein	RasGTPase-activating protein SH3-domain-binding protein (G3BP)	Embryonic specialized cells, such as the ectoderm, mesoderm, and endoderm	Decides the cellular fate during proliferation and differentiation	Diagnosis and evaluation of neural developmental disorders	Schrattenholz et al. (2012) and Kholodenko et al. (2010)
	βIII tubulin, vesicular glutamate transporter (VGLUT1), tyrosine receptor kinase B and C (rtk B, C), BDNF, neurotrophin 3 (NT3), and calretinin.	Ocular fluids and tissues, including tears, aqueous humor, vitreous humor, and ocular fluids/tissue	Provide nourishment and role in apoptosis and neuroprotection	Diagnosis, prognosis, and evaluation of ocular infections and inflammations	Bansal and Gupta (2020)
	Chitinase-3-like protein 1 (CHI3L1), C-X-C motif chemokine 13 (CXCL13), and osteopontin (OPN)	Body fluids such as blood and cerebrospinal fluid (CSF)	Maintaining cytoskeletal and neurofilament integrity throughout the nervous system	Diagnosis and evaluation of neural developmental disorders	Kim (2023)
Lipid	Triacylglycerol (TAG), ether-type phosphatidylcholine (PC-O), sphingomyelin (SM), phosphatidylcholine (PC), phosphatidylinositol (PI), and phosphatidylethanolamine (PE)	Intestinal epithelial cell membrane	Components of various cell membranes	Used to predict the risk of unhealthy aging in a subject	Engel et al. (2021)
	Very long-chain fatty acids, glycolipids, bile acids, and oxidation products of cholesterol	Hepatocyte and endothelial membrane	Localization of ATP-dependent transporter protein on the membrane	Help in the diagnosis of peroxisomal disorders and lysosomal storage disorders	McGranaghan et al. (2021)

(continued)

Table 1.1 (continued)

Nature of biomarker	Target	Samples taken from	Action/effect/function	Uses	References
	Phospholipids, sphingolipids/ ceramides, glycolipids, cholesterol esters, fatty acids, and acylcarnitines	Neurilemma	Provide insights into the underlying pathomechanisms of cardiovascular disease	Diagnosis and prognosis of cardiovascular disease	Mashima and Maekawa (2018)
	Lysophospholipids (LPLs) such as lysophosphatidylcholine (LPC)	Target both soluble and integral membrane-spanning receptors and enzymes, e.g., PAF receptor, G2A sphingophosphorylcholine receptor, LPA receptors (LPA1–6)	Act as signaling molecules and play roles in inflammation, stress, medication, and nutrition	Can be used to detect different pathologies such as infertility, cancer, atherosclerosis, and aging.	Rhee et al. (2011)
Nucleic Acids	8-oxo-7,8-dihydro-2′-deoxyguanosine (8-oxodG)	Target site for oxidative damage to DNA	Correlated with oxidative stress in the human body	Can be used in personalized medicine to detect and determine the cause of an illness, predict disease predisposition, and monitor disease progression	Qvstebq et al. (2009) and Shen et al. (2022)
	8-oxo-7,8-dihydroguanosine (8-oxoG)	Target site for oxidative damage to RNA	Correlated with oxidative stress in the human body	Can be amplified by PCR and related techniques, allowing for the detection and analysis of small amounts of samples with high sensitivity, which is beneficial for disease diagnosis and monitoring	Hanash et al. (2011) and Guo et al. (2017)
Carbohydrate	Mucopolysaccharidoses (MPS)	Cells, blood, and urine	Help build bone, cartilage, tendons, corneas, skin, and connective tissue	Can be used for the detection of mucopolysaccharidoses disease	Bentivegna (2002) and Lawrence et al. (2012)
	Glycoproteomic biomarkers	Target glycan structures on glycoproteins in serum/plasma	Recognizing specific oligosaccharide linkages	Used for the detection of cancer	Cheng et al. (2010) and Ueda (2013)

Nowadays, great deals of studies on nano commodities are conducted in the area of diagnostics and therapeutics. However, commercial nano goods are currently in short supply. The reasons are numerous, including but not limited to; (a) their difficult manufacturing procedures, regardless of "bottom-up" or "top-down" nanofabrication technologies; (b) their low resilience and continuity; (c) the few suitable commercial product evaluation methodologies and standards; and (d) the restricted alternative raw materials. The fourth point is particularly significant in light of the dangerous scenario of ongoing depletion of fossil resources and increasing environmental pollution, a shift toward ecologically friendly and efficient renewable biomolecules as raw materials is an unavoidable trend. The combination of biomolecules is expected to give potentially useful alternatives to these challenges, hence strengthening the clinical uses of biomolecule-based nano products.

This chapter includes an extensive analysis of all the components of biomolecules, including proteins, carbohydrates, nucleic acid, enzymes, vitamins, hormones, and so on, that may be used as valuable and promising biomarkers for the identification of disease, which is one of the most important steps in accurately diagnosing an illness. Thus, this chapter enumerates many benefits and drawbacks of using biomolecules in clinical diagnosis.

1.2 Clinically Important Biomarkers and Related Disorders

"Biomarkers" or "biological markers" are some of the distinguishing features that can be analyzed and evaluated as possible indicators of any ordinary or unusual pathological or pharmaceutical reaction to therapy treatment. Aryutova et al. (2021) describe biomarkers as "substances, structures, or processes that can be quantified in the body or its products and influence or predict the incidences of outcomes or diseases." Furthermore, the significance of a biomarker is determined by its pathophysiological link to a certain clinical objective (Aronson and Ferner 2017). It is an indicator that can detect biological changes associated with illness diagnosis and recurrence (Murdaca et al. 2021). Regardless of their characteristics, the clinical usefulness of biomarkers is mostly determined by their sensitivities, specificities, prediction values, and the precision, with which they can be assessed as well as their dependability, reproducibility, and feasibility (Fig. 1.1). This means that biomarkers play a critical role in clinical diagnosis and research in providing a comprehensive understanding of disease causes and progression (Califf 2018). Clinically important biomarkers which are used in clinical diagnosis are described below.

1.2.1 Carbohydrates

Carbohydrates serve a crucial role in the development of many diabetes-related disorders (Chen et al. 2019; Cui et al. 2019; Ganesan and Xu 2019). Although carbohydrates are directly related to diabetes, however research finding also suggests that it has many pharmacological properties like hypoglycemic, hypolipidemic,

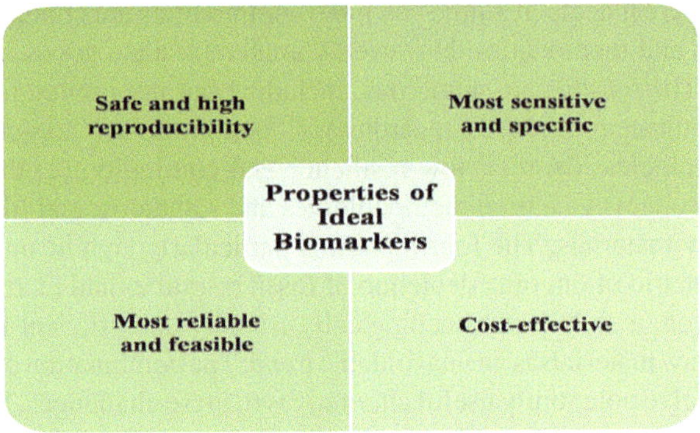

Fig. 1.1 Properties of clinically important biomarkers in disease diagnosis

anticholesterolemic, antioxidant, anti-inflammatory, and detoxifying properties (Savych et al. 2020). Carbohydrates have three distinguishing features: a high density of functional compounds (e.g., hydroxyl), a diversity of structures based on diverse configurations, and excellent biocompatibility due to their ubiquitous presence in the body. To produce carbohydrate-containing therapies, it is critical to use the basic capabilities of carbohydrates for clinical diagnosis. Lactose is a disaccharide that has limited gastrointestinal tract absorption and it breaks down into organic acids by saccharolytic bacteria, enhancing intraluminal gas production and facilitating bowel motions. As a result, it has been utilized to treat persistent constipation. Furthermore, intravenously administered heparin has been utilized in medical applications for decades as an antithrombotic drug (Qiu et al. 2021).

1.2.2 Proteins

Proteins are one of the most important elements of living beings. Investigators have been trying to identify circulating proteins as potential biomarkers in diseases. The goal is to identify and track a wide range of diseases efficiently and easily. However, this search has not been easy, and even when proteins are currently used in medical care, their sensitivity (the ability to detect a disease in patients where the disease is present) and specificity (the ability to rule out the disease in patients where the disease is absent) are relatively low. Thus far, the most prevalent circulating marker proteins can be utilized in the diagnosis of various types of disorders such as cancer, neuroinflammation, hepatocellular injuries, kidney disorders, etc. (Veyssière et al. 2022).

1.2.3 Nucleic Acids

Concerning their stability, quantifiable properties, and capacity to direct human cell reproduction, nucleic acid (NA) holds great promise as a biomarker for severe diseases and disorders. DNA is a biological molecule that is less impacted by external factors than other molecules. NAs have been created to give quality, efficacy, and safety profiles. Oligonucleotides, including single-nucleotide variations, are also quantifiable. The mechanisms of action of NAs differ from other diagnostic markers because they have more specificity to cells and tissues in comparison to other types of drugs. These molecules are difficult to permeate into cells and degrade due to their high molecular weight and very hydrophilic nature. As a result, NAs require high-quality formulations with appropriate drug delivery systems (DDSs) to prevent degradation and assure distribution to specified cells or tissues (Yamada 2021).

1.2.4 Lipids

Lipids are essential components of cells, tissues, organs, and thus the entire organism. They are involved in energy metabolism, cell membrane structure and microenvironment, senescence as well as signaling, and their working range from biological to cellular functions. The absorption of drugs is increased by lipid-based formulations. Bile salts are stimulated to be released when lipids are present in the digestive tract, replicating a partially fed state. Because the medicine is only moderately soluble in the gastrointestinal fluid, bile salts induce it to emulsify, increasing its in vivo solubility. The digestion of the solubilized chemicals is therefore simplified, increasing oral absorption (Ahmed et al. 2023). Lipids serve as an excellent barrier for certain sensitive medications against digestive or external triggers due to their continued stability across a wide range of pH and moisture levels. Lipids are noncytotoxic and sustainable, and they have an extensive record of safety (Angelova et al. 2018).

1.2.5 Enzymes

Recently, the implementation of enzyme markers has been introduced. The enzyme is initially covalently attached to the antibody and the enzyme-labeled antibody is then allowed to react with the cellular antigen, and the locations of enzyme activity are eventually exposed using appropriate cytochemical procedures. Cytochemicals for staining enzymes are available for light and electron microscopy and can be used for microscopic research (de la Fuente et al. 2021). Any enzyme that can be detected using histochemical methods or antibody labeling and that does not alter the structure of the cell can be used as a biomarker. The characteristics of a good enzyme as a biomarker include first, it must have a high specific activity and high turnover number; second, it must be relatively stable at room temperature; third, it must not lose a significant portion of its activity after a coupling procedure; and finally, it must be readily accessible. This clarifies the reason why the number of enzymes that can be utilized as labeling compounds is restricted (Robinson 2015).

1.2.6 Hormones

Hormones are chemical biomarkers synthesized by a particular specific region or organs of the body and delivered into the bloodstream to initiate or regulate specific metabolic processes. Hormone activities vary greatly but might include growth stimulation or inhibition, apoptosis induction or suppression, immune system activation or inhibition, metabolism regulation, etc. Hormones are becoming increasingly important in the genesis and onset of many malignancies (Linkov et al. 2009). Sachdeva et al. (2023) described an emerging approach in the field of contraception that involved detecting biomarkers in urine and saliva samples to know the hormone-induced disorders or hormonal treatments to recover the diseases. Furthermore, the use of urine or saliva to detect the levelof novel biomarkers might be used in public health research and have positive consequences for clinical research and clinical care.

1.2.7 Vitamins

Vitamins are a family of organic substances that are supplied through diet to maintain good physiological functions (Martini et al. 2020). The body normally does not synthesize all vitamins so they must be acquired through diet. Vitamin deficiency is hard to detect, but it might have negative consequences if the shortage is sustained or extended. Nutritional biomarkers like vitamin B9 (folate), vitamin B12 (cobalamin), vitamin A (retinol), and vitamin D were employed and well-established in epidemiological research to show deficiency resulting in health consequences. As a result, nutritional biomarkers that are both accurate and legitimate are necessary to monitor nutritional status and accurately disclose the relationship between nutrition and disease. By detecting the concentration of nutritional biomarkers, it has been found a direct link between vitamin status and disease risk (Chang and Lee 2019; Uusi-Rasi et al. 2019; Murdaca et al. 2021).

1.2.8 Macro and Microelements

Selenium, iodine, copper, zinc, and iron are dietary supplements that are necessary for maintaining physiological function and they are highly desired in healthy diets because they play important functions in cellular metabolism (Kocyłowski et al. 2018). Furthermore, the oxidative or antioxidant characteristics of specific metals may improve cardiovascular health and reduce the incidence of anemia, cancer, and other diseases. Iron deficiency (ID), for example, is a significant contributor to the risk of heart disease and influences total mortality (Naito et al. 2021). To support health, appropriate consumption of vital minerals via diet or supplementation is suggested (Mohammadifard et al. 2019; Pasricha et al. 2021).

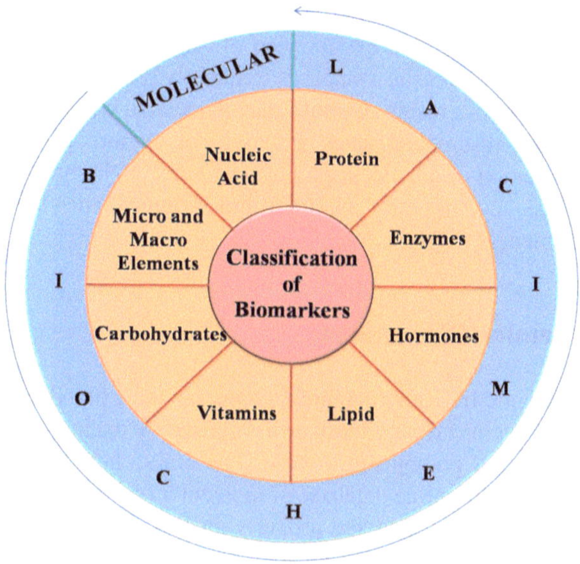

Fig. 1.2 Molecular and important biochemical molecules involved in disease diagnosis

1.3 Classification of Biomarkers

Nucleic acids, proteins, enzymes, lipids, antibodies, antigens, and other biological agents comprise a broad spectrum of bioindicators that are commonly used in the diagnosis and monitoring of diseases (Fig. 1.2) (Budnik et al. 2018). At the cellular or molecular levels, the measured responses can be biochemical, functional, or pathological. Biomolecules can be classified based on how information is gathered. In recent years, several classes of biomarkers, including genetic, biochemical, and protein-based indicators, have been proposed based on their chemical nature. The important classes of biomarkers that are used in clinical diagnosis are described below.

1.3.1 Molecular and Genetic Biomarkers

The European Medicines Agency, in collaboration with the International Council for Harmonization of Technical Requirements for Pharmaceuticals for Human Use, has defined genomic biomarker as "a measurable DNA or RNA characteristic that is an indicator of normal biologic processes, pathogenic processes, and response to therapeutic or other interventions." A molecular and genetic biomarker (DNA or RNA) includes the presence of specific pathogenic moieties, cellular or tissue-level distinguishing traits, and genetic changes (Sarhadi and Armengol 2022). Okeke and Ihekweazu (2021) suggest that molecular markers can be used to assess disease quality and epidemiological outcomes ranging from cardio- and cerebrovascular to other complications. Early diagnosis concepts and techniques, particularly in the case of tumors, have been in desperate need of improvement. The ability to diagnose a diseased state at its earliest stages is frequently required for effective therapy (Ahmad et al. 2023). As a result, biomolecules associated with various cancer types have recently been discovered using a variety of technologies, including two-dimensional evaluation of gel electrophoresis patterns, DNA-based or tissue-based microarrays, mass spectrometric analysis, and bioinformatics-based instruments (Sarhadi and Armengol 2022).

1.3.2 Biochemical Biomarkers

Chemical biomarkers, basically chemical components of blood, urine, saliva, etc. provide information concerning inborn errors caused by the metabolism, metabolic disorders, and onset of pathogenic disease. The 1089 chemical biomarkers were linked to the Marker DB online database of molecular biomarkers. Many biochemical biomarkers can be quantified and properly determined with great sensitivity as well as accuracy (Patron et al. 2019; Wishart et al. 2021). Proteins are cell's important constituents and are frequently altered in disease situations. Proteins can be extracted from practically all biofluids, however, these biomolecules like all analytes, are degraded and altered. Protein biomarkers mainly include protein molecules, protein-derived enzymes and hormones. Human blood and urine include proteases, which cleave proteins into smaller peptides, which can then be cleaved into even smaller bits by peptidases (Sarhadi and Armengol 2022). Surprisingly, the cleavage pattern can be utilized to identify specific tumors (Aster et al. 2017). Protein markers have been traditionally used to quantify a specific protein or isoform. However, evaluation of posttranslational changes is also necessary. Methylation, glycosylation, acetylation, phosphorylation, etc. are examples of posttranslational alterations that can be significant for biomarkers (Gool et al. 2020).

1.4 Omics Science in Clinical Pathology

Omics is the study of biological sciences that includes the analysis of genomics, transcriptomics, proteomics, or metabolomics. The suffix "-ome" refers to the genome, transcriptome, proteome, or metabolome (Singh et al. 2020). The primary goal of omics research is to identify, describe, and quantify all biological molecules involved in the structure, function, and dynamics of a cell, tissue, or organism. Omics have been used in clinical settings to personalize care in a variety of medical sectors, enhancing diagnosis, prognosis, and therapy. Periodic changes in biomarkers will provide time-based trajectories, allowing for the early detection of many types of serious illnesses (Yan et al. 2022). Clinicians have demonstrated that omics-based biomarkers improve patient outcomes, particularly in critical illness management, by highlighting the crucial roles of apoptosis, inflammation, and endothelial dysfunction. The three main fields that comprise the cellular and

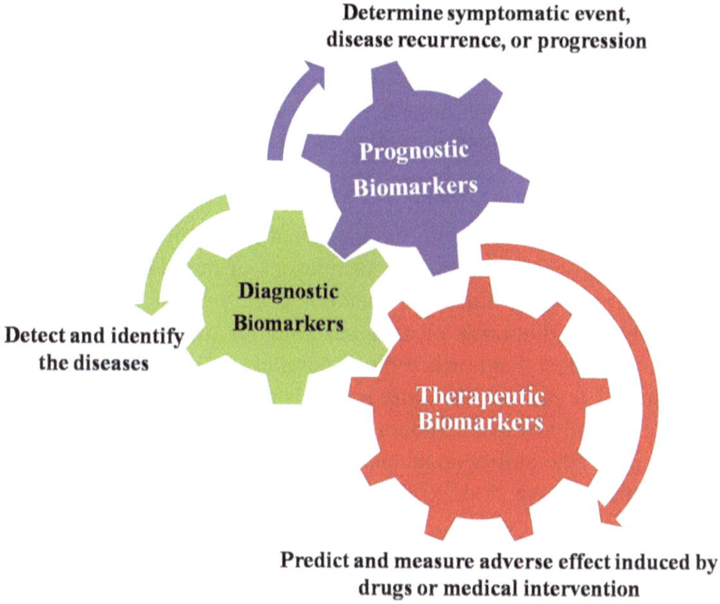

Fig. 1.3 Clinical application of omics science in the diagnosis of diseases with different biomarkers

molecular mechanisms of disease are proteomics, which deals with the proteins that are translated within an organism; transcriptomics, which studies ribonucleic acid molecules within a sample and links genomics and proteomics; and metabolomics, which deals with the small molecules and metabolites found in biological samples. Genomic analysis can be done for both the host and microbes in a sample and involves the study of cellular lipid networks and processes in biological systems, or lipidomics. That is why all of the preceding categories are categorized as therapeutic, prognostic, and diagnostic biomarkers (Fig. 1.3).

1.4.1 Diagnostic Biomarkers

Omics techniques could be used alongside pattern recognition analysis and analytical tools to develop biomarkers for early disease diagnosis. Typical examples of these methods include flux and target analysis, metabolic fingerprinting, metabolic footprinting, metabolic profiling, and understanding disease mechanisms and pathological processes (Qiu et al. 2021). These techniques have made a significant contribution to our understanding of the related metabolic processes. It is still essential for identifying diseases since it can precisely identify alterations in specific metabolic characteristics. Enhancing the prognosis, therapy, and management of diseases requires the ability to recognize metabolite profile alterations during the early stages of diagnosis. More single metabolites or a panel of dysregulated metabolites can be used to create diagnostic models that can distinguish between patients and have diagnostic power. An individual with a disease can be identified or an illness or condition can be detected or confirmed to exist using a diagnostic biomarker. Through both biochemical and clinical validations, biomarkers that are connected with certain illnesses can be identified. Biomarkers verified by substantial biochemical and clinical validations, can be connected with certain illnesses. Not every disease process is fundamentally and directly linked to the biomarkers for that disease. However, evidence of this link with the important illness phenotypes and relationship to the onset, progression, relapse, or relapses of the disease can be explored (Califf 2018). Metabolic profiles and metabolite biomarkers can function as biomarkers for precision diagnosis of different kinds of tuberculosis, with regard to developing early clinical diagnostic tools. A parallel study focused on the metabolic characterization of breast cancer using patient's plasma samples revealed specific metabolic profiles and identified a panel of glutamate, sphingomyelins, and cysteine that showed high predictability about cancer development and hence can be used as diagnostic biomarkers (An et al. 2022).

1.4.2 Prognostic Biomarkers

In patients with an illness or disorder, a prognostic biomarker is used to determine the probability of a symptomatic event, disease recurrence, or disease progression. Early disease diagnosis is challenging because the prognostic stages don't exhibit any symptoms. In order to lower the mortality rate from complex diseases, prognostic biomarkers are crucial. Finding predictive biomarkers that could help clinicians make decisions and advance individual therapy is crucial. Finding practical prognostic indicators in clinical settings is still quite difficult. Regular tests, however, have limited specificity and sensitivity, which make them unsuitable for risk assessment or early diagnosis. Enhancing the identification of risk factors and preventive measures, as well as benefiting from therapy, requires knowledge of several prognostic biomarkers that can be simultaneously quantified in a high-throughput manner. It is interesting to note that a panel of numerous metabolites with notable concentration variations connected with illness status probably benefits greatly from tiny metabolites as biomarkers (Laiakis et al. 2021; Nalbantoglu and Karadag 2021). Surprisingly, the fact that metabolite biomarkers are made up of several metabolites and that variations in their concentrations strongly correlate with the state of the disease may be one of their main findings (Aung et al. 2021; Steinbusch et al. 2021). Changes in tiny metabolites in biological media, such as blood and urine, can be precisely tracked by the way a medicine reacts to it. Qiu et al. (2021) conducted a study that effectively created metabolic phenotyping of tear fluid. The study found that taurine and nicotinic acid represent novel biomarkers that were enhanced in the diabetic group, supporting the prediction of disease development. Additionally, asparagine synthesis was shown to be elevated in metabolite profiles and was linked to a worse prognosis for female patients with colorectal cancer (Shen et al. 2022).

1.4.3 Therapeutic Biomarkers

Therapeutic biomarkers facilitate the potential prediction of patients who may respond to a specific therapy or drug mechanism of action, or who may experience adverse drug reactions. There is not a single treatment that works for every patient. Therefore, the clinic needs to forecast the therapeutic response by matching the correct patients with the proper medication. Further methods are essential to identify viable and useful biomarkers to direct patient care and match the right course of therapy. According to a new study, examining the connection between antivascular endothelial growth factor (Anti-VEGF) therapy and serum metabolome, the differential metabolite lysophosphatidylcholine 18:0 may serve as a biomarker to help choose the best course of treatment for choroidal vasculopathy and macular degeneration (Shen et al. 2022). According to Yan et al. (2022), low levels of kynurenic acid in the CSF fluid and the kynurenic acid/kynurenine ratio are biomarkers of epileptic spasms. Therefore, more treatment approaches to raise kynurenic acid levels should be investigated. Previous research has demonstrated that methyl n-methylaminobenzoate, a response biomarker in the kynurenine pathway, is related to treatment outcomes in schizophrenia patients after 4 weeks of olanzapine monotherapy (Guan et al. 2022). N-methylisoleucine, nornicotine, and 2,3-dihydroxybutanoic acid are three minor metabolites that can be utilized in metabolomic profiling to identify rheumatoid arthritis patients who respond well to methotrexate therapy (Medcalf et al. 2022). These clinical uses of tiny metabolites offer great representations of novel possibilities for targeted therapeutics and the facilitation of precision medicine.

1.5 Promising and Valuable Tools for Identification of Biomolecules

Biomolecules are created by cells of living organisms like metabolites, proteins, carbohydrates, lipids, nucleic acids, and carbohydrates that serve a variety of biological purposes. Biomolecules come in a variety of sizes and configurations. Critical data for hematology, pharmacology, disease diagnosis, and treatment efficacy are obtained through tracking and identifying these biomolecular analytes (Huang et al. 2017). Many detection techniques, such as Surface-Enhanced Raman Spectroscopy (SERS), Surface Plasmon Resonance (SPR), and Gas Chromatography-Mass Spectroscopy (GC-MS), have been developed as a result of the diverse nature of biomolecules (e.g., dimensions, surface charges, mobility, etc.). Surface-Enhanced Raman Scattering (SERS) requires complex optical setup and instrumentation. SERS provides the sensitive detection of target biomolecules (typically at the nM level) by inducing surface plasma resonances that enhance the Raman scattering signals from biomolecules (Han et al. 2022). As a substitute technique, surface-based protein reticulation (SPR) demands the immobilization of sharply directed antibodies on solid surfaces that can withstand nonspecific protein adsorption (Shrivastav et al. 2021; Gao et al. 2022). It is still difficult to manipulate and regulate the polarity of antibodies and remove nonspecific biomolecule adsorption from challenging biological components (Gao et al. 2022). A gas chromatograph analyzer combined with a mass-selective detector is the foundation of the widely used GC-MS technology, which is used for the detection and quantification of volatile compounds. However, the separation and ionization of biomolecules cause damage to them during the measurement. Additionally, GC-MS is not ideal for on-site measurement due to the large instrumentation and drawn-out analysis process.

Nowadays, nucleic acid segments, such as virus genetic material, are frequently detected by reverse transcription-polymerase chain reaction (RT-PCR). Deoxyribonucleic acid (DNA) and ribonucleic acid (RNA) detection are made extremely sensitive and specific using RT-PCR. This is achieved by amplifying certain DNA or RNA segments, which are identified by using molecular fluorophores. However, only RNA or DNA segments with known sequences can be detected by using this method. Amplification inhibitors and inappropriate sample collection are two of the many potential causes of false-positive and false-negative outcomes. The gold standard technique for reliably and sensitively identifying biomolecules, like proteins, is the enzyme-linked immunosorbent test (ELISA). However, antibodies only have a limited surface area, so ELISA may not be used to identify tiny biomolecules (Sakamoto et al. 2018). Additionally, to create a sandwich structure around the target biomolecules in ELISA, labeling antibodies with fluorescent, enzymatic, or radioactive labels is necessary. Typically, fluorescent detection requires the use of an optical reader in a laboratory setting. Transformative advancements like ultra-high sensitivity, less laborious sample preparation, and a larger detection range are widely desired, even if the aforementioned techniques have made impressive developments in recent decades. Recently, for bioanalysis, nanoparticles (such as gold colloids, polymer nanoparticles, magnetic nanoparticles, etc.) are frequently used for the detection of biomolecules (Si et al. 2021). When it comes to enhancing signals transduced from diverse biological systems, nanoparticles exhibit exceptional qualities such as biocompatibility, optical absorption, and emission, or magnetic capabilities and detection limits (Kyriakides et al. 2021).

1.6 Free Radicals and Antioxidants

When oxygen is reduced chemically, photochemically, or biologically, redox-active intermediates known as reactive oxygen species (ROS) are formed. These intermediates, in part, start a series of oxidative reactions. Radical and nonradical species comprise $^1O_2$ and hydrogen peroxide (H_2O_2), while superoxide anions ($O_2^{\bullet -}$), hydroxyl radicals ($HO^{\bullet}$), hydroperoxyl radicals ($HO_2^{\bullet}$), alkoxyl radicals ($RO^{\bullet}$), and peroxyl radicals (ROO•) are more reactive category of ROS (Collin, 2019). Specifically, $^1O_2$ attacks the compound's π-bonds to create hydroperoxide due to its electrophilic qualities. By either a direct reaction or the generation of ROS, $^1O_2$ oxidatively changes biomolecules, including lipids, nucleic acids, and amino acids.

The body uses a variety of antioxidant mechanisms to regulate ROS (Singh et al. 2022). Antioxidant molecules that are present in our body, stop or slow the oxidation of biomolecules like proteins, lipids, and DNA (Singh and Singh 2018; Bouviere et al. 2021). Their mechanisms of action is divided into several defense lines to maintaining a healthy cellular microenvironment: (a) Preventive agents: these include enzymes like superoxide dismutase (SOD), catalase (CAT), and glutathione peroxidase (GPX); proteins that bind metals like ferritin and ceruloplasmin; minerals like selenium (Se), copper (Cu), and zinc (Zn); (b) Radical scavenging agents: these include glutathione, albumin, vitamins C and E, carotenoids, and flavonoids; (c) Repair and de novo enzymes: these include DNA repair enzymes, methionine-sulfoxide reductases, transferases, proteases, and lipases; and (d) Adaptation agents: these produce suitable antioxidant enzymes and transfer them to crucial sites of action.

Antioxidants are classified into two main groups: enzymatic and nonenzymatic antioxidants. The primary enzymes SOD, CAT, and GPX, as well as the secondary enzymes glutathione reductase and glucose-6-phosphate dehydrogenase, are examples of enzymatic antioxidants (Irato and Santovito 2021). Nonenzymatic antioxidants are low-molecular-weight antioxidant molecules such as cellular glutathione and uric acid, which are either lipid-soluble (carotenoids and vitamin E) or water-soluble (vitamin C and phenolic compounds) biomolecules (Flieger et al. 2021). However, dietary antioxidants which are typically found in fruits and vegetables—have also been linked to a healthy balance between antioxidant status and free radicals, which lowers the risk of oxidative stress and, consequently, cancer, cardiovascular disease, and aging (Shukla et al. 2011; Singh et al. 2013; Irato and Santovito 2021).

1.7 Advantages and Disadvantages of Biomolecules in Clinical Diagnosis

Carbohydrates are regarded as one of the most significant biomolecules because these molecules can be utilized to develop new site-specific delivery strategies and act as "blockers/inhibitors" in the development of therapeutics. Specifically, they are used as sensors for the detection of cell types, disease states, protein functions, and developmental stages. However, the primary drawbacks in clinical diagnosis are the identification of carbohydrate biomarkers and the design and synthesis of binders that can precisely recognize the target biomarker with high affinity and specificity.

Lipid-based formulae increase drug bioavailability, prolong half-life, improve pharmacokinetics, are stable across a wide pH range, and act as an excellent barrier for certain sensitive medications to gastrointestinal or environmental conditions. The main limitations include cost of diagnosis, requirement of mass spectrometry, and lack of well-defined methodologies for lipid identification.

Nucleic acid can be used for early screening or susceptibility to the disease, as well as for monitoring the effects of treatment during the disease. This allows clinicians to assess a patient's prognosis and modify treatment plans in response to regular surveillance results. Many patients with early disorders do not receive timely diagnosis. As a result, in the late stages of the disease, delayed diagnosis is one of the most significant causes of death. However, the detection of nucleic acid as a biomarker is the most widely accepted strategy in disease diagnosis but it has some limitations such as sequencing methods have a low sensitivity to mutation detection and is costly and complex for clinical diagnosis. In PCR techniques, the wrong amplification of the targeted sequence limits the power of RT-PCR quantification.

The use of enzymes as biomarkers like amylase, lipase, lactate dehydrogenase, alkaline phosphatase, etc. for clinical diagnosis in humans are used to know the health status of digestive organs, immune system, the contraction of muscles, and other essential organs in humans. There are currently more immunoassays utilized for clinical diagnosis in ELISA due to their easy process, high specificity, and sensitivity and highly efficient, safe and eco-friendly. Still, ELISA displays drawbacks like being expensive and labor-intensive, high risk of erroneous positive or negative results due to inadequate antigen-immobilized microtiter plate surface blocking and antibody instability.

1.8 Future Prospective and Challenges

Categorization of biomarkers by researchers in to several categories was based on different dimension of biomarker's attributes. Through a survey of various pertinent literatures, the qualities of biomarkers are analyzed and summarized. Even with the current biomarkers being used widely, several issues have not yet been solved. These restrictions will hinder the advancement of biomarkers. First, the body's needs for specific molecules may not be adequately reflected by the biomarkers used today, which could lead to a gap between biomarker levels and dietary consumption levels. For instance, low levels of certain mineral nutrients are linked to increased levels of tumor necrosis factor (TNF)-α and interleukin (IL)-1β. These biomarkers can be used to assess health or the likelihood of developing certain diseases, but more precise and sensitive functional biomarkers need to be created (King et al. 2016).

To better understand the molecular-level interacting phenomena correlated with metabolic modifications, future efforts should concentrate on investigating some of the noninvasive, highly specific, and sensitive biomarkers with high diagnostic value; streamlining the evaluation and screening of the candidature of biomarkers for their introduction into clinical diagnosis and developing new algorithm-based approaches and bioinformatics tools and software. Additionally, it would be helpful to demonstrate the efficient and effective affiliations between laboratory and clinical outcomes. Certain critical issues in proteomic studies still need to be addressed. Furthermore, a nonnegligible biological heterogeneity was discovered through the application of methodologies and the employment of many techniques in multiple laboratories even utilizing the same tools which makes it difficult to validate these approaches for use in clinical settings. Metabolomics approaches based on smaller molecular metabolites have significantly changed the landscape of biomedical research. It will soon be possible to successfully validate metabolite-based biomarker molecules before attempting to move them into clinical diagnosis, thus researchers and clinicians must collaborate closely.

1.9 Conclusions

The clinical industries are interested in focusing on biomolecules such as proteins, enzymes, carbohydrates, nucleic acids, vitamins, hormones, and integral components of our body system, as essential tools for making clinical diagnosis. This chapter has summarized the categorization of biomolecule biomarkers for clinical diagnosis based on both nature and function. By extending the engineering development, it has been concluded that omics science also plays important roles in the diagnosis of disease which includes genomics, proteomics, and metabolomics. Omics would become more accessible and economical for frontline practitioners in coming years with the understanding of pathophysiological pathways, and their specific treatments. Furthermore, oxidative stress is emphasized as a possible center of interest for biomolecular diagnosis as oxidative stress is linked to numerous disorders. To diagnose the specific disease, the biomolecule spectrum can provide abundant clues about the structure and/or quantity of interested molecules in samples, the presence of pathogens in clinical samples or disorders of biological processes occurring at the cellular or molecular level with the help of omics science. Finally, the combined development of advanced technologies and biochemical methodology holds great promise in the clinical diagnosis of diseases with the help of biological samples in future.

References

Ahmad A, Imran M, Ahsan H (2023) Biomarkers as biomedical bioindicators: approaches and techniques for the detection, analysis, and validation of novel biomarkers of diseases. Pharmaceutics 15(6):1630

Ahmed S, Shah P, Ahmed O (2023) Biochemistry, lipids'. Stat Pearls [Preprint]. Available at: https://www.ncbi.nlm.nih.gov/books/NBK525952/. Accessed 2 Sept 2023

An R, Yu H, Wang Y, Lu J, Gao Y, Xie X, Zhang J (2022) Integrative analysis of plasma metabolomics and proteomics reveals the metabolic landscape of breast cancer. Cancer Metab 10:13

Angelova MI, Bitbol AF, Seigneuret M, Staneva G, Kodama A, Sakuma Y, Kawakatsu T, Imai M, Puff N (2018) pH sensing by lipids in membranes: the fundamentals of pH-driven migration, polarization and deformations of lipid bilayer assemblies. Biochim Biophys Acta (BBA) Biomembr 1860(10):2042–2063

Aronson JK, Ferner RE (2017) Biomarkers-a general review. Curr ProtocPharmacol 76:9.23.1–9.23.17

Aryutova K, Stoyanov SD, Kandilarova S, Todeva-Radneva A, Kostianev SS (2021) Clinical use of neurophysiological biomarkers and self-assessment scales to predict and monitor treatment response for psychotic and affective disorders. Curr Pharm Des 27(39):4039–4048

Aster JC, Pear WS, Blacklow SC (2017) The varied roles of notch in cancer. Annu Rev Pathol Mech Dis 12:245–275

Aung MT, Ashrap P, Watkins DJ, Mukherjee B, Rosario Z, Vélez-Vega CM, Alshawabkeh AN, Cordero JF, Meeker J (2021) Maternal lipidomic signatures in relation to spontaneous preterm birth and large-for-gestational age neonates. Sci Rep 11(1):8115

Bansal R, Gupta A (2020) Protein biomarkers in uveitis. Front Immunol 11:610428

Bentivegna CS (2002) Advancing monosaccharides as biomarkers: part I. Development of fluorophore-assisted carbohydrate-electrophoresis in *Chironomus riparius*. Aquat Toxicol 61(1–2):95–109

Bouviere J, Fortunato RS, Dupuy C, Werneck-de-Castro JP, Carvalho DP, Louzada RA (2021) Exercise-stimulated ROS sensitive signaling pathways in skeletal muscle. Antioxidants 10(4):537

Budnik LT, Adam B, Albin M et al (2018) Diagnosis, monitoring and prevention of exposure-related non-communicable diseases in the living and working environment: DiMoPEx-project is designed to determine the impacts of environmental exposure on human health. J Occup Med Toxicol 13:6

Califf RM (2018) Biomarker definitions and their applications. Exp Biol Med 243(3): 213–221

Chang SW, Lee HC (2019) Vitamin D and health—the missing vitamin in humans. Pediatr Neonatol 60(3):237–244

Chen X, Qian L, Wang B, Zhang Z, Liu H, Zhang Y, Liu J (2019) Synergistic hypoglycemic effects of pumpkin polysaccharides and puerarin on type II diabetes mellitus mice. Molecules 24(5):955

Cheng Y, Li M, Wang S et al (2010) Carbohydrate biomarkers for future disease detection and treatment. Sci China Chem 53:3–20

Collin F (2019) Chemical basis of reactive oxygen species reactivity and involvement in neurodegenerative diseases. Int J Mol Sci 20(10):2407

Cui J-F, Ye H, Zhu Y-J, Li Y-P, Wang J-F, Wang P (2019) Characterization and hypoglycemic activity of a rhamnan-type sulfated polysaccharide derivative. Mar Drugs 17(1):21

de la Fuente M, Lombardero L, Gómez-González A, Solari C, Angulo-Barturen I, Acera A, Vecino E, Astigarraga E, Barreda-Gómez G (2021) Enzyme therapy: current challenges and future perspectives. Int J Mol Sci 22(17):9181

Engel KM, Schiller J, Galuska CE, Fuchs B (2021) Phospholipases and reactive oxygen species derived lipid biomarkers in healthy and diseased humans and animals—a focus on lysophosphatidylcholine. Front Physiol 12:732319

Flieger J, Flieger W, Baj J, Maciejewski R (2021) Antioxidants: classification, natural sources, activity/capacity measurements, and usefulness for the synthesis of nanoparticles. Materials 14(15):4135

Forman HJ, Zhang H (2021) Targeting oxidative stress in disease: promise and limitations of antioxidant therapy. Nat Rev Drug Discov 20(9):689–709

Ganesan K, Xu B (2019) Anti-diabetic effects and mechanisms of dietary polysaccharides. Molecules 24(14):2556

Gao S, Guisán JM, Rocha-Martin J (2022) Oriented immobilization of antibodies onto sensing platforms—a critical review. Anal Chim Acta 1189:338907

Gool AV, Corrales F, Čolović M, Krstić D, Oliver-Martos B, Martínez-Cáceres E et al (2020) Analytical techniques for multiplex analysis of protein biomarkers. Expert Rev Proteomics 17(4):257–273

Guan X, Xu J, Xiu M, Li X, Liu H, Wu F (2022) Kynurenine pathway metabolites and therapeutic response to olanzapine in female patients with schizophrenia: a longitudinal study. CNS Neurosci Ther 28(10):1539–1546

Guo C, Ding P, Xie C, Ye C, Ye M, Pan C, Cao X, Zhang S, Zheng S (2017) Potential application of the oxidative nucleic acid damage biomarkers in detection of diseases. Oncotarget 8(43):75767–75777

Gupta R, Shukla S, Gupta A, Singh RL, Singh P (2023) Estimation of phytochemical constituents and analysis of antioxidant activity in different parts of *Cassia fistula* plant. Int J Pharma Chem Anal 10(3):194–201

Han XX, Rodriguez RS, Haynes CL et al (2022) Surface-enhanced Raman spectroscopy. Nat Rev Methods Primers 1(1):87
Hanash SM, Baik CS, Kallioniemi O (2011) Emerging molecular biomarkers—blood-based strategies to detect and monitor cancer. Nat Rev Clin Oncol 8(3):142–150
Huang Y, Xu J, Liu J, Wang X, Chen B (2017) Disease-related detection with electrochemical biosensors: a review. Sensors 17(10):2375
Irato P, Santovito G (2021) Enzymatic and non-enzymatic molecules with antioxidant function. Antioxidants 10(4):579
Johnson KB, Wei WQ, Weeraratne D, Frisse ME, Misulis K, Rhee K, Zhao J, Snowdon JL (2021) Precision medicine, AI, and the future of personalized health care. Clin Transl Sci 14(1):86
Kholodenko BN, Hancock JF, Kolch W (2010) Signalling ballet in space and time. Nat Rev Mol Cell Biol 11:414–426
Kim JS (2023) Protein biomarkers in multiple sclerosis. Encephalitis 3(2):54–63
King JC, Brown KH, Gibson RS, Krebs NF, Lowe NM, Siekmann JH, Raitenet DJ (2016) Biomarkers of nutrition for development (BOND)—zinc review. J Nutr 146(4):858S–885S
Kocyłowski R, Lewicka I, Grzesiak M et al (2018) Assessment of dietary intake and mineral status in pregnant women. Arch Gynecol Obstet 297:1433–1440
Kyriakides TR, Raj A, Tseng TH, Xiao H, Nguyen R, Mohammed FS et al (2021) Biocompatibility of nanomaterials and their immunological properties. Biomed Material 16(4):042005
Laiakis EC, Canadell MP, Grilj V, Harken AD, Garty GY, Brenner DJ, Smilenov L, Fornace A (2021) Small molecule responses to sequential irradiation with neutrons and photons for biodosimetry applications: an initial assessment. Rad Res 196(5):468–477
Lawrence R, Brown J, Al-Mafraji K et al (2012) Disease-specific non–reducing end carbohydrate biomarkers for mucopolysaccharidoses. Nat Chem Biol 8:197–204
Linkov F, Yurkovetsky Z, Lokshin A (2009) Hormones as biomarkers: practical guide to utilizing Luminex technologies for biomarker research. In: Tainsky M (ed) Tumor biomarker discovery. Methods in molecular biology, vol 520. Humana Press, Totowa
Martini L, Pecoraro L, Salvottini C, Piacentini G, Atkinson R, Pietrobelli A (2020) Appropriate and inappropriate vitamin supplementation in children. J Nutr Sci 9:E20
Mashima R, Maekawa M (2018) Lipid biomarkers for the peroxisomal and lysosomal disorders: their formation, metabolism and measurement. Biomark Med 12(1):83–95
McGranaghan P, Kirwan JA, Garcia-Rivera MA, Pieske B, Edelmann F, Blaschke F, Appunni S, Saxena A, Rubens M, Veledar E et al (2021) Lipid metabolite biomarkers in cardiovascular disease: discovery and biomechanism translation from human studies. Metabolites 11(9):621
Medcalf MR, Bantis LE, Shi P, Bhadbhade P, Gundry RL, Mikuls TR, England BR et al (2022) Plasma metabolomic profiling as a tool to identify predictive biomarkers of methotrexate efficacy in rheumatoid arthritis. Semin Arthritis Rheum 56:152056
Mohammadifard N, Humphries KH, Gotay C, Mena-Sánchez G, Salas-Salvadó J, Esmaillzadeh A, Ignaszewski A, Sarrafzadegan N (2019) Trace minerals intake: risks and benefits for cardiovascular health. Crit Rev Food Sci Nutr 59(8):1334–1346
Murdaca G, Banchero S, Tonacci A, Nencioni A, Monacelli F, Gangemi S (2021) Vitamin D and folate as predictors of MMSE in Alzheimer's disease: a machine learning analysis. Diagnostics 11(6):940
Naito Y, Masuyama T, Ishihara M (2021) Iron and cardiovascular diseases. J Cardiol 77(2):160–165
Nalbantoglu S, Karadag A (2021) Metabolomics bridging proteomics along metabolites/oncometabolites and protein modifications: paving the way toward integrative multiomics. J Pharm Biomed Anal 199:114031
Okeke IN, Ihekweazu C (2021) The importance of molecular diagnostics for infectious diseases in low-resource settings. Nat Rev Microbiol 19(9):547–548
Pasricha SR, Tye-Din J, Muckenthaler MU, Swinkels DW (2021) Iron deficiency. Lancet 397(10270):233–248
Patron J, Serra-Cayuela A, Han B, Li C, Wishart DS (2019) Assessing the performance of genome-wide association studies for predicting disease risk. PLoS One 14(12):e0220215

Prudent R, Annis DA, Dandliker PJ et al (2020) Exploring new targets and chemical space with affinity selection-mass spectrometry. Nat Rev Chem 5:62–71

Qiu M, Huang S, Luo C, Wu Z, Liang B, Huang H, Ci Z, Zhang D, Han L, Lin J (2021) Pharmacological and clinical application of heparin progress: An essential drug for modern medicine. Biomed Pharmacother 139:111561

Qvstebq R, Kierulf P, Haug KBF (2009) Circulating nucleic acids in blood as biomarkers. Norsk Epidemiologi 16(1):41–49

Rhee EP, Cheng S, Larson MG, Walford GA, Lewis GD, McCabe E, Yang E et al (2011) Lipid profiling identifies a triacylglycerol signature of insulin resistance and improves diabetes prediction in humans. J Clin Invest 121(4):1402–1411

Robinson PK (2015) Enzymes: principles and biotechnological applications. Essays Biochem 59:1–41. https://doi.org/10.1042/bse0590001

Sachdeva R, Kumar N, Brache V et al (2023) New approaches for developing biomarkers of hormonal contraceptive use. Sci Rep 13:245

Sakamoto S, Putalun W, Vimolmangkang S et al (2018) Enzyme-linked immunosorbent assay for the quantitative/qualitative analysis of plant secondary metabolites. J Nat Med 72:32–42

Sarhadi VK, Armengol G (2022) Molecular biomarkers in cancer. Biomolecules 12(8):1021

Savych A, Marchyshyn S, Milian I (2020) Determination of carbohydrates in the herbal antidiabetic mixtures by GC-MC. Acta Pharm 71(3):429–443

Schrattenholz SV, Schopf R, Poznanović S, Manns MK, Groebe K (2012) Protein biomarkers for in vitro testing of toxicology. Mut Res/Gene Toxi Enviro Mut 746(2):113–123

See KC (2023) Personalizing care for critically ill adults using omics: a concise review of potential clinical applications. Cells 12(4):541

Shen Y, Wang H, Xu X, Chen C, Zhu S, Cheng L, Fang J, Liu K, Xu X (2022) Metabolomics study of treatment response to conbercept of patients with neovascular age-related macular degeneration and polypoidal choroidal vasculopathy. Front Pharmacol 13:991879

Shrivastav AM, Cvelbar U, Abdulhalim I (2021) A comprehensive review on plasmonic-based biosensors used in viral diagnostics. Commun Biol 4(1):70

Shukla M, Singh SV, Singh P, Singh U, Vishwakerma SP, Khanna AA, Sexana JK, Singh RL (2011) Antidyslipidimic and antioxidant activity of hydroethanolic fruit extract of *Ficus glomerota*. Asian J Pharm Clin Res 4(2):145–148

Si P, Razmi N, Nur O, Solanki S, Pandey CM, Gupta RK, Malhotra BD et al (2021) Gold nanomaterials for optical biosensing and bioimaging. Nanoscale Adv 3(10):2679–2698

Singh P, Singh RL (2018) Quantification of phytochemicals imparting antioxidant activities in commonly used vegetables. Int J Appl Sci Biotech 6(2):107–112

Singh P, Vishwakarma SP, Singh RL (2013) Evaluation of antioxidant, oxidative DNA damage protective and antimicrobial activities of *Foeniculum vulgare* plant. J Med Plant Res 7(35):2551–2563

Singh RL, Vishwakarma SP, Singh P (2014) Vitamins and minerals: roles and plant sources. In: Prakash D (ed) Phytochemicals of nutraceutical importance and their role in human health. CABI-UK, pp 248–265

Singh P, Monda S, Singh RL (2020) Introduction. In: Mondal S, Singh RL (eds) Advances in animal genomics. Elsevier, pp 1–12

Singh P, Singh RL, Pathak N, Singh PK, Tripathi M, Mondal S (2022) Phytochemistry and nutraceutical properties of *Carica papaya* (Linn.): a review. Diet Suppl Nutr 1(9):1–15

Singh U, Singh P, Singh PK, Shukla S, Singh RL (2023) Evaluation of antioxidant, oxidative DNA damage and hepatoprotective activities of ethanolic leaf extract of *Lagenaria siceraria* plant against acetaminophen-induced hepatotoxiocity. Funct Foods Health Dis 13(3):117–134

Steinbusch LKM, Wang P, Waterval HWAH, Stassen FAPM, Coene KLM, Engelke UFH, Habets DDJ, Bierau J, Körver-Keulartset IMLW (2021) Targeted urine metabolomics with a graphical reporting tool for rapid diagnosis of inborn errors of metabolism. J Inherit Metab Dis 44(5):1113–1123

Ueda K (2013) Glycoproteomic strategies: from discovery to clinical application of cancer carbohydrate biomarkers. Proteomics Clin Appl 7(9–10):607–617

Uusi-Rasi K, Patil R, Karinkanta S et al (2019) Serum 25-hydroxyvitamin D levels and incident falls in older women. Osteoporos Int 30:93–101
Veyssière H, Bidet Y, Penault-Llorca F et al (2022) Circulating proteins as predictive and prognostic biomarkers in breast cancer. Clin Proteomics 19:25
Wishart DS, Bartok B, Oler E, Liang KYH, Budinski Z, Berjanskii M, Guo A, Cao X, Wilson M (2021) Marker DB: an online database of molecular biomarkers. Nucleic Acids Res 49(D1):D1259–D1267
Yamada Y (2021) Nucleic acid drugs—current status, issues, and expectations for exosomes. Cancers 13(19):5002
Yan Y, Chen J, Liang Q et al (2022) Metabolomics profile in acute respiratory distress syndrome by nuclear magnetic resonance spectroscopy in patients with community-acquired pneumonia. Respir Res 23:172
Yang S, Lian G (2019) ROS and diseases: role in metabolism and energy supply. Mol Cell Biochem 467(1):1–12

Carbohydrate Metabolism in Health and Diseases

2

Mohd. Waiz, Kehkashan Rehman, Parvej Ahmad, and M. Salman Khan

Abstract

Disruption of metabolic pathways that play a central role in the metabolism of carbohydrates, lipids, and proteins is responsible for several metabolic diseases, including hyperglycemia, which is characterized by increased blood glucose levels due to the dysregulation of carbohydrate hydrolysis. Chronic hyperglycemia causes several chemical modifications to nucleic acids, proteins, and lipids through a nonenzymatic process called glycation, leading to the formation of advanced glycation end products (AGEs). AGEs can promote oxidative stress, inflammation, and apoptosis, which cause damage to tissues and muscles and contribute to several pathophysiological conditions, including retinopathy, nephropathy, neuropathy, atherosclerosis, and cancer. Moreover, recent studies have also reported that carbohydrates are involved in the progression of hypercholesterolemia by influencing HMG-CoA reductase activity, which is a rate-limiting enzyme in cholesterol synthesis. This chapter provides a basic concept of carbohydrate metabolism and AGEs formation, accumulation, and receptor interaction as well as their role in the progression of metabolic diseases.

Keywords

Carbohydrate metabolism · Metabolic disorder · Advanced glycation end products · Hyperglycemia

M. Waiz · K. Rehman · P. Ahmad
IIRC-5, Clinical Biochemistry & Natural Product Research Lab, Department of Biosciences, Integral University, Lucknow, U.P., India

M. S. Khan (✉)
IIRC-5, Clinical Biochemistry & Natural Product Research Lab, Department of Biosciences, Integral University, Lucknow, U.P., India

Department of Biotechnology, Era University, Lucknow, U.P., India

R. L. Singh et al. (eds.), *Clinical Applications of Biomolecules in Disease Diagnosis*, https://doi.org/10.1007/978-981-97-4723-8_2

2.1 Introduction

The metabolic processes that occur in living organisms to break down carbohydrates and use them for energy production or storage are referred to as carbohydrate metabolism (Pösö and Puolanne 2005). Carbohydrates are one of the main sources of energy in the human diet, and are essential for various physiological functions (Mann et al. 2007). Altered carbohydrate metabolism is involved in the initiation, progression, and development of several metabolic diseases, including hyperglycemia and cardiovascular diseases. There are three main forms of carbohydrates: monosaccharides, disaccharides, and polysaccharides. Monosaccharides, including glucose, fructose, and galactose, are the most basic carbohydrates. Sucrose (glucose + fructose) and lactose (glucose + galactose) are monosaccharides that combine to generate disaccharides (Niaz et al. 2020; Qi and Tester 2019). On the other hand, Polysaccharides, are complex carbohydrates composed of many monosaccharide units, such as starch and glycogen. Carbohydrate metabolism begins with digestion in the gastrointestinal tract, where complex carbohydrates are broken down into monosaccharides through enzymatic action (Goodman 2010; Evans 2013). Once absorbed into the bloodstream, monosaccharides are transported to cells throughout the body where they can be used for several purposes (Evans 2013). Glucose, the principal monosaccharide generated from carbohydrate breakdown, is essential for energy metabolism (Tirone and Brunicardi 2001). In a process known as glycolysis, glucose is broken down in the cytoplasm of cells to produce adenosine triphosphate (ATP), the cell's primary energy currency. This process generates a small quantity of ATP and pyruvate, which can then enter various metabolic pathways, depending on the cellular state (Chandel 2021).

In aerobic circumstances, pyruvate is transported into the mitochondria where it undergoes oxidative phosphorylation. This process results in the formation of ATP through the electron transport chain and the Krebs cycle. This process converts glucose to carbon dioxide and water to efficiently generate energy (Nath and Villadsen 2015). In times of limited oxygen availability, pyruvate can undergo fermentation, producing lactate and other byproducts and reducing ATP compared to aerobic metabolism. Excess glucose can be converted into glycogen in the liver and muscle cells through glycogenesis. When energy is needed, glycogen can be broken down into glucose through glycogenolysis, facilitating rapid glucose release into the bloodstream (Zhang et al. 2021). Overall, carbohydrate metabolism is a complex and tightly regulated process that ensures efficient utilization and storage of carbohydrates for energy production in the body. It is crucial for the proper functioning of the body and is controlled by factors such as nutritional intake, hormone regulation, and metabolic needs (Maughan 2009).

2.2 Overview of Metabolic Pathways

2.2.1 Glycolysis

Glycolysis is the metabolic mechanism that converts glucose, a six-carbon sugar, into pyruvate, a three-carbon molecule. It occurs in the cytoplasm and is regulated by multiple enzymatic processes (Campbell 2014). The overall process can be summarized as follows: Glucose Phosphorylation: The enzyme hexokinase (EC 2.7.1.1) phosphorylates glucose using one ATP molecule, resulting in glucose-6-phosphate. This phosphorylation phase traps glucose within the cell and prepares it for subsequent metabolism (Fothergill-Gilmore and Michels 1993). Rearrangement and Phosphorylation: The isomerization converts glucose-6-phosphate into its isomer, fructose-6-phosphate (Guo et al. 2012a). Utilizing another ATP molecule, phosphofructokinase-1 (PFK-1) (EC 2.7.1.11) catalyzes the phosphorylation of fructose-6-phosphate to create fructose-1,6-bisphosphate. Glucose molecules are now known to be involved in glycolysis (Dixon and Rees 1980). Cleavage and Sugar Activation: The cleavage of fructose-1,6-bisphosphate results in two three-carbon compounds, dihydroxyacetone phosphate (DHAP) and glyceraldehyde-3-phosphate (G3P). Aldolase (EC 4.1.2.13) was responsible for this cleavage. DHAP is a simple intermediate that easily transforms into G3P (Silvestri et al. 2003). Energy Generation and NADH Production: Glyceraldehyde-3-phosphate dehydrogenase (GAPDH) oxidizes G3P to produce NADH and 1,3-bisphosphoglycerate. This process transfers electrons to NAD^+, reduces them to NADH, and uses energy to phosphorylate ADP to ATP, resulting in glycolysis's initial ATP molecule (du Plessis et al. 2015). After that1,3-bisphosphoglycerate converted into 3-phosphoglycerate, ATP production, and 2-phosphoglycerate, with phosphoglycerate mutase assisting in the latter and enolase catalyzing the latter to PEP. ATP Generation and Pyruvate Production: In the final stage of glycolysis, pyruvate kinase catalyzes the conversion of PEP to pyruvate, accompanied by the production of one ATP molecule through substrate-level phosphorylation (Dixon and Rees 1980). Following glycolysis, a single molecule of glucose is converted into two molecules of pyruvate. This process also results in a net gain of two ATP molecules (produced directly in steps 4 and 6) and two NADH molecules (generated in step 4) (Fig. 2.1). Pyruvate molecules can further enter other metabolic pathways, such as the Krebs cycle, depending on cellular conditions and energy demands (Gray et al. 2014).

2.2.2 Kreb's Cycle

The Krebs cycle is also known as the citric acid cycle or the tricarboxylic acid (TCA) cycle (Gupta and Gupta 2021). It begins with the introduction of acetyl-CoA, which is produced from pyruvate and fatty and amino acids. Each acetyl-CoA molecule reacts with oxaloacetate, a four-carbon chemical, to form citrate with the help of citrate synthase. Citrate Isomerization: Citrate undergoes an isomerization step catalyzed by aconitase, converting it into its isocitrate isomer. Isocitrate

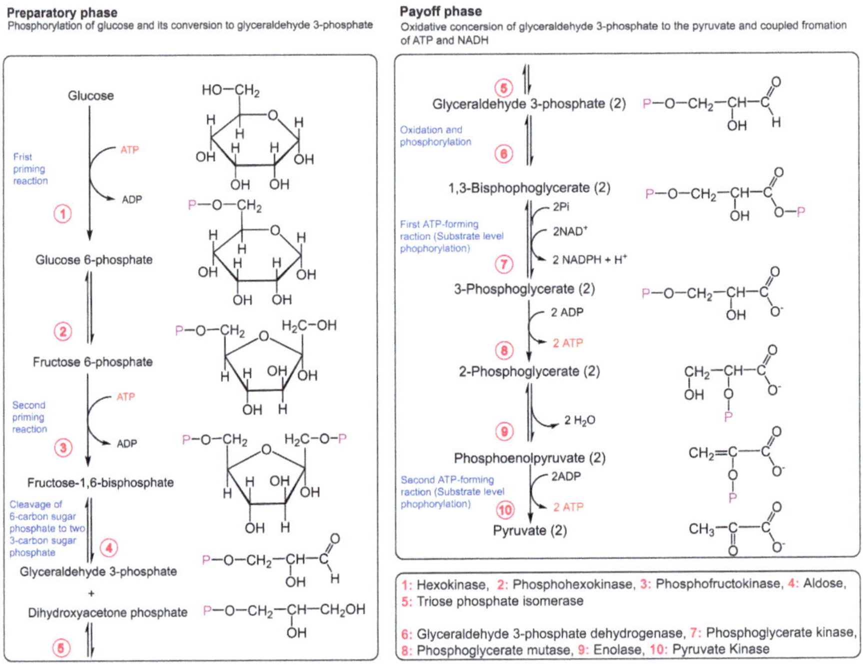

Fig. 2.1 This figure delineates the glycolytic pathway, depicting the preparatory and payoff phases. In the preparatory phase, glucose undergoes phosphorylation and conversion to glyceraldehyde-3-phosphate. Notably, hexokinase and phosphofructokinase catalyze the first and third reactions, consuming two ATP molecules. The payoff phase, represented in the right panel, involves the conversion of glyceraldehyde-3-phosphate to pyruvate and concurrent ATP and NADH generation. Phosphoglycerate kinase facilitates the first ATP-yielding reaction, converting 1,3-phosphoglycerate to 3-phosphoglycerate. The second ATP-yielding reaction ensues as phosphoenolpyruvate is converted to pyruvate by pyruvate kinase. This depiction elucidates the discrete stages of glycolysis, accentuating the enzymatic events and energy transactions integral to this metabolic pathway

Oxidation: Isocitrate dehydrogenase oxidizes isocitrate, resulting in the production of alpha-ketoglutarate. One molecule of nicotinamide adenine dinucleotide (NAD) + hydrogen (H) (NADH) is produced during this process (Akram 2014). α-Ketoglutarate Oxidation:α-Ketoglutarate is further oxidized by α-ketoglutarate dehydrogenase (also known as the oxoglutarate dehydrogenase complex (OGDC), resulting in the formation of succinyl-CoA, one molecule of NADH, and one molecule of carbon dioxide (CO_2). Succinyl-CoA Production: Succinyl-CoA synthetase (EC 6.2.1.4) (also known as succinate thiokinase or succinate-CoA ligase) catalyzes the conversion of succinyl-CoA to succinate, producing one molecule of ATP via substrate-level phosphorylation. Succinate Dehydrogenation: The Succinate dehydrogenase oxidizes succinate to fumarate (EC 1.3.5.1). This step involves the reduction of FAD to $FADH_2$, forming a single molecule of $FADH_2$ (Bodner 1986; Akram 2014). Fumarate Hydration: The enzyme fumarase (or fumarate hydratase) (EC

4.2.1.2) converts fumarate to malate. Malate Oxidation: Malate dehydrogenase (EC 1.1.1.37) oxidizes malate to oxaloacetate by reducing NAD+ to NADH, yielding one molecule of NADH (Gupta and Gupta 2021). The Krebs cycle, which involves the oxidation of one acetyl-CoA molecule, generates three NADH molecules, one $FADH_2$ molecule, one ATP molecule, and two CO_2 molecules. Oxaloacetate was then regenerated in the next cycle. High-energy electron carriers from the Krebs cycle enter the electron transport chain in the mitochondria, where they produce ATP via oxidative phosphorylation. This cycle is a major factor in carbon fuel oxidation (Riera et al. 2008).

2.2.3 Oxidative Phosphorylation

The process of oxidative phosphorylation represents the final step in cellular respiration, taking place on the inner layer of the mitochondrial membrane (Meyer et al. 2022). Electron Transport Chain (ETC): The ETC, where NADH and $FADH_2$ transport high-energy electrons to the inner mitochondrial membrane, initiates oxidative phosphorylation. NADH dehydrogenase, succinate dehydrogenase, cytochrome b-c1 complex, cytochrome c, and cytochrome c oxidase are all components of ETC (Osellame et al. 2012). Electron Transfer and Proton Pumping: The electrons' energy is progressively released as they pass through the ETC protein complexes. The aforementioned energy facilitates the movement of protons (H^+) from the mitochondrial matrix to the area between the inner and outer mitochondrial membranes, thereby establishing a difference in electric charge across the inner mitochondrial membrane. Complexes I, III, and IV actively participate in the transportation of protons during the process of electron transfer (Ramsay 2009). Chemiosmosis and ATP Synthesis: the electron transport chain is used in chemiosmosis, the mechanism by which ATP is synthesized (Ahmad et al. 2023). Protons enter the mitochondrial matrix via ATP synthase, an inner-membrane enzyme. Their energy is utilized to transform ADP and Pi into ATP via oxidative phosphorylation, a process that combines electron transport and ATP synthesis (Neupane et al. 2019). Oxygen as the Terminal Electron Acceptor: The ultimate electron acceptor at the end of the ETC was molecular oxygen (O_2). Water (H_2O) is formed when O_2 interacts with electrons and protons. As the last electron sinks, oxygen is required for oxidative phosphorylation to occur (Schmidt-Rohr 2020). Oxidative phosphorylation generates a substantial amount of ATP, with the yield varying depending on the molecule. One NADH molecule creates three ATP molecules on average, whereas one $FADH_2$ molecule produces two. Shuttle systems and the Krebs cycle have an impact on real ATP yield (Fig. 2.2) (Zheng 2012).

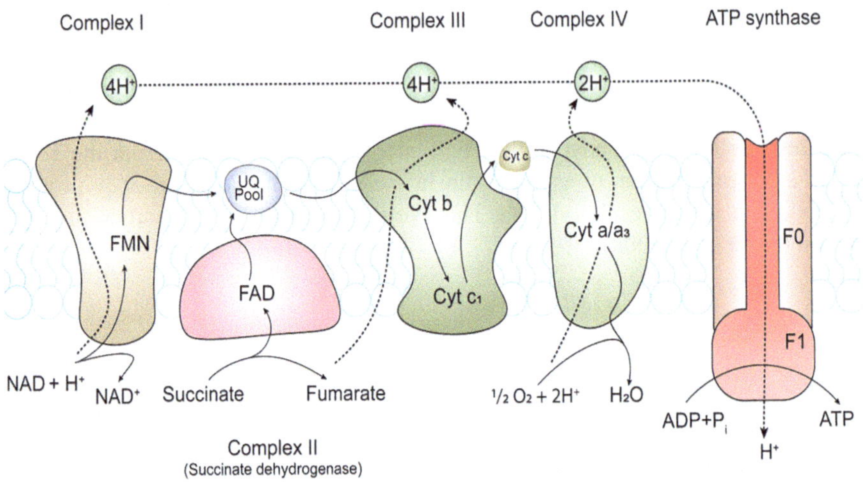

Fig. 2.2 The schematic diagram delineates the electron transport chain (ETC), comprising Complexes I, II, III, IV, and ATP synthase. The ETC facilitates the conversion of reducing equivalents, derived from NADH and FADH2 generated in glycolysis, fatty acid oxidation, and the Tricarboxylic Acid (TCA) cycle, into ATP. Electron flow commences with NADH binding to Complex I or succinate/FADH2 binding to Complex II, initiating a cascade that culminates in ATP synthesis by ATP synthase. Dysregulation in these pathways compromises ATP production, imposing cellular stress. This depiction underscores the intricate interplay of metabolic processes in sustaining mitochondrial electron flow and ATP generation

2.3 Impaired Glucose Metabolism in Metabolic Diseases

Impaired glucose metabolism is a common feature of various metabolic diseases, including diabetes mellitus (Abdul-Ghani et al. 2006). Type-2 diabetes mellitus (T2DM) is characterized by insulin resistance, in which the body's cells become less receptive to insulin action (Goldstein 2002). This impairs cellular glucose uptake, resulting in increased blood glucose levels. Furthermore, the pancreas may produce inadequate levels of insulin. IR often arises from obesity, physical inactivity, or genetic factors. This results in reduced glucose transport into cells, particularly in the muscle, liver, and adipose tissue. Consequently, glucose levels remain elevated in the bloodstream (Pérez-Martin et al. 2001). In T2DM, the β-cells in the pancreas responsible for insulin production may become dysfunctional over time. This leads to decreased insulin secretion, exacerbating glucose dysregulation (Kaiser et al. 2003). Moreover, chronic low-grade inflammation is commonly observed in patients with insulin resistance and metabolic diseases. Inflammatory cytokines disrupt the insulin signaling pathways, leading to decreased glucose uptake and utilization in cells (de Luca and Olefsky 2008). Furthermore, impaired mitochondrial function results in reduced cellular energy production and oxidative stress, leading to disrupted glucose metabolism (Bhatti et al. 2017). Overall, impaired glucose metabolism in metabolic diseases stems from insulin resistance, beta-cell dysfunction, adipose tissue dysfunction, inflammation, and mitochondrial dysfunction (Gonzalez-Franquesa and Patti 2017). Understanding the underlying mechanisms is crucial for developing effective interventions and therapies to manage and treat these conditions. Lifestyle changes such as regular exercise, a well-balanced diet, and weight management are key components in improving glucose metabolism and preventing or managing metabolic diseases (Strasser 2013).

2.4 Dysregulation of Glucose Uptake and Utilization

Dysregulation of glucose uptake and utilization refers to impaired or interrupted mechanisms involved in the uptake, storage, and utilization of glucose by cells (Arnold et al. 2018). Here, we provide an overview of the key aspects affected by dysregulation. Glucose transporters are proteins that facilitate glucose transport across cell membranes (Klip and Pâquet 1990). Glucose transporter 4 (GLUT4) is the most well-known transporter and is principally responsible for insulin-regulated glucose absorption in the skeletal muscle and adipose tissue (Watson et al. 2004). Under dysregulated conditions, the translocation and activation of GLUT4 to the cell membrane are impaired, resulting in reduced glucose uptake (Giorgino et al. 2000). However, glucose can be stored as glycogen to meet future energy demand. Dysregulation of glycogen synthesis occurs when the enzymes involved in glycogen synthesis are dysfunctional or when the signaling pathways controlling glycogen synthesis are impaired. This leads to reduced glycogen synthesis and storage capacity, contributing to elevated blood glucose levels and impaired glucose utilization (Grimes and Jope 2001). Furthermore, glucose is synthesized in the liver from noncarbohydrate substances such as amino acids and glycerol is known as gluconeogenesis (Maughan 2009). Dysregulation of gluconeogenesis can occur when there is an aberrant increase in glucose synthesis, even in the presence of high blood glucose levels. This can result in higher blood glucose levels and disturbed glucose homeostasis (Gadupudi et al. 2016). When cells are unable to efficiently take up and utilize glucose, they rely on alternative energy sources such as fatty acids. This can lead to a shift toward fatty acid metabolism and a decrease in glucose oxidation, affecting cellular energy balance and potentially leading to metabolic complications (Santiago and Potashkin 2013). Altered glucose uptake and utilization is typically related to metabolic illnesses such as type 2 diabetes, obesity, metabolic syndrome, and insulin resistance (Tang et al. 2021). These disorders affect the physiological mechanisms involved in glucose metabolism, resulting in increased blood glucose levels, decreased energy metabolism, and increased risk of developing chronic hyperglycemia consequences (Bolk et al. 2001). The proper management of dysregulation typically involves lifestyle modifications (diet and exercise), medication, and other interventions targeted at improving glucose uptake and utilization, restoring glucose homeostasis, and preventing long-term complications (Bianchi et al. 2007).

2.5 Influence of Altered Carbohydrate Metabolism on Metabolic Pathways

The alteration of carbohydrate metabolism can have significant influences on various metabolic pathways in the body (Poznyak et al. 2020). Here's an overview of how changes in carbohydrate metabolism can impact other metabolic pathways including such as glycolysis, gluconeogenesis, lipid metabolism, insulin signaling, and ketogenesis. Glycolysis is responsible for the breakdown of glucose to produce energy in the form of ATP. Altered carbohydrate metabolism can affect glycolysis in several ways (Wu et al. 2005). For instance, if glucose availability is reduced, such as during fasting or a low-carbohydrate diet, the rate of glycolysis may decrease as there is limited substrate available. Conversely, under conditions of excessive glucose, such as uncontrolled diabetes or excessive carbohydrate intake, glycolysis may be upregulated to meet increased energy demands (Cherkas et al. 2020). However, gluconeogenesis serves as a crucial pathway for maintaining blood glucose levels during fasting or prolonged periods of carbohydrate restriction. Altered carbohydrate metabolism, particularly reduced carbohydrate intake, can stimulate gluconeogenesis to provide glucose for energy requirements. However, excessive carbohydrate intake can suppress gluconeogenesis (Fig. 2.3) (Hatting et al. 2018).

On the other hand, carbohydrate metabolism can influence lipid metabolism through various mechanisms. When carbohydrate intake is insufficient, the body may switch to utilizing the stored fat as an energy source. This can lead to increased lipolysis (breakdown of fats) and subsequent fatty acid oxidation, thereby contributing to weight loss and ketone body production (Galgani and Ravussin 2008). Conversely, a high-carbohydrate diet, especially one rich in refined carbohydrates and sugars, can stimulate de novo lipogenesis and conversion of excess carbohydrates into fatty acids for storage as triglycerides (Softic et al. 2016). However, one of the major pathways, insulin signaling, that involved in the progression of metabolic disorder affected by altered carbohydrate metabolism. Elevated blood glucose levels prompt the pancreas to release insulin, which stimulates glucose uptake into cells (Mul et al. 2015). IR, in conditions such as T2DM or MetS can disrupt insulin signaling and impair glucose uptake. This leads to dysregulation of other metabolic pathways involved in energy production and storage (Withers and White 2000). Furthermore, the body reaches a condition of ketosis when there are few or no carbon atoms available. This occurs when the liver transforms fatty acids into ketone bodies such as acetoacetate and beta-hydroxybutyrate, which are used as alternative fuel sources by the brain and other tissues (Herdt 2000). Dysregulated carbohydrate metabolism, particularly in conditions such as uncontrolled diabetes or prolonged fasting, can increase ketone production and affect the balance between carbohydrate and fat utilization signaling and ketogenesis (Abdul Kadir et al. 2020). The body adapts to changes in carbohydrate availability to maintain its energy balance and glucose homeostasis. However, dysregulation of carbohydrate metabolism, such as in diabetes or other metabolic disorders, can disrupt these pathways, leading to metabolic imbalances and associated complications (Longo et al. 2019).

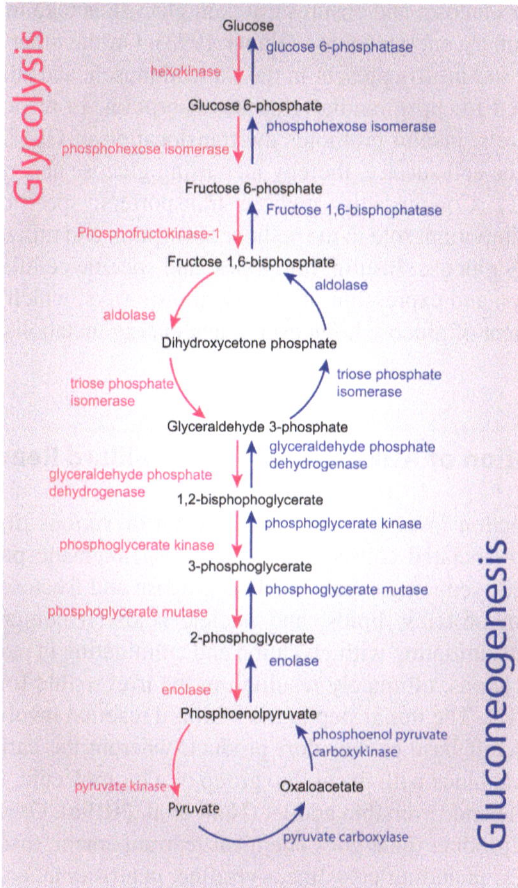

Fig. 2.3 Schematic representation of gluconeogenesis

2.6 Role of Glucose Transporters in Cellular Uptake and Utilization of Carbohydrates

Glucose transporters (GLUTs) are essential for the uptake and utilization of carbohydrates, particularly glucose, by cells (Herman 2006). GLUTs that are present in a variety of isoforms (GLUT1–4) allow glucose to enter cells from the extracellular fluid by facilitating its transit across cell membranes (Watson et al. 2004). GLUT1, which is found in various tissues and cells, is responsible for basal glucose uptake and is not tightly regulated by insulin. This ensured that cells had a consistent supply of glucose for key metabolic functions (Chadt and Al-Hasani 2020). GLUT2, located predominantly in the liver, pancreas, and small intestine, facilitates bidirectional glucose transport. It allows glucose uptake into hepatocytes and pancreatic β-cells, and mediates glucose release from hepatocytes into the bloodstream (Leturque et al. 2009). Moreover, GLUT3 present in neurons and other tissues, has a high affinity for glucose, and ensures efficient glucose uptake in the brain, where glucose is the primary energy source (Haber 1993). On the other hand, GLUT4 is insulin-regulated and mostly present in the skeletal muscle and adipose tissue. This process is required for postprandial glucose absorption. In response to increased blood glucose levels, insulin promotes the translocation of GLUT4 from intracellular vesicles to the cell surface, thereby increasing glucose absorption (Thurmond and Pessin 2001). Conclusively, glucose transporters, specifically GLUT1 to GLUT4, play an important role in the cellular absorption and utilization of carbohydrates, principally glucose. Insulin, hormones, and specific cellular demands influence the regulation and expression of glucose transporters, which contribute to the overall management of glucose homeostasis and energy metabolism (Arévalo-Ruiz et al. 2017).

2.7 Formation of AGEs Through the Maillard Reaction

The Maillard reaction involves in the progression of various diseases, including diabetes and its associated consequences. This nonenzymatic process entails the interaction between reducing sugars, such as glucose and fructose, and free amino groups present in proteins, lipids, and nucleic acids. It comprises an intricate sequence of events, initiating with glycation and culminating in rearrangements and chemical modifications, ultimately resulting in the irreversible formation of AGEs (Ahmad et al. 2018). The initial step in the Maillard reaction involves the formation of a reversible Schiff base or Amadori product, wherein the carbonyl group of a reducing sugar combines with the amino group of a biomolecule, such as a protein, to produce a labile and reversible adduct (Nabi et al. 2019a). Over time, the Schiff base or Amadori product undergoes chemical rearrangement, resulting in structurally diverse AGEs, including crossline, pyrraline, pentosidine, and others (Ahmad et al. 2022). AGE production is influenced by various parameters, including lowering sugar concentration and the extent of glycation, as well as the existence of oxidative stress and reactive oxygen species (ROS) (Alvi et al. 2021). Diabetes promotes the development and accumulation of AGEs owing to elevated glucose levels and prolonged exposure to hyperglycemic settings. AGE accumulation contributes to a number of disorders, most notably diabetes and its complications (Waiz et al. 2023). AGEs cause cellular dysfunction by activating receptors for advanced glycation end products (RAGE) and causing oxidative stress (Hu et al. 2020; Nabi et al. 2021a). When AGEs bind to RAGE, they activate intracellular signaling pathways that promote inflammation, oxidative stress, and pro-inflammatory cytokines, ultimately leading to insulin resistance, poor pancreatic beta-cell function, and the progression of diabetic complications (Fig. 2.4) (Waiz et al. 2024). Understanding the mechanisms driving AGE formation and developing strategies to minimize their detrimental consequences could have a significant impact on diabetes prevention and control (Rungratanawanich et al. 2021).

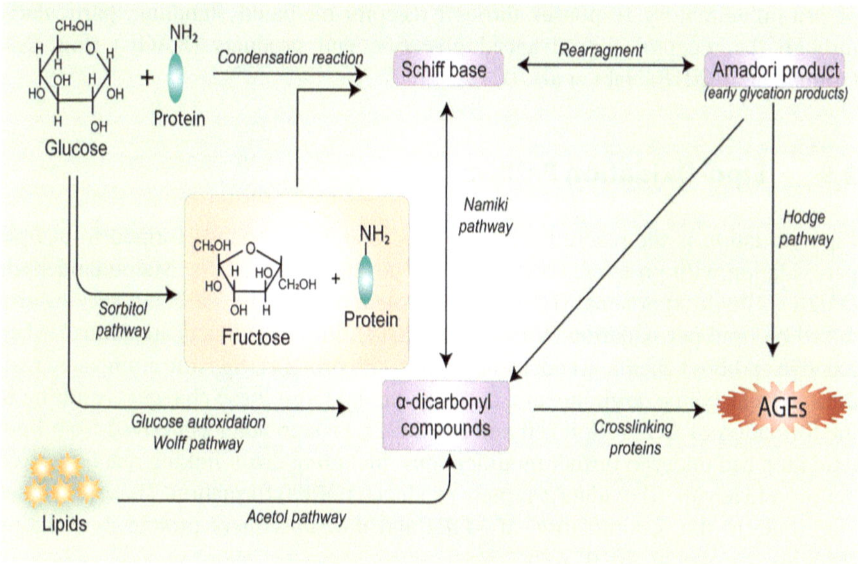

Fig. 2.4 The schematic representation of AGEs formation via different pathways. Nonenzymatic reactions between carbonyl groups of reducing sugars and primary amino groups of proteins result in the formation of Schiff bases. Subsequent Amadori rearrangements transform these Schiff bases into protein-derived aminomethyl ketones. Additionally, 1,2-dicarbonyl compounds, originating from autoxidations of glucose or Schiff bases (and their Amadori products), react with protein amino groups, leading to glycoxidation. This process yields various protein cross-links and modifications, collectively known as Advanced Glycation End Products (AGEs). AGEs are formed through glycoxidative reactions, contributing to protein alterations and potentially impacting cellular function

2.8 Glycoxidation and Lipo-Oxidation Pathways Leading to AGEs Formation

Two key mechanisms that contribute to the production of AGEs are glycoxidation and lipo-oxidation (Ahmad et al. 2018). These pathways involve the interaction of sugars or lipid-derived reactive carbonyl compounds with proteins, leading to AGE formation. Glycoxidation is the process by which reducing carbohydrates such as glucose reacts with proteins to form early glycation products, which then undergo further oxidative modifications to form AGEs (Ahmad et al. 2014). The key steps in the glycoxidation pathway are as follows: The first stage involves a nonenzymatic reaction between a reducing sugar and a protein, which produces reversible Schiff bases or Amadori products. These early glycation products can be rearranged and transformed (Nabi et al. 2021b). These early glycation products are sensitive to oxidative changes mediated by reactive oxygen species (ROS) and carbonyl species (RCS). Further glycation, oxidation, and cross-linking events may occur, resulting in the development of irreversible AGEs (Nabi et al. 2019a). These AGEs can cause protein cross-linking, alteration of protein structure and function, and the induction of pro-inflammatory responses through receptor-mediated signaling, particularly through the receptor for advanced glycation end products (RAGE) (Fig. 2.4) (Ahmad et al. 2018; Nabi et al. 2021a).

2.9 Lipo-Oxidation Pathway

Lipo-oxidation is the reaction of reactive carbonyl compounds formed from lipid peroxidation with proteins, resulting in the production of AGEs. Malondialdehyde (MDA), 4-hydroxynonenal (HNE), and acrolein are reactive carbonyl species produced by lipid peroxidation (Vistoli et al. 2013). Protein adducts are formed when reactive carbonyl chemicals covalently bond with proteins. Specific amino acid residues such as lysine, arginine, and cysteine can undergo these changes, resulting in the formation of AGEs (Spickett and Pitt 2020). Protein adducts derived from lipo-oxidation can undergo further modifications, including cross-linking reactions, oxidation, and advanced oxidation protein product (AOPPs) formation. These processes contribute to the accumulation of AGEs and the subsequent protein dysfunction (Moldogazieva et al. 2019).

2.10 Receptor-Mediated Mechanisms AGEs Uptake and Signaling

RAGE is a cell surface receptor of the immunoglobulin superfamily with numerous ligands. It is expressed by endothelial, immune, neuronal, and pancreatic beta cells, among other cell types (Madonna et al. 2018). Binding of AGEs to RAGE initiates a series of intracellular activities. AGEs can bind to the extracellular domain of RAGE, inducing conformational changes in the receptor and increasing downstream signaling (Bongarzone et al. 2017). Many signaling pathways are activated by this connection, including mitogen-activated protein kinases (MAPKs), nuclear factor-kappa B (NF-kB), and phosphoinositide 3-kinase (PI3K)/Akt (Kierdorf and Fritz 2013). These pathways control gene expression, inflammation, cell survival, and the production of pro-inflammatory cytokines, chemokines, and adhesion molecules that contribute to oxidative stress, inflammation, and cellular dysfunction. Collectively, these pathways contribute to cellular dysfunction (Waghela et al. 2021). Increased AGE build-up and RAGE activation contribute to the development and progression of metabolic disorder including diabetes and cardiovascular diseases. Insulin resistance, reduced pancreatic beta-cell function, and the development of diabetic nephropathy, retinopathy, neuropathy, and cardiovascular disease are all caused by RAGE-mediated signaling pathways (Litwinoff et al. 2015). Targeting the interactions between AGEs and RAGE has been shown to be a potential therapeutic strategy for reducing the negative effects of AGE-RAGE signaling. Inhibition of RAGE activation with specific antagonists or blocking antibodies has shown promise in reducing inflammation, improving insulin sensitivity, and attenuating diabetic complications. Understanding these mechanisms can provide insights into potential therapeutic interventions for improving disease outcomes (Briceno Noriega et al. 2022).

2.11 Impact of Hyperglycemia and Oxidative Stress on AGEs Accumulation

The formation of advanced glycation end products (AGEs) is influenced by hyperglycemia and oxidative stress. Hyperglycemia is defined as high glucose levels in the bloodstream, which is common in people with diabetes or impaired glucose tolerance (Waiz et al. 2023). Excess glucose reacts with proteins, lipids, and nucleic acids, resulting in AGEs formation via glycation. Under prolonged hyperglycemic conditions, AGE accumulation is accelerated because glucose availability boosts glycation processes and the creation of reactive carbonyl species (RCS), which are significant precursors of AGE formation (Nabi et al. 2018; Waiz et al. 2022). However, variety of factors, such as mitochondrial malfunction, increased flux via the polyol pathway, enzyme activation, oxidative stress, a situation characterized by an imbalance between reactive oxygen species (ROS) generation and antioxidant defense systems, can directly influence AGE buildup (Juan et al. 2021).

2.12 Receptor for AGE (RAGE) and Its Expression in Various Tissues

RAGE (receptor for advanced glycation end products) is a cell surface receptor that plays a key role in the pathophysiology of metabolic disorders. RAGE is expressed in various tissues, including the vasculature, kidneys, nervous system, retina, adipose tissue, and pancreas. RAGE activation in the vasculature leads to endothelial dysfunction, inflammation, and increased permeability, contributing to diabetic vascular complications (Alvi et al. 2021). Moreover, it also contributes to diabetic nephropathy, characterized by glomerular hypertrophy, albuminuria, and renal fibrosis (Nabi et al. 2019b, 2021a). RAGE activation also affects the nervous system leading to diabetic neuropathy, cognitive impairment, and neurodegenerative disorders (Wendt et al. 2003). On the contrary, the activation of RAGE within retinal cells is implicated in the pathogenesis of diabetic retinopathy, characterized by retinal inflammation, vascular permeability, and the formation of cellular capillaries. In adipose tissue, RAGE activation is associated with chronic low-grade inflammation, adipose dysfunction, and insulin resistance, collectively fostering the onset of metabolic disorders linked to obesity. Moreover, within the pancreatic milieu, RAGE activation is recognized for its contributory role in beta-cell dysfunction and diminished insulin synthesis, thereby substantiating the progression toward diabetes (Grigsby et al. 2014). Understanding the expression pattern and functional significance of RAGE in diverse tissues provides insight into the mechanisms underlying the development and progression of metabolic disorders. Targeting RAGE and its downstream signaling pathways has emerged as a viable treatment method to attenuate the negative consequences of RAGE activation and to lower the burden of metabolic problems associated with diabetes.

2.13 Signaling Pathways Activated by RAGE-AGEs Interaction

The interaction between the RAGE receptor and advanced glycation end products (AGEs) initiates a cascade of intracellular signaling pathways that play a critical role in the pathogenesis of various metabolic ailments (Rungratanawanich et al. 2021). RAGE-AGE association activates the NF-kB pathway, which translocates NF-kB from the cytoplasm to the nucleus and upregulates pro-inflammatory cytokines and adhesion molecules. This activation promotes inflammation and enhances the cellular response to the interaction between RAGE and AGEs, resulting in increased inflammation (Ganesh Yerra et al. 2013). It also influences the MAPK pathway, phosphorylate transcription factors, and produce pro-inflammatory cytokines and matrix metalloproteinases, contributing to inflammation, cell proliferation, apoptosis, and tissue remodeling (El-Far et al. 2020). Another most affected pathway is the PI3K/Akt pathway, which is essential for cell survival, growth, and metabolism. This route influences metabolic disease pathogenesis by regulating downstream effectors, inflammatory mediators, oxidative stress, and insulin signaling (Fig. 2.5) (Ru and Wang 2020).

Moreover, RAGE activation also stimulates the JAK/STAT pathway and phosphorylates STAT transcription factors. These proteins modulate inflammation, immune responses, and cell proliferation, contributing to the amplification of inflammatory responses induced by RAGE–AGE interaction (Malemud and Pearlman 2009). RAGE activation triggers reactive oxygen species (ROS) through NADPH oxidase activation and mitochondrial dysfunction, leading to oxidative stress and cellular damage (Wendt et al. 2002). These pathways, including NF-κB, amplify the inflammatory response and contribute to the development and progression of diseases, such as diabetes (Fig. 2.5). Moreover, AGEs play a significant role in progression and development of several diseases including atherosclerosis (Table 2.1). These pathways can provide insight into the pathogenesis of metabolic diseases and offer potential therapeutic strategies to mitigate the detrimental effects of RAGE activation and reduce disease-associated complications. Targeting of these pathways may provide potential therapeutic solutions (Teshima et al. 2014).

2.14 Cross Talk Between RAGE and Other Receptors in Metabolic Diseases

In metabolic disorders, the receptor for advanced glycation end products (RAGE) interacts with various receptors and signaling pathways. This interaction increases the inflammatory response, oxidative stress, and cellular dysfunction, all of which contribute to the pathophysiology of many metabolic diseases including diabetes

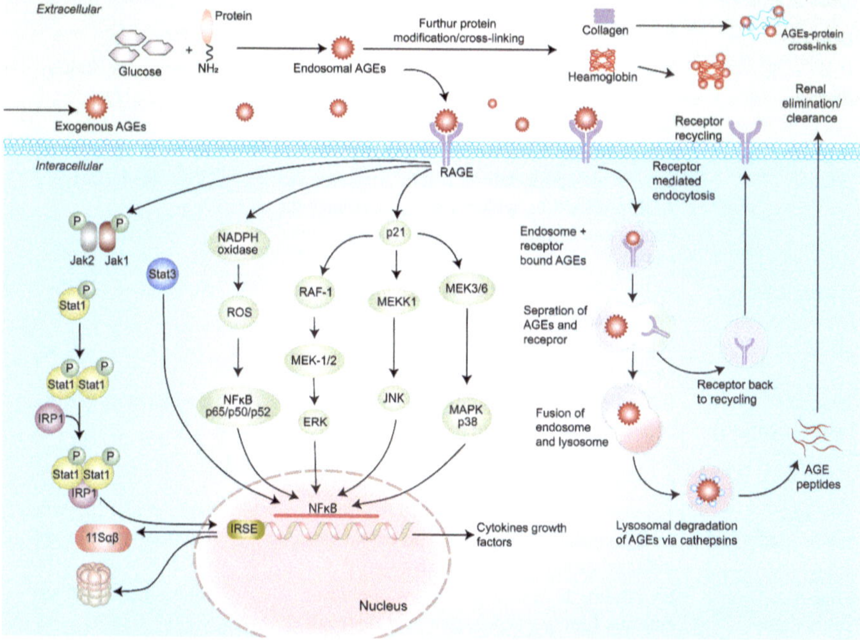

Fig. 2.5 The interaction between AGEs and their receptor RAGE initiates diverse signaling pathways, encompassing the activation of protein kinase C (PKC), tyrosine phosphorylation of JAK-STAT, PI3K-Akt, MAPK, and calcium signaling pathways. The activation of MAPK and PI3K-Akt signaling pathways subsequently triggers the activation of NADPH, fostering the generation of reactive oxygen species. This activation further stimulates the NF-κB signaling pathway, leading to heightened production of cytokines and growth factors. Ultimately, these molecular events contribute to cellular and tissue damage, illustrating the intricate and interconnected nature of signaling pathways involved in the response to AGE–RAGE binding

(Tenório et al. 2019). RAGE and TLRs exchange ligands, resulting in the synergistic activation of inflammatory signaling pathways, such as NF-kB and MAPKs (Rosenberg et al. 2017). The interaction between RAGE and the receptor for interleukin-1 (IL-1R) forms a complex that enhances inflammatory response. RAGE can act as a co-receptor for IL-1R, amplifying the downstream signaling pathways triggered by IL-1β, a potent pro-inflammatory cytokine. This cross talk potentiates the activation of NF-κB and MAPKs, leading to the increased production of inflammatory mediators and exacerbation of tissue damage (Chen and Nuñez 2010). Soluble RAGE (sRAGE), a shortened RAGE receptor, functions as a decoy receptor by competitively binding to RAGE ligands, such as AGEs. sRAGE inhibits ligand binding to cellular RAGE and reduces RAGE-mediated signaling pathways. Reduced sRAGE levels and a decreased sRAGE-to-RAGE ratio have been observed in metabolic diseases, leading to enhanced RAGE signaling and increased inflammation (Guo et al. 2012b).

Furthermore, RAGE activation can interfere with insulin signaling by inhibiting IR autophosphorylation and phosphorylation of downstream insulin receptor

Table 2.1 Role of AGEs in the development of diseases

Diseases	Role of AGEs	References
Parkinson's disease	The potential of AGEs to cross-link alpha synuclein may also account for the association between AGEs and Parkinson's disease.	Rungratanawanich et al. (2021)
Alzheimer's disease	A significant characteristic of the disease, tau protein phosphorylation, can be triggered by AGE production, which also up-regulates the expression of the amyloid precursor protein (APP).	Ko et al. (2010) and Byun et al. (2017)
Atherosclerosis	AGEs bind to RAGE on endothelial, smooth muscle, and monocytes, activating pathways causing inflammation, atherogenesis, vasoconstriction, and atherosclerosis.	Singh et al. (2022)
Diabetic cardiomyopathy	It has been demonstrated that AGEs cross connect the Ryanodine receptor and SERCA2a domains in cardiomyocytes, changing the homeostasis of calcium.	Bodiga et al. (2014)
Diabetic nephropathy	Engagements of RAGE with AGEs elicit oxidative stress generation and evoke vascular inflammation and thrombogenesis.	Nabi et al. (2019a) and Parwani and Mandal (2023)
Diabetic retinopathy	By binding to several cell types, including endothelium, pericytes, glia, and microglia, AGEs can initiate intracellular signaling via RAGE.	Sharma et al. (2012)
Osteoarthritis (OA)	RAGE on chondrocytes is activated by AGEs, hardening the matrix and decreasing synthetic capability. This could make it more difficult for articular chondrocytes to preserve matrix integrity, which would raise the risk of osteoarthritis.	DeGroot et al. (1999, 2001)
Cancer	AGE-RAGE interaction causing a rise in the production of ROS and inflammatory processes.	Schröter and Höhn (2019)

substrates (IRS). This disrupts insulin-stimulated glucose uptake and metabolism, contributing to insulin resistance, a hallmark of metabolic diseases, such as T2DM (Khalid et al. 2021). PPARs, which regulate lipid and glucose metabolism can be disrupted by RAGE interaction and lead to insulin resistance and metabolic disorders. This disruption of PPAR signaling amplifies inflammatory responses, oxidative stress, and cellular dysfunction in metabolic diseases (Yamagishi et al. 2009). Deregulation of these processes contributes to the pathophysiology of metabolic diseases and their consequences. Understanding the interplay between RAGE and other receptors offers insight into the complex molecular mechanisms underlying metabolic diseases, and targeting these pathways may offer potential therapeutic strategies (Saad et al. 2015).

2.15 Role of AGEs in the Progression of Diabetes

The glycation process, with its intricate nature and gradual unfolding, is subject to the presence of substrates and can lead to unprompted harm to proteins within physiological systems. Under normal physiological circumstances, there exists a baseline production of AGEs; however, this rate accelerates substantially when persistent hyperglycemic conditions persist due to an increase in glucose availability. Initially, it was believed that advanced glycation primarily impacted long-lived proteins, such as collagen and elastin; however, extended periods of elevated hyperglycemia can also cause alterations in short-lived proteins, including circulating plasma proteins and lipids (Nabi et al. 2018, 2020; Ahmad et al. 2022). Diabetic patients, particularly those with impaired renal function, typically exhibit significantly elevated levels of circulating AGEs. AGEs induce harm by forming intermolecular and intramolecular cross-links, leading to irreversible disruption of protein structure and function, ultimately resulting in deleterious effects (Alenazi et al. 2022). This covalent cross-link formation deactivates biologically active proteins and enzymes, rendering them resistant to proteolytic digestion, creating catalytic sites for the formation of reactive oxygen species (ROS), instigating pro-inflammatory events, disrupting intracellular signaling, and contributing to various metabolic and biochemical perturbations, including those associated with diabetes (Rafi et al. 2023; Ashraf et al. 2015; Nabi et al. 2021b). Furthermore, advancing age is an important risk factor for the development of diabetes mellitus, and this is likely due to the accumulation of AGEs in tissues. The interaction of AGEs with a variety of cell surface receptors triggers a range of cell-mediated pathophysiological responses, which may help to explain the link between AGE accumulation and the onset of diabetes. When AGEs bind to their respective receptors, they initiate a cascade of signals that directly affect cellular function and metabolism by increasing inflammation and oxidative stress (Asadipooya and Uy 2019).

2.16 Future Perspective

AGEs represent a complex mixture of compounds that can arise either endogenously or through dietary sources. Over the past five decades, there has been a marked shift in dietary patterns toward heat-processed, tasty, and appetite-stimulating foods rich in AGEs, which has significantly contributed to increased oxidative stress and inflammation, correlating with the rising prevalence of diabetes. AGEs are formed through nonenzymatic glycation and cross-linking of proteins, lipids, and nucleic acids, leading to structural and functional modifications and resulting in detrimental effects on health. These effects include the release of pro-inflammatory molecules, compromise of mitochondrial oxidative stress, activation of the AGEs/Receptor for AGEs (RAGE) signaling cascade, and induction of transcription factors that upregulate genes implicated in diabetic complications. Therefore, investigating the role of AGEs in diabetic complications is a critical area of contemporary research, aimed at elucidating the critical biochemical mechanisms involved in the development of type 2 diabetes and its complications. Recent studies evaluating circulating AGE concentrations in individuals with chronic type 2 diabetes undergoing oral hypoglycemic treatment aim to establish connections between AGE levels, diabetes pathobiology, and complication severity. Modifying dietary AGE intake through lifestyle modifications such as restricting dietary AGE intake, engaging in moderate exercise, and ceasing cigarette smoking emerges as a crucial intervention to modulate the AGEs/RAGE axis and promote health. However, a challenge lies in developing a specific, sensitive, cost-effective method for measuring circulatory AGEs, given the limitations of sophisticated laboratory techniques for widespread monitoring. Advancing our understanding of various AGEs and their associated cellular responses is essential for informing logical intervention strategies.

Acknowledgments The authors would like to acknowledge Integral University for providing necessary facilities.

References

Abdul Kadir A, Clarke K, Evans RD (2020) Cardiac ketone body metabolism. Biochim Biophys Acta Mol Basis Dis 1866(6):165739

Abdul-Ghani MA, Tripathy D, DeFronzo RA (2006) Contributions of β-cell dysfunction and insulin resistance to the pathogenesis of impaired glucose tolerance and impaired fasting glucose. Diabetes Care 29(5):1130–1139

Ahmad S, Khan MS, Akhter F et al (2014) Glycoxidation of biological macromolecules: a critical approach to halt the menace of glycation. Glycobiology 24(11):979–990

Ahmad S, Akhter F, Shahab U et al (2018) Do all roads lead to the Rome? The glycation perspective! Semin Cancer Biol 49:9–19

Ahmad S, Nabi R, Alvi SS et al (2022) Carvacrol protects against carbonyl osmolyte-induced structural modifications and aggregation to serum albumin: insights from physicochemical and molecular interaction studies. Int J Biol Macromol 213:663–674

Ahmad M, Wolberg A, Kahwaji CI (2024) Biochemistry, Electron Transport Chain. [Updated 2023 Sep 4]. In: StatPearls [Internet]. Treasure Island (FL): StatPearls Publishing https://www.ncbi.nlm.nih.gov/books/NBK526105/

Akram M (2014) Citric acid cycle and role of its intermediates in metabolism. Cell Biochem Biophys 68(3):475–478

Alenazi F, Saleem M, Khaja ASS et al (2022) Metformin encapsulated gold nanoparticles (MTF-GNPs): a promising antiglycation agent. Cell Biochem Funct 40(7):729–741

Alvi SS, Nabi R, Khan MS, Akhter F, Ahmad S, Khan MS (2021) Glycyrrhizic acid scavenges reactive carbonyl species and attenuates glycation-induced multiple protein modification: an in vitro and in Silico study. Tatone C, ed. Oxid Med Cell Longev 2021:1–14

Arévalo-Ruiz M, Doria F, Belmonte-Reche E et al (2017) Synthesis, binding properties, and differences in cell uptake of G-quadruplex ligands based on carbohydrate naphthalene diimide conjugates. Chem—A Eur J 23(9):2157–2164

Arnold SE, Arvanitakis Z, Macauley-Rambach SL et al (2018) Brain insulin resistance in type 2 diabetes and Alzheimer disease: concepts and conundrums. Nat Rev Neurol 14(3):168–181

Asadipooya K, Uy EM (2019) Advanced glycation end products (AGEs), receptor for AGEs, diabetes, and bone: review of the literature. J Endocr Soc 3(10):1799–1818

Ashraf JM, Ahmad S, Choi I et al (2015) Recent advances in detection of AGEs: immunochemical, bioanalytical and biochemical approaches. IUBMB Life 67(12):897–913

Bhatti JS, Bhatti GK, Reddy PH (2017) mitochondrial dysfunction and oxidative stress in metabolic disorders—a step towards mitochondria based therapeutic strategies. Biochim Biophys Acta Mol Basis Dis 1863(5):1066–1077

Bianchi C, Penno G, Romero F, Del Prato S, Miccoli R (2007) Treating the metabolic syndrome. Expert Rev Cardiovasc Ther 5(3):491–506

Bodiga VL, Eda SR, Bodiga S (2014) Advanced glycation end products: role in pathology of diabetic cardiomyopathy. Heart Fail Rev 19(1):49–63

Bodner GM (1986) Metabolism Part II: the tricarboxylic acid (TCA), citric acid, or Krebs cycle. J Chem Educ 63(8):673

Bolk J, van der Ploeg T, Cornel JH, Arnold AER, Sepers J, Umans VAWM (2001) Impaired glucose metabolism predicts mortality after a myocardial infarction. Int J Cardiol 79(2–3):207–214

Bongarzone S, Savickas V, Luzi F, Gee AD (2017) Targeting the receptor for advanced glycation endproducts (RAGE): a medicinal chemistry perspective. J Med Chem 60(17):7213–7232

Briceno Noriega D, Zenker HE, Croes C-A et al (2022) Receptor-mediated effects of advanced glycation end products (AGEs) on innate and adaptative immunity: relevance for food allergy. Nutrients 14(2):371

Byun K, Yoo Y, Son M et al (2017) Advanced glycation end-products produced systemically and by macrophages: a common contributor to inflammation and degenerative diseases. Pharmacol Ther 177:44–55

Campbell I (2014) Intermediary metabolism. Anaesth Intensive Care Med 15(7):350–352

Chadt A, Al-Hasani H (2020) Glucose transporters in adipose tissue, liver, and skeletal muscle in metabolic health and disease. Pflügers Arch Eur J Physiol 472(9):1273–1298

Chandel NS (2021) Glycolysis. Cold Spring Harb Perspect Biol 13(5):a040535

Chen GY, Nuñez G (2010) Sterile inflammation: sensing and reacting to damage. Nat Rev Immunol 10(12):826–837

Cherkas A, Holota S, Mdzinarashvili T, Gabbianelli R, Zarkovic N (2020) Glucose as a major antioxidant: when, what for and why it fails? Antioxidants 9(2):140

de Luca C, Olefsky JM (2008) Inflammation and insulin resistance. FEBS Lett 582(1):97–105

DeGroot J, Verzijl N, Bank RA, Lafeber FPJG, Bijlsma JWJ, TeKoppele JM (1999) Age-related decrease in proteoglycan synthesis of human articular chondrocytes: the role of nonenzymatic glycation. Arthritis Rheum 42(5):1003–1009

DeGroot J, Verzijl N, Budde M, Bijlsma JWJ, Lafeber FPJG, TeKoppele JM (2001) Accumulation of advanced glycation end products decreases collagen turnover by bovine chondrocytes. Exp Cell Res 266(2):303–310

Dixon WL, ap Rees T (1980) Identification of the regulatory steps in glycolysis in potato tubers. Phytochemistry 19(7):1297–1301

du Plessis S, Agarwal A, Mohanty G, van der Linde M (2015) Oxidative phosphorylation versus glycolysis: what fuel do spermatozoa use? Asian J Androl 17(2):230

El-Far AH, Sroga G, Al Jaouni SK, Mousa SA (2020) Role and mechanisms of RAGE-ligand complexes and RAGE-inhibitors in cancer progression. Int J Mol Sci 21(10):3613

Evans A (2013) Digestion resistant carbohydrates. In: Resistant starch. Wiley, Chichester, pp 95–110

Fothergill-Gilmore LA, Michels PAM (1993) Evolution of glycolysis. Prog Biophys Mol Biol 59(2):105–235

Gadupudi GS, Klingelhutz AJ, Robertson LW (2016) Diminished phosphorylation of CREB is a key event in the dysregulation of gluconeogenesis and glycogenolysis in PCB126 hepatotoxicity. Chem Res Toxicol 29(9):1504–1509

Galgani J, Ravussin E (2008) Energy metabolism, fuel selection and body weight regulation. Int J Obes 32(S7):S109–S119

Ganesh Yerra V, Negi G, Sharma SS, Kumar A (2013) Potential therapeutic effects of the simultaneous targeting of the Nrf2 and NF-κB pathways in diabetic neuropathy. Redox Biol 1(1):394–397

Giorgino F, de Robertis O, Laviola L et al (2000) The sentrin-conjugating enzyme mUbc9 interacts with GLUT4 and GLUT1 glucose transporters and regulates transporter levels in skeletal muscle cells. Proc Natl Acad Sci 97(3):1125–1130

Goldstein BJ (2002) Insulin resistance as the core defect in type 2 diabetes mellitus. Am J Cardiol 90(5):3–10

Gonzalez-Franquesa A, Patti M-E (2017) Insulin resistance and mitochondrial dysfunction. In: Mitochondrial dynamics in cardiovascular medicine. Springer, Cham, pp 465–520

Goodman BE (2010) Insights into digestion and absorption of major nutrients in humans. Adv Physiol Educ 34(2):44–53

Gray LR, Tompkins SC, Taylor EB (2014) Regulation of pyruvate metabolism and human disease. Cell Mol Life Sci 71(14):2577–2604

Grigsby JG, Cardona SM, Pouw CE et al (2014) The role of microglia in diabetic retinopathy. J Ophthalmol 2014:1–15

Grimes CA, Jope RS (2001) The multifaceted roles of glycogen synthase kinase 3β in cellular signaling. Prog Neurobiol 65(4):391–426

Guo WA, Knight PR, Raghavendran K (2012a) The receptor for advanced glycation end products and acute lung injury/acute respiratory distress syndrome. Intensive Care Med 38(10):1588–1598

Guo X, Li H, Xu H et al (2012b) Glycolysis in the control of blood glucose homeostasis. Acta Pharm Sin B 2(4):358–367

Gupta R, Gupta N (2021) Tricarboxylic acid cycle. In: Fundamentals of bacterial physiology and metabolism. Springer Singapore, pp 327–346

Haber RS (1993) Tissue distribution of the human GLUT3 glucose transporter. Endocrinology 132(6):2538–2543

Hatting M, Tavares CDJ, Sharabi K, Rines AK, Puigserver P (2018) Insulin regulation of gluconeogenesis. Ann N Y Acad Sci 411(1):21–35

Herdt TH (2000) Ruminant adaptation to negative energy balance. Vet Clin North Am Food Anim Pract 16(2):215–230

Herman MA (2006) Glucose transport and sensing in the maintenance of glucose homeostasis and metabolic harmony. J Clin Invest 116(7):1767–1775

Hu R, Wang M, Ni S et al (2020) Salidroside ameliorates endothelial inflammation and oxidative stress by regulating the AMPK/NF-κB/NLRP3 signaling pathway in AGEs-induced HUVECs. Eur J Pharmacol 867:172797

Juan CA, Pérez de la Lastra JM, Plou FJ, Pérez-Lebeña E (2021) The chemistry of reactive oxygen species (ROS) revisited: outlining their role in biological macromolecules (DNA, lipids and proteins) and induced pathologies. Int J Mol Sci 22(9):4642

Kaiser N, Leibowitz G, Nesher R (2003) Glucotoxicity and β-cell failure in type 2 diabetes mellitus. J Pediatr Endocrinol Metab 16(1):5–22

Khalid M, Alkaabi J, Khan MAB, Adem A (2021) Insulin signal transduction perturbations in insulin resistance. Int J Mol Sci 22(16):8590

Kierdorf K, Fritz G (2013) RAGE regulation and signaling in inflammation and beyond. J Leukoc Biol 94(1):55–68

Klip A, Pâquet MR (1990) Glucose transport and glucose transporters in muscle and their metabolic regulation. Diabetes Care 13(3):228–243

Ko S-Y, Lin Y-P, Lin Y-S, Chang S-S (2010) Advanced glycation end products enhance amyloid precursor protein expression by inducing reactive oxygen species. Free Radic Biol Med 49(3):474–480

Leturque A, Brot-Laroche E, Le Gall M (2009) GLUT2 mutations, translocation, and receptor function in diet sugar managing. Am J Physiol Endocrinol Metab 296(5):E985–E992

Litwinoff E, Hurtado del Pozo C, Ramasamy R, Schmidt A (2015) Emerging targets for therapeutic development in diabetes and its complications: the RAGE signaling pathway. Clin Pharmacol Ther 98(2):135–144

Longo R, Peri C, Cricrì D et al (2019) Ketogenic diet: a new light shining on old but gold biochemistry. Nutrients 11(10):2497

Madonna R, Pieragostino D, Balistreri CR et al (2018) Diabetic macroangiopathy: pathogenetic insights and novel therapeutic approaches with focus on high glucose-mediated vascular damage. Vascul Pharmacol 107:27–34

Malemud C, Pearlman E (2009) Targeting JAK/STAT signaling pathway in inflammatory diseases. Curr Signal Transduct Ther 4(3):201–221

Mann J, Cummings JH, Englyst HN et al (2007) FAO/WHO scientific update on carbohydrates in human nutrition: conclusions. Eur J Clin Nutr 61(S1):S132–S137

Maughan R (2009) Carbohydrate metabolism. Surgery 27(1):6–10

Meyer EH, Letts JA, Maldonado M (2022) Structural insights into the assembly and the function of the plant oxidative phosphorylation system. New Phytol 235(4):1315–1329

Moldogazieva NT, Mokhosoev IM, Mel'nikova TI, Porozov YB, Terentiev AA (2019) Oxidative stress and advanced lipoxidation and glycation end products (ALEs and AGEs) in aging and age-related diseases. Oxid Med Cell Longev 2019:1–14

Mul JD, Stanford KI, Hirshman MF, Goodyear LJ (2015) Exercise and regulation of carbohydrate metabolism. Prog Mol Biol Transl Sci 135:17–37

Nabi R, Alvi SS, Khan RH, Ahmad S, Ahmad S, Khan MS (2018) Antiglycation study of HMG-R inhibitors and tocotrienol against glycated BSA and LDL: a comparative study. Int J Biol Macromol 116:983–992

Nabi R, Alvi SS, Saeed M, Ahmad S, Khan MS (2019a) Glycation and HMG-CoA reductase inhibitors: implication in diabetes and associated complications. Curr Diabetes Rev 15(3):213–223

Nabi R, Alvi SS, Shah A et al (2019b) Modulatory role of HMG-CoA reductase inhibitors and ezetimibe on LDL-AGEs-induced ROS generation and RAGE-associated signalling in HEK-293 cells. Life Sci 235:116823

Nabi R, Alvi SS, Shah MS et al (2020) A biochemical & biophysical study on in-vitro anti-glycating potential of iridin against d-Ribose modified BSA. Arch Biochem Biophys 686:108373

Nabi R, Alvi S, Alouffi S et al (2021a) Amelioration of neuropilin-1 and RAGE/matrix metalloproteinase-2 pathway-induced renal injury in diabetic rats by rosuvastatin. Arch Biol Sci 73(2):265–278

Nabi R, Alvi SS, Shah A, et al (2021b) Ezetimibe attenuates experimental diabetes and renal pathologies via targeting the advanced glycation, oxidative stress and AGE-RAGE signalling in rats. Arch Physiol Biochem 129(4):831–846

Nath S, Villadsen J (2015) Oxidative phosphorylation revisited. Biotechnol Bioeng 112(3):429–437

Neupane P, Bhuju S, Thapa N, Bhattarai HK (2019) ATP synthase: structure, function and inhibition. Biomol Concepts 10(1):1–10

Niaz K, Khan F, Shah MA (2020) Analysis of carbohydrates (monosaccharides, polysaccharides). In: Recent advances in natural products analysis. Elsevier, Amsterdam, pp 621–633

Osellame LD, Blacker TS, Duchen MR (2012) Cellular and molecular mechanisms of mitochondrial function. Best Pract Res Clin Endocrinol Metab 26(6):711–723

Parwani K, Mandal P (2023) Role of advanced glycation end products and insulin resistance in diabetic nephropathy. Arch Physiol Biochem 129(1):95–107

Pérez-Martin A, Raynaud E, Mercier J (2001) Insulin resistance and associated metabolic abnormalities in muscle: effects of exercise. Obes Rev 1:47–59

Pösö AR, Puolanne E (2005) Carbohydrate metabolism in meat animals. Meat Sci 70(3):423–434

Poznyak A, Grechko AV, Poggio P, Myasoedova VA, Alfieri V, Orekhov AN (2020) The diabetes mellitus–atherosclerosis connection: the role of lipid and glucose metabolism and chronic inflammation. Int J Mol Sci 21(5):1835

Qi X, Tester RF (2019) Fructose, galactose and glucose—in health and disease. Clin Nutr ESPEN 33:18–28

Rafi Z, Khan M, Khan S et al (2023) Glycation derived AuNPs bioconjugated novel herbal drug isoferulic acid: as a potential anti-glycation, anti-diabetic and antineoplastic agent. Colloids Surf A Physicochem Eng Asp 676:132148

Ramsay RR (2009) Electron carriers and energy conservation in mitochondrial respiration. ChemTexts 5(2):9

Riera JJ, Schousboe A, Waagepetersen HS, Howarth C, Hyder F (2008) The micro-architecture of the cerebral cortex: functional neuroimaging models and metabolism. Neuroimage 40(4):1436–1459

Rosenberg JH, Rai V, Dilisio MF, Agrawal DK (2017) Damage-associated molecular patterns in the pathogenesis of osteoarthritis: potentially novel therapeutic targets. Mol Cell Biochem 434(1–2):171–179

Ru J, Wang Y (2020) Osteocyte apoptosis: the roles and key molecular mechanisms in resorption-related bone diseases. Cell Death Dis 11(10):846

Rungratanawanich W, Qu Y, Wang X, Essa MM, Song B-J (2021) Advanced glycation end products (AGEs) and other adducts in aging-related diseases and alcohol-mediated tissue injury. Exp Mol Med 53(2):168–188

Saad MI, Abdelkhalek TM, Saleh MM et al (2015) Insights into the molecular mechanisms of diabetes-induced endothelial dysfunction: focus on oxidative stress and endothelial progenitor cells. Endocrine 50(3):537–567

Santiago JA, Potashkin JA (2013) Shared dysregulated pathways lead to Parkinson's disease and diabetes. Trends Mol Med 19(3):176–186

Schmidt-Rohr K (2020) Oxygen is the high-energy molecule powering complex multicellular life: fundamental corrections to traditional bioenergetics. ACS Omega 5(5):2221–2233

Schröter D, Höhn A (2019) Role of advanced glycation end products in carcinogenesis and their therapeutic implications. Curr Pharm Des 24(44):5245–5251

Sharma Y, Saxena S, Mishra A, Saxena A, Natu SM (2012) Advanced glycation end products and diabetic retinopathy. J Ocul Biol Dis Infor 5(3–4):63–69

Silvestri MG, Desantis G, Mitchell M, Wong C-H (2003) Asymmetric aldol reactions using aldolases. Topics in stereochemistry 23:267–342. https://doi.org/10.1002/0471224499.ch5

Singh S, Siva BV, Ravichandiran V (2022) Advanced glycation end products: key player of the pathogenesis of atherosclerosis. Glycoconj J 39(4):547–563

Softic S, Cohen DE, Kahn CR (2016) Role of dietary fructose and hepatic De novo lipogenesis in fatty liver disease. Dig Dis Sci 61(5):1282–1293

Spickett CM, Pitt AR (2020) Modification of proteins by reactive lipid oxidation products and biochemical effects of lipoxidation. Essays Biochem 64(1):19–31

Strasser B (2013) Physical activity in obesity and metabolic syndrome. Ann N Y Acad Sci 1281(1):141–159

Tang C, Cao G, Zhao W et al (2021) Lactobacillus acidophilus NX2-6 improved high-fat diet-induced glucose metabolism disorder independent of promotion of insulin secretion in mice. J Agric Food Chem 69(51):15598–15610

Tenório MB, Ferreira RC, Moura FA, Bueno NB, de Oliveira ACM, Goulart MOF (2019) Cross-talk between oxidative stress and inflammation in preeclampsia. Oxid Med Cell Longev 2019:1–26

Teshima Y, Takahashi N, Nishio S et al (2014) Production of reactive oxygen species in the diabetic heart. Circ J 78(2):300–306

Thurmond DC, Pessin JE (2001) Molecular machinery involved in the insulin-regulated fusion of GLUT4-containing vesicles with the plasma membrane. Mol Membr Biol 18(4):237–245

Tirone TA, Brunicardi FC (2001) Overview of glucose regulation. World J Surg 25(4):461–467

Vistoli G, De Maddis D, Cipak A, Zarkovic N, Carini M, Aldini G (2013) Advanced glycoxidation and lipoxidation end products (AGEs and ALEs): an overview of their mechanisms of formation. Free Radic Res 47(sup1):3–27

Waghela BN, Vaidya FU, Ranjan K, Chhipa AS, Tiwari BS, Pathak C (2021) AGE-RAGE synergy influences programmed cell death signaling to promote cancer. Mol Cell Biochem 476(2):585–598

Waiz M, Alvi SS, Salman Khan M (2022) Potential dual inhibitors of Pcsk-9 and Hmg-R from natural sources in cardiovascular risk management. EXCLI J 21:47–76

Waiz M, Alvi SS, Khan MS (2023) Association of circulatory PCSK-9 with biomarkers of redox imbalance and inflammatory cascades in the prognosis of diabetes and associated complications: a pilot study in the Indian population. Free Radic Res 57:1–14

Waiz M, Alvi SS, Ahmad S, Khan MS (2024) Association of PCSK-9 with the biomarkers of Type-2 diabetes and its complications in the Indian population: a pilot study. Clinical Laboratory 70(2), p286

Watson RT, Kanzaki M, Pessin JE (2004) Regulated membrane trafficking of the insulin-responsive glucose transporter 4 in adipocytes. Endocr Rev 25(2):177–204

Wendt T, Bucciarelli L, Qu W et al (2002) Receptor for advanced glycation endproducts (RAGE) and vascular inflammation: insights into the pathogenesis of macrovascular complications in diabetes. Curr Atheroscler Rep 3:228–237

Wendt TM, Tanji N, Guo J et al (2003) RAGE drives the development of glomerulosclerosis and implicates podocyte activation in the pathogenesis of diabetic nephropathy. Am J Pathol 162(4):1123–1137

Withers DJ, White M (2000) Perspective: the insulin signaling system—a common link in the pathogenesis of type 2 diabetes. Endocrinology 141(6):1917–1921

Wu C, Khan SA, Lange AJ (2005) Regulation of glycolysis—role of insulin. Exp Gerontol 40(11):894–899

Yamagishi S, Nakamura K, Matsui T (2009) Regulation of advanced glycation end product (AGE)-receptor (RAGE) system by PPAR-gamma agonists and its implication in cardiovascular disease. Pharmacol Res 60(3):174–178

Zhang H, Ma J, Tang K, Huang B (2021) Beyond energy storage: roles of glycogen metabolism in health and disease. FEBS J 288(12):3772–3783

Zheng J (2012) Energy metabolism of cancer: glycolysis versus oxidative phosphorylation (review). Oncol Lett 4(6):1151–1157

Protein Metabolism and Its Profiling for the Diagnosis of Metabolic Disorders

3

Sameen Shafi, Hafizurrahman Khan, and Preeti Bajpai

Abstract

Proteins play a pivotal role as essential macromolecules, constructed from amino acids, and are vital for the intricate functions within cells. In maintaining cellular well-being, a delicate equilibrium prevails between the synthesis and breakdown of proteins. Yet, when subjected to heightened catabolic activity triggered by epigenetic influences, disturbances in metabolism can emerge. These disruptions manifest as Hereditary Metabolic Disorders (HMDs) and genetic disease conditions predominantly inherited through autosomal recessive traits. HMDs, characterized by inherent errors in protein metabolism enzymes, impair the body's capacity to appropriately process proteins and other crucial substances. Diagnosing these disorders can be challenging due to varied and nonspecific symptoms. Early detection is vital and involves basic metabolic tests in blood or urine. Advanced techniques, such as ion exchange chromatography, mass spectrometry, and next-generation sequencing (NGS), aid in identifying specific biomarkers and genetic mutations associated with these conditions. The use of advanced diagnostic methods and promising metabolomic approaches with biomarkers holds the potential in improving the management of HMDs and provides better outcomes for affected individuals. The chapter will center on protein metabolic inherited disorders, exploring the diverse diagnostic techniques employed for their identification and management.

S. Shafi · P. Bajpai (✉)
Department of Zoology, Mahatma Gandhi Central University, Motihari, Bihar, India
e-mail: preetibajpai@mgcub.ac.in

H. Khan
Department of Zoology, Maharani Janki Kunwar College, Bettiah, Bihar, India

R. L. Singh et al. (eds.), *Clinical Applications of Biomolecules in Disease Diagnosis*, https://doi.org/10.1007/978-981-97-4723-8_3

Keywords
Inborn metabolic disorders · Amino acid disorders · Ion exchange chromatography · Tandem mass spectrometry · Reverse-phase high-performance liquid chromatography

3.1 Introduction

Proteins are the building blocks of life, essential for the structure and function of cells. Protein catabolism into amino acids is a finely orchestrated process essential for maintaining cellular function, energy balance, and overall homeostasis. The catabolism unfolds the proteins through a meticulously regulated enzymatic cascade. Proteases, the chief enzyme, catalyze the hydrolysis of peptide bonds, culminating in the fragmentation of proteins. Endopeptidases are released by the mucosal lining of the stomach and the exocrine pancreas to cleave the polypeptide chain. The last amino acids from the C or N terminus of a dipeptide or tripeptide are removed by the exopeptidase aiding absorption at the microvilli. The process is intricately regulated by various cellular signals that respond to nutritional cues, environmental stresses, and hormonal fluctuations. This regulatory network fine-tunes the pace and intensity of protein catabolism. Protein catabolism not only facilitates the replenishment of amino acid pools but also provides fuels for critical cellular functions (Bröer and Gauthier-Coles 2022).

Amino acids, serving as the fundamental constituents of proteins, are essential for sustaining life. The delicate harmony between protein synthesis and degradation is crucial for overall well-being. Amino acids exhibit diverse roles in biological systems, playing integral parts in biosynthetic pathways that contribute to the formation of various biomolecules. Beyond their role in biosynthesis, these compounds are pivotal in neurotransmission processes and contribute significantly to numerous essential physiological functions. Upon assimilation from dietary sources, amino acids serve dual purposes. They not only act as building blocks for protein and biomolecule synthesis but also serve as potential sources of energy. Through oxidative pathways, amino acids undergo catabolism, yielding urea and carbon dioxide, valuable as energy substrates (Blanco and Blanco 2022). The upcoming chapter will delve into a comprehensive exploration of amino acid metabolism, encompassing various metabolic disorders and the diagnostic methodologies utilized for their detection.

3.1.1 Amino Acid Metabolism

Amino acids represent a class of organic compounds characterized by the presence of amino (-NH2) and carboxyl (-COOH) functional groups. They can be categorized into α-, β-, γ-, and δ-amino acids based on the relative positioning of these functional groups within their core molecular structure. The 22 alpha-amino acids serve as the essential building blocks of proteins, with nine of them classified as essential amino acids. Their essentiality arises from the fact that the human body cannot autonomously synthesize them, hence, they are acquired through dietary sources. In contrast, nonessential amino acids are synthesized through various metabolic pathways, often derived from intermediates of glycolysis, the citric acid cycle (Krebs cycle), or the pentose phosphate pathway. These pathways involve enzymatic reactions converting precursor molecules into specific amino acids, crucial for maintaining overall health (Litwack 2021).

Proteins are constructed by linking amino acids through peptide bonds, forming polypeptide chains. Postsynthesis, these chains frequently undergo posttranslational modifications, contributing to the diversity of their structure and function. Additionally, complex protein structures emerge when multiple polypeptide chains combine (Nakamoto 2011). The proteins obtained from our diet undergo metabolism facilitated by a variety of digestive enzymes in the stomach and intestine. Following digestion, the resulting free amino acid monomers are transported via the portal vein to the liver. In the liver, immunoglobulins, albumin, and other nonessential amino acids are synthesized. Amino acids enter cells through facilitated active diffusion, relying on specific amino acid transporters (AAT). These transporters function not just as conduits, but also as monitors of amino acid levels, offering early indications in the event of nutritional deficiencies (Bröer and Gauthier-Coles 2022). The metabolic fate of amino acids is determined by the body's metabolic state. Once proteins have fulfilled their function, they are no longer needed and can be degraded by lysosomal enzymes. Protein catabolism is intricately regulated to maintain cellular protein homeostasis. Regulatory mechanisms, such as ubiquitination and autophagy, respond to cellular signals related to nutrient availability, energy status, and stress (Díaz-Villanueva et al. 2015). The ubiquitin-proteasome system (UPS) attaches ubiquitin molecules to target proteins, marking them for degradation by the proteasome. This complex breaks down ubiquitin-tagged proteins into short peptides, further degraded into free monomers. These amino acids then serve as substrates for synthesizing other proteins required in metabolic processes (Tanaka 2009).

The catabolism of proteins is essential for cellular turnover and is particularly crucial during malnutrition. Amino acids released can enter the citric acid cycle to produce energy, meeting the body's demand. In a healthy state, oxidative deamination breaks down amino acids into carbon and nitrogen skeletons. This process starts with transferring the amino group to alpha-ketoglutaric acid, forming glutamate, which enters the urea cycle. The resulting carbon can be utilized for energy generation, while the nitrogen, in the form of ammonia, is toxic. The liver's urea cycle converts toxic ammonia into urea, a less-toxic compound easily eliminated through the kidneys (Barmore et al. 2023). Broadly, multiple enzymes are intricately connected to the metabolism of particular amino acids. Consequently, any impairment in an enzyme corresponds to a malfunction in the metabolism of the associated amino acid, which serves as its substrate. Such dysfunction can result in the accumulation and/or deficiency of specific metabolites, as well as the generation of harmful metabolites. These alterations thereby exhibit the potential to inflict significant damage to the health of affected individuals.

3.2 Protein Metabolic Disorders

Protein metabolic disorders or more commonly called amino acid disorders (Table 3.1) are a group of inborn errors of metabolism caused by the inherited defects in pathways involved in amino acid metabolism. These disorders typically follow an autosomal recessive mode of inheritance, i.e., both parents must carry a mutated gene for the disorder to be expressed in their child. The inherited defect can be seen in the pathway as enzyme insufficiency or by the diminished biological activity of the enzyme.

Amino acid disorders are indicated with its varied and broad clinical symptoms. Some of the clinical manifestations may include developmental delays, intellectual disabilities, failure to thrive, seizures, and various neurological and metabolic abnormalities. The management of amino acid disorders typically involves a combination of medical support and dietary interventions. These interventions aim to limit the consumption of specific amino acids that the individual's metabolism cannot properly process. This may involve dietary restrictions, the use of supplements, and the incorporation of medical foods specially formulated to meet the nutritional

Table 3.1 List of amino acid disorders

Concerned amino acid	Disorders name	Enzyme involved
Aromatic amino acid disorder	Phenylketonuria; tyrosinemia I; tyrosinemia II; tyrosinemia III	Phenylalanine hydroxylase (Muntau et al. 2019) Fumarylacetoacetate hydrolase (FAH) (Bas et al. 2003) Tyrosine aminotransferase (TAT) (Antonini et al. 2020) 4-hydroxyphenylpyruvate dioxygenase (HPPD) (Wilson et al. 2021)
Branched-chain amino acid disorder	Maple syrup urine disease (MSUD)	Branched-chain keto acid dehydrogenase complex (BCKDC) (Odièvre et al. 2005)
Sulfur amino acid disorder	Homocystinuria S-adenosylhomocysteine hydroxylase deficiency; cystathioniuria; cystinuria	Cystathionine beta-synthase (CBS) (Morris et al. 2017) S-adenosylhomocysteine hydroxylase (AdoHcyas) (Coşar et al. 2014) Cystathione gamma lyase (CTH) (Shirozu et al. 2013) Mutation in SLC3A1 amino acid transport gene (Bisceglia et al. 2001)
Urea cycle disorder	Citrullinemia I; CPS-I deficiency; ornithine transcarbamylase deficiency; hyperornithinemia-hyperammonemia-homocitrullinuria (HHH)	Argininosuccinate synthase (ASS) (Husson et al. 2003) Carbamoyl phosphate-I synthase (CPSi) (Yeldandi et al. 1990) Ornithine transcarbamylase(OTC) (Prieve et al. 2018) Ornithine translocase (ORCi) (Zanatta et al. 2016)

needs of individuals with these disorders (Van Vliet et al. 2014). Regular monitoring of amino acid levels through quantitative amino acid analysis is crucial for assessing the effectiveness of dietary treatment in individuals who have already been diagnosed with amino acid disorder. The following section discusses in detail about the various amino acid disorders, the enzymes involved, and clinical symptoms and methods to detect and mitigate them.

3.2.1 Aromatic Amino Acid Disorders

Aromatic amino acids including phenylalanine, tryptophan, and tyrosine are obtained through dietary intake as humans lack metabolic pathways to synthesize them. Aromatic amino acids play a crucial role as precursors in the synthesis of various biologically active compounds. Specifically, phenylalanine serves as a precursor for the synthesis of several other amino acids. The conversion of dietary phenylalanine to tyrosine is facilitated by phenylalanine hydroxylase. Tyrosine, in turn, contributes to the biosynthesis of neurotransmitters and hormones such as dopa, dopamine, octopamine, norepinephrine, and epinephrine (Vavricka et al. 2014). Additionally, tyrosine is involved in melanin synthesis across diverse organisms (Whitten and Coates 2017). Tryptophan, another aromatic amino acid, acts as a precursor for the biosynthesis of tryptamine, serotonin, auxin, kynurenines, and melatonin (Hardeland and Poeggeler 2003; Mukherjee and Maitra 2015). Various enzymes involved in the metabolism of aromatic amino acids have been identified as drug targets for diseases, including neurodegenerative conditions, schizophrenia, and cancers (Selvan et al. 2016; Stone and Darlington 2013). The subsequent discussion provides a detailed exploration of the disorders arising from these metabolic processes.

3.2.1.1 Phenylketonuria (PKU)

Phenylketonuria was first described by Asbjorn Folling in the year 1934. Dr. Folling found increased levels of phenyl ketones in the urine of two brothers suffering from a mental disability and emitted peculiar odor (Williams et al. 2008). PKU is an inborn error of amino acid metabolism and is inherited in autosomal recessive fashion. The WHO guidelines recommend the phenylalanine levels in blood to be maintained at 120–360 μmol throughout life (Muntau et al. 2019). One out of 20,000 new born children in the US suffers from this disease. The deficiency of phenylalanine hydroxylase (PAH) is one of the major causes of PKU. PAH uses tetrahydrobiopterin as a cofactor to convert phenylalanine to tyrosine (Fig. 3.1). However, in individuals with PKU, PAH is deficient or nonfunctional, leading to a build-up of phenylalanine in the body. There are cases that have also reported a lack of biopterin to be one of the contributors to the disease (Davis et al. 1991). The clinical diagnosis of PKU shows low intelligence quotient (IQ), eczema, strange, odor, and seizures. The pregnant female should maintain optimum levels of phenylalanine in blood as higher levels cause teratogenic defects in developing fetus. The existing treatment plan incorporates dietary restrictions on phenylalanine along with the

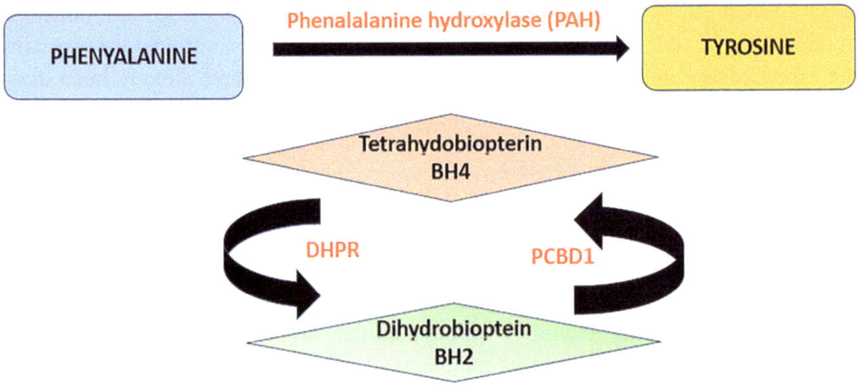

Fig. 3.1 Reaction catalyzed by phenylalanine hydroxylase (PAH) for conversion of phenylalanine to tyrosine. Tetrahydrobiopterin (BH4) is a cofactor of PAH. DHPR dihydropteridine reductase, PCBD1 pterin-4-α-carbinolamine dehydratase

supplementation of tyrosine. A more recent and promising approach involves introducing sapropterin, a biopterin cofactor, into the diets of individuals with restricted hydroxylase activity (Muntau et al. 2021). Additionally, subcutaneous injections of phenylalanine in the form of Pegvaliase are administered to maintain optimal blood levels. The effectiveness of this treatment is heightened by early detection through a newborn screening program, which identifies hyperphenylalaninemia within 24 h of birth (Rocha et al. 2021).

3.2.1.2 Alkaptonuria

Alkaptonuria is a disorder caused by a defect in a gene called homogentisate dioxygenase. The darker shade of urine is an indicator of the disease in children with growing age (Kisa et al. 2021). Accumulation of homogentisic acid within the cartilage can lead to arthritis in the joints and spine, as well as the formation of kidney stones and dysfunction of cardiac valves. (Galderisi et al. 2021). Elevated levels of homogentisic acid in urine serve as a quantifiable marker for diagnosing the disorder, with allelic variants confirming its presence. Dietary restrictions targeting tyrosine and phenylalanine have limited efficacy in reversing the disorder. However, the drug nitisinone shows promise in reducing plasma and urinary homogentisic acid levels. Moreover, nitisinone administration has been linked to a reduction in optic ocular ochronosis, a common complication of this disease. Administered orally in capsule form, nitisinone generally lacks severe complications. However, caution is advised regarding dietary intake, as nitisinone may elevate blood tyrosine levels, emphasizing the preference for a low-tyrosine and low-phenylalanine diet alongside medication (Hughes et al. 2020).

3.2.1.3 Tyrosinemia Type I

Tyrosinemia type I is a rare metabolic disorder caused by a deficiency of the enzyme fumarylacetoacetate hydrolase (FAH), which is involved in the breakdown of tyrosine (Tang and Kong 2021). The diagnosis of this condition can be established by the elevated levels of succinylacetone in blood and urine. Methionine and phenylalanine levels in the blood plasma are higher than normal in the blood in the affected patients. Tyrosine metabolites and the compound δ-aminolevulinic acid (δ-ALA) are found in elevated amounts in the urine. Furthermore, molecular genetic testing involving identification of biallelic pathogenic variants in the FAH gene, can also be used as diagnostic tool (Bliksrud et al. 2012). The primary treatment for tyrosinemia type I involves the use of a medication called Nitisinone (Orfadin®). Nitisinone works by blocking the action of para-hydroxyphenylpyruvic acid dioxygenase (p-HPPD), the enzyme responsible for the second step in the tyrosine degradation pathway (Philip et al. 2012). This inhibition prevents the accumulation of fumarylacetoacetate and its conversion to succinylacetone. Treatment should begin as soon as the diagnosis is confirmed. However, it is important to note that nitisinone treatment can lead to an increase in blood tyrosine levels (Dardevet 2016). Therefore, dietary management is also a crucial component of therapy. A controlled intake of phenylalanine and tyrosine should be initiated immediately after diagnosis to prevent the formation of tyrosine crystals in the cornea (Chinsky et al. 2017). In cases where tyrosinemia type I is severe and unresponsive to nitisinone therapy or if there is evidence of malignant changes in hepatic tissue, liver transplantation may be considered. However, with the availability of nitisinone, liver transplantation is now typically reserved for severe cases or those who do not respond to medical therapy. Early diagnosis and intervention are crucial in managing this condition effectively and preventing severe complications (Bartlett et al. 2014).

3.2.1.4 Tyrosinemia Type II

Tyrosinemia type II known as Richner Hanhart syndrome occurs due to the deficiency in hepatic cytosolic enzyme, tyrosine aminotransferase (TAT). TAT catalyzes the conversion of L-tyrosine to p-hydroxyphenylpyruvate in the multienzyme process required for tyrosine metabolism. The disease is characterized by oculocutaneous disorder which includes sensitivity to light, inflammation of sclera, hyperkeratosis of skin, lesions on palms and soles (Bas et al. 2003). Tandem mass spectrometry is used to screen the newborns. The techniques used minimal amount of blood sample to detect accumulated metabolite. However, it is crucial to differentiate between the types of tyrosinemia as all the types are usually accompanied with liver failure. The treatment comprises of low protein intake with a recommended levels at 200–500 μmol/l of tyrosine to be maintained (Stauss et al. 2020). During pregnancy, it is advisable to take balance number of amino acids so that the fetal risk of the disease could be evaded.

3.2.1.5 Tyrosinemia Type III

Tyrosinemia type III, also known as 3-Hydroxykynureninuria, is a rare autosomal recessive genetic metabolic disorder characterized by a deficiency in the enzyme 4-hydroxyphenylpyruvate dioxygenase (4-HPPD) (Szymanska et al. 2015). This enzyme, encoded by the mutated gene HPPD, plays a crucial role in breaking down the amino acid tyrosine. Individuals affected by Tyrosinemia III inherit two copies of the mutated HPPD gene, leading to a lack of functional 4-HPPD. The consequence of this enzyme deficiency is the accumulation of tyrosine and its metabolites within the body. This accumulation results in the distinctive symptoms associated with Tyrosinemia III. The disorder manifests due to the disrupted metabolic process, highlighting the importance of 4-HPPD in the breakdown of tyrosine. The severity of Tyrosinemia III symptoms varies among individuals. Key manifestations include developmental delays, affecting both motor and intellectual skills in children (Scott 2006). Neurological symptoms comprise intellectual disability, seizures, and behavioral issues. Ocular symptoms such as cataracts and corneal opacities are common, alongside heightened skin sensitivity to sunlight and unusual rashes. Some cases may involve psychiatric problems like depression and anxiety, and less commonly, kidney issues.

Although a cure for Tyrosinemia III remains elusive, current management strategies offer promising avenues for enhancing patients' well-being. Central to this approach is a strict low-tyrosine diet, which involves avoiding high-protein foods to mitigate symptoms. Complementing this dietary regimen are specialized medical formulas designed to provide essential amino acids while excluding tyrosine and phenylalanine. Nitisinone, a commonly prescribed medication, plays a pivotal role in this regimen. Vigilant monitoring of blood tyrosine levels and other metabolic markers is essential for gauging the efficacy of dietary and medical interventions, enabling healthcare providers to make necessary adjustments promptly. Although Tyrosinemia III presents challenges, a comprehensive strategy encompassing dietary management, medication, and supportive therapies holds significant promise in improving patients' quality of life. Timely diagnosis and intervention are crucial elements contributing to better outcomes for individuals grappling with this condition (Heylen et al. 2012).

3.2.1.6 Albinism

Albinism, stemming from the Latin word meaning "white," manifests as a group of disorders characterized by a lack of pigmentation in the skin due to deficient or nearly absent melanin production in the ectodermal tissue. Its occurrence is estimated to be about 1 in 40,000 individuals (Lasseaux et al. 2022). Melanocytes, originating from the neural crest ectoderm during embryonic development, migrate to various tissues like the skin, eyes, hair, and inner ear. Despite the unaffected number of melanocytes in individuals with albinism, these cells house specialized organelles called melanosomes, crucial for melanin production (Hayashi and Suzuki 2017). Melanin, renowned for its UV protection properties, also plays vital roles in the embryological development of ocular structures and oculoneural pathways.

Two primary forms of melanin exist: eumelanin, responsible for black or brown skin pigmentation and UV protection, and pheomelanin, lacking UV protective qualities. The activation of melanocortin one receptors, encoded by the MC1R gene, on melanocytes stimulates the synthesis of eumelanin over pheomelanin. Notably, individuals with predominantly pheomelanin are categorized as Fitzpatrick scale 1, displaying red or blond hair and light-colored, ruddy skin. Besides the skin, eyes, and hair, the retinal pigment epithelium represents the only other tissue capable of melanin synthesis, albeit without a neural crest origin.

The consequences of hypomelanosis can be broadly categorized into dermatological and ophthalmologic outcomes, each with distinct implications. Albinism, characterized by a deficiency in eumelanin, presents an elevated susceptibility to sun-related issues, including solar lentigines, actinic keratoses, and solar erythema. Moreover, individuals with albinism face an increased risk of UV-associated malignancies, particularly squamous cell carcinomas, due to the photo-protective role of eumelanin. In the realm of ophthalmology, melanin plays a crucial role during embryonic development, influencing the formation of key structures such as the fovea, optic nerves, optic tracts, and visual cortex. Whereas, in albinism the fovea exhibits compromised development, appearing hypoplastic or even absent (Chong et al. 2009). Melanin's role in guiding the migration of developing optic nerve fibers is pivotal for the proper establishment of binocular vision. However, in albinism, misrouting occurs, with some axons from retinal ganglion cells projecting erroneously to the contralateral hemisphere to dorsal lateral geniculate nucleus, resulting in an increased optic fiber crossover. This misdirection of nerve fibers contribute to binocular vision deficits, often presenting as strabismus. Additionally, the occipital lobe in individuals with albinism demonstrates decreased gyrification, thought to be linked to a lack of foveal input, along with localized increases in cortical thickness. Similar findings have been observed in congenitally blind subjects. The absence or partial deficiency of melanin during embryonic development leads to structural abnormalities in ocular components and misrouting of optic tracts. These alterations give rise to a spectrum of irreversible intraocular (refractive) and extraocular (oculomotor or nonrefractive) challenges. Typically, individuals with albinism exhibit visual acuity ranging from 20/60 to 20/400. The most studied albinism is oculocutaneous albinism (OCA). OCA is phenotypically similar disorder due to error in the pathway synthesizing melanin. Seven types of nonsyndromic albinism (OCA1 to OCA7) have been described (Teramae et al. 2019).

3.2.1.6.1 Ocular Albinism (OA1)

OA1 is an X-linked condition involving the GPR143 gene product, a G-protein coupled receptor. Mutations in this gene result in dysfunctional melanosome biogenesis, characterized by the presence of "macromelanosomes."

OCA1 OCA1 is an autosomal recessive genetic condition characterized by the disruption of melanin synthesis. The TYR gene, responsible for encoding the tyrosinase enzyme, plays a crucial role in this process. Tyrosinase normally regulates hydroxylates L-tyrosine to L-DOPA and oxidates L-DOPA to DOPA quinine conversion and serves as the rate-limiting step in melanin synthesis. Loss of function in TYR leads to an inability to produce melanin.

OCA2 OCA2 is associated with the OCA2 gene product, an enigmatic melanosome transmembrane protein P. The exact function of this protein remains unknown.

OCA3 OCA3, following an autosomal recessive pattern, involves the TYRP1 gene product, tyrosinase-related protein 1. This protein is believed to stabilize and modulate tyrosinase activity, contributing to melanosome integrity.

OCA4 Autosomal recessive OCA4 involves the SLC45A2 gene, coding for a membrane-associated transport protein (MATP). MATP is thought to transport substances essential for melanin biosynthesis into the melanosome.

OCA5 OCA5 is an autosomal recessive condition, although the specific gene responsible is yet to be identified.

OCA6 In a similar autosomal recessive fashion, OCA6 involves the SLC24A5 gene, which codes for a Na/K/Ca cation exchange protein with a structure akin to that seen in OCA4. Its function is believed to be similar to that of OCA4.

OCA7 OCA7 is another autosomal recessive condition, involving the LRMDA gene, which codes for a protein speculated to play a role in melanocyte differentiation.

3.2.1.6.2 Hermansky-Pudlak Syndrome (HPS)

HPS, an autosomal recessive syndrome with ten identified genotypes, affects various genes, including HPS1, AP3B1, HPS3, HPS4, HPS5, HPS6, DTNBP1, BLOC1S3, BLOC1S6, and AP3D1. Subjects with HPS exhibit oculocutaneous albinism, accumulation of ceroid lipofuscin in tissues, and bleeding diathesis due to a lack of dense platelet granules. Some subtypes present with immunodeficiency and hemophagocytic syndrome.

3.2.1.6.3 Chediak-Higashi Syndrome (CHS)

CHS, following an autosomal recessive pattern, involves the LYST gene, coding for a protein directing material into lysosomes. Mutations result in giant cytoplasmic granules (lysosomes), leading to increased susceptibility to infections, neutropenia, peripheral neuropathy, coagulopathy, and hypomelanosis.

3.2.1.6.4 Angelman Syndrome (AS) and Prader-Willi Syndrome (PWS)

AS and PWS both result from spontaneous partial deletions of chromosome 15q, causing albinism due to haploinsufficiency for the OCA2 gene product. AS arises from maternal chromosome 15q deletion or uniparental disomy with both copies of paternal origin, while PWS is associated with paternal chromosome 15q deletion or uniparental disomy with both copies of maternal origin.

Early detection of albinism is crucial for effective management of ocular symptoms, optimizing visual potential, and influencing broader aspects of safety, education, self-esteem, and cognitive development. Skin and hair hypopigmentation may not always raise concerns, particularly in families with diverse pigmentation where the contrast is less apparent. Iris hypopigmentation, typical in infants, might be initially overlooked as melanin deposits take months to reach adult coloration. However, ocular manifestations, including nystagmus and photophobia within the first 3–6 months should be taken seriously.

3.2.2 Branched-Chain Amino Acid (BCAA) Disorders

Branched-Chain Amino Acids (BCAA) constitute a cluster of amino acids distinguished by their branched fatty side chains, viz. Leucine (Leu), Isoleucine, and Valine. Together, these amino acids account for 35% of the essential amino acids found in the muscle tissues of the human body. The catabolic breakdown of BCAA involves a pivotal step regulated by the branched-chain α-keto acid dehydrogenase (BCKDH) complex, serving as a rate-limiting factor (Shimomura et al. 2006). The resulting products play integral roles in diverse physiological activities. Notably, leucine activates the mammalian target of rapamycin (mTOR), fostering protein synthesis in the liver (de Bandt 2016). Additionally, BCAAs contribute to glycogen synthesis by enhancing glycogen absorption in both liver and skeletal tissues.

Expanding their influence, BCAAs also contribute to the regulation of the immune system. They stimulate lymphocyte proliferation and activate T cells, showcasing their involvement in immune response modulation (Calder 2006). In summary, BCAA's multifaceted functions encompass protein synthesis, glycogen regulation, and immune system modulation, underscoring their essential role in sustaining various physiological processes within the human body. The primary conditions associated with disruptions in BCAA metabolism encompass maple syrup urine disease (MSUD), isovaleric academia (IVA), methylmalonic academia (MMA), propionic academia (PA), and 3-methylcrotonyl carboxylase deficiency (3-MCC) (Table 3.2). These disorders exhibit diverse manifestations, ranging from metabolic decompensation in newborns to acute intermittent episodes in older children or adults, as well as chronic neurodevelopmental childhood illnesses (Holeček 2018). Among these, MSUD stands out as a rarity in most populations, with estimated incidence rates of 1:185,000 live births (Packman et al. 2012). It occurs due to the aberrant activity of the branched-chain alpha-keto acid dehydrogenase (BCKAD) complex (Ali and Ngu 2018). The BCKAD complex is made up of four catalytic subunits; E1α and E1β are encoded by branched-chain alpha-keto acid (BCAA) dehydrogenase complex A (BCKDHA), branched-chain alpha-keto acid (BCAA) dehydrogenase complex B (BCKDHB) gene, respectively. Dihydrolipoamide branched-chain transacylase (DBT) codes the E2 subunit while the E3 is coded by dihydrolipoamide dehydrogenase (DLD). The multienzyme complex located in mitochondria is responsible for the breakdown of branched amino acids. The catabolism of these amino acids is required for the synthesis of protein, fatty acids, and cholesterol. The clinical characteristics of the disease include urine emitting odor similar to maple syrup during few hours of birth followed by ketoacidosis, development delays, and neurological failure leading to death (Frazier et al. 2014).

The MSUD can be prenatally diagnosed by examining the activity of BCKAD enzyme in the chorionic villi cells. The amniotic fluid can also be measured for the branched-chain amino acids. The newborn screening for MSUD is performed within 24–48 h after birth. Tandem mass spectrometry is used to measure the concentration of leucine and isoleucine (Heldt et al. 2005). It readily detects classic MSUD but for other variants of the disease, other tests are required. Liquid chromatography is used to distinguish amino acids with same molecular mass like allo-isoleucine, leucine, and isoleucine. The liquid chromatography also helps in differentiating from hydroxyprolinemia. Another more specific diagnostic tool used for MSUD is plasma amino acid testing. The detection of allo-isoleucine by this tool provides more specificity and sensitivity to the diagnosis (Schadewaldt et al. 1999). Further techniques

Table 3.2 List of disorders caused due to defects in branched-chain amino acid metabolism

S.no.	Branched-chain amino acid disorders	Characteristic feature
1.	Propionic acidemia	Increase levels of propionic acid, 3-hydroxypropionate, methyl citrate, and propionyl carnitine (Forny et al. 2021)
2.	Methylmalonic aciduria	Mutation in methylmalonyl CoA mutase; cobalamin A; cobalamin-B (Ghoraba et al. 2015)
3.	Isovaleric acidemia (IVA)	Deficiency of isovaleryl-CoA dehydrogenase (IVD) (Khan et al. 2020)
4.	3-methylcrotonylcarboxylase deficiency (3-MCC)	Mutation in α and β subunits of MCC enzyme (Baumgartner et al. 2001)
5.	Maple syrup urine disease	
	Classical	Neonatal; mutation in E1α; E1β; E2 subunit of branched-chain alpha-keto acid dehydrogenase complex (genes: BCAKDHA; BCAKDHB; and DBT); <2% of BCKAD enzymatic activity (Blackburn et al. 2017)
	Intermediate	Variable; mutation in E1α; E1β; E2 subunit of branched-chain alpha-keto acid dehydrogenase complex (genes: BCAKDHA; BCAKDHB; and DBT); > 30% of BCKAD residual activity (Sajeev et al. 2021)
	Intermittent	Variable; mutation in E1α; E1β; E2 subunit of branched-chain alpha-keto acid dehydrogenase complex (genes: BCAKDHA; BCAKDHB; and DBT); completely asymptomatic (Pode-Shakked et al. 2020)
	Thiamine responsive	Mutation in E2 subunit of branched-chain alpha-keto acid dehydrogenase (BCAKDH) complex (Scriver et al. 1971)
	E3 deficient	Mutation in E3 subunit of branched-chain alpha-keto acid dehydrogenase (BCAKDH) complex (Chuang et al. 2020)

to confirm the MSUD include gas chromatography and dinitrophenylhydrazine (DNPH) test which detects the presence of branched-chain amino acids. DNPH is also used to detect these amino acids in urine. Equal volume of urine and DNPH reagent is mixed together. A yellow white precipitate develops on detection of amino acids. Ketonuria is the major symptom which develops from the disease and this can be tested with standard urine (Stauss et al. 2020).

3.2.3 Sulfur-Containing Amino Acid Disorders

Methionine and cysteine are the two sulfur-containing amino acids. They are found in various metabolic pathways and also act as catalyst for various enzymatic reactions. Both of these amino acids contribute to the cellular pool of organic sulfur, maintain sulfur homeostasis, and regulate one carbon metabolism.

Disorders related to sulfur amino acids primarily result from abnormalities in methionine demethylation, homocysteine remethylation (Hcy), cobalamin and folate metabolism, Hcy transsulfuration, as well as catabolism of cysteine (Cys) and hydrogen sulfide (H_2S) (Sacharow et al. 2004). Among these homocystinuria, a complex metabolic disorder characterized by cystathionine beta-synthase deficiency, leads to the accumulation of homocysteine (Kožich et al. 2016). It can be characterized by disorders of skeletal system (increased extremities, scoliosis, and stiff joints) delay in CNS development, optic disorders like ecotopia lens, and progressive myopia. This abnormal buildup of homocysteine can lead to serious health consequences.

One of the critical pathways in methionine metabolism is transmethylation, which converts methionine into homocysteine (Fig. 3.2). Subsequently, homocysteine undergoes two main metabolic routes: remethylation and transsulfuration. In the remethylation pathway, homocysteine is converted back into methionine with the help of vitamin B12 (cobalamin) and folate. This pathway is crucial for maintaining a balance between methionine and homocysteine levels in the body (Padmanabha et al. 2021). Disorders related to homocystinuria can be categorized into two groups: remethylation disorders and transsulfuration disorders. Remethylation disorders are associated with defects in enzymes like methylenetetrahydrofolate reductase (MTHFR) and cobalamin (Cbl) metabolism (Brustolin et al. 2010). In these cases, the body struggles to effectively convert homocysteine back into methionine, leading to elevated homocysteine levels. On the other hand, classic homocystinuria is a transsulfuration disorder primarily caused by a deficiency in an enzyme called cystathionine β-synthase (CBS) (Gerrard and Dawson 2022). CBS plays a pivotal role in converting homocysteine into cystathionine, a necessary step for the transsulfuration of homocysteine into cysteine. When CBS is deficient or dysfunctional, this pathway is blocked, resulting in the accumulation of both homocysteine and methionine.

Diagnosing homocystinuria can be challenging, especially through newborn screening, which typically measures methionine levels (Morris et al. 2017). In remethylation disorders and some cobalamin-related cases, methionine levels may

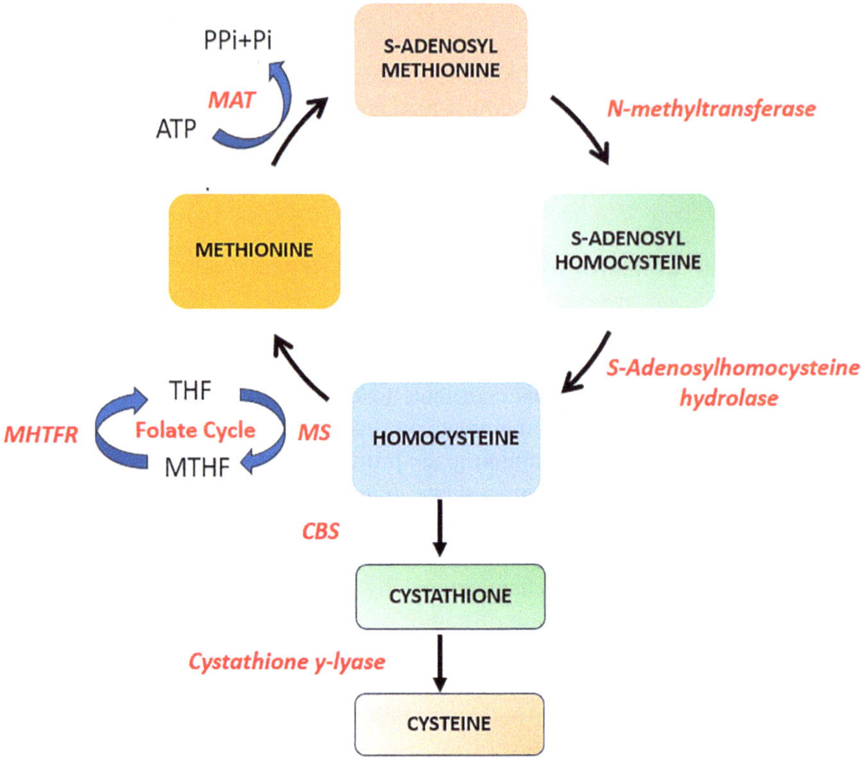

Fig. 3.2 Sulfur-containing amino acids metabolism. CBS cystathionine β synthase, MAT methionine adenosyl transferase, MS methionine synthase, MTHFR methylene tetrahydrofolate reductase, THF tetrahydrofolate

appear normal, leading to potential under diagnosis. To improve detection rates, some studies advocate for including the analysis of total homocysteine in diagnostic workflows. Total homocysteine analysis measures all forms of homocysteine in plasma/serum, providing a more comprehensive view of the metabolic abnormalities (Alcaide et al. 2015). Early diagnosis and appropriate management are crucial to mitigate the health risks associated with homocystinuria.

3.2.3.1 Cystathioninuria

Cystathioninuria occurs due to a defect in the cystathionase gamma-lyase (CTH) gene which is involved in the breakdown of cystathione to cysteine. This leads to an increase in excretion of cystathione in urine due to its elevated concentration in plasma. The mutation changes the conformation of CTH enzyme which inhibits its capacity to bind to its cofactor, pyridoxal phosphate (Frimpter 1965). The disease is autosomal recessive with low incidences of 1 per 14,000 live births (Wong et al. 1979). The disorder is not clinically associated with consistent or particular pathologic symptoms; however, some individuals may develop convulsions, thrombocytopenia, and low intelligence quotient.

3.2.3.2 Cystinuria

Cystinuria arises from a mutation affecting the transport protein responsible for structurally related amino acids. This condition was first documented by Wollaston in the early 1800s when he observed yellow stones composed of sulfur-containing amino acids in patients' urine, coining the term "cystic oxide" and naming the syndrome cystinuria ("XIII. On Cystic Oxide, a New Species of Urinary Calculus," 1810). Pathologically, cystinuria is characterized by diminished tubular reabsorption of lysine, arginine, ornithine, and cystine, leading to their excessive excretion in urine. The low solubility of cystine in water fosters the accumulation of stones in the kidneys, ureters, and urinary bladder. Individuals with cystinuria excrete approximately 600–1800 mg of cysteine per day, exacerbating stone formation (Townsend et al. 2004).

Cystinuria manifests in three distinct types, namely Type I, Type II, and Type III, each characterized by variations in the excretion levels of cystine and dibasic amino acids in the urine of affected individuals. In Type I, individuals exhibit normal excretion levels of cystine and dibasic amino acids, and this type is associated with mutations in the SLC3A1 amino acid transport gene (Sen Lee et al. 1993; Calonge et al. 1994). On the other hand, Type II is identified by a substantial increase, specifically 9–15 times more, in the excretion of these substances. The causative factor for Type II is linked to mutations in genes SLC7A9 and SLC7A3, both encoding amino acid transporters. Whereas, Type III is characterized by individuals excreting twice the normal range of cystine, and it is also attributed to mutations in the SLC7A9 and SLC7A3 genes, further underscoring the role of amino acid transporters in the development of Types II and III cystinuria. Adequate water intake plays a crucial role in preventing the formation of crystals and facilitates the dissolution of existing ones. While alkalinizing urine can effectively inhibit stone formation, it comes with the drawback of increasing the likelihood of complications related to the development of calcium stones in renal pathology. In addressing cystinuria, a treatment approach involves the use of penicillamine, which acts by replacing cysteine to form disulfides of penicillamine and cysteine. This resulting disulfide is notably more soluble than cystine, contributing to a reduction in the excessive accumulation of amino acids (Townsend et al. 2004).

3.2.4 Urea Cycle Disorders (UCDs)

Protein catabolism involves breaking down amino acids into carbon skeletons which can serve as energy sources or metabolic intermediates, and ammonia, a toxic compound. The liver's urea cycle converts ammonia into nontoxic urea for excretion. Urea cycle disorders (UCDs) occur due to the defects in the enzymes or transporters of the urea cycle which is involved in the removal of toxic ammonia from the bloodstream by converting it to less toxic urea which is excreted by the kidneys (Braissant

2010). Newborns with mutation in any of the four enzymes of the cycle can fall seriously ill within 36–48 h after birth. Urea cycle disorders are rare genetic conditions with an estimated incidence of 1 in 8000 births (Ah Mew et al. 2003). UCD is complex mechanism due to the involvement of multiple enzymes and cellular organelle, i.e., mitochondrial matrix and cytoplasm (Summar and Mew 2018). The first two reactions of the urea cycle occur in the matrix of mitochondria. Carbamoyl phosphate synthetase I (CPS1) catalyzes the first step which involves addition of ammonia to bicarbonate forming carbamoyl phosphate which enters the urea cycle (Fig. 3.3). Various stress effectors molecules can easily initiate defects in CPS1 resulting in hyperammonemia. The ammonia released gets accumulated in the bloodstream, posing a severe threat to the central nervous system. Hyperammonemia can result in neurological damage, seizures, coma, and even death if not promptly treated (Summar and Mew 2018).

In second step, ornithine transcarbamylase (OTC), a liver enzyme, catalyzes the combination of carbamoyl phosphate and ornithine to produce citrulline. OTC gene defects, inherited in an X-linked fashion due to its location on the X chromosome, are the most common cause of UCDs. Males typically exhibit more prominent OTC deficiency, while females may range from asymptomatic to severe based on the inactivated X chromosome. Additionally, defects in the mitochondrial ornithine translocase (ORNT1) protein, responsible for transporting ornithine from the cytoplasm to the mitochondrial matrix, can result in hyperammonemia (Buerger et al. 2019). The remaining enzymatic steps of the urea cycle occur in the cytoplasm. Argininosuccinate synthase 1 (ASS1) catalyzes the ligation of citrulline and aspartic acid to form argininosuccinic acid. Defects in ASS1 lead to both citrullinemia and hyperammonemia (Ye et al. 2000). Aspartic acid, crucial for the urea cycle in the cytoplasm, is transported from the mitochondria by the citrin transporter. This transporter also plays a role in transporting glutamate from the cytoplasm into the mitochondria. Defects in the citrin transporter protein can result in both hyperammonemia and citrullinemia. In the subsequent phase of the urea cycle, argininosuccinate undergoes cleavage facilitated by argininosuccinate lyase (ASL), resulting in the formation of fumaric acid and arginine (Erez et al. 2011). Deficiencies in ASL are characterized by elevated levels of citrulline and argininosuccinate, manifesting as persistent hyperammonemia in infants and occasional induction of hyperammonemia in adults. In the final stage of the urea cycle, arginine is cleaved into urea and ornithine by the arginase enzyme, encoded by the ARG1 gene (Erez et al. 2011). Arginase deficiency leads to both argininemia and hyperammonemia. Symptoms of arginase deficiency typically emerge in children around the age of three, with hyperammonemia developing later in life.

3.2.4.1 Hyperornithinemia-Hyperammonemia-Homocitrullinuria (HHH) Syndrome

Hyperornithinemia-Hyperammonemia-Homocitrullinuria (HHH) syndrome is a specific disorder within the urea cycle, stemming from mutations in the SLC25A15 or ORNT1 gene, causing a deficiency in ornithine translocase (ORNT1). This essential protein is pivotal in transporting ornithine, lysine, and arginine across the

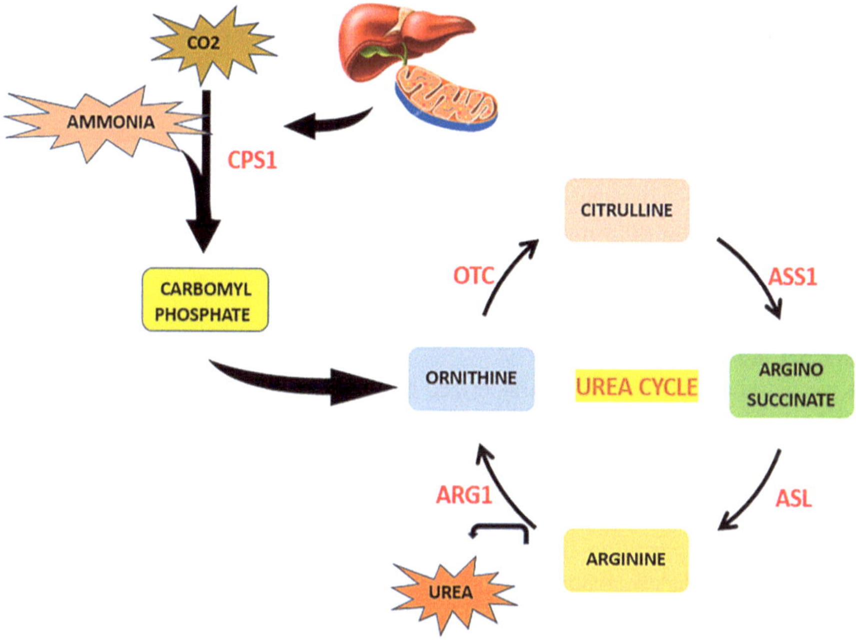

Fig. 3.3 Urea cycle disorders. ARG1 arginase, ASS1 argininosuccinate synthase, ASL argininosuccinate lyase, CPS1 carbamoyl phosphate I synthase, OTC ornithine transcarbamylase

inner mitochondrial membrane in both peripheral tissues and pericentral hepatocytes (Lee et al. 2014). In HHH syndrome, the inadequacy of ORNT1 disrupts the transport of ornithine into the mitochondria, resulting in an accumulation of ornithine in the cytosol (hyperornithinemia). In the liver, mitochondrial ornithine is a crucial substrate for ornithine transcarbamylase (OTC), an enzyme integral to the urea cycle (Lee et al. 2014). The diminished availability of mitochondrial ornithine decelerates the urea cycle, leading to an elevation in ammonia and carbamoyl-phosphate levels (culminating in hyperammonemia). Concurrently, surplus carbamoyl-phosphate can engage with lysine, giving rise to homocitrulline (resulting in homocitrullinuria) or enter in the pyrimidine pathway, triggering the formation of orotic acid, subsequently excreted in urine. As is the case with other urea cycle disorders, early diagnosis during infancy is imperative for enhancing the clinical outcome of HHH syndrome.

3.2.5 Protein Deficient Disorders

Protein deficiency disorders, also known as protein-energy malnutrition (PEM), result from inadequate intake of protein and calories necessary for proper growth and maintenance of bodily functions. These disorders primarily affect vulnerable populations such as children, pregnant women, and the elderly, particularly in regions with limited access to nutritious food. Proteins are essential for the body's structural integrity, immune function, and enzymatic processes (Singh and Singh 2018). In protein deficiency, the body prioritizes essential functions, sacrificing muscle mass and other nonessential processes. Children with PEM may experience stunted growth, developmental delays, and increased susceptibility to infections. There are two main types of protein deficiency disorders: marasmus and kwashiorkor. Marasmus is characterized by severe overall malnutrition, leading to muscle wasting and a gaunt appearance. Kwashiorkor, on the other hand, results from a lack of protein despite sufficient caloric intake, causing edema and a swollen abdomen. The following sections give a more detailed insight of the two disorders.

3.2.5.1 Kwashiorkor

Kwashiorkor is a condition characterized by severe protein malnutrition and bilateral swelling of extremities, primarily affecting infants and children, typically between the weaning age and 5 years. It is prevalent in regions marked by extreme poverty and starvation. In the 1950s, the World Health Organization recognized it as a significant public health crisis (Marks et al. 2019). Peripheral edema is a hallmark of kwashiorkor, resulting from an imbalance in fluid dynamics across capillary walls due to reduced albumin levels. Albumin, essential for maintaining osmotic pressure, is markedly low in children with kwashiorkor, leading to intravascular depletion. The subsequent rise in antidiuretic hormone (ADH) due to hypovolemia contributes to edema, along with increased plasma renin causing sodium retention (Obonyo et al. 2017). Kwashiorkor is further characterized by depleted body fat stores, low weight for height, and reduced mid-upper arm circumference in affected children. Additional features include thin, dry skin, an enlarged head relative to the body, an emaciated appearance, bradycardia, hypotension, hypothermia, and thin, shrunken arms, thighs, and buttocks with redundant skin folds. Low glutathione levels, indicative of heightened oxidant stress, are associated with the disease, and nutritional improvements and antioxidants are considered potential interventions (Singh et al. 2013; Gupta et al. 2023). Experimental theories suggesting the involvement of microbiome/virone alterations require further investigation for a comprehensive understanding.

The treatment of individuals admitted with kwashiorkor adheres to ten universal principles applied across various phases, beginning with emergency stabilization and progressing through rehabilitation. The first phase focuses on addressing and preventing hypocalcemia, managing hypothermia, and addressing dehydration, laying the foundation for stabilizing the patient's immediate health. Subsequent phases involve correcting electrolyte imbalances, managing and preventing infections, and addressing micronutrient deficiencies, aiming to restore the body's overall nutritional equilibrium. Cautious feeding is initiated, ensuring a gradual and systematic approach to meet nutritional needs. The principle of achieving catch-up growth becomes pivotal in the rehabilitation phase, promoting the recovery of physical health and overall well-being. Simultaneously, the provision of sensory stimulation and emotional support plays a crucial role in enhancing the psychosocial aspects of recovery. The treatment approach also emphasizes preparing for postrecovery follow-up, ensuring continued monitoring and support for sustained health improvement. These principles collectively constitute a holistic and structured framework, guiding healthcare professionals in providing comprehensive care to individuals affected by kwashiorkor, encompassing both the immediate challenges and the long-term recovery process (Sadler et al. 2008).

3.2.5.2 Marasmus

Marasmus is a severe form of malnutrition resulting from an inadequate intake of total calories, leading to significant loss of both adipose tissue and muscle. Children affected by marasmus often exhibit a weight-for-height value more than three standard deviations below the average for their age or sex. The root cause is often linked to poverty, which hampers a household's ability to provide an ample and consistent food supply for children (Arthur et al. 2015). This condition contributes to a substantial number of deaths, with 300,000 directly attributed to malnutrition annually, and half of these occurring in young children. Globally, an estimated 852 million people suffer from undernourishment, predominantly in less developed regions. Approximately 18 million children in low- to middle-income countries, primarily in Asia, are believed to be affected by marasmus (Titi-Lartey and Gupta 2023).

Physiologically, marasmus is marked by an increase in total body water relative to body weight (Barltrop and Sandhu 1985). Weight loss in marasmus correlates with higher total body water, and children with severe wasting may experience a 10–33% reduction in total body potassium. Diarrhea contributes to potassium loss, leading to intracellular deficits in this essential mineral, as well as depletion of sodium, phosphorus, and calcium (Garrow et al. 1965). Severe malnutrition, including marasmus, is associated with villous atrophy, loss of brush border enzymes, crypt hypoplasia, and impaired absorption across the gastric mucosa. Reduced gastric acid secretion can contribute to bacterial overgrowth (Grover and Ee 2009). In infants with marasmus, observable symptoms include irritability, apathy, sunken fontanelles due to dehydration, and a shrunken, wasted appearance with reduced subcutaneous fat. Weight loss is initially noticeable in the groin or axilla, progressing to the buttocks, face, and thighs, resulting in a characteristic "old man" appearance (Trehan and Manary 2015). Marasmus may also manifest with hypotension, hypothermia, and bradycardia. Micronutrient deficiencies often accompany marasmus, with dry eyes and Bitot spots indicating vitamin A deficiency. Spoon-shaped nails (koilonychia) suggest iron deficiency anemia, while hypocalcemia may present with Chvostek or Trousseau signs. Prolonged deficiencies in calcium and vitamin D can lead to associated conditions like rickets and other bone deformities over time.

Anthropometry plays a crucial role in diagnosing marasmus, although its reliable and repeatable application can be challenging in low-income countries. Accurate measurements are essential, including vertical length with a precision of 0.5 cm, weight with a precision of 100 g, and middle-upper arm circumference (MUAC) with a precision of at least 2 mm or less (Wootton et al. 2014). In addition to these measurements, the presence of pitting edema should be carefully examined to rule out the possibility of kwashiorkor, another severe form of malnutrition. This thorough assessment helps healthcare professionals evaluate the nutritional status of individuals, especially in regions where resources may be limited. Despite the challenges, precise anthropometric measurements remain a valuable tool in identifying and managing cases of marasmus and other nutritional disorders. Research indicates that community-based treatment for children with uncomplicated severe malnutrition yields better results than hospital care. The management of marasmus typically involves three primary phases: (1) Resuscitation and stabilization, (2) Nutritional rehabilitation, and (3) Follow-up and prevention of recurrence.

3.3 Diagnostic Techniques for Amino Acid Metabolic Disorders

3.3.1 Quantitative Analysis

Quantitative analysis of amino acids in various biological fluids serves as a diagnostic tool for monitoring individuals diagnosed with disorders and those experiencing deficiencies in specific amino acids. This analysis involves assessing blood, plasma, urine, and sometimes cerebrospinal fluid in certain cases. Notably, each disorder is distinguished by deviating levels of particular amino acids. The interpretation of the analysis is based on a panel of amino acid ratios. For example, in phenylketonuria (PKU), the diagnosis considers the levels of both phenylalanine and tyrosine, emphasizing the importance of understanding specific amino acid profiles for accurate assessments (Eastman et al. 2000).

3.3.2 Newborn Screening

Newborn screening is a crucial public health initiative that aims to identify congenital disorders in infants shortly after birth (Urv and Parisi 2017). Timely detection of these disorders through newborn screening can significantly improve the prognosis and quality of life for affected infants. Robert Guthrie in early 1960 was the first to analyze phenylketonuria, an amino acid disorder in children using bacterial inhibition assay (Vockley et al. 2014). Since then, screening for Inborn Metabolic Disorder (IMD) has been performed worldwide. Newborn screening typically involves a simple blood test, often performed within the first 24–48 h after birth. A small blood sample is collected from the baby's heel and sent to a laboratory for analysis. In the case of amino acid disorders, the test measures the levels of amino acids in the blood. Elevated or decreased levels of specific amino acids can be indicative of underlying disorders. Newborn screening plays a vital role in identifying amino acid disorders early in an infant's life. Early detection enables healthcare professionals to initiate interventions promptly, preventing or mitigating the devastating effects of these genetic disorders (Blumenfeld et al. 1966). As advancements in genetic testing continue to evolve, newborn screening will remain a cornerstone of preventive healthcare, ensuring that every child has the opportunity to thrive from the very beginning of life.

3.3.3 Ion Exchange Chromatography

In the early 1950s, a significant breakthrough occurred in the field of quantitative amino acid analysis with the work of Moore and Stein. They introduced a method for separating amino acids using a resin column made up of polystyrene, enabling accurate diagnostics (Moore et al. 1958a). This pioneering technique, known as ion exchange chromatography (IEC) with postcolumn reaction has been considered a gold standard for clinical amino acid analysis for many years. Even with the continuous evolution of analytical methods, ion exchange chromatography, paired with a lithium buffer system, along with postcolumn visualization, carried out using ninhydrin which is detected by UV, remains a prevalent and trusted approach (Moore et al. 1958b). Ion exchange chromatography (IEC) relies on the principle of ionic interactions between amino acids in the sample and the functional groups attached to the column's stationary phase (Adhikari et al. 2010). Within the chromatographic column, which is packed with negatively charged resins, the sample is initially introduced to the column under low acidic conditions, all amino acids carry a positive charge, leading to strong interactions with the negatively charged column. In the course of chromatography, the adjustment of pH and salt concentrations is orchestrated using a lithium buffer. This manipulation induces a transformation in the charge status of amino acids. As the pH approaches the isoelectric point of an amino acid, it loses its net charge, resulting in weaker interactions with the charged column. This dynamic alteration in charge interactions plays a crucial role in the separation and elution of amino acids during the Ion Exchange Chromatography (IEC) process. The intensity of absorbance from the eluted amino acids analyte is directly proportional to the concentration of the amino acid in the biological sample (Ullah 2012).

Despite its notable strengths, including high reproducibility and excellent linearity across a broad range, ion exchange chromatography (IEC) for amino acid analysis does have certain limitations. One prominent drawback is the considerable time investment required to achieve a comprehensive amino acid profile, often spanning around 150 min. Additionally, the method lacks specificity in identifying amino acids since it relies on retention. Moreover, a common challenge encountered in standard Ion Exchange Chromatography (IEC) methods is the co-elution of amino acids. For example, homocitrulline may co-elute with methionine, presenting diagnostic complexities in conditions such as HHH syndrome. Similarly, allo-isoleucine, a crucial marker for diagnosing maple syrup urine disease (MSUD), might co-elute with cystathionine. Additionally, argininosuccinic acid, accumulated in individuals with arginine-succinate lyase deficiency, shares the same retention time as leucine.

Another limitation pertains to the instability of ninhydrin, a reagent utilized for postcolumn derivatization. It is advisable to store ninhydrin for a duration of less than or equal to 1 month, contributing to an increase in the overall cost of the test.

Despite these challenges, IEC remains a valuable tool in clinical amino acid analysis. However, researchers and clinicians need to exercise caution and awareness regarding these limitations when interpreting results.

3.3.4 Reverse-Phase High-Performance Liquid Chromatography (RP-HPLC)

In recent years, reverse-phase high-performance liquid chromatography (RP-HPLC) and ultra-high-performance liquid chromatography (UPLC) have emerged as compelling alternatives to ion exchange chromatography for amino acid analysis. RP-HPLC, is particularly well-suited for the separation and analysis of proteins based on their hydrophobicity. The principle behind RP-HPLC involves the use of a nonpolar stationary phase, typically composed of hydrocarbon chains bound to solid supports, and a polar mobile phase. In this technique, proteins are eluted based on their affinity for the stationary phase, with more hydrophobic proteins interacting strongly and eluting later than hydrophilic ones. For optical detection in RP-HPLC, the utilization of o-phthalaldehyde (OPA) for derivatization can be implemented either as a precolumn or a postcolumn reaction (Cunico and Schlabach 1983). In this reaction, in the presence of a thiol compound such as 2-mercaptoethanol, a stable fluorescent product is formed (Dorresteijn et al. 1996). This resulting product can be identified using fluorimetry (with excitation at 340 nm and emission at 410 nm) or UV detection (at 340 nm). It is noteworthy that while OPA derivatization is highly reproducible and automated, it may not be the optimal choice for detecting proline/hydroxyproline and sulfur-containing amino acids.

As an alternative, other reagents for RP-HPLC with precolumn derivatization include 4-(4-Dimethylaminophenylazo) benzene sulfonyl chloride (DABS-Cl) and 9-fluorenylmethyl chloroformate (FMOC-Cl) (Schneider 1989; Bütikofer et al. 1991; Sherwood 2000). These reagents provide flexibility and versatility for amino acid detection, catering to a broader range of amino acids, including those not ideally suited for OPA derivatization.

RP-HPLC, capable of detecting and quantifying proteins at very low concentrations, emerges as a valuable tool for diagnosing protein disorders, even in their early stages. Its exceptional resolution facilitates the separation of closely related proteins with distinct hydrophobic properties, such as mutated or variant forms associated with protein disorders. RP-HPLC enables precise quantification of protein fractions, offering valuable insights into the extent of protein abnormalities and disease severity. Despite these advantages, certain considerations must be taken into account. The biological structure of proteins can be highly complex, with various proteins sharing similar hydrophobic properties, making separation and identification challenging in such cases. The requisite nonpolar stationary phase in RP-HPLC can lead to denaturation of proteins, altering their structure and function. These factors underscore the need for careful consideration of the method's limitations when applied to complex protein structures.

3.3.5 Tandem Mass Spectrometry

Tandem mass spectrometry (TMS) stands out as a potent analytical technique for amino acid analysis, and it is frequently preferred for its heightened sensitivity, capability for high-throughput analysis, and cost-effectiveness (Shuken 2023). In the case of newborn screening, blood samples are collected on filter paper, and specific-sized disks, typically 3 mm in diameter, are extracted from the paper for subsequent analysis (Lehotay et al. 2011). The TMS process involves two fundamental steps: ionization and mass analysis. In the initial step, a sample containing amino acids undergoes ionization, commonly through electrospray ionization or laser ablation, resulting in the formation of charged molecules (ions). The subsequent step involves mass analysis. In MS/MS, the ions undergo further fragmentation, often through collision-induced dissociation (CID) or alternative fragmentation techniques, generating a series of daughter ions with distinct mass-to-charge ratios (Chace and Kalas 2005). Unlike early assays that required derivatization, modern TMS allows for simultaneous analysis of multiple amino acids and their metabolites in a single sample. This advancement facilitates high-throughput screening and comprehensive profiling. The measurement of mass-to-charge ratios for these daughter ions enables clinicians to identify and quantify amino acids with remarkable accuracy, allowing for the early detection of amino acid disorders even before clinical symptoms manifest. Tandem MS/MS is adept at detecting subtle variations in amino acid concentrations within normal and pathological ranges. This capability proves invaluable in the realm of personalized medicine, aiding clinicians in prescribing tailored treatment strategies.

While tandem mass spectrometry (MS/MS) has revolutionized the realm of amino acid disorder detection, offering clinicians and researchers a potent instrument for early diagnosis and monitoring, it is imperative to recognize its inherent constraints. MS/MS relies on discerning specific mass-to-charge ratios of ions to identify and quantify amino acids, yet it confronts challenges such as potential false-positive or false-negative outcomes due to interferences from isobaric compounds like leucine, isoleucine, hydroxyproline, or matrix effects. These complexities may lead to misdiagnoses, underscoring the necessity for meticulous data interpretation and validation (Neagu et al. 2022). The inherent instability of cysteine and homocysteine renders their detection impossible via tandem MS/MS, as the ionization process generates cystine and homocysteine, further complicating the analysis. The handling, preparation, and processing of samples are pivotal yet intricate and time-consuming. Positive MS/MS screening results mandate additional confirmatory testing through techniques such as gas chromatography-mass spectrometry (GC-MS) or genetic testing, introducing complexity and cost to the diagnostic process. While tandem mass spectrometry boasts heightened sensitivity, specificity, and the ability to analyze multiple amino acids simultaneously, its susceptibility to false results and the challenges associated with sample preparation, costs, and the requirement for specialized expertise may restrict its utility in particular healthcare environments. Therefore, it is paramount to acknowledge these limitations and advocate for a comprehensive approach, integrating MS/MS screening with confirmatory tests to ensure precision in diagnosis.

3.4 Conclusions

In summary, amino acid disorders represent a heterogeneous group of inborn errors of metabolism, characterized by diverse clinical and biochemical parameters. To facilitate early diagnosis and intervention, most amino acid disorders are incorporated into newborn screening programs. The analysis of physiological amino acid levels plays a pivotal role in diagnosing and managing the metabolic disorders. Tandem mass spectrometry is the most used tool in analyzing some of the amino acids in affected neonates even in the absence of clinical symptoms. Over the past six decades, various analytical methods have been developed, with notable advancements in derivatization, chromatography, and mass spectrometry. Nevertheless, ion exchange chromatography remains the gold standard technique in this field. Methodological enhancements include streamlined sample pretreatment, wider linear concentration ranges for the analysis of over 40 amino acids, increased automation, enhanced sensitivity, reduced run times, and improved specificity. These improvements will expedite diagnosis and therapy monitoring for amino acid disorders. The diagnostic field is expanding with the use of whole-genome sequencing in detecting anomalies. Whole-genome sequencing contributes significantly to the identification of novel genes, biomarker discovery, and the elucidation of disease associations, thus advancing the field. The primary objective of treating amino acid disorders is to restore the balance between substrates and downstream products while preventing the accumulation of toxic substances. Simultaneously, nutritional management must meet the basic dietary requirements for growth and normal development. While many amino acid disorders lack curative treatments, current interventions significantly enhance patients' quality of life. Over the forthcoming years, continued methodological advancements are expected to deepen our understanding of inborn errors of metabolism, particularly those related to amino acids, further improving disease outcomes.

References

Adhikari S, Manthena PV, Sajwan K et al (2010) A unified method for purification of basic proteins. Anal Biochem 400(2):203–206

Ah Mew N, Simpson KL, Gropman AL et al (2003) Urea cycle disorders overview. In: GeneRev®. University of Washington, Seattle, Seattle

Alcaide P, Krijt J, Ruiz-Sala P et al (2015) Enzymatic diagnosis of homocystinuria by determination of cystathionine-ß-synthase activity in plasma using LC-MS/MS. Clin Chim Acta 438:261–265

Ali EZ, Ngu LH (2018) Fourteen new mutations of BCKDHA, BCKDHB and DBT genes associated with maple syrup urine disease (MSUD) in Malaysian population. Mol Genet Metab Rep 17:22–30

Antonini R, Scaini G, Michels M et al (2020) Effects of omega-3 fatty acids supplementation on inflammatory parameters after chronic administration of L-tyrosine. Metab Brain Dis 35(2):295–303

Arthur SS, Nyide B, Soura AB et al (2015) Tackling malnutrition: a systematic review of 15-year research evidence from INDEPTH health and demographic surveillance systems. Glob Health Action 8(1):28298

Barltrop D, Sandhu BK (1985) Marasmus-1985. Postgrad Med J 61(720):915

Barmore W, Azad F, Stone WL (2023) Physiology, urea cycle. In: StatPearls. StatPearls Publishing, Treasure Island

Bartlett DC, Lloyd C, McKiernan PJ et al (2014) Early nitisinone treatment reduces the need for liver transplantation in children with tyrosinaemia type 1 and improves post-transplant renal function. J Inherit Metab Dis 37(5):745–752

Bas AY, Kunak B, Ertan U et al (2003) Tyrosinemia type 1: a case report. Int Pediatr 18(1):183–191

Baumgartner MR, Almashanu S, Suormala T et al (2001) The molecular basis of human 3-methylcrotonyl-CoA carboxylase deficiency. J Clin Invest 107(4):495–504

Bisceglia L, Purroy J, Jiménez-Vidal M et al (2001) Cystinuria type I: identification of eight new mutations in SLC3A1. Kidney Int 59(4):1250–1256

Blackburn PR, Gass JM, Pinto e Vairo F et al (2017) Maple syrup urine disease: mechanisms and management. Appl Clin Genet 10:57–66

Blanco A, Blanco G (2022) Medical biochemistry. Academic, London

Bliksrud YT, Brodtkorb E, Backe PH et al (2012) Hereditary tyrosinaemia type i in Norway: incidence and three novel small deletions in the fumarylacetoacetase gene. Scand J Clin Lab Invest 72(5):369–373

Blumenfeld CM, Wallace MJ, Anderson R (1966) Phenylketonuria-the Guthrie screening test-a method of quantitation, observations on reliability and suggestions for improvement. Calif Med 105(6):429

Braissant O (2010) Current concepts in the pathogenesis of urea cycle disorders. Mol Genet Metab 100:S3–S12

Bröer S, Gauthier-Coles G (2022) Amino acid homeostasis in mammalian cells with a focus on amino acid transport. J Nutr 152(1):16–28

Brustolin S, Giugliani R, Félix TM (2010) Genetics of homocysteine metabolism and associated disorders. Braz J Med Biol Res 43(1):1–7

Buerger C, Garbade SF, Dietrich Alber F (2019) Impairment of cognitive function in ornithine transcarbamylase deficiency is global rather than domain-specific and is associated with disease onset, sex, maximum ammonium, and number of hyperammonemic events. J Inherit Metab Dis 42(2):243–253

Bütikofer U, Fuchs D, Bosset JO et al (1991) Automated HPLC-amino acid determination of protein hydrolysates by precolumn derivatization with OPA and FMOC and comparison with classical ion exchange chromatography. Chromatographia 31(9–10):441–447

Calder PC (2006) Branched-chain amino acids and immunity. J Nutr 136(1):288–293

Calonge MJ, Gasparini P, Chillarón J (1994) Cystinuria caused by mutations in rBAT, a gene involved in the transport of cystine. Nat Genet 6(4):420–425

Chace DH, Kalas TA (2005) A biochemical perspective on the use of tandem mass spectrometry for newborn screening and clinical testing. Clin Biochem 38(4):296–309

Chinsky JM, Singh R, Ficicioglu C et al (2017) Diagnosis and treatment of tyrosinemia type I: a US and Canadian consensus group review and recommendations. Genet Med 19(12):1380–1395

Chong GT, Farsiu S, Freedman SF et al (2009) Abnormal foveal morphology in ocular albinism imaged with spectral-domain optical coherence tomography. Arch Ophthalmol 127:37–44

Chuang DT, Cox RP, Max Wynn R (2020) Maple syrup urine disease: biochemical, clinical and therapeutic considerations. Rosenberg's Mol Genet Basis Neurol Psychiatr Dis 1:865–875

Coşar A, Ipçioğlu OM, Özcan Ö et al (2014) Folate and homocysteine metabolisms and their roles in the biochemical basis of neuropsychiatry. Turk J Med Sci 44(1):1–9

Cunico RL, Schlabach T (1983) Comparison of ninhydrin and o-phthalaldehyde post-column detection techniques for high-performance liquid chromatography of free amino acids. J Chromatogr A 266(C):461–470
Dardevet D (2016) The molecular nutrition of amino acids and proteins. Elsevier, London
Davis MD, Kaufman S, Milstien S (1991) Conversion of 6-substituted tetrahydropterins to 7-isomers via phenylalanine hydroxylase-generated intermediates. Proc Natl Acad Sci U.S.A 88(2):385–389
de Bandt JP (2016) Leucine and mammalian target of rapamycin-dependent activation of muscle protein synthesis in aging. J Nutr 146(12):2616–2624
Díaz-Villanueva JF, Díaz-Molina R, García-González V (2015) Protein folding and mechanisms of proteostasis. Int J Mol Sci 16(8):17193–17230
Dorresteijn RC, Berwald LG, Zomer G et al (1996) Determination of amino acids using o-phthalaldehyde-2-mercaptoethanol derivatization: effect of reaction conditions. J Chromatogr A 724(1–2):159–167
Eastman JW, Sherwin JE, Wong R et al (2000) Use of the phenylalanine: tyrosine ratio to test newborns for phenylketonuria in a large public health screening programme. J Med Screen 7(3):131–135
Erez A, Nagamani SCS, Lee B (2011) Argininosuccinatelyase deficiency-Argininosuccinic aciduria and beyond. Am J Med Genet C Semin Med Genet 157(1):45–53
Forny P, Hörster F, Ballhausen D et al (2021) Guidelines for the diagnosis and management of methylmalonic acidaemia and propionic acidaemia: first revision. J Inherit Metab Dis 44(3):566–592
Frazier DM, Allgeier C, Homer C et al (2014) Nutrition management guideline for maple syrup urine disease: an evidence- and consensus-based approach. Mol Genet Metab 112(3):210–217
Frimpter GW (1965) Cystathioninuria: nature of the defect. Science 149(3688):1095–1096
Galderisi S, Cicaloni V, Milella MS et al (2021) Homogentisic acid induces cytoskeleton and extracellular matrix alteration in alkaptonuric cartilage. J Cell Physiol 236(8):6011–6024
Garrow JS, Fletcher K, Halliday D (1965) Body composition in severe infantile malnutrition. J Clin Invest 44:417–425
Gerrard A, Dawson C (2022) Homocystinuria diagnosis and management: it is not all classical. J Clin Pathol 75(11):745–750
Ghoraba DA, Mohammed MM, Zaki OK (2015) Mutation analysis of methylmalonyl CoA mutase gene exon 2 in Egyptian families: identification of 25 novel allelic variants. Meta Gene 3:71–88
Grover Z, Ee LC (2009) Protein energy malnutrition. Pediatr Clin North Am 56(5):1055–1068
Gupta R, Shukla S, Gupta A, Singh RL, Singh P (2023) Estimation of phytochemical constituents and analysis of antioxidant activity in different parts of *Cassia fistula* plant. J Pharma Chem Ana 10(3):194–201
Hardeland R, Poeggeler B (2003) Non-vertebrate melatonin. J Pineal Res 34(4):233–241
Hayashi M, Suzuki T (2017) Oculocutaneous albinism type 4. GeneRev®. University of Washington, Seattle, Seattle
Heldt K, Schwahn B, Marquardt I et al (2005) Diagnosis of MSUD by newborn screening allows early intervention without extraneous detoxification. Mol Genet Metab 84(4):313–316
Heylen E, Scherer G, Vincent MF et al (2012) Tyrosinemia type III detected via neonatal screening: management and outcome. Mol Genet Metab 107(3):605–607
Holeček M (2018) Branched-chain amino acids in health and disease: metabolism, alterations in blood plasma, and as supplements. Nutr Metab 15(1):1–12
Hughes JH, Wilson PJM, Sutherland H et al (2020) Dietary restriction of tyrosine and phenylalanine lowers tyrosinemia associated with nitisinone therapy of alkaptonuria. J Inherit Metab Dis 43(2):259–268
Husson A, Brasse-Lagnel C, Fairand A et al (2003) Argininosuccinatesynthetase from the urea cycle to the citrulline-NO cycle. Eur J Biochem 270(9):1887–1899
Khan A, Zahid B, Khan S et al (2020) Isovaleric acidemia: a rare case of an inborn error of metabolism. Cureus 12(2):e7150

Kisa PT, Gunduz M, Dorum S et al (2021) Alkaptonuria in Turkey: clinical and molecular characteristics of 66 patients. Eur J Med Genet 64(5):104197
Kožich V, Morris AAM, Blom HJ (2016) Disorders of sulfur amino acid metabolism. In:Inborn Metabolic Diseases, 2016, p. 309–320
Lasseaux E, Neveu MM, Fiore M et al (2022) Albinism. In: Clinical ophthalmic genetics and genomics. Academic, London, pp 339–402
Lee W, Sen WRG et al (1993) Cloning and chromosomal localization of a human kidney cDNA involved in cystine, dibasic, and neutral amino acid transport. J Clin Invest 91(5):1959–1963
Lee HHC, Poon KH, Lai CK et al (2014) Hyperornithinaemia-hyperammonaemiahomocitrullinuria syndrome: a treatable genetic liver disease warranting urgent diagnosis. HKMJ 20(1):63
Lehotay DC, Hall P, Lepage J et al (2011) LC-MS/MS progress in newborn screening. Clin Biochem 44(1):22–31
Litwack G (2021) Human biochemistry. Academic, London. https://doi.org/10.1016/C2019-0-03231-1
Marks RR, Burgy JR, Davis LS (2019) Acute kwashiorkor in the setting of cerebral palsy and pancreatic insufficiency. Cutis 103(1):10–12
Moore S, Spackman DH, Stein WH (1958b) Automatic recording apparatus for use in the chromatography of amino acids. Fed Proc 17(4):1107–1115
Moore S, Spackman DH, Stein WH (1958a) Chromatography of amino acids on sulfonated polystyrene resins: an improved system. Anal Chem 30(7):1185–1190
Morris AAM, Kožich V, Santra S et al (2017) Guidelines for the diagnosis and management of cystathionine beta-synthase deficiency. J Inherit Metab 40(1):49–74
Mukherjee S, Maitra SK (2015) Gut melatonin in vertebrates: chronobiology and physiology. Front Endocrinol 6:131836
Muntau AC, Adams DJ, Bélanger-Quintana A et al (2019) International best practice for the evaluation of responsiveness to sapropterindihydrochloride in patients with phenylketonuria. Mol Genet Metab 127(1):1–11
Muntau AC, Burlina A, Eyskens F et al (2021) Long-term efficacy and safety of sapropterin in patients who initiated sapropterin at < 4 years of age with phenylketonuria: results of the 3-year extension of the SPARK open-label, multicentre, randomised phase IIIb trial. Orphanet J Rare Dis 16(1):341
Nakamoto T (2011) Mechanisms of the initiation of protein synthesis: in reading frame binding of ribosomes to mRNA. Mol Biol Rep 38(2):847–855
Neagu AN, Jayathirtha M, Baxter E et al (2022) Applications of tandem mass spectrometry (MS/MS) in protein analysis for biomedical research. Molecules 27(8):2411
Obonyo N, Brent B, Olupot-Olupot P (2017) Myocardial and haemodynamic responses to two fluid regimens in African children with severe malnutrition and hypovolaemic shock (AFRIM study). Crit Care 21(1):1–3
Odièvre MH, Chretien D, Munnich A et al (2005) A novel mutation in the dihydrolipoamide dehydrogenase E3 subunit gene (DLD) resulting in an atypical form of α-ketoglutarate dehydrogenase deficiency. Hum Mutat 25(3):323–324
Packman W, Mehta I, Rafie S et al (2012) Young adults with MSUD and their transition to adulthood: psychosocial issues. J Genet Couns 21(5):692–703
Padmanabha H, Mahale R, Christopher R et al (2021) Clinical, biochemical, radiological, and genetic profile of patients with homocysteine remethylation pathway defect and spastic paraplegia. Ann Indian Acad Neurol 24(6):908–916
Philip LP, Keith H, Chiles J et al (2012) Partial pyridoxine responsiveness in PNPO deficiency. JIMD Rep 4:113–116
Pode-Shakked N, Korman SH, Pode-Shakked B et al (2020) Clues and challenges in the diagnosis of intermittent maple syrup urine disease. Eur J Med Genet 63(6):103901
Prieve MG, Harvie P, Monahan SD et al (2018) Targeted mRNA therapy for ornithine transcarbamylase deficiency. Mol Ther 26(3):801–813

Rocha JC, Bausell H, Bélanger-Quintana A et al (2021) Development of a practical dietitian road map for the nutritional management of phenylketonuria (PKU) patients on pegvaliase. Mol Genet Metab Rep 28:100771

Sacharow SJ, Picker JD, Levy HL (2004) Homocystinuria caused by cystathionine Beta-synthase deficiency. In: GeneReviews®. University of Washington, Seattle

Sadler K, Kerac M, Collins S et al (2008) Improving the management of severe acute malnutrition in an area of high HIV prevalence. J Trop Pediatr 54(6):364–369

Sajeev M, Chin S, Ho G et al (2021) Challenges in diagnosing intermediate maple syrup urine disease by newborn screening and functional validation of genomic results imperative for reproductive family planning. Int J Neonatal Screen 7(2):25

Schadewaldt P, Bodner-Leidecker A, Hammen HW et al (1999) Significance of L-alloisoleucine in plasma for diagnosis of maple syrup urine disease. Clin Chem 45(10):1734–1740

Schneider HJ (1989) Amino acid analysis using DABS-CL. Chromatographia 28(1–2):45–48

Scott CR (2006) The genetic tyrosinemias. Am J Med Genet Seminars Med Genet 142C(2):121–126

Scriver CR, Clow CL, Mackenzie S et al (1971) Thiamine-responsive maple-syrup-urine disease. Lancet 297(7694):310–312

Selvan SR, Dowling JP, Kelly WK et al (2016) Indoleamine 2,3-dioxygenase (IDO): biology and target in cancer immunotherapies. Curr Cancer Drug Targets 16(9):755–764

Sherwood RA (2000) Amino acid measurement in body fluids using PITC derivatives. Methods Mol Biol (Clifton, N.J.) 159:169–175

Shimomura Y, Honda T, Shiraki M et al (2006) Branched-chain amino acid catabolism in exercise and liver disease. J Nutr 136(1):250–253

Shirozu K, Tokuda K, Marutani E (2013) Cystathione-γ-lyase deficiency protects mice from galactosamine/LPS-induced acute liver failure. FASEB J 27(S1):1161–1167

Shuken SR (2023) An introduction to mass spectrometry-based proteomics. J Proteome Res 22(7):2151–2171

Singh P, Singh RL (2018) Quantification of phytochemicals imparting antioxidant activities in commonly used vegetables. Int J Appl Sci Biotechnol 6(2):107–112

Singh P, Vishwakarma SP, Singh RL (2013) Evaluation of antioxidant, oxidative DNA damage protective and antimicrobial activities of *Foeniculum vulgare* plant. J Med Plant Res 7(35):2551–2563

Stauss K, Puffenberger E, Morton D (2020) Maple syrup urine disease. 2006 Jan 20 [updated 2013 May 9]. In Gene Reviews

Stone TW, Darlington LG (2013) The kynurenine pathway as a therapeutic target in cognitive and neurodegenerative disorders. Br J Pharmacol 169(6):1211–1227

Summar ML, Mew NA (2018) Inborn errors of metabolism with hyperammonemia: urea cycle defects and related disorders. Pediatr Clin North Am 65(2):231–246

Szymanska E, Sredzinska M, Ciara E et al (2015) Tyrosinemia type III in an asymptomatic girl. Mol Genet Metab Rep 5:48–50

Tanaka K (2009) The proteasome: overview of structure and functions. Proc Jpn Acad Ser B Phys Biol Sci 85(1):12–36

Tang Y, Kong Y (2021) Hereditary tyrosinemia type I: newborn screening, diagnosis and treatment. Zhejiang Da XueXueBao. Yi Xue Ban = J Zhejiang Univ Med Sci 50(4):514

Teramae A, Kobayashi Y, Kunimoto H et al (2019) The molecular basis of chemical chaperone therapy for Oculocutaneous albinism type 1A. J Invest Dermatol 139(5):1143–1149

Titi-Lartey OA, Gupta V (2023). Marasmus. [Updated 2023 Jul 24]. In: StatPearls [Internet]. StatPearls Publishing, Treasure Island; 2024 Jan

Townsend DM, Tew KD, Tapiero H (2004) Sulfur-containing amino acids and human disease. Biomed Pharmacother 58(1):48–55

Trehan I, Manary MJ (2015) Management of severe acute malnutrition in low-income and middle-income countries. Arch Dis Child 100(3):283–287

Ullah H (2012) The role of ion exchange chromatography in purification and characterization of molecules. In: Ion exchange technologies. IntechOpen, London, pp 331–342

Urv TK, Parisi MA (2017) Newborn screening: beyond the spot. Adv Exp Med Biol 1031:323–346

Van Vliet D, Derks TG, Van Rijn M et al (2014) Single amino acid supplementation in aminoacidopathies: a systematic review. Orphanet J Rare Dis 9(1):7
Vavricka CJ, Han Q, Mehere P et al (2014) Tyrosine metabolic enzymes from insects and mammals: a comparative perspective. Insect Sci 21(1):13–19
Vockley J, Andersson HC, Antshel KM et al (2014) Phenylalanine hydroxylase deficiency: diagnosis and management guideline. Genet Med 16(2):188–200
Whitten MMA, Coates CJ (2017) Re-evaluation of insect melanogenesis research: views from the dark side. Pigment Cell Melanoma Res 30(4):386–401
Williams RA, Mamotte CDS, Burnett JR (2008) Phenylketonuria: an inborn error of phenylalanine metabolism. Clin Biochem Rev 29(1):31
Wilson PJM, Ranganath LR, Bou-Gharios G et al (2021) Expression of tyrosine pathway enzymes in mice demonstrates that homogentisate 1,2-dioxygenase deficiency in the liver is responsible for homogentisic acid-derived ochronotic pigmentation. JIMD Rep 58(1):52–60
Wong LTK, Hardwick DF, Applegarth DA et al (1979) Review of metabolic screening program of Children's Hospital, Vancouver, British Columbia. 1971–1977. Clin Biochem 12(5):167–172
Wootton S, Durkin K, Jackson A (2014) Quality control issues related to assessment of body composition. Food Nutr Bull 35:579–585
XIII (1810) On cystic oxide a new species of urinary calculus. Philos Trans R Soc Lond 100:223–230
Ye X, Whiteman B, Jerebtsova M et al (2000) Correction of argininosuccinatesynthetase (AS) deficiency in a murine model of citrullinemia with recombinant adenovirus carrying human AS cDNA. Gene Ther 7(20):1777–1782
Yeldandi AV, Tan X, Dwivedi RS et al (1990) Co expression of glutamine synthetase and carbamoyl phosphate synthaoe I genes in pancreatic hepatocytes of rat. Proc Natl Acad Sci U.S.A 87(3):881–885
Zanatta Â, Rodrigues MDN, Amaral AU et al (2016) Ornithine and homocitrulline impair mitochondrial function, decrease antioxidant defenses and induce cell death in menadione-stressed rat cortical astrocytes: potential mechanisms of neurological dysfunction in HHH syndrome. Neurochem Res 41(9):2190–2198

Disorders of Lipid Metabolism

4

Shireen Masood, Osaid Masood, Syed Faiz Mujtaba, and Monisha Banerjee

Abstract

Lipids are organic molecules readily soluble in organic solvents, insoluble in water due to biochemical transformations, and metabolized through several enzymes of β-oxidation pathway. They play a vital role in metabolic functions, such as structural organization, energy depository, cellular signaling, energy metabolism, transmission of nerve impulse, and transport of nutrients. The dysfunction in lipid metabolism leads to a number of complications like hyperlipidemia, hypercholesterolemia, triglyceridemia, hypertriglyceridemia, and hyperlipoproteinemia. Moreover, secondary metabolites of lipid irregularities result in a wide range of disorders like hypothyroidism, obesity, diabetes mellitus, liver cholestasis, Gaucher disease, Tay-Sachs disease, and renal ailments. They are also reported in varied conditions like lymphoma, porphyria, and Cushing disease. Disorders of lipid metabolism often cause atherosclerotic vascular diseases including thromboembolic cerebrovascular disease, ischemic

S. Masood · O. Masood
Molecular and Human Genetics Laboratory, Department of Zoology, University of Lucknow, Lucknow, Uttar Pradesh, India

S. F. Mujtaba
Department of Zoology, Faculty of Science, Shia Post Graduate College, Lucknow, Uttar Pradesh, India

M. Banerjee (✉)
Molecular and Human Genetics Laboratory, Department of Zoology, University of Lucknow, Lucknow, Uttar Pradesh, India

Institute of Advanced Molecular Genetics and Infectious Diseases (IAMGID), ONGC Centre for Advanced Studies (OCAS), University of Lucknow, Lucknow, Uttar Pradesh, India

R. L. Singh et al. (eds.), *Clinical Applications of Biomolecules in Disease Diagnosis*, https://doi.org/10.1007/978-981-97-4723-8_4

heart disease, and peripheral vascular disease. Lipid metabolism disorders have become a widespread global issue, necessitating increased research to develop prognostic/diagnostic strategies. The chapter will mainly deal with the genetic and epigenetic studies in lipid metabolism disorders.

Keywords

Epigenetics · Genetic studies · Lipids disorders · Metabolism · Primary disorders · Secondary disorders

4.1 Introduction

Lipids are organic molecules that are soluble in nonpolar and organic solvents but are not soluble in polar solvent like water. Lipids consist of triglycerides (oil and fats), phospholipids, steroid, and waxes. Glycerol, a polyol with three carbons and three fatty acids are esters that make up fats and oils. Fatty acids are chains of hydrocarbons with differing degrees of saturation along with carboxylic acid groups at end chains. Phospholipids are necessary components of cell membranes that are made up of a glycerol backbone, two tails of fatty acid that are hydrophobic, and a phosphate group that is hydrophilic. Cholesterol, which is produced in the liver, is an important form of steroid. Steroids are made up of four fused rings. Biochemical lipids are stored in tissues all over the body, specifically in adipose tissue, a type of connective tissue. Endogenous triglycerides and cholesterol esters are synthesized to further formulate very low-density lipid (VLDL) that binds with low-density lipid receptors (LDLR) via surface apolipoproteinB-100 (APOB) and apolipoprotein E (APOE) and converts to intermediate-density lipoprotein via lipoprotein lipase (LPL). The liver subsequently alters intermediate-density lipoprotein to produce low-density lipid (LDL), which transports hepatic cholesterol to cells through apoB/E receptor-mediated endocytosis for probable steroid or hormone synthesis and integrity of cell (Panickar and Bhathena 2010). Lipids protect the human organs including spleen, liver, heart, and kidneys against harm (Church et al. 2012).

The lipid metabolism is controlled by a family of proteins known as sterol regulatory element-binding proteins (SREBS). The disorders related to lipid metabolism which arises primarily due to genetic predisposition followed by environmental changes and lifestyle can be primary or secondary. The primary lipid metabolism disorders are the underlying cause of secondary disorders.

4.2 Normal Lipid Metabolism

All physiological and cellular functions carried out in the body require energy, source of which are fatty acids (FA). They are reserved as triglycerides in the form of small sized lipid droplets (LD) predominantly in the liver and adipose tissue. The mobilization of FA from triglyceride reserves in adipose tissue requires the action of triglyceride lipases, which are subsequently released into the circulation and drawn into hepatocytes, where it is reintroduced into triglyceride (Gibbons et al. 2004) (Fig. 4.1).

A part of the re-esterified triglyceride binds APOB to create VLDL, which is then released into the bloodstream. This activity is modulated through microsomal triglyceride transfer protein (MTTP) and is followed by the encapsulation with a phosphatidylcholine enriched monolayer of the phospholipid of neutral lipid core (Sundaram and Yao 2010). APOB, APOC, cholesterol, and other apolipoproteins can also be identified on the exterior of triglyceride VLDL particles. FA is utilized to form fructose and glucose through de novo. The majority of triglyceride required for VLDL triglyceride formation is also synthesized from FA from adipose tissue entering the liver (Mato et al. 2019). Since cholesterol and triglycerides are insoluble in water, they must be delivered in conjunction with proteins. Lipoproteins are complex particles that comprise a central core of cholesterol esters and triglycerides surrounded by phospholipids, apolipoproteins, and free cholesterol which aid in its production and function. Plasma lipoproteins are classified into seven categories depending on their size, lipid composition, and apolipoproteins (chylomicrons, VLDL, chylomicron remnants, intermediate density lipid (IDL), LDL, high-density lipid (HDL), and lipoprotein A (LPA)) (Feingold 2022).

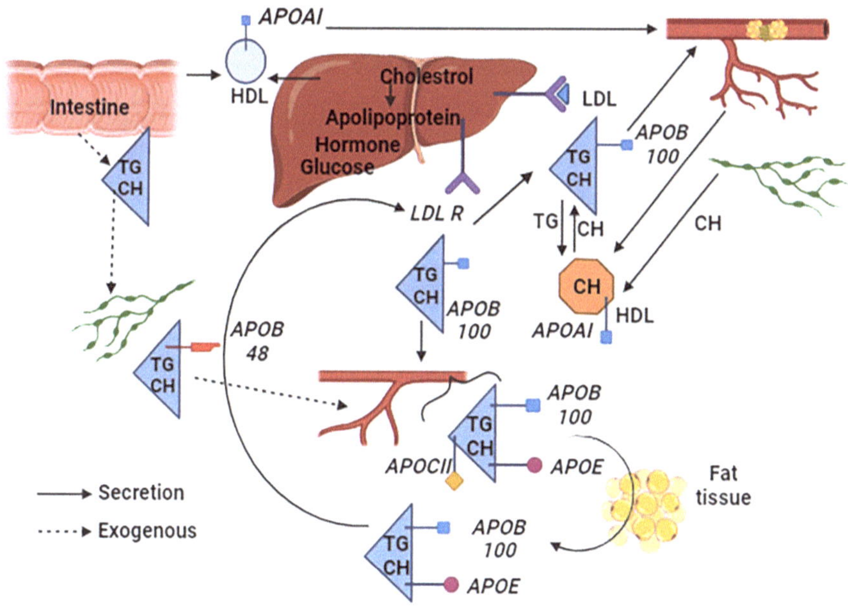

Fig. 4.1 Normal lipid metabolism involving liver, intestine, blood vascular system, lymphatic system, and adipocytes. APOAI apolipoprotein AI, HDL high-density lipid, LDL low-density lipid, TG triglycerides, CH cholesterol, LDL R low-density lipid receptor, APOB apolipoprotein B, APOCII apolipoprotein CII, APOE apolipoprotein E

Sterol regulatory element-binding proteins (SREBPs) are a family of membrane-bound transcription factors that regulate lipid homeostasis in animals (Horton et al. 2002). SREBPs regulate the expression of about 30 genes directly, that regulate the production and absorption of cholesterol, FAs, phospholipids, and triglycerides, along with the cofactor NADPH (nicotinamide adenine dinucleotide phosphate hydrogen) which is necessary for their synthesis (Osborne 2000; Fernandez and Esteller 2010). Isoforms of SREBP such as SREBP-1a, SREBP-1c, and SREBP-2 are encoded by the mammalian genome. SREBPs differ in terms of tissue specific expression, target gene selectivity, and relative potency of their transactivation regions. Individual isoform analyses indicate that SREBP-1 is more selective for genes associated with fatty acid metabolics and de novo lipid genesis, whilst SREBP-2 is more sensitive for genes engaged primarily in cholesterol homeostasis. In mice, abnormal SREBP expression causes metabolic disorders with physiological effects analogous to particular lipid metabolism illnesses in humans (Fernandez and Esteller 2010). Genome-wide association studies (GWAS) have recognized over 500 SNPs (single-nucleotide polymorphisms) linked with altered lipid levels describing ~40% of the individual differences (Garcia et al. 2019). The disorders in normal lipid metabolic processes have led to a number of disorders which can be classified into primary and secondary lipid metabolic disorders. Primary lipid metabolic disorders result from the primary biochemical alterations while secondary results from manifestations of the primary alterations (Fig. 4.2).

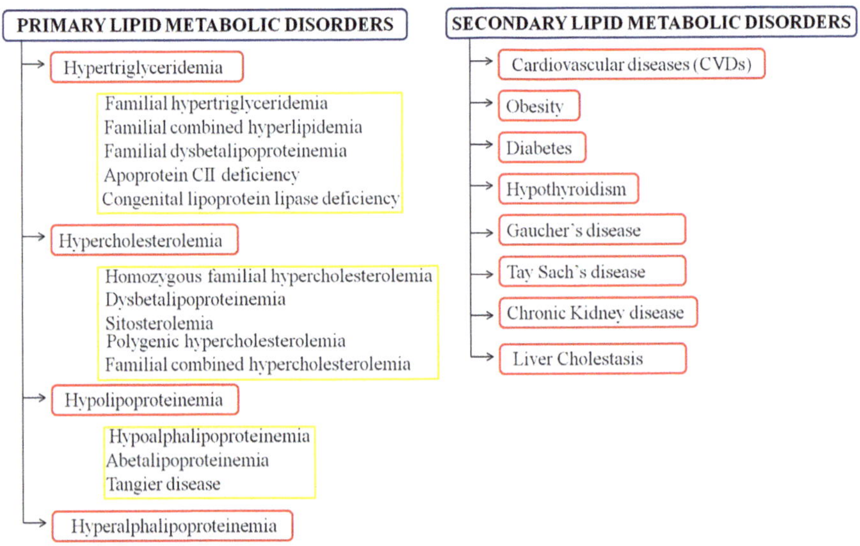

Fig. 4.2 Primary and secondary lipid metabolic disorders

4.3 Primary Lipid Metabolic Disorders

Lipid metabolism is a crucial and complex biological process of digestion, absorption, assimilation, decomposition, and metabolism of lipids, such as triglyceride, cholesterol, and FAs. A lipid metabolic disorder that manifests at a young age is frequently caused by monogenic factors. However, in addition to nongenetic influences, a large number of these disorders has a complex polygenic background where the cumulative effect of several mutations boosts the triglyceride levels (Miller et al. 2011). Primary lipid disorders have specific characteristics and risks associated with it as described below (Table 4.1).

4.3.1 Hypertriglyceridemia

It can be characterized by plasma triglyceride levels greater than 150 mg/dL, impacts around 10% of the human population, making it the most frequent type of dyslipidemia globally (Simha 2020). Associated syndromes are described below:

Table 4.1 Characteristics and risks associated with primary lipid metabolic disorders

Disorders	Lipid effect	Inheritance	Risk
Familial hypertriglyceridemia	VLDL ↑ HDL-C↓ LDL-C↑	Autosomal dominant	Pancreatitis
Familial combined hyperlipidemia	Cholesterol ↑ Triglyceride ↑	Autosomal dominant	CVD
Familial dysbetalipoproteinemia	Chylomicrons↑ VLDL↑	Autosomal recessive	Atherosclerotic CVD
Apoprotein CII deficiency	Chylomicrons↑ VLDL ↑	Autosomal recessive	–
Congenital lipoprotein lipase deficiency	Triglyceride↑ Chylomicrons↑	Autosomal recessive	Pancreatitis
Homozygous familial hypercholesterolemia	LDL-C↑	Autosomal dominant	Atherosclerotic CVD
Dysbetalipoproteinemia	VLDL↑	–	
Sitosterolemia	LDL-C↑	Autosomal recessive	Xanthomas, arthralgia, and premature atherosclerosis
Polygenic hypercholesterolemia	LDL-C↑	Autosomal dominant	CVD
Familial combined hypercholesterolemia	LDL-C↑	Autosomal dominant	CVD
Hypoalphalipoproteinemia	HDL-C↓	Autosomal dominant	CVD
Abetalipoproteinemia	LDL-C↓	Autosomal recessive	Fat malabsorption, spinocerebellar degeneration, acanthocytosis, and retinitis pigmentosa
Tangier disease	HDL-C↓	Autosomal recessive	Premature cardiovascular events
Hyperalphalipoproteinemia	HDL-C ↑	–	–

4.3.1.1 Familial Hypertriglyceridemia

Familial hypertriglyceridemia (type IV familial dyslipidemia) is an autosomal dominant condition characterized by the overproduction of very low-density lipoproteins (VLDL) from the liver. As a result, the patient will have an excessive number of triglycerides and VLDL on the lipid profile that can cause acute pancreatitis. HDL cholesterol levels are also lowered; LDL cholesterol and ApoB levels can be low or normal. There is a high variation in the phenotypic expression involving hypercholesterolemia and hypertriglyceridemia. The heterozygous variants in the genes implicated in familial hypertriglyceridemia include *LPL*, *APOA5*, *APOC2*, *GPIHBP1* (glycosylphosphatidylinositol-anchored HDL-binding protein), and *LMF1* (lipase maturation factor) that affect LPL maturation and function (Pingitore et al. 2016; Bello-Chavolla et al. 2018).

4.3.1.2 Familial Combined Hyperlipidemia

An autosomal dominant disorder, polygenic in nature, characterized by the excessive synthesis of lipoproteins containing apolipoprotein B. The gene-cluster *APOA1-C3-A4-A5, LPL, APOE, USF1* (upstream transcription factor), *TNFRSF1B* (tumor necrosis factor receptor superfamily), and *LDLR* are involved in familial combined hyperlipidemia and alter LPL maturation and function, RLP (remnant like particle) lipid clearance, or VLDL secretion from the liver (Geurts 2000; Groenendijk et al. 2001; Pajukanta et al. 2004; Civeira et al. 2008; Liu et al. 2010; Solanas-Barca et al. 2012). It tends to be more atherogenic than familial hypertriglyceridemia and is linked to an increased risk of cardiovascular disease (Carrasquilla et al. 2021).

4.3.1.3 Congenital Lipoprotein Lipase Deficiency

Lipoprotein lipase deficiency (LPLD), type I hyperlipoproteinemia or familial chylomicronemia syndrome is an autosomal recessive metabolism disease characterized by severe elevations in circulating triglyceride concentrations in addition to an increased risk of pancreatitis, persistent pancreatitis, and fatalities from acute pancreatitis. Biallelic mutations in the LPL gene and four other causative genes that generate enzymes and cofactors that communicate with LPL, including *APOC2*, *GPIHBP1*, *APOA5*, and *LMF1*, cause monogenic chylomicronemia (Baass et al. 2020).

4.3.1.4 Apoprotein C-II Deficiency

An autosomal recessive condition is characterized by a deficiency of *APO C-II*, a necessary cofactor for lipoprotein lipase action; leading to a plasma buildup of chylomicron and VLDL (Ginsberg et al. 2021). The mutations in *APOC2* lead to impaired binding of lipoprotein with APOC-II which further interferes with LPL activity. Of all the mutations, some were found to be present in majority of the population with apoprotein C-II deficiency, for example, a nonsense mutation in R4 position, missense mutations, lysine to threonine, and lysine to glutamic acid at K19T and K55Q respectively (Wolska et al. 2017).

4.3.1.5 Familial Dysbetalipoproteinemia

The presence of mixed hyperlipidemia, with significantly elevated residual chylomicron and very low-density lipoprotein remnants (cholesterol- VLDL), as well as pathogenic mutations in the *APOE* gene, is a reliable marker for the condition. It is a recessively inherited disease due to homozygosity in *APOE2* (rs7412; Arg176Cys). Other genes that are found to be associated are *LDLR, APOB, PCSK9 (*Proprotein convertase subtilisin/kexin type*)*, and other hypertriglyceridemia-associated genes (Boot et al. 2020).

4.3.2 Hypercholesterolemia

It can be described as a disorder in which there are high cholesterol levels in the blood which include cholesterol transported within the apoB-containing lipoproteins represented by VLDL, LDL, IDL, and LPA (Civeira et al. 2022). Three main conditions result in hypercholesterolemia:

4.3.2.1 Homozygous Familial Hypercholesterolemia

It is an autosomal dominant hereditary condition that can be homozygous or heterozygous that affects one in every 250 persons. Pathogenic mutations in *LDLR, APOB,* and *PCSK9* genes cause this disorder, which hinders the removal of LDLs from the blood, increasing the possibility of early atherosclerotic cardiovascular disease.

4.3.2.2 Dysbetalipoproteinemia

It is an aggressive form of mixed hyperlipidemia characterized by an accumulation of triglyceride-rich residual particles in the blood, commonly known as VLDL. It is most commonly caused by a tendency of aberrant particle catabolism caused by homozygosity for receptor-defective binding APOE2 isoforms expressed by *APOE* gene variant (Civeira et al. 2022).

4.3.2.3 Sitosterolemia

It is also known as phytosterolemia, is an uncommon autosomal recessive illness caused by homozygous or compound heterozygous mutations in the *ABCG5* (ATP-binding cassette subfamily G member) or *ABCG8* genes, which prevent the formation of heterodimers resulting in the drain of free sterols from enterocytes and hepatocytes into the intestine and bile (Escolà-Gil et al. 2014). Clinical manifestations include xanthomas, arthralgia, and premature atherosclerosis.

4.3.2.4 Polygenic Hypercholesterolemia

It is an autosomal dominant and most common disorder to raise cholesterol levels. It occurs by polymorphisms in the *LDLR*, where over 1200 variants have been reported and sometimes by mutations in *APOB100* or *PCSK9* genes. It may lead to early and recurring cardiovascular disease, if increased LDL-C levels are not addressed or managed insufficiently (Stein and Raal 2013).

4.3.2.5 Familial Combined Hyperlipidemia

As discussed in triglyceridemia.

4.3.3 Hyperalphalipoproteinemia (HALP)

HALP is a disease characterized by raised plasma HDL-C levels, as well as increased plasma total cholesterol, whereas LDL and VLDL levels are generally normal (Zhou et al. 2020). Primary causes of HALP consist of *CETP* (Cholesteryl ester transfer protein) (splicing defect in intron and missense mutation in exon), hepatic lipase (LIPC) (polymorphisms in promoter), *APO-CIII* (c.13-2A > G and c.55 + 1G > A), scavenger receptor class B Type I (SR-BI) (point mutation), and endothelial lipase (LIPG) deficiency. These are associated with increased plasma HDL-C levels (Giammanco et al. 2021).

4.3.4 Hypolipoproteinemia

Low blood lipid levels are referred to as hypolipoproteinemia. This issue could have a genetic component, or it could be linked to other conditions like anemia or an overactive thyroid. It includes three primary conditions:

4.3.4.1 Hypoalphalipoproteinemia

It is an autosomal dominant disorder and can be defined as an extreme reduction in HDL cholesterol levels in plasma that is associated with a greater risk of coronary heart disease. Rare heterozygous missense, nonsense, or splicing variations in the *ABCA1, APOA1,* and *LCAT* (lecithin:cholesterol acyl transferase) genes, which encode ABCA1, APOA-I, and LCAT, are present in 10–35% of those who are afflicted (Aguilar-Salinas et al. 2009).

4.3.4.2 Hypobetalipoproteinemia

LDL-C, typically ranged between 20 and 50 mg/dL, and apoB are two features of the autosomal co-dominant condition. Familial hypobetalipoproteinemia has been linked to more than 60 mutations that result in truncations in the *APOB* gene, which spans from *APOB2* to *APOB-89*. Truncations are named according to how much of the original APOB-100 molecule remains; for example, apoB-67 is the amino terminus of 67% of APOB-100 (Welty 2014). Increased degradation of VLDL and extremely low release of the truncated APOB are caused by decreased LDL APOB-100 synthesis. Reduced triglyceride outflow from the liver as a result of this decreased APOB secretion also contributes to the emergence of fatty liver (Tanoli et al. 2004). This condition has also been associated with hepatic steatosis, insulin sensitivity, hepatic cirrhosis, and carcinoma.

4.3.4.3 Abetalipoproteinemia (ABL)

Rare autosomal recessive illness ABL is characterized by undetectable levels of LDL and APOB and exceptionally low-serum levels of triglycerides and cholesterol (sub 30 mg/dl). It is a lipid transport protein that is produced by the microsomal triglyceride transfer protein (*MTP*) gene and found in the endoplasmic reticulum of enterocytes and hepatocytes. When VLDL triglyceride is assembled in the rough endoplasmic reticulum, MTP catalyzes the transfer of triglyceride to the developing APOB particle. APOB is co-translationally transported through the endoplasmic reticulum membrane. The *MTP* gene has more than 30 known mutations (Welty 2014). Hallmark symptoms include fat malabsorption, spinocerebellar degeneration, acanthocytosis, and retinitis pigmentosa.

4.3.4.4 Tangier Disease

A hereditary condition known as Tangier illness is characterized by a lack of or incredibly low levels of HDL-cholesterol. It has an autosomal recessive genetic profile and is caused by a defective mutation in the *ABCA1* gene, which is required for the production of particles of HDL from cholesterol and phospholipids. ABCA1 deficiency in peripheral cell populations, including macrophages and Schwann cells, hampers cellular cholesterol export. Due to the buildup of cholesterol in these cells, symptoms such as pharyngeal tonsillitis, ocular opacity, hepatosplenomegaly, lymphadenopathy, and peripheral neuropathy might develop (Koseki et al. 2021).

Defects in genes associated with lipid metabolism thus result in primary disorders (Table 4.2). However, recent studies have demonstrated the role of epigenetics in the irregular lipid metabolic machinery. The carnitine palmitoyltransferase-1A (CPT1A) converts long-chain acyl-CoAs into long-chain acylcarnitines which makes it easier for them to pass the mitochondrial membrane and enter the mitochondrial matrix, where fatty acid oxidation occurs. DNA hypermethylation in this gene (cg00574958) has been linked to a decreased risk of metabolic disorders, including low triglycerides, reduced obesity risk, type 2 diabetes, hypertension, and metabolic syndrome (Lai et al. 2020). The gene *ABCG*1 possessed many methylation sites which were found to be associated with triglyceride level out of which two sites cg06500161 and cg27243685

Table 4.2 Genes involved in primary lipid metabolic disorders

Genetic disorders	Protein (gene) defect
Familial hypertriglyceridemia	LPL, APOA5, APOC2, GPIHBP1, and LMF1
Familial dysbetalipoproteinemia	APOE
Apoprotein CII deficiency	APOC2
Congenital lipoprotein lipase deficiency	LPL
Homozygous familial hypercholesterolemia	LDLR, APOB, and PCSK9
Dysbetalipoproteinemia	APOE
Sitosterolemia	ABCG5 or ABCG8
Polygenic hypercholesterolemia	LDLR
Hypoalphalipoproteinemia	ABCA1, APOA1, and LCAT
Abetalipoproteinemia	MTTP
Tangier disease	ABCA1
Hyperalphalipoproteinemia	CETP, LIPC, APOC3, SR-BI, and LIPG

were found to be positively correlated (Mazaheri-Tehrani et al. 2022). Three SREBF1 gene loci, including cg11024682, cg20544516, and cg08129017 have a substantial and favorable relationship with blood triglyceride level in terms of methylation level (Pfeiffer et al. 2015; Sayols-Baixeras et al. 2016; Hedman et al. 2017). The expression of SREBF1 and its methylation are inversely correlated. As a result, hypertriglyceridemia may be caused by decreased SREBF1 expression. A meta-analysis demonstrated that cg11024682 has a bigger impact on serum triglyceride levels than other SREBF1 locations (Mazaheri-Tehrani et al. 2022). In the third exonic region of *APOA5*, the CGI (cytosine guanine islands) methylation was found to be associated with high triglyceride levels as well as a high risk of dyslipidemia (Oliva et al. 2016). The genetic studies were performed in various ethnic populations to draw conclusive evidence. The epigenetic studies are limited which open up future possibilities to work on lipid metabolism-related genes. Patients with familial hypercholesterolemia found a significant correlation between HDL-C, LDL-C, and triglyceride levels and ABCG1 DNA methylation (Guay et al. 2014).

4.4 Secondary Lipid Metabolic Disorders

Secondary dyslipidemia results from other underlying conditions and medications (Fig. 4.3). According to Vodnala et al. (2012), secondary dyslipidemia accounts for roughly 30–40% of dyslipidemia cases. The best line of treatment is to identify and treat the underlying diseases or drugs that are the primary causes of secondary dyslipidemia.

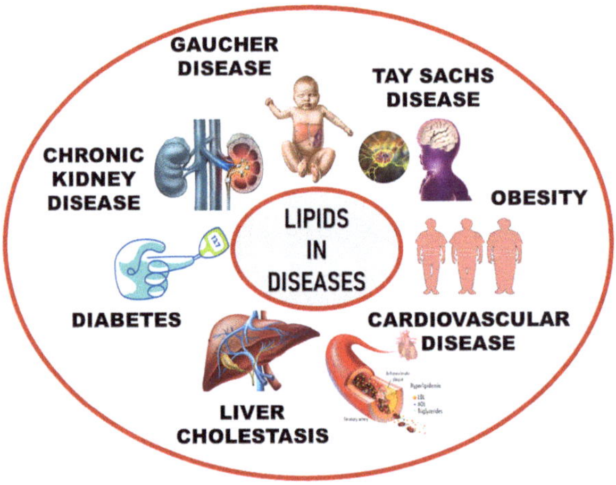

Fig. 4.3 The representation of various diseases caused by disruption in the physiological lipid metabolism

4.4.1 Cardiovascular Diseases (CVDs)

With an estimated 17.9 million deaths per year, CVDs are among the top causes of death worldwide. Heart and blood vessel problems known as CVDs include rheumatic heart disease, coronary heart disease, cerebrovascular disease, and other ailments (Singh et al. 2016; Singh et al. 2023). Heart attacks and strokes account for more than four out of every five CVD deaths, and one-third of these deaths happen before the age of 70 (Upadhyay 2015; Yu et al. 2019). Due to their cellular synthesis, transportation, assembly, degradation, oxidation, and plasma concentrations, lipids and lipoproteins have a significant impact on the initiation and progression of CVD (Upadhyay 2015; Yu et al. 2019). High plasma levels of LDL, VLDL, IDL, and LPA increase the risk of CVD by encouraging the development of thrombus and plaque (Fig. 4.4) (Krauss 2010; Nordestgaard et al. 2010). The factors that cause the development of atherosclerotic plaques include native LDL and its posttranslationally oxidized form, oxLDL (Ference et al. 2017). Moreover, a methylation site in AHRR at cg05575921 was additionally directly and inversely related to triglyceride levels, LDL-C, and HDL-C. This CpG was found linked to both the incidence and prevalence of coronary heart disease (CHD) (Ligthart et al. 2016; Sano et al. 2018). The hematopoietic DNA demethylating enzyme TET2 plays a significant role in avoiding atherosclerosis. It was demonstrated that cardiac dysfunction in heart failure is exacerbated by hematopoietic or myeloid cell-specific TET2 deletion (Sano et al. 2018). This depletion enhances the formation of atherosclerosis in ApoE defective (ApoE−/−) mice by modifying Beclin1-dependent autophagic processes, whereas TET2 overexpression prevents this development, suggesting that TET2 may be a useful therapeutic target in the management of atherosclerosis (Peng et al.

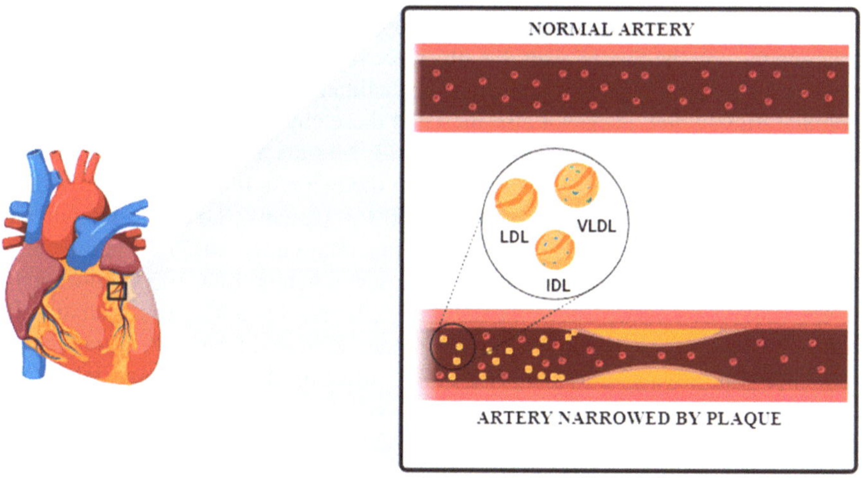

Fig. 4.4 Involvement of lipoproteins (VLDL very low-density lipoprotein, LDL low-density lipoprotein, IDL intermediate-density lipoprotein) in plaque formation

2016; Li et al. 2022). In ApoE−/− mice, enhancer of zeste homolog 2 (EZH2) overexpression stimulates the development of atherosclerosis through DNMT1-mediated ABCA1 methylation, which lowers macrophage cholesterol export. This further leads to increased production of the foam cells and plaque formation (Lv et al. 2016). A strong correlation between the incidence of CAD and hypomethylation in the APOC3, CETP (cholesteryl ester transfer protein), and APOC1 gene promoters was found. The promoters of the APOA5 and LIPC (Lipase C, Hepatic type) genes were more methylated in those at a higher risk of developing CAD (Li et al. 2022). In another study, it was found that cg06500161 in *ABCG1* was hypermethylated in myocardial infarction patients leading to a decrease in the expression of *ABCG1* and thereby increasing the levels of HDL-C and triglyceride (Bays et al. 2013; Pfeiffer et al. 2015).

4.4.2 Obesity

Insulin resistance, abnormalities in lipid metabolism, and metabolic syndrome are frequently linked to obesity. Elevated triglyceride, VLDL, APOB, and non-HDL-C values are among the lipid changes frequently found in obese people and HDL-C/ApoA-I levels are typically low. Although LDL-C values are typically in the normal to mildly raised range, a rise in small dense LDL is frequently observed, leading to an elevation in the amount of LDL particles (Grundy 2004; Franssen et al. 2011; Bays et al. 2013; Xiao et al. 2016).

For various reasons, these small dense LDL particles are thought to be more pro-atherogenic than larger LDL particles (Berneis and Krauss 2002). Small dense LDL particles spend more time in the bloodstream because they have a lower affinity for the LDL receptor. Additionally, these small particles enter the artery wall more readily than large particles and bind to intra-arterial proteoglycans more strongly, trapping them in the arterial wall. Finally, since small dense LDL particles are more easily oxidized, macrophage uptake may be facilitated. In people with obesity, post-prandial triglyceride levels are also high and these chylomicron remnants are pro-atherogenic (Klop et al. 2013; Bays et al. 2013; Masuda and Yamashita 2017). The severity of lipid abnormalities increases with the increase in BMI (basal metabolic index). According to Bays et al. (2013), 50–60% of patients who are overweight and 60–70% of obese individuals have dyslipidemia (Bays et al. 2013).

In a study, it was found that there were seven genes; LOC105378740 (RNA gene), LINC02506 (RNA gene), CSMD1 (CUB and Sushi Multiple Domains 1), MELK (Maternal Embryonic Leucine Zipper Kinase), FAM81A (family with sequence similarity 81 member A), ERAL1 (Era Like 12S Mitochondrial RRNA Chaperone 1), and miRNA144 that may have been connected to lipid levels and BMI. The biological pathways of interest were the control of reverse transport of cholesterol and the control of particle clearance of high-density lipoproteins. Genetic variables influence BMI and blood lipid levels, and there was a genetic correlation between the two. Their genetic correlation may differ according to gender. The MIR144 gene and associated metabolic pathways may have an impact on lipid levels and obesity, according to bivariate GWAS research (Ke et al. 2022).

The agouti mouse model is one of the most widely used instances of how epigenetics affects obesity. The agouti viable yellow (Avy) gene, which determines the color of the mouse coat and is typically imprinted by DNA methylation, is mutated in this model. This gene exhibits aberrant hypomethylation, leading to ectopic production of its protein. This protein binds to the hypothalamic melanocortin 4 receptor (MC4R), impairing its function and inducing hyperphagic obesity. It is interesting to note that this phenotype was diminished in mice raised on a diet high in methyl donors, suggesting that epigenetic changes can be reversed in response to environmental stimuli like nutrition (Mahmoud 2022). Adiponectin (ADIPOQ) and leptin (LEP) are the top genes related to obesity. A study found a correlation between the promoter region hypomethylation of LEP gene and obese individuals in the Indian population (Sadashiv et al. 2021). Another study demonstrated the relation between hypomethylation of LEP gene and hypermethylation of ADIPOQ gene with obesity in a different study population (Houde et al. 2015). It has also been demonstrated that histone alterations control the expression of key genes involved in adipogenesis, including preadipocyte factor-1 (Pref-1), adipocyte protein 2 (aP2), peroxisome proliferator-activated receptor γ (PPARG), and CCAAT enhancer-binding protein β (C/EBPB), C/EBPA (Zhang et al. 2012; Dhana et al. 2018). Obesity was also found to be associated with altered DNA methylation in the lipid metabolism-related genes ABCG1, carnitine palmitoyl-transferase 1 A (CPT1A), and SREBF1 (Dhana et al. 2018).

4.4.3 Diabetes

Chronically high blood glucose is a hallmark of type 2 diabetes mellitus (T2DM), which is brought on by both decreased insulin production and insulin resistance. Insulin resistance in target tissues such as skeletal muscle, liver, and adipose tissue is known to be influenced by aging, obesity, and sedentary lifestyles. After prolonged exposure to high glucose and fat concentrations, pancreatic islet function declines. Globally, the number of T2DM patients is rising at alarming rates due to aging populations and rising obesity prevalence (Hall et al. 2018). Diabetic dyslipidemia is characterized by hypertriglyceridemia, high levels of VLDL, small dense LDL, and reduced levels of HDL cholesterol. Patients with diabetes have low LPL activity, which inhibits the breakdown of triglyceride-rich lipoproteins and absorption of free fatty acids. The enhanced triglyceride intake of the liver stimulates the synthesis of VLDL. The interchange of cholesteryl ester from HDL to VLDL or LDL to triglycerides is facilitated by hypertriglyceridemia.

As a result, the buildups of chylomicrons and VLDL in blood circulation, and the levels of serum triglycerides are significantly increased in type 1 diabetes mellitus (T1DM) patients with poor management of glucose. Diabetic patients have high levels of cholesteryl ester in VLDL (Rivellese et al. 1988; James and Pometta 1990; Bagdade et al. 1991). In T1DM patients, LDL cholesterol levels often stay normal, while triglyceride-rich LDL or small dense LDL is frequently elevated. In T1DM, the amount of HDL-cholesterol or the ratio of cholesterol to triglycerides in HDL is frequently decreased. Through glucose control, the levels of triglycerides, VLDL, LDL, and HDL cholesterol are typically brought back to normal ranges in T1DM patients with no nephropathy. Treatment for hypoglycemia frequently fails to normalize alterations in lipoprotein composition (Bagdade and Subbaiah 1989; Dullaart 1995).

In type 2 diabetes mellitus (T2DM) with inadequate glucose control, the triglyceride or chylomicronemia increase is not as obvious as they are in T1DM. The level of LDL-cholesterol in T2DM patients is often increased or subnormal. T2DM patients have commonly shown higher levels of small dense LDL (Stewart et al. 1994) and frequently shown reduced HDL cholesterol levels. This is connected to increased APOB levels and reduced APOA1 levels. In T2DM patients, hypoglycemic treatment alone is unable to normalize dyslipoproteinemia (Laakso 1997). Gestational diabetes mellitus (GDM) is a form of carbohydrate intolerance that was initially noticed during pregnancy. Mothers with GDM and their children have been observed to have a higher incidence of T2DM (Metzger and Coustan 1998). During an average pregnancy, cholesterol and triglyceride levels start to decline at around 7 weeks of pregnancy and gradually rise up to term. After delivery, the cholesterol levels often come back to normal levels (Darmady and Postle 1982). GDM frequently exhibits elevated triglycerides and reduced HDL cholesterol (Knopp et al. 1980).

In a meta-analysis consisting of a total of more than 26,000 diabetic individuals, it was found that the genetic variant in PPARγ rs1801282 (C > G) for lowering vulnerability to T2DM (Li et al. 2015, p. 201). In another study, it was determined that the ABCA1 C69T gene mutation dramatically decreased the probability of hypertriglyceridemia in patients with diabetes (Yan et al. 2019). The APOB gene's rs1042031 SNP and the ABCC8 gene's rs6544718 SNP were protective because they are linked to a notably lower incidence of T2DM. The SNPs for rs676210 (APOB), rs13702 (LPL), and rs1800961 hepatocyte nuclear factor-4 (HNF4A), were linked to a higher risk of T2DM (Nicchio et al. 2021). Hyper DNA methylation of PPARGC1A and reduced expression of these important genes were seen in the islets from T2DM donors, and these findings were linked to compromised production of insulin. Furthermore, elevated glucose and glycated hemoglobin (HbA1c) appeared to directly enhance these genes' DNA methylation (Hall et al. 2013, 2018). In Indian and European T2D patients, a study found that methylation in SREBF1 and ABCG1 was linked to a higher incidence of T2D and a fourfold increased risk of exhibiting the pathology in cases (Chambers et al. 2015). Patients with obesity and diabetes had lower expression of insulin receptor substrate 2 (IRS2) in their livers than individuals who were obese but did not have diabetes. DNA methylation at CpG5 in IRS2 and elevated miRNA hsa-let-7e-5p (let-7e-5p) in liver are associated with decreased IRS2 expression (Krause et al. 2020).

4.4.4 Hypothyroidism

Thyroid dysfunction has a great impact on lipids as well as a number of other cardiovascular risk factors. Hypothyroidism is relatively common and is associated with an unfavorable effect on lipids. The initial step in the production of cholesterol is the 3-hydroxy-3-methylglutaryl-coenzyme A (HMG-CoA) reductase, which is induced by thyroid hormones. Triiodothyronine (T3) also increases LDL receptors by regulating the activity of the *LDLR* gene. The direct binding of T3 to particular thyroid hormone-responsive elements (TREs) is how this T3-mediated gene activation occurs (Bakker et al. 1998). Additionally, T3 modulates the expression of the *LDLR* gene through the SREBP-2 (Shin and Osborne 2003). In addition, T3 has been linked to preventing LDL from oxidation (Faure et al. 2004). CETP activity, which swaps cholesteryl esters from HDL2 to the VLDL and triglycerides in the opposite way, is activated by thyroid hormones, which can affect HDL metabolism (Lagrost 1994). Furthermore, thyroid hormones activate the LPL, which breakdowns the triglyceride-rich lipoproteins, and the hepatic lipase which hydrolyzes HDL2 to HDL3 and contributes to the conversion of IDL to LDL and in turn LDL to small dense LDL (sdLDL) (Kuusi et al. 1980; Santamarina-Fojo et al. 2004). APOAV, which is significantly involved in the control of triglyceride, is upregulated as a result of T3 (Prieur et al. 2005). A study by Rensen et al. (2005) found a correlation between higher levels of APOAV and lower levels of triglycerides (Rensen et al. 2005).

According to the proposed processes, this impact is caused by a decrease in hepatic VLDL-triglyceride production and an increase in plasma LPL levels and activity. This further increases the creation of lipoprotein remnants through increased LPL-mediated lipolysis of VLDL-triglyceride (Rensen et al. 2005). Additionally, APOAV has also been linked to a greater clearance of lipoprotein core remnants. This was brought on by a rise in hepatic intake further caused by an increased affinity for the LDL receptor (Rensen et al. 2005). When treating dyslipidemia brought on by hypothyroidism, we should take into account both overt hypothyroidism, in which thyroid hormone levels are low, and subclinical hypothyroidism, in which thyroid-stimulating hormone (TSH) levels are high despite normal thyroid hormone levels. A rise in total cholesterol, LDL-cholesterol, APOB, and LPA are seen in overt hypothyroidism (Pearce 2004). According to him, LDL-cholesterol has significantly increased by roughly 30%. The increased LDL-cholesterol levels in hypothyroidism are explained by the fact that thyroid hormone drives LDL-cholesterol breakdown and the conversion of cholesterol to bile acids by promoting the expression of the LDL-receptor and 7 alpha-hydroxylase, respectively (Duntas and Brenta 2017).

Subclinical hypothyroidism (SH), defined as the clinical status of elevated serum TSH levels with normal levels of free triiodothyronine (FT3) and free thyroxine (FT4), is a far more common disorder than overt hypothyroidism. SH is associated with increased levels of total cholesterol and LDL-cholesterol (Tanis et al. 1996; Danese et al. 2000; Luboshitzky et al. 2002; Jung et al. 2003; Monzani et al. 2004; Walsh et al. 2005; Erem 2006; Duman et al. 2007). In addition, some studies have shown that SH dyslipidemia may also be accompanied by increased triglycerides (Milionis et al. 2005; Toruner et al. 2008) and decreased HDL-cholesterol levels.

4.4.5 Gaucher Disease

An autosomal recessive inborn metabolic error called Gaucher disease, results from the harmful buildup of glucocerebroside lipids in multiple organs. It is brought on by *GBA1* (glucocerebrosidase) gene mutations that cause a lack of glucocerebrosidase activity in the lysosomes. The most common mutations in the GBA gene, which is found on chromosome 1q21, are N370S and L444P. The mutation N370S is unique to the Caucasian group, which could eventually support the founder effect theory. The absence of neurological symptoms is ensured by the existence of this mutation, even on a single allele, as it leaves enough enzyme activity for proper glucocerebroside catabolism in neurons. The mutation L444P is panethnic; more than 70% of patients have homozygosity for this mutation, which causes the neuronopathic version of the illness (Riboldi and Di Fonzo 2019). Hepatosplenomegaly (an abnormal enlargement of the liver and/or spleen), anemia (low amounts of circulating red blood cells), thrombocytopenia (low levels of platelets), and skeletal deformities are common signs of Gaucher disease.

4.4.6 Tay-Sachs Disease

Tay-Sachs disease is an uncommon, inherited metabolic condition that mostly impacts young children and causes cells, particularly in the brain, to gradually deteriorate and die. It is brought on by a mutation in the hexosaminidase A enzyme, which permits the unfavorable accumulation of lipids (fatty substances like oils and acids) in cells. To have a child with Tay-Sachs disease, both parents must have the defective gene. Tay-Sachs disease belongs to a class of genetic conditions known as GM2 (monosialotetrahexosylganglioside) gangliosidoses. The most prevalent and severe types of GM2 gangliosidosis are frequently observed to have a total Hexa deficiency. More than 100 different variants in HEXA have been identified which include deletions, splicing abnormalities, nonsense, and missense mutations. These mutations can cause defects in translation and transcription as well as improper protein folding, dimerization, and catalytic malfunction (Triggs-Raine et al. 2001; Cachon-Gonzalez et al. 2018). Gemfibrozil, a lipid-lowering FDA-approved drug, when administered orally greatly reduced glial activation and inflammation in addition to decreasing the buildup of GM2 gangliosides/glycoconjugates in the Tay-Sachs mice's motor cortex. Additionally, it enhanced behavioral function and prolonged the life span of Tay-Sachs mice by elevating PPARα levels in their brains. These findings point to a useful role for gemfibrozil, which stops the advancement of TSD and lengthens the life span of Tay-Sachs mice via a PPARα-dependent mechanism (Raha et al. 2023).

4.4.7 Chronic Kidney Disease (CKD)

The dyslipidemia that CKD is associated includes hypertriglyceridemia, raised LDL cholesterol, a buildup of APOB-containing lipoproteins, higher concentrations of LPA particles, and decreased HDL levels (Cohen and Fisher 2013). Patients on dialysis and those with nephrotic syndrome have elevated plasma triglyceride concentrations in the beginning stages of CKD (Batista et al. 2004; Kwan et al. 2007).

Patients having CKD also have higher ratios of triglycerides to cholesterol in LDL and HDL particles. In CKD, hypertriglyceridemia is brought on by the delayed breakdown of triglyceride-rich lipoproteins, including VLDL particles and chylomicron remnants, along with decreased LCAT activity (Guarnieri et al. 1978; Kwan et al. 2007). In CKD, ApoC-III, a competitive inhibitor of LPL, is expressed at higher levels and LPL expression is lowered (Cheung et al. 1996; Reiss et al. 2015). Triglycerides carried by chylomicrons and VLDL particles are broken down by LPL (Mead et al. 2002; Ginsberg et al. 2021). Because adipocytes and myocytes express less VLDL receptors, the removal of VLDL particles from the circulation and their conversion into IDL particles are similarly compromised in CKD (Liang and Vaziri 1997; Liang et al. 1998; Vaziri 2006). Fatty acid-binding proteins (FABPs) have a preference for recognizing long-chain fatty acids as substrates. Elevated levels of L-FABP exacerbated interstitial inflammation and encouraged the growth of rat tubular epithelial cells in the rat kidney disease model (Watanabe et al. 2018).

4.4.8 Liver Cholestasis

Cholestasis is a reduction or stoppage of bile flow. Bile is the digestive fluid produced by the liver and it is the route for cholesterol elimination from the body. Plasma cholesterol is taken up by the liver in the form of APOB-containing lipoproteins (i.e., remnant lipoproteins and LDL) by receptor-mediated endocytosis or by selective uptake of HDL cholesterol (Cohen and Fisher 2013). By converting to bile salts and by biliary secretion, cholesterol gets eliminated. Hypercholesterolemia can result from biliary obstruction, which is especially common when it is brought on by cholestasis disorders, which interfere with the removal of cholesterol.

Hypercholesterolemia associated with cholestasis is largely attributable to the formation of lipoprotein X, an atypical lipoprotein particle. Lipoprotein X comprises principally unesterified cholesterol and phospholipids (Jahn et al. 1985), resembling the cholesterol-phospholipid vesicles that are secreted by the liver into bile (Cohen 1999). The principal proteins associated with lipoprotein-X are APOC and albumin contained within the core (Narayanan 1984; Hussain et al. 2015). The lipids of the particle comprise a sphere, with an aqueous core. Lipoprotein-X is devoid of APOB. It appears to be formed due to the secretion of biliary-type particles into plasma in the setting of obstruction to bile flow (Elferink et al. 1998), although defects in plasma cholesterol esterification may also contribute (Jahn et al. 1985). According to recent research, the maternal intake of a high-fat or high-cholesterol western diet might cause the pathogenesis of fatty liver disease in male children by modifying the expression of the ApoB gene. They discovered that there is enhanced methylation of CpG dinucleotides in the promoter region of the ApoB gene based on DNA methylation studies (Chen et al. 2020). The methylation amount within the Nrf2 promoter in the mice liver induced by HFD can be alleviated by resveratrol, an inhibitor of glucose transporter 9 (GLUT9). This is associated with reduced levels of triglycerides and decreased expression of lipogenic genes, such as SREBP-1c and Fas cell surface death receptor (FAS) in fatty liver disease (Hosseini et al. 2020). It was discovered that barrier-to-autointegration factor 60a (BAF, chromatin decondensation protein) regulates hepatic fatty acid oxidation and stimulates the transcriptional activation of mitochondrial and peroxisomal fat-oxidation genes through its interactions with PPARα and PGC-1α (Peroxisome proliferator-activated receptor-gamma coactivator). Furthermore, BAF60a functions as a diet-sensitive subunit and stimulates the expression of genes involved in the absorption of cholesterol and the metabolism of hepatic bile acid (Meng et al. 2015).

4.4.9 Other Diseases

Several other diseases, including Cushing's syndrome and acute porphyria, have been linked to lipid metabolism abnormalities. Dyslipidemia is one of the features of human Cushing's syndrome including—in association with central obesity, hypertension, impaired glucose tolerance or diabetes, insulin resistance, hypercoagulability, and it determines an increased cardiovascular risk. There is an increase in VLDL and LDL levels in the blood, but not HDL levels, which causes levels of triglycerides and total cholesterol to rise in people with Cushing's syndrome (Sun et al. 2021).

Acute porphyrias (AP) are autosomal dominant inherited diseases, associated with neurovisceral attacks. Hypercholesterolemia was described some decades ago in patients with AP (Taddeini et al. 1964; Hollander et al. 1967; Lees et al. 1970; Stein and Raal 2013). It was seen more frequently in patients with symptomatic illness and in cases with latent AP (Hollander et al. 1967; Lees et al. 1970). However, there are few lipoprotein investigations of AP patients, and the outcomes have been conflicting. LDL-C levels have been reported to be higher by some authors (Stein and Tschudy 1970; Lees et al. 1970), but others only describe an increase in HDL-cholesterol (Mustajoki and Nikkilä 1984).

4.5 Conclusions

Genetic testing has become a frequent factor in the clinical treatment of individuals with lipid disorders. Familial hypercholesterolemia, familial chylomicronemia syndrome, sitosterolemia, lysosomal acid lipase deficiency, and other dyslipidemias are among the dyslipidemias where genetic diagnosis may be clinically beneficial. In the last decade, pharmacogenetics has emerged in the personalized medicine era. Studies have discovered variations that may be pertinent to fibrate monotherapy, most of them have to do with lipid metabolism. Genetics of Lipid Lowering Drugs and Diet Network (GOLDN) study showed many SNPs in PPARA linked with LDL and triglyceride response to fenofibrate (agonists of PPAR (peroxisome proliferator-activated receptor)), encoded by *PPARA*. Through observational and genetic investigations, important lipoprotein metabolism-related proteins, including PCSK9, apolipoprotein(a), cholesteryl ester transfer protein, and angiopoietin-related protein 3, have been identified as feasible candidates for therapeutic intervention.

Numerous strategies can be employed to target these proteins, including base editing, antisense oligonucleotides, and short interfering RNA, as well as protein inhibition or interference and the insertion of loss-of-function mutations. These cutting-edge and unique approaches offer previously unheard possibilities for the prevention of lipid-associated disorders by complementing and possibly even replacing current medications. Genetic testing along with the emerging pharmaco-epigenetics will lead to enhanced effectiveness of lipid-lowering therapy, lessen its negative effects, boost patient compliance, and improve the diagnostic procedure and therapeutic approach.

References

Aguilar-Salinas CA, Canizales-Quinteros S, Rojas-Martínez R et al (2009) Hypoalphalipoproteinemia in populations of Native American ancestry: an opportunity to assess the interaction of genes and the environment. Curr Opin Lipidol 20:92–97

Baass A, Paquette M, Bernard S, Hegele RA (2020) Familial chylomicronemia syndrome: an under-recognized cause of severe hypertriglyceridaemia. J Intern Med 287:340–348

Bagdade JD, Subbaiah PV (1989) Whole-plasma and high-density lipoprotein subfraction surface lipid composition in IDDM men. Diabetes 38:1226–1230

Bagdade JD, Ritter MC, Subbaiah PV (1991) Accelerated cholesteryl ester transfer in patients with insulin-dependent diabetes mellitus. Eur J Clin Investig 21:161–167

Bakker O, Hudig F, Meijssen S, Wiersinga WM (1998) Effects of triiodothyronine and amiodarone on the promoter of the human LDL receptor gene. Biochem Biophys Res Commun 249:517–521

Batista MC, Welty FK, Diffenderfer MR et al (2004) Apolipoprotein A-I, B-100, and B-48 metabolism in subjects with chronic kidney disease, obesity, and the metabolic syndrome. Metabolism 53:1255–1261

Bays HE, Toth PP, Kris-Etherton PM et al (2013) Obesity, adiposity, and dyslipidemia: a consensus statement from the National Lipid Association. J Clin Lipidol 7:304–383

Bello-Chavolla OY, Kuri-García A, Ríos-Ríos M et al (2018) Familial combined hyperlipidemia: current knowledge, perspectives, and controversies. RIC 70:1094

Berneis KK, Krauss RM (2002) Metabolic origins and clinical significance of LDL heterogeneity. J Lipid Res 43:1363–1379

Boot CS, Luvai A, Neely RDG (2020) The clinical and laboratory investigation of dysbetalipoproteinemia. Crit Rev Clin Lab Sci 57:458–469

Cachon-Gonzalez MB, Zaccariotto E, Cox TM (2018) Genetics and therapies for GM2 gangliosidosis. Curr Gene Ther 18:68–89

Carrasquilla GD, Christiansen MR, Kilpeläinen TO (2021) The genetic basis of hypertriglyceridemia. Curr Atheroscler Rep 23:39

Chambers JC, Loh M, Lehne B et al (2015) Epigenome-wide association of DNA methylation markers in peripheral blood from Indian Asians and Europeans with incident type 2 diabetes: a nested case-control study. Lancet Diabetes Endocrinol 3:526–534

Chen H-C, Chen Y-Z, Wang C-H, Lin F-J (2020) The nonalcoholic fatty liver disease-like phenotype and lowered serum VLDL are associated with decreased expression and DNA hypermethylation of hepatic ApoB in male offspring of ApoE deficient mothers fed a with Western diet. J Nutr Biochem 77:108319

Cheung AK, Parker CJ, Ren K, Iverius P-H (1996) Increased lipase inhibition in uremia: identification of pre-β-HDL as a major inhibitor in normal and uremic plasma. Kidney Int 49:1360–1371

Church C, Horowitz M, Rodeheffer M (2012) WAT is a functional adipocyte? Adipocyte 1:38–45

Civeira F, Jarauta E, Cenarro A et al (2008) Frequency of low-density lipoprotein receptor gene mutations in patients with a clinical diagnosis of familial combined hyperlipidemia in a clinical setting. J Am Coll Cardiol 52:1546–1553
Civeira F, Arca M, Cenarro A, Hegele RA (2022) A mechanism-based operational definition and classification of hypercholesterolemia. J Clin Lipidol 16:813–821
Cohen DE (1999) Hepatocellular transport and secretion of biliary lipids. Curr Opin Lipidol 10:295–302
Cohen DE, Fisher EA (2013) Lipoprotein metabolism, dyslipidemia, and nonalcoholic fatty liver disease. Semin Liver Dis 33:380–388
Danese MD, Ladenson PW, Meinert CL, Powe NR (2000) Effect of thyroxine therapy on serum lipoproteins in patients with mild thyroid failure: a quantitative review of the literature*. J Clin Endocrinol Metabol 85:2993–3001
Darmady J, Postle A (1982) Lipid metabolism in pregnancy. BJOG 89:211–215
Dhana K, Braun KVE, Nano J et al (2018) An epigenome-wide association study of obesity-related traits. Am J Epidemiol 187:1662–1669
Dullaart RP (1995) Plasma lipoprotein abnormalities in type 1 (insulin-dependent) diabetes mellitus. Neth J Med 46:44–54
Duman D, Demirtunc R, Sahin S, Esertas K (2007) The effects of simvastatin and levothyroxine on intima-media thickness of the carotid artery in female normolipemic patients with subclinical hypothyroidism: a prospective, randomized-controlled study. J Cardiovasc Med 8:1007–1011
Duntas LH, Brenta G (2017) Thyroid hormones: a potential ally to LDL-cholesterol-lowering agents. HJ 15:500–510
Elferink RP, Ottenhoff R, Van Marle J et al (1998) Class III P-glycoproteins mediate the formation of lipoprotein X in the mouse. J Clin Invest 102:1749–1757
Erem C (2006) Blood coagulation, fibrinolytic activity and lipid profile in subclinical thyroid disease: subclinical hyperthyroidism increases plasma factor X activity. Clin Endocrinol 64:323–329
Escolà-Gil JC, Quesada H, Julve J et al (2014) Sitosterolemia: diagnosis, investigation, and management. Curr Atheroscler Rep 16:424
Faure P, Oziol L, Artur Y, Chomard P (2004) Thyroid hormone (T3) and its acetic derivative (TA3) protect low-density lipoproteins from oxidation by different mechanisms. Biochimie 86:411–418
Feingold KR (2022) Lipid and lipoprotein metabolism. Endocrinol Metab Clin N Am 51:437–458
Ference BA, Ginsberg HN, Graham I et al (2017) Low-density lipoproteins cause atherosclerotic cardiovascular disease. 1. Evidence from genetic, epidemiologic, and clinical studies. A consensus statement from the European Atherosclerosis Society Consensus Panel. Eur Heart J 38:2459–2472
Fernandez AF, Esteller M (2010) Viral epigenomes in human tumorigenesis. Oncogene 29:1405–1420
Franssen R, Monajemi H, Stroes ESG, Kastelein JJP (2011) Obesity and dyslipidemia. Med Clin North Am 95:893–902
Garcia JA, Kalacas NA, Sy Ortin T et al (2019) XRCC4 c.1394G>T Single-nucleotide polymorphisms and breast cancer risk among Filipinos. Asian Pac J Cancer Prev 20:1097–1101
Geurts JMW (2000) Identification of TNFRSF1B as a novel modifier gene in familial combined hyperlipidemia. Hum Mol Genet 9:2067–2074
Giammanco A, Noto D, Barbagallo CM et al (2021) Hyperalphalipoproteinemia and beyond: the role of HDL in cardiovascular diseases. Life 11:581
Gibbons GF, Wiggins D, Brown A-M, Hebbachi A-M (2004) Synthesis and function of hepatic very-low-density lipoprotein. Biochem Soc Trans 32:59–64
Ginsberg HN, Packard CJ, Chapman MJ et al (2021) Triglyceride-rich lipoproteins and their remnants: metabolic insights, role in atherosclerotic cardiovascular disease, and emerging therapeutic strategies—a consensus statement from the European atherosclerosis society. Eur Heart J 42:4791–4806

Groenendijk M, Cantor RM, Funke H, Dallinga-Thie GM (2001) Two newly identified SNPs in the APO AI-CIII intergenic region are strongly associated with familial combined hyperlipidaemia. Eur J Clin Invest 31:852–859
Grundy SM (2004) Obesity, metabolic syndrome, and cardiovascular disease. J Clin Endocrinol Metabol 89:2595–2600
Guarnieri GF, Moracchiello M, Campanacci L et al (1978) Lecithin-cholesterol acyltransferase (LCAT) activity in chronic uremia. Kidney Int Suppl 13:S26–S30
Guay S-P, Brisson D, Lamarche B et al (2014) Epipolymorphisms within lipoprotein genes contribute independently to plasma lipid levels in familial hypercholesterolemia. Epigenetics 9:718–729
Hall E, Dayeh T, Kirkpatrick CL et al (2013) DNA methylation of the glucagon-like peptide 1 receptor (GLP1R) in human pancreatic islets. BMC Med Genet 14:76
Hall E, Dekker Nitert M, Volkov P et al (2018) The effects of high glucose exposure on global gene expression and DNA methylation in human pancreatic islets. Mol Cell Endocrinol 472:57–67
Hedman ÅK, Mendelson MM, Marioni RE et al (2017) Epigenetic patterns in blood associated with lipid traits predict incident coronary heart disease events and are enriched for results from genome-wide association studies. Circ Cardiovasc Genet 10:e001487
Hollander CS, Scott RL, Tschudy DP et al (1967) Increased protein-bound iodine and thyroxine-binding globulin in acute intermittent porphyria. N Engl J Med 277:995–1000
Horton JD, Goldstein JL, Brown MS (2002) SREBPs: activators of the complete program of cholesterol and fatty acid synthesis in the liver. J Clin Invest 109:1125–1131
Hosseini H, Teimouri M, Shabani M et al (2020) Resveratrol alleviates non-alcoholic fatty liver disease through epigenetic modification of the Nrf2 signaling pathway. Int J Biochem Cell Biol 119:105667
Houde A-A, Légaré C, Biron S et al (2015) Leptin and adiponectin DNA methylation levels in adipose tissues and blood cells are associated with BMI, waist girth and LDL-cholesterol levels in severely obese men and women. BMC Med Genet 16:29
Hussain I, Ahmad Z, Garg A (2015) Extreme hypercholesterolemia presenting with pseudohyponatremia—a case report and review of the literature. J Clin Lipidol 9:260–264
Jahn CE, Schaefer EJ, Taam LA et al (1985) Lipoprotein abnormalities in primary biliary cirrhosis. Association with hepatic lipase inhibition as well as altered cholesterol esterification. Gastroenterology 89:1266–1278
James RW, Pometta D (1990) Differences in lipoprotein subfraction composition and distribution between type I diabetic men and control subjects. Diabetes 39:1158–1164
Jung CH, Sung KC, Shin HS et al (2003) Thyroid dysfunction and their relation to cardiovascular risk factors such as lipid profile, hsCRP, and waist hip ratio in Korea. Korean J Intern Med 18:146–153
Ke J, Gao W, Wang B et al (2022) Exploring the genetic association between obesity and serum lipid levels using bivariate methods. Twin Res Hum Genet 25:234–244
Klop B, Elte JWF, Cabezas MC (2013) Dyslipidemia in obesity: mechanisms and potential targets. Nutrients 5:1218–1240
Knopp RH, Chapman M, Bergelin R et al (1980) Relationships of lipoprotein lipids to mild fasting hyperglycemia and diabetes in pregnancy. Diabetes Care 3:416–420
Koseki M, Yamashita S, Ogura M et al (2021) Current diagnosis and Management of Tangier Disease. J Atheroscler Thromb 28:802–810
Krause C, Geißler C, Tackenberg H et al (2020) Multi-layered epigenetic regulation of IRS2 expression in the liver of obese individuals with type 2 diabetes. Diabetologia 63:2182–2193
Krauss RM (2010) Lipoprotein subfractions and cardiovascular disease risk. Curr Opin Lipidol 21:305–311
Kuusi T, Saarinen P, Nikkilä EA (1980) Evidence for the role of hepatic endothelial lipase in the metabolism of plasma high density lipoprotein2 in man. Atherosclerosis 36:589–593
Kwan BCH, Kronenberg F, Beddhu S, Cheung AK (2007) Lipoprotein metabolism and lipid management in chronic kidney disease. J Am Soc Nephrol 18:1246–1261
Laakso M (1997) Dyslipidemia, morbidity, and mortality in non-insulin-dependent diabetes mellitus. Lipoproteins and coronary heart disease in non-insulin-dependent diabetes mellitus. J Diabetes Complicat 11:137–141

Lagrost L (1994) Regulation of cholesteryl ester transfer protein (CETP) activity: review of in vitro and in vivo studies. Biochim Biophys Acta (BBA) Lipids Lipid Metab 1215:209–236
Lai C-Q, Parnell LD, Smith CE et al (2020) Carbohydrate and fat intake associated with risk of metabolic diseases through epigenetics of CPT1A. Am J Clin Nutr 112:1200–1211
Lees RS, Song CS, Levere RD, Kappas A (1970) Hyperbeta-lipoproteinemia in acute intermittent porphyria: preliminary report. N Engl J Med 282:432–433
Li T, Shi Y, Yin J et al (2015) The association between lipid metabolism gene polymorphisms and nephropathy in type 2 diabetes: a meta-analysis. Int Urol Nephrol 47:117–130
Li W, Wang Y, Huang R et al (2022) Association of lipid metabolism-related gene promoter methylation with risk of coronary artery disease. Mol Biol Rep 49:9373–9378
Liang K, Vaziri ND (1997) Acquired VLDL receptor deficiency in experimental nephrosis. Kidney Int 51:1761–1765
Liang K, Oveisi F, Vaziri ND (1998) Role of secondary hyperparathyroidism in the genesis of hypertriglyceridemia and VLDL receptor deficiency in chronic renal failure. Kidney Int 53:626–630
Ligthart S, Marzi C, Aslibekyan S et al (2016) DNA methylation signatures of chronic low-grade inflammation are associated with complex diseases. Genome Biol 17:255
Liu Z-K, Hu M, Baum L et al (2010) Associations of polymorphisms in the apolipoprotein A1/C3/A4/A5 gene cluster with familial combined hyperlipidaemia in Hong Kong Chinese. Atherosclerosis 208:427–432
Luboshitzky R, Aviv A, Herer P, Lavie L (2002) Risk factors for cardiovascular disease in women with subclinical hypothyroidism. Thyroid 12:421–425
Lv Y-C, Tang Y-Y, Zhang P et al (2016) Histone methyltransferase enhancer of zeste homolog 2-mediated ABCA1 promoter DNA methylation contributes to the progression of atherosclerosis. PLoS One 11:e0157265
Mahmoud AM (2022) An overview of epigenetics in obesity: the role of lifestyle and therapeutic interventions. Int J Mol Sci 23:1341
Masuda D, Yamashita S (2017) Postprandial hyperlipidemia and remnant lipoproteins. J Atheroscler Thromb 24:95–109
Mato JM, Alonso C, Noureddin M, Lu SC (2019) Biomarkers and subtypes of deranged lipid metabolism in non-alcoholic fatty liver disease. WJG 25:3009–3020
Mazaheri-Tehrani S, Khoshhali M, Heidari-Beni M et al (2022) A systematic review and meta-analysis of observational studies on the effects of epigenetic factors on serum triglycerides. Arch Endocrinol Metab 66:407. https://doi.org/10.20945/2359-3997000000472
Mead J, Irvine S, Ramji D (2002) Lipoprotein lipase: structure, function, regulation, and role in disease. J Mol Med 80:753–769
Meng Z-X, Wang L, Chang L et al (2015) A diet-sensitive BAF60a-mediated pathway links hepatic bile acid metabolism to cholesterol absorption and atherosclerosis. Cell Rep 13:1658–1669
Metzger BE, Coustan DR (1998) Summary and recommendations of the fourth international workshop-conference on gestational diabetes mellitus. The Organizing Committee. Diabetes Care 21(Suppl 2):B161–B167
Milionis HJ, Tambaki AP, Kanioglou CN et al (2005) Thyroid substitution therapy induces high-density lipoprotein–associated platelet-activating factor–acetylhydrolase in patients with subclinical hypothyroidism: a potential antiatherogenic effect. Thyroid 15:455–460
Miller M, Stone NJ, Ballantyne C et al (2011) Triglycerides and cardiovascular disease: a scientific statement from the American Heart Association. Circulation 123:2292–2333
Monzani F, Caraccio N, Kozàkowà M et al (2004) Effect of levothyroxine replacement on lipid profile and intima-media thickness in subclinical hypothyroidism: a double-blind, placebo-controlled study. J Clin Endocrinol Metabol 89:2099–2106
Mustajoki P, Nikkilä EA (1984) Serum lipoproteins in asymptomatic acute porphyria: no evidence for hyperbetalipoproteinemia. Metabolism 33:266–269
Narayanan S (1984) Biochemistry and clinical relevance of lipoprotein X. Ann Clin Lab Sci 14:371–374

Nicchio IG, Cirelli T, Nepomuceno R et al (2021) Polymorphisms in genes of lipid metabolism are associated with type 2 diabetes mellitus and periodontitis, as comorbidities, and with the subjects' periodontal, glycemic, and lipid profiles. J Diabetes Res 2021:1–21
Nordestgaard BG, Chapman MJ, Ray K et al (2010) Lipoprotein(a) as a cardiovascular risk factor: current status. Eur Heart J 31:2844–2853
Oliva I, Guardiola M, Vallvé J-C et al (2016) *APOA5* genetic and epigenetic variability jointly regulate circulating triacylglycerol levels. Clin Sci 130:2053–2059
Osborne TF (2000) Sterol regulatory element-binding proteins (SREBPs): key regulators of nutritional homeostasis and insulin action. J Biol Chem 275:32379–32382
Pajukanta P, Lilja HE, Sinsheimer JS et al (2004) Familial combined hyperlipidemia is associated with upstream transcription factor 1 (USF1). Nat Genet 36:371–376
Panickar KS, Bhathena SJ (2010) Control of fatty acid intake and the role of essential fatty acids in cognitive function and neurological disorders. Fat detection: taste, texture, and post ingestive effects 470–474
Pearce EN (2004) Hypothyroidism and dyslipidemia: modern concepts and approaches. Curr Cardiol Rep 6:451–456
Peng J, Yang Q, Li A-F et al (2016) Tet methylcytosine dioxygenase 2 inhibits atherosclerosis via upregulation of autophagy in ApoE−/− mice. Oncotarget 7:76423–76436
Pfeiffer L, Wahl S, Pilling LC et al (2015) DNA methylation of lipid-related genes affects blood lipid levels. Circ Cardiovasc Genet 8:334–342
Pingitore P, Lepore SM, Pirazzi C et al (2016) Identification and characterization of two novel mutations in the LPL gene causing type I hyperlipoproteinemia. J Clin Lipidol 10:816–823
Prieur X, Huby T, Coste H et al (2005) Thyroid hormone regulates the Hypotriglyceridemic gene APOA5. J Biol Chem 280:27533–27543
Raha S, Dutta D, Paidi RK, Pahan K (2023) Lipid-lowering drug gemfibrozil protects mice from Tay-Sachs disease via peroxisome proliferator-activated receptor α. Cells 12:2791
Reiss AB, Voloshyna I, De Leon J et al (2015) Cholesterol metabolism in CKD. Am J Kidney Dis 66:1071–1082
Rensen PCN, van Dijk KW, Havekes LM (2005) Apolipoprotein AV: low concentration, high impact. Arterioscler Thromb Vasc Biol 25:2445–2447
Riboldi GM, Di Fonzo AB (2019) GBA, Gaucher disease, and Parkinson's disease: from genetic to clinic to new therapeutic approaches. Cells 8:364
Rivellese A, Riccardi G, Romano G et al (1988) Presence of very low density lipoprotein compositional abnormalities in type 1 (insulin-dependent) diabetic patients; effects of blood glucose optimisation. Diabetologia 31:884. https://doi.org/10.1007/BF00265371
Sadashiv MA, Khokhar M et al (2021) Leptin DNA methylation and its association with metabolic risk factors in a northwest Indian obese population. J Obes Metab Syndr 30:304–311
Sano S, Oshima K, Wang Y et al (2018) Tet2-mediated clonal hematopoiesis accelerates heart failure through a mechanism involving the IL-1β/NLRP3 Inflammasome. J Am Coll Cardiol 71:875–886
Santamarina-Fojo S, González-Navarro H, Freeman L et al (2004) Hepatic lipase, lipoprotein metabolism, and atherogenesis. Arterioscler Thromb Vasc Biol 24:1750–1754
Sayols-Baixeras S, Subirana I, Lluis-Ganella C et al (2016) Identification and validation of seven new loci showing differential DNA methylation related to serum lipid profile: an epigenome-wide approach. The REGICOR study. Hum Mol Genet 25:4556–4565
Shin D-J, Osborne TF (2003) Thyroid hormone regulation and cholesterol metabolism are connected through Sterol Regulatory Element-Binding Protein-2 (SREBP-2). J Biol Chem 278:34114–34118
Simha V (2020) Management of hypertriglyceridemia. BMJ 371:m3109
Singh P, Kakkar P, Singh RL (2016) Protective effect of *Trigonella foenum-graecum* and *Foeniculum vulgare* mature leaf against t-BHP induced toxicity in primary rat hepatocytes. J Exp Food Chem 2(2):111
Singh P, Pandey P, Singh PK, Tripathi M, Singh RP, Shukla S, Pathak N, Singh RL (2023) A comprehensive review on phytochemistry, nutritional and pharmacological properties of *Momordica charantia*. IP Int J Compr Adv Pharmacol 8(2):73–79

Solanas-Barca M, De Castro-Orós I, Mateo-Gallego R et al (2012) Apolipoprotein E gene mutations in subjects with mixed hyperlipidemia and a clinical diagnosis of familial combined hyperlipidemia. Atherosclerosis 222:449–455
Stein EA, Raal FJ (2013) Polygenic familial hypercholesterolaemia: does it matter? Lancet 381:1255–1257
Stein JA, Tschudy DP (1970) Acute intermittent porphyria. A clinical and biochemical study of 46 patients. Medicine (Baltimore) 49:1–16
Stewart JM, Kilpatrick ES, Cathcart S, Small M, Dominiczak MH (1994) Low-density lipoprotein particle size in type 2 diabetic patients and age matched controls. Annals of clinical biochemistry 31(2):153–9
Sun X, Feng M, Lu L et al (2021) Lipid abnormalities in patients with Cushing's disease and its relationship with impaired glucose metabolism. Front Endocrinol 11:600323
Sundaram M, Yao Z (2010) Recent progress in understanding protein and lipid factors affecting hepatic VLDL assembly and secretion. Nutr Metab (Lond) 7:35
Taddeini L, Nordstrom KL, Watson CJ (1964) Hypercholesterolemia in experimental and human hepatic porphyria. Metabolism 13:691–701
Tanis BC, Westendorp RGJ, Smelt AHM (1996) Effect of thyroid substitution on hypercholesterolaemia in patients with subclinical hypothyroidism: a reanalysis of intervention studies. Clin Endocrinol 44:643–649
Tanoli T, Yue P, Yablonskiy D, Schonfeld G (2004) Fatty liver in familial hypobetalipoproteinemia. J Lipid Res 45:941–947
Toruner F, Altinova AE, Karakoc A et al (2008) Risk factors for cardiovascular disease in patients with subclinical hypothyroidism. Adv Ther 25:430–437
Triggs-Raine B, Mahuran DJ, Gravel RA (2001) 17. Naturally occurring mutations in GM2 gangliosidosis: a compendium. In: Advances in genetics. Elsevier, San Diego, pp 199–224
Upadhyay RK (2015) Emerging risk biomarkers in cardiovascular diseases and disorders. J Lipids 2015:971453
Vaziri ND (2006) Dyslipidemia of chronic renal failure: the nature, mechanisms, and potential consequences. Am J Physiol Renal Physiol 290:F262–F272
Vodnala D, Rubenfire M, Brook RD (2012) Secondary causes of dyslipidemia. The American journal of cardiology 110(6):823–825
Walsh JP, Bremner AP, Bulsara MK et al (2005) Thyroid dysfunction and serum lipids: a community-based study. Clin Endocrinol 63:670–675
Watanabe K, Tanaka M, Yuki S et al (2018) How is edaravone effective against acute ischemic stroke and amyotrophic lateral sclerosis? J Clin Biochem Nutr 62:20–38
Welty FK (2014) Hypobetalipoproteinemia and abetalipoproteinemia. Curr Opin Lipidol 25:161–168
Wolska A, Dunbar RL, Freeman LA et al (2017) Apolipoprotein C-II: new findings related to genetics, biochemistry, and role in triglyceride metabolism. Atherosclerosis 267:49–60
Xiao C, Dash S, Morgantini C et al (2016) Pharmacological targeting of the atherogenic dyslipidemia complex: the next frontier in CVD prevention beyond lowering LDL cholesterol. Diabetes 65:1767–1778
Yan H, Cheng L, Jia R et al (2019) ATP-binding cassette sub-family a member1 gene mutation improves lipid metabolic abnormalities in diabetes mellitus. Lipids Health Dis 18:103
Yu X-H, Zhang D-W, Zheng X-L, Tang C-K (2019) Cholesterol transport system: an integrated cholesterol transport model involved in atherosclerosis. Prog Lipid Res 73:65–91
Zhang Q, Ramlee MK, Brunmeir R et al (2012) Dynamic and distinct histone modifications modulate the expression of key adipogenesis regulatory genes. Cell Cycle 11:4310–4322
Zhou H, Gong Y, Wu Q et al (2020) Rare diseases related with lipoprotein metabolism. Adv Exp Med Biol 1276:171–188

Nucleic Acid Metabolism and Disorders

5

Saba Ubaid and Shivani Pandey

Abstract

Nucleic acid metabolism refers to the entire spectrum of nucleotide synthesis and degradation activities. Anabolic mechanisms such as phosphate, pentose sugar, and a nitrogenous base react chemically to produce nucleotide which polymerizes to form either deoxyribonucleic acid or ribonucleic acid. A catabolic process occurs when nucleic acid is degraded to its monomeric subunit nucleotides. To produce new nucleotides, portions of the nucleotides or nucleobases can also be rescued. Metabolism studies must take into account salvage pathways, which allow partially catabolic products to be redirected into an anabolic pathway for incorporation into new macromolecular structures, in addition to the relatively simple catabolism pathways of nucleic acids. Several modifications in purine and pyrimidine metabolism cause many inborn errors of metabolism, however, understanding these faults is also essential for the effective treatment of numerous life-threatening disorders. These comprise gout, infectious disorders, inflammatory conditions including rheumatoid arthritis, cardiovascular disease, and a variety of cancers such as leukemia, pancreatic cancer, nonsmall cell lung cancer, and colon cancer. This chapter explores the multifaceted role of nucleic acid metabolism, elucidating its pivotal role in cellular functions and disease pathology. Insights derived from these intricate pathways will not only unravel the origins of severe conditions but also pave the way for innovative therapeutic strategies targeting a spectrum of challenging medical afflictions.

S. Ubaid · S. Pandey (✉)
Department of Biochemistry, King George's Medical University, Lucknow, Uttar Pradesh, India
e-mail: shivanipandey@kgmcindia.edu

R. L. Singh et al. (eds.), *Clinical Applications of Biomolecules in Disease Diagnosis*, https://doi.org/10.1007/978-981-97-4723-8_5

Keywords

Nucleic acids · Purine · Pyrimidine · Catabolism · Anabolism

5.1 Introduction

Nucleic acids are crucial biomolecules found in both cells and viruses, serving vital functions in storing and expressing genetic information (Vlassov et al. 2009). DNA, or deoxyribonucleic acid, carries the instructions necessary for protein production within cells. RNA, another type of nucleic acid, exists in various molecular forms and participates in numerous cellular processes, such as facilitating protein synthesis (Elson 1965). Nucleic acid metabolism encompasses the processes of synthesizing and breaking down nucleic acids. Polymers of nucleotides are known as nucleic acids. Phosphate, pentose sugar, and a nitrogenous base combine chemically to produce nucleotides, which is an anabolic process (Hartman and Buchanan 1959). Conversely, the degradation of nucleic acids is a catabolic reaction. Moreover, components of nucleotides or nucleobases can be salvaged for the regeneration of new nucleotides. Enzymes are essential for both the synthesis and degradation reactions, and any deficiencies or defects in these enzymes can lead to various diseases (Hartman and Buchanan 1959).

A series of seminars conducted in Tel Aviv, Israel (1974) emphasized the research of purine and pyrimidine metabolism (Peters et al. 2006). Excessive phosphoribosyl pyrophosphate synthetase activity, gout linked to hypoxanthine-guanine phosphoribosyltransferase deficiency, and severe combined immunodeficiency disease (SCID) due to adenosine deaminase (ADA) deficiency were identified as potentially life-threatening and debilitating conditions associated with significant metabolic disorders in that particular decade (Kamatani et al. 2021). Gene therapy was initially applied to treat the SCID in humans. But at the same time, number of purine and pyrimidine analogs emerged, and these are still the mainstays of treatment for some of the diseases listed above. Some treatments used for these conditions include allopurinol for gout, arabinosyl-cytosine (ara-C) for acute myeloid leukemia, 6-mercaptopurine (6MP) for childhood leukemia, uracil analogs like 5-fluorouracil (5FU) for colon cancer, and azathioprine for inflammatory diseases, such as rheumatoid arthritis and inflammatory bowel disease (Peters et al. 2006). Another surge in the development of novel analogs occurred in the 1980s. Among them were ganciclovir and azidothymidine (AZT), employed as antiviral agents; doxifluridine, later formulated as capecitabine; and deoxynucleoside analogs such as fludarabine, cladribine, and gemcitabine, utilized as anticancer agents (Peters et al. 2006).

The advancement of these compounds hinged on the insights gained from the previous studies of inborn errors of metabolism. These investigations provided an understanding of the functions and control of crucial enzymes in the metabolism of purine and pyrimidine, highlighting how modifying these enzymes could treat the mentioned diseases. Moreover, they revealed reasons behind varying treatment responses among patients, often attributed to genetic polymorphisms. Numerous studies have emphasized on the significance of genetic variations in determining treatment outcomes. The characterization of the human genome spurred research into leveraging the genetic foundations of these diseases to enhance therapeutic strategies. This chapter aims to elucidate the critical role of nucleic acid metabolism in human health and disease, exploring its biochemical intricacies and translational implications. From the synthesis and degradation of nucleic acids to the development of novel therapies, this chapter delves into the fundamental processes underlying genetic information storage and expression.

5.2 Purine Metabolism

5.2.1 Purine Nucleotide Biosynthesis

The synthesis of purine nucleotides starts with 5-Phosphoribosyl-α-pyrophosphate (PRPP), which produces the nucleotide IMP via a sequence of events. Ribose 5-phosphate pyrophosphokinase catalyzes the first step, which produces PRPP from ribose 5-phosphate and the two high-energy phosphates of ATP. Purine ring structure building takes place on the scaffold of ribose 5-phosphate. In addition to its function in purine synthesis, PRPP is also involved in the biosynthesis of pyrimidines and histidines. In the second stage, the enzyme PRPP glutamylamidotransferase converts PRPP to Phosphoribosyl-β-amine (PRA) (Smith 1998; Zalkin 1993). Two half-reactions are involved in this transformation: first, glutamine is hydrolyzed to produce glutamate and ammonia, and then ammonia is transferred to PRPP to produce PRA while releasing pyrophosphate (Messenger and Zalkin 1979). The next step is to use the enzyme GAR synthetase to convert PRA to Glycinamide ribonucleotide (GAR). In this stage, atoms C(4), C(5), and N(7) of the purine base are provided by glycine, which ligates to the PRA amino group in an ATP-dependent manner. Two distinct enzymes are involved in the production of N-formylglycinamide ribonucleotide (FGAR). Most organisms include GAR transformylase, which gets its purine C(8) carbon atom from the cofactor N10-formylTHF. In *E. Coli* and similar microbes, FGAR synthetase and GAR transformylase are present. The ATP-dependent ligation of GAR and formate to produce FGAR is catalyzed by FGAR synthetase enzyme. The ATP-dependent amidation of FGAR to FGAM, the fifth step in the purine biosynthesis pathway, is catalyzed by FGAM synthetase I. The γ-phosphate of an ATP molecule activates the amide oxygen of FGAR, leading to the production of a putative iminophosphate intermediate (Westheimer 1981). Next, nucleophilic ammonia produced from a glutamine molecule attacks this intermediate. The results of this transformation are glutamate, ADP, and FGAM (Buchanan 1982; Schendel et al. 1989). The sixth step of purine biosynthesis involves the ATP-dependent transformation of FGAM into AIR (5-Aminoimidazole ribonucleotide), facilitated by AIR synthetase. This process results in the formation of the five-membered ring of the purine base. Essentially, the formyl oxygen of FGAR is activated by ATP's γ-phosphate, allowing it to be attacked by the N1 of FGAR, leading to ring closure (Schrimsher et al. 1986). Furthermore, AIR undergoes direct carboxylation at the C(4) position by CO_2 with the help of class II AIR carboxylase, resulting in the formation of CAIR. The eight-step biosynthesis process of N-succinocarboxamide-5-aminoimidazole ribonucleotide involves the ligation of the carboxylate group of CAIR to the amino group of aspartate, which is catalyzed by SAICAR synthetase. The remaining aspartate is removed as succinate in the following step, and the amino group supplies the N(1) nitrogen atom of the last purine base. Adenylosuccinate lyase catalyzes the biosynthesis of aminoimidazole-4-carboxamide ribonucleotide (AICAR) in the ninth step. Adenylosuccinate lyase stands out in the pathway due to its ability to accommodate two different substrates, performing two distinct reactions within the same active site. Firstly, it converts SAICAR to AICAR in step 9, and secondly, it catalyzes the second step in the conversion of IMP to AMP. Both reactions result in the release of fumarate through β-elimination, suggesting a shared mechanism and likely employing the same catalytic residues. In the penultimate step of de novo purine biosynthesis, AICAR is transformed by AICAR transformylase, transferring a formyl group to produce FAICAR. Finally, the pathway concludes with IMP synthase catalyzing a straightforward cyclization reaction of FAICAR to generate IMP, eliminating water in the process.

The synthesis of IMP requires a total of two moles of formate, five moles of ATP, glutamine, and one mole each of glycine, CO_2, and aspartate. The formyl moieties are carried by tetrahydrofolate (THF) in the form of N10-formyl-THF (Chandel 2021) (Fig. 5.1)

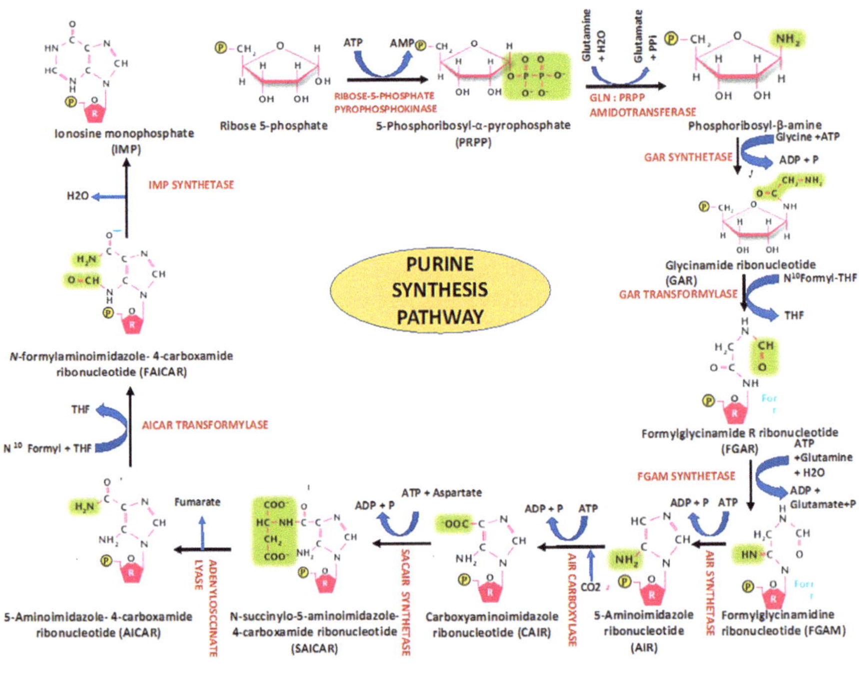

Fig. 5.1 The purine synthesis pathway. Synthesis of purine nucleotide begins with PRPP and through a series of reactions, it generates the IMP

Two moles of formate and glutamine with one mole of glycine, five moles of ATP, CO_2, and aspartate are required in total for IMP synthesis. Tetrahydrofolate (THF) carries the formyl moieties in the form of N10-formyl-THF (Chandel 2021). The initial two steps of the purine biosynthesis pathway are the ones that limit the pace. The initial step, which is catalyzed by ribose-5-pyrophosphokinase, is an ADP and GDP-inhibited process. PRPP amidotransferase catalyzes the second reaction. By binding to AMP, GMP, and IMP, this process is also feedback suppressed allosterically. On the other hand, PRPP increases amidotransferase activity. Different processes can be used to convert IMP into GMP or AMP. GTP is crucial for AMP development, and ATP is needed for GMP production. This permits regulation of AMP or GMP in response to GTP or ATP availability, respectively. At the expense of producing GMP, the accumulation of GTP leads to the synthesis of AMP from IMP. On the other hand, GMP synthesis is favored over AMP synthesis when there is an excess of ATP.

Additionally, the accumulation of GMP or AMP inhibits the activity of IMP dehydrogenase and adenylosuccinate synthetase, which stops the synthesis of additional GMP or AMP, respectively. Following this, adenylate kinase or guanylate kinase can use ATP to convert AMP or GMP to ADP or GDP, respectively. Free ATP is typically utilized as the phosphate donor in these reactions since it is more abundant than any other nucleotide (Chandel 2021).

5.2.2 Salvage Pathway for Purine Nucleotide Synthesis

Salvage pathway is a set of procedures that can be used to synthesize nucleotides from purine bases and purine nucleosides. The free purine bases guanine, adenine, and hypoxanthine are converted to their respective nucleotides by phosphoribosylation. Adenosine phosphoribosyltransferase (APRT), which transforms adenine into AMP, and hypoxanthine-guanine phosphor ribosyl transferase (HGPRT), which uses PRPP as a substrate to convert hypoxanthine to IMP or guanine to GMP, are two important transferases that are involved in the salvage pathway of purines. The salvage route lowers PRPP levels and, as a result, slows down the pace of de novo purine synthesis by using PRPP as a substrate. Depending on PRPP availability, the de novo purine synthesis and salvage processes are closely linked (Chandel 2021).

The purine nucleotide cycle is a significant salvage pathway that increases the trichloroacetic acid (TCA) cycle's capacity to create additional nicotinamide adenine dinucleotide (NADH) for ATP synthesis by generating the TCA-cycle intermediate fumarate. An increase in muscular activity raises the need for ATP metabolically. A rise in AMP levels indicates that the body is not producing enough energy from the electron transport chain and TCA cycle to meet its needs. Because muscle cells lack the majority of the enzymes involved in the crucial anaplerotic reactions that is, the enzymes that replenish pathway intermediates, this cycle is crucial to muscle cells. Therefore, fumarate produced by the purine nucleotide cycle allows the muscle to replenish TCA-cycle intermediates. Deficiencies in myoadenylate deaminase, the muscle-specific isoenzyme of AMP deaminase, lead to cramping and fatigue during exercise (Chandel 2021).

5.2.3 Purine Nucleotide Degradation

Purine rings are insoluble in human tissue. Purine nucleotide catabolism produces uric acid, which is discharged in the urine as sodium urate crystals and has low water solubility. The fundamental purine structure is unaffected by purine catabolism. The primary pathway begins with the conversion of AMP to IMP, which results in the production of hypoxanthine. It is possible to convert GMP into guanosine and then guanine again. Xanthine is produced by both hypoxanthine and guanine. In the last stage, xanthine oxidase transforms xanthine into uric acid, which results in the production of hydrogen peroxide (Chandel 2021).

5.3 Disorders of Purine Metabolism

5.3.1 Gout

Gout is an inflammatory disease that develops due to monosodium urate (MSU) crystal deposition which is the end product of purine metabolism (De Miguel et al. 2012; Dalbeth et al. 2015). Genome-wide association studies (GWAS) showed that variations in the serum urate level are associated with single-nucleotide polymorphisms. The urate transporter genes SLC2A9 (encoding GLUT9), SLC17A1 (encoding NPT1), SLC22A12 (encoding URAT1), and ABCG2 are strongly associated with variation in serum urate levels.

5.3.1.1 Treatment

Incident of gout disease can be avoided by employing strategies that lower serum urate concentrations. Intervention research in obese adults found that, when compared to conventional treatment, bariatric surgery reduced serum urate levels and lowered the incidence of gout by 34% over a 26-year follow-up. The uricosuric effects of fenofibrate, a PPARα agonist, cause very slight drops in serum urate levels. Fenofibrate medication decreased the incidence of gout in people with type 2 diabetes mellitus by 50% when compared with a placebo in a post hoc analysis of the FIELD (Fenofibrate Intervention and Event Lowering in Diabetes) trial (Waldman et al. 2018). In a previous study, it was found that the urate-lowering drug febuxostat reduces the incidence of gout in patients with Chronic Kidney Disease (CKD) stage 3 and asymptomatic hyperuricemia (Kimura et al. 2018). An alternate method for preventing incident gout could be to administer anti-inflammatory medication to delay the onset of an acute inflammatory reaction to deposited MSU crystals (Table 5.1).

5.3.2 Lesch-Nyhan's Disease/Syndrome

The purine bases hypoxanthine and guanine are converted to their corresponding nucleotides, IMP and GMP, via HPRT, the location of the Lesch-Nyhan illness defect (Seegmiller et al. 1967). The HPRT gene is located in the X chromosome and

Table 5.1 Inborn errors of purine and pyrimidine metabolism and their treatments

Purine and pyrimidine disorders	Enzyme defect	Treatment
Gout	Deficiencies in two enzymes, hypoxanthine:guanine phosphoribosyltransferase and adenylosuccinate lyase.	Fenofibrate, febuxostat, (Waldman et al. 2018; Kimura et al. 2018)
Lesch-Nyhan's disease	Complete deficiency of hypoxanthine-guanine phosphoribosyl-transferase (HPRT)	Allopurinol (or febuxostat), low purine intake and high fluid intake, alkalinization of urine (Fu et al. 2014)
Adenine phosphoribosyltransferase deficiency	Accumulated adenine transforms into the highly insoluble 2, 8 dihydroxyadenine.	Allopurinol, low purine, alkalinization of urine not recommended (Cartier 1974)
Phosphoribosyl pyrophosphate (PRPP) synthetase superactivity	"Superactivity" of phosphoribosyl pyrophosphate (PRPP) synthetase	Allopurinol (or febuxostat), urine alkalinization, high fluid intake, low-purine/low-fructose diet. (Kranen et al. 1985)
Adenosine deaminase (ADA) deficiency	Deficiency of adenosine deaminase	None established yet, but potential strategies include fresh-frozen plasma, recombinant ADA2, and bone marrow transplantation. (Giblett et al. 1972; Fu et al. 2014; Glick 2006)
Purine nucleoside phosphorylase deficiency	Purine nucleoside phosphorylase deficiency	Bone marrow transplantation, Repeated infusions of normal, irradiated erythrocytes (Markert 1991)
Xanthine oxidase deficiency—hereditary xanthinuria	Lack of xanthine oxidase	Low-purine diet and increasing fluid consumption, including low-sulfur diets, sulfate, and molybdenum supplementation (Wadman et al. 1983)
Adenylosuccinate lyase deficiency	Deficiency of adenylosuccinate lyase	Injection of uridine, oral D-ribose, and D-glucose (Salerno et al. 1998), oral supplementation of adenine, allopurinol (Boria et al. 2010)
Myoadenylate deaminase deficiency ATIC deficiency	Lack of AMP deaminase activity	D-ribose therapy (Patten 1982)
Thiopurine methyltransferase deficiency	Deficiency of TPMT	Azathioprine, 6-mercaptopurine, and 6-thioguanine (Hitchings and Elion 1969)
Deoxyguanosine kinase deficiency	Mitochondrial deoxyguanosine kinase deficiency	Liver transplantation (Dimmock et al. 2008a, b), formulas with an enriched medium-chain triglyceride (Feranchak et al. 2007)

(continued)

Table 5.1 (continued)

Purine and pyrimidine disorders	Enzyme defect	Treatment
UMP synthase deficiency	Overproduction of orotic acid and a shortage of pyrimidine nucleotides (Smith et al. 1973)	Administration of uridine
Dihydropyrimidine dehydrogenase deficiency	DPD deficiency	No available treatment
Dihydropyrimidinase deficiency	Reduced amounts of the neurotransmitter alanine	No available treatment
β-ureidopropionase deficiency	Ureidopropionase deficiency	No available treatment
Pyrimidine 5′-nucleotidase deficiency	Erythrocyte-specific deficiency	Blood transfusions (Chou and De Loughery 2001)
Cytosolic 5′-nucleotidase superactivity	Increases in the activity of cytosolic 5-nucleotidase	No available treatment
Thymidine phosphorylase deficiency	Thymidine phosphorylase deficiency	No available treatment
Dihydroorotate dehydrogenase deficiency	DHODH deficiency	No available treatment
Cytidine deaminase deficiency	CDA deficiency causes a reduction in the basal activity of poly(ADP-ribose) polymerase 1 (PARP-1) (Gemble et al. 2015)	No available treatment
Thymidine kinase 2 deficiency	Thymidine kinase-2 deficiency (Saada et al. 2001)	No available treatment

is expressed in a recessive manner. Females who have inherited the defective gene exhibit nonrandom inactivation of X chromosome (De Gregorio et al. 2000). Variations in the quantity of residual activity in the intact cell assay are correlated with phenotypic variation in clinical expression (Page et al. 1981).

The neurologic hallmarks of delayed motor development, dystonia and involuntary movements, and self-injurious behavior are combined with symptoms that are the result of uric acid buildup to form the classic clinical picture. Without assistance, the patient cannot sit or walk. These kids are typically seen sitting in a small wheelchair that is securely fastened around their waist and chest. Dystonia, chorea, and athetosis are examples of extrapyramidal symptoms that emerge. The back's opisthotonic arching is a consistent observation. Signs of a pyramidal tract include positive Babinski reflexes and spasticity hyperreflexia. Most notably, the condition is characterized by self-injurious behavior (Puig et al. 2001). Patients do not have a pain threshold. Additionally, aggressive behavior is directed at other people. Both uricosuria and hyperuricemia are typical. For this and other metabolic hyperuricemias, a high uric acid to creatinine ratio is a helpful first screening test (Kaufman et al. 1968; Ohdoi et al. 2003). Apart from Lesch-Nyhan disease, there exist additional HPRT mutations exhibiting distinct characteristics. Gout sufferers and those with calculi in the urinary tract are among them (Kelley et al. 1969; Kogut et al. 1970). Certain variations have some residual activity in the erythrocyte assay, making it simple to differentiate them from the patient described in the classic case.

5.3.2.1 Treatment

The cornerstone medication for treating hyperuricemia is allopurinol, which inhibits the xanthine oxidase enzyme to prevent xanthine and hypoxanthine from being converted to uric acid. Reducing excessive production of uric acid lessens the consequences of gouty arthritis, urolithiasis, and nephrolithiasis. Since hyperuricemia is a congenital defect, it is best to begin medication as soon as possible to avoid consequences. Studies on the long-term safety and effectiveness of allopurinol have shown no adverse effects, provided that the dosage is kept low to prevent the development of xanthine calculi. On neurodevelopmental and cognitive results, however, allopurinol therapy has little effect. Toxic epidermal necrolysis, Stevens-Johnson syndrome, and drug hypersensitivity syndrome are uncommon side effects of allopurinol therapy. In these circumstances, another xanthine oxidase inhibitor called febuxostat may be employed (Fu et al. 2014). Therapies such as exchange transfusion and bone marrow transplantation have not proven to be effective (Pralong et al. 2005). Studies on the use and effectiveness of therapies such as gabapentin usage, deep brain stimulation (Pralong et al. 2005), dopamine replacement therapy, and local botulinum injection are now underway (Dabrowski et al. 2005) (Table 5.1).

5.3.3 Adenine Phosphoribosyltransferase Deficiency

Adenine phosphoribosyltransferase (APRT) deficiency is another abnormality of purine metabolism that causes renal stone disease in children (Barratt et al. 1979). APRT is a purine salvage enzyme because it catalyzes the conversion of adenine to its mononucleotide (AMP). Due to the lack of APRT, accumulated adenine is converted to 2, 8 dihydroxyadenine, which is highly insoluble. APRT gene is located at the chromosome 16q24.3 (Kahan et al. 1974). Previous studies conducted in Japan show three mutations in 96% of patient alleles (Kamatani et al. 1992). According to Kamatani et al. (1992), the most frequent mutation is a T-to-C mutation that converts methionine 136 to threonine (M136T).

5.3.3.1 Treatment

To prevent the synthesis of 2,8-dihydroxyadenine in symptomatic individuals, allopurinol should be administered in accordance with the guidelines provided under PRPP synthetase superactivity. Dietary purine limitation and increased fluid intake are advised for patients with or without stones. However, alkalinizing the urine is not recommended because 2,8-dihydroxyadenine's solubility does not increase to pH 9 like uric acid does (Cartier 1974) (Table 5.1).

5.3.4 Phosphoribosyl Pyrophosphate (PRPP) Synthetase Superactivity

Phosphoribosyl pyrophosphate (PRPP) synthetase variation results in more than normal enzymatic activity, which produces hyperuricemia (Nyhan et al. 1969). Nephropathy, calculi in the urinary tract, and gouty arthritis are among the complications. Sensorineural deafness was also present in several kindreds (Simmonds et al. 1984; Becker et al. 1988). Females can also exhibit symptoms of this X-linked illness (Becker et al. 1988). By evaluating the in vivo conversion of [14C] glycine to urinary uric acid, the overproduction of purine de novo was confirmed (Simmonds et al. 1984). Purines, adenine, guanine, and hypoxanthine are incorporated into nucleotides by intact cultivated fibroblasts in this disease more quickly than by controls (Becker et al. 1980). The elevated levels of PRPP enzyme can be used as a screening method but due to its aberrant nature, measuring its activity is less accurate (Becker et al. 1980). The X chromosome has two genes called PRPS1 and S2 that code for PRPP synthetase, located at Xq22–24 and Xp22.2–22.3 (Becker et al. 1995). A tiny number of point mutations in PRPS1 have been identified in people who have changed the enzyme's allosteric characteristics and overactivity. No mutation was detected in the PRPS1 or S2 DNA of the six patients with overactive PRPP synthase. Rather, higher concentrations of the S1 isoform were observed, whose physicochemical and catalytic characteristics were normal (Taira et al. 1987; Roessler et al. 1993).

5.3.4.1 Treatment

Allopurinol, which inhibits xanthine oxidase, the final enzyme in the purine degradation process, is the recommended treatment for patients. As a result, less uric acid is produced, and its position is taken up by the somewhat more soluble compounds xanthine and hypoxanthine, which are about ten times more soluble than uric acid. Xanthine calculi may develop following allopurinol medication in a small percentage of patients who have a significant increase in de novo synthesis (Kranen et al. 1985). As a result, further precautions against crystallization are advised. As uric acid and xanthine are more soluble at alkaline pH than at acid pH, these include a low-purine diet (free of organ meats, fish such as tuna, anchovy, herring, salmon, mackerel, and sardines, peas, and dried beans), increasing fluid intake, and administration of potassium citrate, or citrate mixtures, sodium bicarbonate, to bring urinary pH to 6.0–6.5 (Bhasin et al. 2014) (Table 5.1).

5.3.5 Adenosine Deaminase (ADA) Deficiency

Purine nucleotides have now been shown to play a significant role in immunodeficiency. The irreversible deamination of adenosine to inosine and deoxyadenosine to deoxyinosine is catalyzed by ADA. According to Giblett et al. (1972), the primary metabolic cause of severe combined immunodeficiency disease (SCID) is adenosine deaminase (ADA, EC 3.5.4.4) deficiency. ADA gene is present on chromosome 20q13.11. Many mutations have been identified (Hirschhorn et al. 1992; Markert

1994). The majority of which are single-amino acid alterations. Numerous unrelated patients have been identified to have both R211H and the comparatively common mutation 329V (Hirschhorn et al. 1992).

5.3.5.1 Treatment

Adenosine deaminase deficiency (ADA) is a purine salvage pathway deficiency that results in buildup of toxic metabolites causing death in rapidly dividing cells, especially lymphocytes. The most complete form of ADA leads to severe combined immune deficiency (SCID). Treatment with enzyme replacement therapy (ERT) was developed in the 1970s and became the treatment for ADA SCID by the 1980s (Markert et al. 1987; Hershfield 1995). It remains an option for some infants with SCID, and a stop-gap measure for others awaiting curative therapy. For some infants with ADA SCID who have matching family donors, a hematopoietic stem cell transplant (HSCT) is an option for cure. Gene therapy for ADA SCID, approved in some countries and in trials in others, is becoming possible for more infants with this disorder (Table 5.1).

5.3.6 Purine Nucleoside Phosphorylase Deficiency

Giblett (1985) have found that purine nucleoside phosphorylase (PNP, EC 2.4.2.1) deficiency results in a clinical syndrome of SCID that is indistinguishable from that of ADA deficiency. The reversible cleavage of inosine and guanosine into their corresponding bases, hypoxanthine and guanine, is catalyzed by PNP. Additional substrates are deoxyguanosine and deoxyinosine. Among immune deficiency illnesses, PNP deficiency is distinct in that it causes hypouricemia and decreased uric acid excretion in the urine (Chu et al. 1989). Cultured fibroblasts or erythrocyte lysates can be used to demonstrate a lack of enzymatic activity (Dooley et al. 1987).

5.3.6.1 Treatment

Up until recently, the majority of patients passed away from severe bacterial or viral infections, albeit later in life than children with untreated ADA deficiencies. Bone marrow transplantation and repeated infusions of normal, irradiated erythrocytes were the methods of treatment (Markert 1991). A successful bone marrow transplant with a matched donor has been reported more recently (Carpenter et al. 1996). Gene and enzyme therapy could be made available soon (Table 5.1).

5.3.7 Xanthine Oxidase Deficiency: Hereditary Xanthinuria

Hereditary xanthinuria, which is caused by a lack of xanthine oxidase (EC 1.2.3.2), is a common cause of hypouricemia. Xanthine oxidase is required to break down the purine base xanthine to uric acid. Many patients are diagnosed with hypouricemia; however, because xanthine is very insoluble, patients also run the risk of developing calculus and its associated renal problems (Carpenter et al. 1986). Patients experiencing cramps or soreness in their muscles may also have xanthine crystals in their muscles (Chalmers et al. 1969). Isolated deficiency of xanthine oxidase has been referred to as Type I. Aldehyde oxidase catalyzes the conversion of allopurinol to oxypurinol, which patients can oxidize. Patients classified as type II are unable to metabolize allopurinol due to deficiencies in both oxidase activities (Reiter et al. 1990). The assay of the xanthine in the urine is typically used to make the diagnosis.

5.3.7.1 Treatment

Although isolated xanthine oxidase insufficiency is often benign, renal stones can be avoided by following a low-purine diet and increasing fluid consumption. When xanthine oxidase and sulfite oxidase deficiencies coexist, the prognosis is quite bad. All treatment approaches have failed thus far, including low-sulfur diets, sulfate, and molybdenum supplementation (Wadman et al. 1983), and attempts to bind sulfite with thiol-containing medications (Table 5.1).

5.3.8 Adenylosuccinate Lyase (ADSL) Deficiency

Adenylosuccinate lyase (ADSL) deficiency is the other purine metabolism illness for which the Bratton-Marshall reaction is positive (Georges and Berghe 1984). One of the main symptoms is mental impairment, and most people experience seizures. Following the first report on autistic behavior, there has been an ongoing, largely ineffective search among autistic communities for aberrant purine metabolism (Georges and Berghe 1984). In the eighth step of the de novo synthesis of purines, succinyl amino imidazole carboxamide ribotide (SAICAR) (SAICAMP) is converted to AICAR by the defective enzyme adenylosuccinate lyase (adenylosuccinase, EC 4.3.2.2). In purine nucleotide interconversion, the enzyme also catalyzes the conversion of adenylosuccinate to AMP. Human chromosome 22q1.3.1.–1.3.2 is where the gene is located, and a point mutation was found in the index family. There are more missense mutations discovered, with R426H being the most prevalent (Georges and Berghe 1984; Stone et al. 1992).

5.3.8.1 Treatment

Even while the number of individuals with ADSL deficiency is rising, there are still very few reports of therapeutic considerations and attempts. Only two of them, the injection of uridine and D-ribose, had some positive benefits (Salerno et al. 1998). Some ADSL patients have received D-ribose injection, which promotes de novo purine synthesis and increases the availability of phosphoribosyl pyrophosphate (PRPP). S-adenosyl-l-methionine (SAMe) was recently assessed by Werkhoven et al. (2013) as a possible therapy for ADSL deficiency (Table 5.1).

5.3.9 Myoadenylate Deaminase Deficiency (MADD)

Catalyzed by myoadenylate deaminase (also known as AMP deaminase, AMP aminohydrolase, EC 3.5.4.6), AMP is converted to IMP and NH3. Some affected patients may not exhibit any symptoms at all, whereas others report muscle soreness or cramping after exercising (Fishbein et al. 1978). The enzyme activity assay conducted on the biopsied muscle confirms the diagnosis. The AMPD1 gene on chromosome 1p13–21 is mutated. There can be an increase in blood creatine kinase. Myoglobinuria is an uncommon side effect.

The forearm ischemia exercise test is a diagnostic tool used to differentiate between different metabolic disorders. In patients with adenylosuccinate lyase deficiency, the test reveals a lack of the normal rise in venous ammonia levels, contrasting with McArdle glycogen storage disease where the typical rise in lactic acid levels is absent. This test helps in making the diagnosis and distinguishing between the two conditions. Furthermore, a homozygous C to T transition at nucleotide 34 has been consistently observed in all affected patients (Morisaki et al. 1992). By eliminating a *MaeII* restriction site, the mutation offers a straightforward diagnostic procedure.

5.3.9.1 Treatment

Pharmacological and/or dietary therapy may be beneficial, even essential, for MADD patients whose lifestyle requires significant physical activity for their well-being (Engel 1986). Accordingly, D-ribose therapy has been found to be both successful and unsuccessful (Patten 1982; Lecky 1983). But recently, a 55-year-old man with MADD has shown significant clinical improvement after taking high doses of D-ribose orally in accordance with a detailed treatment regimen (Zöllner et al. 1986). The treatment's rationale arises from the fact that D-ribose provides an extra energy source that might potentially counteract the reduced energy supply caused by defective glycogenolysis and glycolysis, as well as improve the synthesis of purine nucleotides from scratch (Fishbein et al. 1978) (Table 5.1).

5.3.10 Thiopurine Methyltransferase (TPMT) Deficiency

According to Weinshilboum (1989), TPMT is a cytoplasmic enzyme that preferentially catalyzes the S-methylation of thiopurines, thiopyrimidines, and thiophenols. The metabolism of the immunosuppressive and antileukemic drugs azathioprine, mercaptopurine (MP), and thioguanine (TG) is significantly influenced by TPMT. Apart from xenobiotic metabolism, the endogenous function of TPMT is unknown. A common genetic polymorphism regulates the amount of TPMT activity in human tissue, and inherited variations in TPMT activity play a significant role in individual variations in thiopurine medication toxicity and therapeutic efficacy (Lennard and Lilleyman 1996). A well-established example of the therapeutic utility of pharmacogenetics and a prime example of pharmacogenomics research is the TPMT genetic polymorphism (Weinshilboum and Wang 2006). Variations in the TPMT gene cause TPMT deficiency (Lennard 2014). The TPMT enzyme is essential for metabolizing thiopurine medications. Once within the body, these medications are transformed into harmful substances that destroy bone marrow-derived immune system cells. The TPMP enzyme effectively breaks medications into harmless, inactive molecules. The stability and activity of the TPMT enzyme are decreased by alterations in the TPMT gene (Wennerstrand et al. 2012). Without sufficient amounts of this enzyme, the medications cannot be "turned off," which means they remain in the body for longer and continue to damage cells (Barreto et al. 2014).

5.3.10.1 Treatment

The decrease activity of TPMP inhibits the processing of thiopurine-containing medications (Booth et al. 2011). These medications block, azathioprine, 6-mercaptopurine, and 6-thioguanine suppress the immune system of the body. Certain autoimmune diseases, such as Crohn's disease and rheumatoid arthritis, are brought on by immune system abnormalities and are treated using thiopurine medicines. Additionally, these medications are used to treat a variety of cancer types, with a focus on lymphomas and leukemias, which are malignancies of the immune system's cells. Furthermore, organ transplant recipients take thiopurine medications to help stop their immune systems from attacking their newly donated organ (Hitchings and Elion 1969).

Hematopoietic toxicity, or damage to the bone marrow, is a possible side effect of thiopurine medication treatment (De Boer et al. 2007). While anyone taking these medications has the potential to experience this problem, those who lack TPMT are most vulnerable. Normally, the bone marrow produces multiple blood cell types, such as oxygen-carrying red blood cells, immune-stimulating white blood cells, and platelets that aid in blood coagulation. Myelosuppression is a disorder when the bone marrow is unable to produce enough of these cells as a result of damage to the bone marrow (Carey 2003). Anemia, or a lack of red blood cells, can result in weakness, pallor, shortness of breath, and excessive exhaustion. Neutropenia, or low white blood cell counts, can cause recurrent infections that could be fatal. Thrombocytopenia, or low-platelet counts, can easily bruise and bleed (Table 5.1).

5.3.11 Deoxyguanosine Kinase (DGUOK) Deficiency

A mitochondrial deoxyguanosine kinase deficiency was found in multiple patients with the hepatocerebral form of mitochondrial DNA depletion syndrome, which is characterized by early progressive liver failure, neurological abnormalities, hypoglycemia, and increased lactate (Schwartzberg 2006). This enzyme is crucial for the supply of precursors of mitochondrial DNA, especially in the liver and brain where there is no cytosolic form of the enzyme. It phosphorylates the deoxy counterpart of guanosine into deoxyGMP. In 19 patients from three families, a single-nucleotide loss in the mitochondrial deoxyguanosine kinase gene was associated with the condition (Mandel et al. 2001). Additional mutations have been found since then.

5.3.11.1 Treatment

In the context of rigorous clinical studies, no therapeutic therapies have been shown to date to enhance the outcomes of children with DGUOK deficiency. Nonetheless, following supportive care for liver failure, case reports have shown recovery of liver disease parameters (Mousson de Camaret et al. 2007). As a result, it is advised that a person with a DGUOK deficit be assessed by a medical professional experienced in treating pediatric liver failure. A dietician with experience in managing children with hepatic and renal failure should be involved in their care, as there is anecdotal evidence of the benefits of supportive care and data showing children with cholestatic liver disease or renal disease are at significant risk of nutritional insufficiency (Feranchak et al. 2007). Infants with cholestasis may benefit more nutritionally from formulas with enriched medium-chain triglyceride content than from those that are mostly made of long-chain triglycerides (Feranchak et al. 2007). In people with isolated hepatic illness, cornstarch may lessen symptomatic hypoglycemia (Dimmock et al. 2008a, b). Good nutritional control can be achieved with fractional meals and enteral nourishment at night (Ducluzeau et al. 2002). It is important to prevent nutritional deficiencies like those caused by a lack of fat-soluble vitamins or an insufficiency of vital fatty acids. Consequently, it could be necessary to supply fat-soluble vitamins and essential fatty acids to affected patients (Feranchak et al. 2007).

Orthotopic liver transplantation has little survival benefit for children with multisystem disease (Dimmock et al. 2008a, b). Orthotopic liver transplantation is a promising therapeutic option because it has demonstrated remarkable 10-year survival in several children with solitary hepatic or hepatorenal illness. But given that at least one child with isolated liver illness before to transplant developed neurologic symptoms following liver transplantation, this choice needs to be carefully discussed with parents (Dimmock et al. 2008a, b). Liver transplantation may be necessary for this condition despite the problems surrounding hepatic dysfunction due to the possibility of hepatocellular carcinoma developing into a malignant transformation (Dimmock et al. 2008a, b). The established long-term dangers associated with liver transplantation must be considered while deciding on orthotopic liver transplantation.

A sensitive but nonspecific marker called Alpha-fetoprotein (AFP) is utilized to distinguish between nonmalignant liver disease and hepatocellular carcinoma. The possibility of hepatocellular carcinoma should be taken into consideration in individuals with a solid tumor found by abdominal ultrasound examination and a highly elevated serum AFP concentration, even though the value of a highly elevated serum concentration of AFP in the detection of hepatocellular carcinoma in DGUOK deficiency is unknown (Freisinger et al. 2006).

Urine examination and evaluation for urine amino acids is advised due to the hazards associated with renal illness and its implications for dietary management (Dimmock et al. 2008a, b). Even in those who have survived for a long time, hypotonia could be a major cause of reduced life quality. Thus, it is appropriate to assess these children's gross motor abilities and development continuously while providing appropriate therapy (Dimmock et al. 2008a, b). Ursodeoxycholic acid has little proof of effectiveness in treating this disease, despite being utilized in other cholestatic disorders (Feranchak et al. 2007). According to Lee et al. (2009), the use of this medicine for treatment did not prove to be beneficial for one infant who suffered from chronic illness. According to Dimmock et al. (2008a, b), routine immunizations, including the influenza vaccine, are currently advised for all DGUOK-deficient individuals and their household contacts. This is because routine immunizations have not been linked to clinical decompensation in individuals with DGUOK deficiency, but intercurrent infections have been linked to acute decompensation in patients with this and other mtDNA deletion disorders (Table 5.1).

5.4 Pyrimidine Metabolism

5.4.1 Pyrimidine Nucleotide Biosynthesis

Pyrimidine and purine synthesis differ in that the pyrimidine ring is completely formed before being connected, whereas the ring in purine is constructed utilizing the ribose sugar (PRPP) as the base (Chandel 2021). Pyrimidine synthesis is frequently an easier process than purine synthesis due to its simpler six-membered ring base structure. Aspartate and carbamoyl phosphate are components of the ring structure. One mole of glutamine, two moles of ATP, and one mole of bicarbonate are transformed into carbamoyl phosphate in the cytoplasm. Although carbamoyl phosphate is produced in the mitochondrion from ammonia and bicarbonate, it is also used in the urea cycle. Carbamoyl phosphate synthetase I catalyzes the urea cycle process, while carbamoyl phosphate synthetase II (CPS-II) produces the pyrimidine nucleotide precursor (Chandel 2021). Aspartate transcarbamoylase, the rate-limiting step in pyrimidine biosynthesis, then acts to funnel carbamoyl phosphate into the pyrimidine nucleotide biosynthesis pathway, forming the N-carbamoylaspartate. Next, dihydroorotase cycles the pyrimidine ring. The first complete pyrimidine base, orotate, is constructed in the following steps. OMP is produced when PRPP binds to orotate and is then decarboxylated to produce UMP. While UMP is converted to UDP by an ATP-dependent NMP kinase, UTP is produced by the NDP kinase. UTP is converted to CTP from ATP and glutamine by the enzyme CTP synthase (Chandel 2021).

The de novo production of thymine nucleotides also begins with uridine nucleotides. Thymidine is exclusively found in DNA. The initial conversion involves UDP being transformed into dUDP, a process catalyzed by Ribonucleotide reductase (RR). This conversion eliminates the 2′-hydroxyl group, resulting in the formation of the deoxynucleotide. Subsequently, dUMP, a precursor to the development of dTMP, is created from dUDP. UTP diphosphatase quickly converts any phosphorylated dUDP to dUTP into dUMP, lowering the amount of dUTP available and avoiding the incorrect incorporation of uracil into DNA. 5-methyl uracil serves as the nitrogen base for thymine. Thymidylate synthase employs N5, N10-methylene THF as a methyl provider to facilitate the methylation process of dUMP, resulting in the production of dTMP. During this process, N5,N10-methylene THF is transformed into dihydrofolate (DHF).

Dihydrofolate reductase (DHFR) utilizes NADPH to convert DHF back into THF, thus replenishing the necessary cofactor for thymidylate synthase to continue its catalytic activity. The enzyme serine hydroxymethyl transferase then uses serine as a substrate to convert THF to N5, N10-methylene THF. DHFR's significance in thymidine nucleotide biosynthesis renders it an attractive target for cancer therapy due to the heightened demand for thymidine nucleotides in rapidly dividing cancer cells, essential for DNA synthesis (Chandel 2021).

Eventually, damaged DNA synthesis and cell death result from an inability of the cell to regenerate THF. Therefore, inhibitors of thymidylate synthase, such 5-fluorouracil, can be used therapeutically to target rapidly reproducing cancer

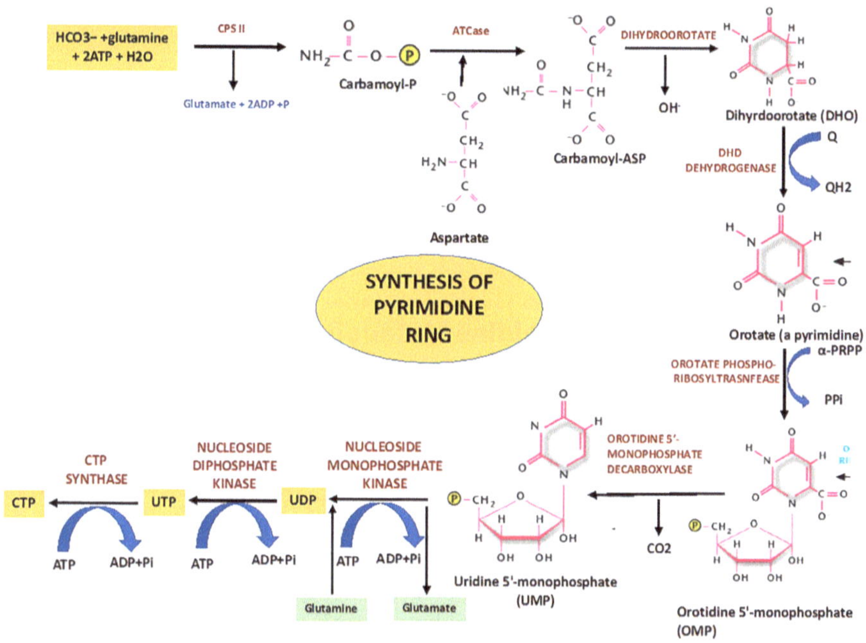

Fig. 5.2 Pyrimidine synthesis pathway. Synthesis of pyrimidine nucleotide begins with the carbamoyl phosphate and aspartate that leads to the generation of CTP

cells. 5-fluorouracil is metabolically transformed to 5-fluorodeoxyuridine monophosphate, which is then irreversibly linked to thymidylate synthase. Folate analogs, including methotrexate, reversibly block the enzyme DHFR, reducing the availability of THF to reduce purine production and methylation of dUMP to dTMP (Chandel 2021). Animals primarily regulate pyrimidine synthesis in the first phase, which is catalyzed by CPS-II. ATP activates it, and UDP and UTP limit feedback. Additional allosteric activation occurs via PRPP, a substrate of UMP synthase. Because PRPP is also a substrate in the initial stage of purine biosynthesis, it functions as a feed-forward activator to assist in coordinating the creation of pyrimidines and purines. UMP and, to a lesser extent, CMP competitively block orotidine 50-phosphate decarboxylase, another enzyme that is regulated. Lastly, CTP feedback inhibits CTP synthase (Chandel 2021) (Fig. 5.2).

5.4.2 Catabolism of Pyrimidine Nucleotides

The pyrimidine ring is degradable, in contrast to the purine ring. Water-soluble pyrimidine bases can degrade the pyrimidine ring into metabolites that can enter metabolic pathways like the TCA cycle. Purine salvage pathways are therefore more important than pyrimidine salvage pathways. Dihydrouracil dehydrogenase catalyzes the conversion of thymine and uracil into ureidoisobutyrate and ureidopropionate, respectively. Ureidopropionase catalyzes the conversion of ureidopropionate and ureidoisobutyrate into β-alanine or β-aminoisobutyrate, NH_3, and CO_2. Following conversion to acetyl-CoA, β-alanine is utilized in the production of fatty acids. Furthermore, β-aminoisobutyrate is transformed into methylmalonyl-CoA, which is subsequently utilized in the TCA cycle, and succinyl-CoA. Additionally, excretion of β-alanine and β-aminoisobutyrate is possible (Chandel 2021).

5.5 Disorders of Pyrimidine Metabolism

5.5.1 UMP Synthase Deficiency

Uridine monophosphate (UMP) synthase is a dual-function enzyme essential for synthesizing pyrimidines. In its first reaction, it catalyzes the conversion of orotic acid to orotidine monophosphate (OMP) via orotate phosphoribosyltransferase (OPRT). In its second reaction, it decarboxylates OMP into UMP through orotidine decarboxylase (ODC). When there's a defect in this enzyme, it causes a cascade of effects. Firstly, there's an excessive buildup of orotic acid and a shortage of pyrimidine nucleotides. This overproduction is due to reduced feedback inhibition of cytosolic carbamoyl phosphate synthetase II by pyrimidine nucleotides. Secondly, the lack of pyrimidine nucleotides hampers cell division, leading to conditions like megaloblastic anemia and growth delays.

5.5.1.1 Treatment

Administration of uridine is beneficial for the enzyme defect as it is involved in the conversion of the uridine kinase into UMP.

5.5.2 Dihydropyrimidine Dehydrogenase (DPD) Deficiency

DPD catalyzes the degradation of uracil and thymine into dihydrouracil and dihydrothymine, respectively. Its deficiency causes the accumulation of uracil (Berger et al. 1984). It is unknown why some pediatric patients develop a substantial DPD deficiency while others do not. Common symptoms of DPD deficiency include delayed cognitive and motor development, along with convulsive disorders (Chen et al. 2014). Some patients may also experience growth retardation, microcephaly, dysmorphia, autism, hypotonia, and ocular abnormalities, although these are less frequent (Chen et al. 2014; Van Kuilenburg et al. 2004). Moreover, individuals with DPD deficiency have a significantly diminished ability to break down the chemotherapy drug 5-fluorouracil, which can lead to severe and potentially fatal multiorgan toxicity (Van Kuilenburg et al. 2004).

5.5.2.1 Treatment

For pediatric patients, there is no available treatment. The symptoms often don't change, although some cases have resulted in a more severely damaged child dying in their early years of life.

5.5.3 Dihydropyrimidinase (DHP) Deficiency

Dihydrouracil and dihydrothymine are cleaved into ureidoisobutyrate and ureidoipropionate, respectively, by dihydropyrimidinase (DHP). As a result, significant levels of dihydrouracil and dihydrothymine, which are typically detected in trace amounts, are eliminated through the urine (Van Gennip et al. 1997). Thymine and uracil excretion are also somewhat elevated. Reduced amounts of the neurotransmitter alanine may be involved, although the causes and mechanisms of the symptoms remain unknown, just like with DPD deficiency. There have also been reports of increased sensitivity to 5-fluorouracil, which can result in severe toxicity (van Kuilenburg et al. 2003).

5.5.3.1 Treatment

The prognosis appears uncertain, and there is no treatment.

5.5.4 β-Ureidopropionase Deficiency

In vitro H-NMR spectroscopy of urine revealed elevated levels of ureidoisobutyric acid (also known as N-carbamylaminoisobutyric acid) and ureidopropionic acid (also called N-carbamylalanine) in a female infant born to consanguineous parents who presented with muscle hypotonia, severe developmental delay, and dystonic movements (Assmann 1998). These findings led to the identification of ureidopropionase deficiency, also known as β-alanine synthase deficiency, in the liver (Van Kuilenburg et al. 2004).

5.5.5 Pyrimidine 5′-Nucleotidase Deficiency

This erythrocyte-specific deficiency causes pyrimidine nucleotide buildup, which in turn causes chronic hemolytic anemia and basophilic stippling (Valentine et al. 1974). It is still unclear how the elevated pyrimidine nucleotides produce hemolysis.

5.5.5.1 Treatment

Blood transfusions may be required on an intermittent basis to treat episodes of severe anemia brought on by fever, infections, pregnancy, parvovirus B19 infection, accidents, and surgeries. Patients with this illness are likely to require blood transfusions in the long run, and it is advisable to transfuse blood that is matched for all Rhesus and Kell antigens. As it has been demonstrated for other chronic hemolytic disorders, especially after splenectomy (Chou and De Loughery 2001), the prolonged hemolysis is likely to cause a prothrombotic state, hence prophylactic antivenous thromboembolism seems reasonable when patients are bedridden. Although splenomegaly is frequently encountered, splenectomy has been documented in a number of individuals whose cases were published in the 1970s. However, the procedure has generally been found to be of little value, with the best result being a temporary improvement in the blood count (Paglia and Valentine 1980). Nonetheless, a small number of individuals are said to have seen notable clinical improvement after splenectomy (McMahon et al. 1981). Folic acid supplements could be beneficial, especially in cases of inadequate diet or malnourishment. The patient should also be informed about the heightened risk of gallstones and should be properly evaluated and should any suspicious signs appear.

5.5.6 Cytosolic 5′-Nucleotidase Superactivity

A condition involving developmental delay, recurrent infections, seizures, growth retardation, ataxia, autistic characteristics, and hypouricosuria has been reported in four unrelated children (Page et al. 1997). Research conducted on the fibroblasts of the patients revealed 6- to 20-fold increases in the activity of cytosolic 5-nucleotidase, determined by using a purine (AMP) or pyrimidine (UMP) as the substrate. In light of the potential for this heightened catabolism to result in pyrimidine nucleotide deficit, uridine at a daily dose of 1 g/kg was administered to the patients. Notable improvements in development were observed, along with a reduction in the incidence of infections and seizures.

5.5.7 Thymidine Phosphorylase Deficiency

Thymidine phosphorylase deficiency has been observed in individuals with mitochondrial neurogastrointestinal encephalomyopathy (MNGIE), an autosomal recessive disorder linked to numerous deletions of skeletal muscle mitochondrial DNA, due to a variety of mutations (Nishino et al. 2000). Thymidine builds up significantly as a result of the enzyme deficit, which most likely causes an imbalance in the nucleotides within the mitochondria and compromises the replication of mitochondrial DNA.

5.5.8 Dihydroorotate Dehydrogenase (DHO) Deficiency

Dihydroorotate (DHO) is converted to OA by Dihydroorotate dehydrogenase (DHODH). Miller syndrome, which is characterized by severe micrognathia, cleft lip, eyelid colobomas, supernumerary nipples, and developmental anomalies of the ulnar ray, is brought on by DHODH deficiency (Ng et al. 2010). Urine orotic acid and plasma dihydroorotate are examples of diagnostic indicators.

5.5.9 Cytidine Deaminase (CDA) Deficiency

According to Demontis et al. (1998), CDA is an enzyme of the pyrimidine salvage pathway that catalyzes the hydrolytic deamination of cytidine and deoxycytidine to uridine and deoxyuridine, respectively. Previous research demonstrated that the disequilibrium in the pyrimidine pool brought on by CDA deficiency causes a reduction in the basal activity of poly(ADP-ribose) polymerase 1 (PARP-1), an enzyme with multiple functions involved in various cellular processes. This, in turn, contributes to the development of genetic instability in CDA-deficient cells (Gemble et al. 2015).

5.5.10 Thymidine Kinase 2 Deficiency

Saada et al. (2001) reported the discovery of two mutations within the gene encoding thymidine kinase-2, the mitochondrial variant of the thymidine salvage enzyme, in four distinct patients presenting with severe, isolated myopathy accompanied by depletion of mitochondrial DNA in muscle tissue. Similar to the inadequacies of thymidine phosphorylase and deoxyguanosine kinase, the problem most likely results in an imbalance of mitochondrial nucleotides, which impedes mitochondrial DNA replication.

5.6 Diagnosis of Inborn Errors of Purine and Pyrimidine Metabolism

Pyrimidine metabolism disorders have a wide range of phenotypic manifestations and little clinical understanding; they are frequently misdiagnosed or go untreated. The clinical manifestations linked to deficits in ureidopropionase (UP), dihydropyrimidinase (DPH), and dihydropyrimidine dehydrogenase (DPD) have been seen to vary significantly. All three of the pyrimidine catabolic faults have been reported in asymptomatic individuals, including those who belong to the same family as the affected patients. It is plausible that ascertainment bias led to the detection of these deficits. However, overlapping clinical characteristics across all three abnormalities point to the possibility that the neuropathologies are caused by the route being blocked. In this instance, the related clinical symptoms might be influenced by the existence of a contemporaneous secondary polymorphic gene variant. Currently, there aren't any specific therapy is known.

According to Duran et al. (1997), screening for these illnesses shouldn't be limited to a select set of patients with clear diagnoses. Some of these disorders have the potential to be fatal and it is essential to identify these flaws as soon as possible

(Simmonds and Van Gennip 2003). There are no easily detectable particular end products that can be used to screen for pyrimidine diseases, unlike purine metabolism, which includes uric acid as an end product that can be used to assess overall activity. Urine testing specifically is the most effective method for diagnosing the most known hereditary abnormalities in pyrimidine metabolism. Although pyrimidine buildup in blood and cerebrospinal fluid is less noticeable because of their high renal clearance, these bodily fluids can nevertheless be employed for screening (Duran et al. 1997).

Prior to this, only a limited number of specialized laboratories were able to detect and diagnose purine and pyrimidine diseases. While orotic acid, orotidine, and dihydropyrimidines are weakly identified by Gas chromatography-mass spectrometry (GC-MS), pyrimidine degradation deficiencies have long been identifiable by GC-MS analysis of urinary trimethysilylated organic acid extracts (Wadman et al. 1984). Nowadays, tandem mass spectrometry (MS/MS), high-performance liquid chromatography (HPLC) in conjunction with UV/diode-array detection, or high-performance nuclear magnetic resonance (H-NMR) are frequently used techniques for the identification of pyrimidine metabolites (Hartmann et al. 2006). Some nucleotide metabolites have been added to the screening panel by neonatal screening laboratories through the use of rapid multiple metabolite HPLC-MS/MS screening technologies. As a result, these abnormalities are now being identified more frequently (van Werkhoven et al. 2013). Many pyrimidine illnesses can have their deficiencies confirmed using enzyme assays with erythrocytes, leukocytes, or fibroblasts, as well as through mutational investigation of the corresponding genes. Molecular genetic approaches are expected to become more prominent in enabling prompt diagnosis with the introduction of more cost-effective and efficient next-generation sequencing techniques, especially in situations where access to advanced biochemical analysis is not easily available. Surprisingly wide gaps still exist in our understanding of pyrimidine metabolism disorders; for instance, no errors in Carbamoyl-phosphate synthetase 2, Aspartate transcarbamoylase, and Dihydroorotase (CAD) function have been reported. As was the case for the pyrimidine ailment DHODH deficiency, gene discovery techniques employing whole-exome or whole-genome sequencing are likely to identify novel disease genes involved in pyrimidine metabolism, sometimes with unexpected phenotypes (Ng et al. 2010). The creation of a "pyrimidine disorder gene panel" will significantly increase the speed at which known illness genes may be screened, hence improving the accuracy of genetic diagnosis.

Due to the fact that nucleotides are normally found inside cells, defects in the purine and pyrimidine pathways are first screened for by detecting the related nucleosides or bases. For example, adenylosuccinate lyase deficiency results in intracellular accumulation of succinyl-AMP and SAICA-ribotide (nucleotides), but these eventually break down to form the nucleosides succinyl-adenosine and SAICA-riboside, which then seep into extracellular fluids and can be detected in urine. Urine excretion profiles can therefore be used to diagnose the most known hereditary abnormalities of purine and pyrimidine metabolism. Urine excretion is high (except for uric acid), making it the preferable sample for screening for these abnormalities. As a result, screening for purine and pyrimidine diseases rarely uses plasma.

The nucleotide profile of erythrocyte extracts can be useful for the diagnosis and therapeutic monitoring of certain purine and pyrimidine diseases. For instance, blood transfusions, a typical treatment for SCID, might temporarily remove the metabolite deoxyadenosine, which can confuse the diagnosis of ADA-SCID, which is defined as a buildup of this metabolite in plasma and urine. Nevertheless, deoxy-ATP is highly persistent and accumulates in ADA-deficient erythrocytes, offering a distinct diagnostic characteristic (Simmonds and Van Gennip 2003). In particular, erythrocyte nucleotides are the only way to diagnose certain nucleotide disorders when urine and plasma metabolite patterns are normal. For example, a deficiency in inosine triphosphate pyrophosphohydrolase (ITPA) causes an abnormal accumulation of the purine nucleotide ITP in erythrocytes (Sumi et al. 2002).

Serum and urine uric acid concentrations that are abnormally high or low can be used to diagnose a variety of purine abnormalities, including hereditary renal hypouricemia, PRS-1 super-activity, xanthine oxidase (XOD), purine nucleoside phosphorylase (PNP) deficits, HPRT, and familial juvenile hyperuricemic nephropathy (FJHN). Urate can be normal, hence the uric acid assay is not a reliable way to screen for PRS-1 deficiency (de Brouwer et al. 2010). Many purine and pyrimidine illnesses can have their deficiencies confirmed using enzyme assays with erythrocytes, leukocytes, or fibroblasts. An exercise test using the forearm muscles is typically used to check for myeloadenylate deaminase (MADA) deficiency. Since the majority of patients have the c.34C > T mutation, molecular investigation of the AMPD1 gene is helpful for diagnosis (Morisaki et al. 1992).

5.7 Conclusions

The study of nucleic acid metabolism and its disorders represents a critical area of research with profound implications for human health. A deep understanding of the mechanism involved in the nucleic acid metabolism will provide us with the opportunity to develop potential therapeutic interventions for addressing associated disorders. Looking ahead, future research holds promising prospects for the development of novel diagnostic tools and targeted therapies aimed at mitigating the impact of nucleic acid metabolism disorders. Advances in technologies such as gene editing, RNA interference, and nucleic acid-based therapeutics offer unprecedented opportunities for precise intervention at the molecular level, potentially transforming the landscape of treatment for these conditions. However, despite significant progress, there are notable gaps in our understanding of nucleic acid metabolism and its dysregulation in disease states. Further elucidating the intricate pathways and regulatory mechanisms involved, as well as uncovering the full spectrum of genetic and environmental factors contributing to these disorders, will be essential for realizing the full potential of personalized medicine approaches. Moreover, addressing the ethical, legal, and social implications of emerging technologies in this field will be paramount to ensuring equitable access to innovative treatments while safeguarding patient rights and privacy.

References

Assmann B (1998) Presumptive ureidopropionase deficiency as a new defect in pyrimidine catabolism found with in vitro H-NMR spectroscopy. J Inher Metab Dis 21(2):1

Barratt TM, Simmonds HA, Cameron JS et al (1979) Complete deficiency of adenine phosphoribosyltransferase: a third case presenting as renal stones in a young child. Arch Dis Child 54(1):25

Barreto JN, McCullough KB, Ice LL et al (2014) Antineoplastic agents and the associated myelosuppressive effects: a review. J Pharm Pract 27(5):440–446

Becker MA, Raivio KO, Bakay BOHDAN et al (1980) Variant human phosphoribosylpyrophosphate synthetase altered in regulatory and catalytic functions. J Clin Invest 65(1):109–120

Becker MA, Puig JG, Mateos FA et al (1988) Inherited superactivity of phosphoribosylpyrophosphate synthetase: association of uric acid overproduction and sensorineural deafness. Am J Med 85(3):383–390

Becker MA, Smith PR, Taylor W et al (1995) The genetic and functional basis of purine nucleotide feedback-resistant phosphoribosylpyrophosphate synthetase superactivity. J Clin Invest 96(5):2133–2141

Berger R, Stoker-de Vries SA, Wadman SK et al (1984) Dihydropyrimidine dehydrogenase deficiency leading to thymine-uraciluria. An inborn error of pyrimidine metabolism. Clin Chim Acta 141(2–3):227–234

Bhasin B, Stiburkova B, De Castro-Pretelt M et al (2014) Hereditary renal hypouricemia: a new role for allopurinol? Am J Med 127(1):e3–e4. epub

Booth RA, Ansari MT, Loit E et al (2011) Assessment of thiopurine S-methyltransferase activity in patients prescribed thiopurines: a systematic review. Ann Intern Med 154(12):814–823

Boria I, Garelli E, Gazda HT et al (2010) The ribosomal basis of diamond- blackfan anemia: mutation and database update. Hum Mutat 34:1269–1279

Buchanan JM (1982) Covalent reaction of substrates and antimetabolites with formylglycinamide ribonucleotide amidotransferase. In: Methods in enzymology, vol 87. Academic Press, pp 76–84

Carey PJ (2003) Drug-induced myelosuppression: diagnosis and management. Drug Saf 26(10):691–706

Carpenter TO, Lebowitz RL, Nelson D et al (1986) Hereditary xanthinuria presenting in infancy with nephrolithiasis. J Pediatr 109(2):307–309

Carpenter PA, Ziegler JB, Vowels MR (1996) Late diagnosis and correction of purine nucleoside phosphorylase deficiency with allogeneic bone marrow transplantation. Bon Marrow Transpl 17(1):121–124

Cartier P (1974) Une nouvelle maladiemetabolique: le deficit en adenine-phosphoribosyltransferase avec lithiasa de 2, 8-dihydroxyadenine. CR Acad Sci 279:883–886

Chalmers RA, Watts RWE, Bitensky L et al (1969) Microscopic studies on crystals in skeletal muscle from two cases of xanthinuria. J Pathol 99(1):45–56

Chandel NS (2021) Nucleotide metabolism. CSH Perspect Biol 13(7):a040592

Chen BC, MohdRawi R, Meinsma R et al (2014) Dihydropyrimidine dehydrogenase deficiency in two Malaysian siblings with abnormal MRI findings. Mol Syndromol 5(6):299–303

Chou R, De Loughery TG (2001) Recurrent thromboembolic disease following splenectomy for pyruvate kinase deficiency. Am J Hematol 67(3):197–199

Chu SY, Cashion P, Jiang M (1989) Purine nucleoside phosphorylase in erythrocytes: determination of optimum reaction conditions. Clin Biochem 22(1):3–9

Dabrowski E, Smathers SA, Ralstrom CS et al (2005) Botulinum toxin as a novel treatment for self-mutilation in Lesch—Nyhan syndrome. Dev Med Child Neurol 47(9):636–639

Dalbeth N, House ME, Aati O et al (2015) Urate crystal deposition in asymptomatic hyperuricaemia and symptomatic gout: a dual energy CT study. Ann Rheum Dis 74(5):908–911
De Boer NK, Van Bodegraven AA, Jharap B et al (2007) Drug insight: pharmacology and toxicity of thiopurine therapy in patients with IBD. Nat Clin Pract Gastroenterol Hepatol 4(12):686–694
de Brouwer AP, van Bokhoven H, Nabuurs SB et al (2010) PRPS1 mutations: four distinct syndromes and potential treatment. Am J Hum Genet 86(4):506–518
De Gregorio L, Nyhan WL, Serafin E et al (2000) An unexpected affected female patient in a classical Lesch–Nyhan family. Mol Genet Metab 69(3):263–268
De Miguel E, Puig JG, Castillo C et al (2012) Diagnosis of gout in patients with asymptomatic hyperuricaemia: a pilot ultrasound study. Ann Rheum Dis 71(1):157–158
Demontis S, Terao M, Brivio M et al (1998) Isolation and characterization of the gene coding for human cytidine deaminase. Biochim Biophys Acta Gene Struct Exp 1443(3):323–333
Dimmock DP, Dunn JK, Feigenbaum A et al (2008a) Abnormal neurological features predict poor survival and should preclude liver transplantation in patients with deoxyguanosine kinase deficiency. Liver Transpl 14(10):1480–1485
Dimmock DP, Zhang Q, Dionisi-Vici C et al (2008b) Clinical and molecular features of mitochondrial DNA depletion due to mutations in deoxyguanosine kinase. Hum Mutat 29(2):330–331
Dooley T, Fairbanks LD, Simmonds HA et al (1987) First trimester diagnosis of adenosine deaminase deficiency. Prenat Diagn 7(8):561–565
Ducluzeau PH, Lachaux A, Bouvier R et al (2002) Progressive reversion of clinical and molecular phenotype in a child with liver mitochondrial DNA depletion. J Hepatol 36(5):698–703
Duran M, Dorland L, Meuleman EEE et al (1997) Inherited defects of purine and pyrimidine metabolism: laboratory methods for diagnosis. J Inherit Metab Dis 20:227–236
Elson DAVID (1965) Metabolism of nucleic acids (macromolecular DNA and RNA). Annu Rev Biochem 34(1):449–486
Engel AG (1986) Carnitine deficiency syndromes and lipid storage myopathy. Myology:1663–1696
Feranchak AP, Sokol RJ, Suchy F (2007) Medical and nutritional management of cholestasis in infants and children. Liver Dis Child 3:190–231
Fishbein WN, Armbrustmacher VW, Griffin JL (1978) Myoadenylate deaminase deficiency: a new disease of muscle. Science 200(4341):545–548
Freisinger P, Fütterer N, Lankes E et al (2006) Hepatocerebral mitochondrial DNA depletion syndrome caused by deoxyguanosine kinase (DGUOK) mutations. Arch Neurol 63(8):1129–1134
Fu R, Ceballos-Picot I, Torres RJ, Group L-NDIS et al (2014) Genotype- phenotype correlations in neurogenetics: Lesch-Nyhan disease as a model disorder. Brain 137(5):1282–1303
Fu R, Chen CJ, Jinnah HA (2014) Genotypic and phenotypic spectrum in attenuated variants of Lesch–Nyhan disease. Mol Genet Metab 112(4):280–285
Gemble S, Ahuja A, Buhagiar-Labarchède G et al (2015) Pyrimidine pool disequilibrium induced by a cytidine deaminase deficiency inhibits PARP-1 activity, leading to the under replication of DNA. PLoS Genet 11(7):e1005384
Georges J, Berghe V (1984) An infantile autistic syndrome characterised by the presence of succinylpurines in body fluids. Lancet 324(8411):1058–1061
Giblett ER (1985) ADA and PNP deficiencies: how it all began. Ann New York Acad Sci 451(1):1–8
Giblett E, Anderson J, Cohen F et al (1972) Adenosine-deaminase deficiency in two patients with severely impaired cellular immunity. Lancet 300(7786):1067–1069
Glick N (2006) Dramatic reduction in self-injury in Lesch–Nyhan disease following S-adenosylmethionine administration. J Inherit Metab Dis 29:687
Hartman SC, Buchanan JM (1959) Nucleic acids, purines, pyrimidines (nucleotide synthesis). Annu Rev Biochem 28(1):365–410
Hartmann S, Okun JG, Schmidt C et al (2006) Comprehensive detection of disorders of purine and pyrimidine metabolism by HPLC with electrospray ionization tandem mass spectrometry. Clin Chem 52(6):1127–1137
Hershfield MS (1995) PEG-ADA replacement therapy for adenosine deaminase deficiency: an update after 8.5 years. Clin Immunol Immunopathol 76(3):S228–S232

Hirschhorn R, Ellenbogen A, Tzall S (1992) Five missense mutations at the adenosine deaminase locus (ADA) detected by altered restriction fragments and their frequency in ADA–patients with severe combined immunodeficiency (ADA–SCID). Am J Med Genet 42(2):201–207

Hitchings GH, Elion GB (1969) Role of antimetabolites in immunosuppression and transplantation. Acc Chem Res 2(7):202–209

Kahan B, Held KR, DeMars R (1974) The locus for human adenine phosphoribosyltransferase on chromosome no. 16. Genetics 78(4):1143–1156

Kamatani N, Hakoda M, Otsuka S et al (1992) Only three mutations account for almost all defective alleles causing adenine phosphoribosyltransferase deficiency in Japanese patients. J Clin Invest 90(1):130–135

Kamatani N, Jinnah HA, Hennekam RC et al (2021) Purine and pyrimidine metabolism. In: Emery and Rimoin's principles and practice of medical genetics and genomics. Academic Press, pp 183–234

Kaufman JM, Greene ML, Seegmiller JE (1968) Urine uric acid to creatinine ratio—a screening test for inberited disorders of purine metabolism: Phosphoribosyltransferase (PRT) deficiency in X-linked cerebral palsy and in a variant of gout. J Pediatr 73(4):583–592

Kelley WN, Greene ML, Rosenbloom FM et al (1969) Hypoxanthine-guanine phosphoribosyltransferase deficiency in gout. Ann Intern Med 70(1):155–206

Kimura K, Hosoya T, Uchida S et al (2018) Febuxostat therapy for patients with stage 3 CKD and asymptomatic hyperuricemia: a randomized trial. Am J Kidney Dis 72(6):798–810

Kogut MD, Donnell GN, Nyhan WL et al (1970) Disorder of purine metabolism due to partial deficiency of hypoxanthine-guanine phosphoribosyltransferase: a study of a family. Am J Med 48(2):148–161

Kranen S, Keough D, Gordon RB et al (1985) Xanthine-containing calculi during allopurinol therapy. J Urol 133(4):658–659

Lecky BRF (1983) Failure of D-ribose in myoadenylate deaminase deficiency. Lancet 321(8317):193

Lee NC, Dimmock D, Hwu WL et al (2009) Simultaneous detection of mitochondrial DNA depletion and single-exon deletion in the deoxyguanosine gene using array-based comparative genomic hybridisation. Arch Dis Child 1:55–58

Lennard L (2014) Implementation of TPMT testing. Br J ClinPharmacol 77(4):704–714

Lennard L, Lilleyman JS (1996) Individualizing therapy with 6-mercaptopurine and 6-thioguanine related to the thiopurine methyltransferase genetic polymorphism. Ther Drug Monit 18(4):328–334

Mandel H, Szargel R, Labay V et al (2001) The deoxyguanosine kinase gene is mutated in individuals with depleted hepatocerebral mitochondrial DNA. Nat Genet 29(3):337–341

Markert ML (1991) Purine nucleoside phosphorylase deficiency. Immunodef Rev 3(1):45–81

Markert ML (1994) Molecular basis of adenosine deaminase deficiency. Immunodeficiency 5(2):141–157

Markert ML, Hershfield MS, Schiff RI et al (1987) Adenosine deaminase and purine nucleoside phosphorylase deficiencies: evaluation of therapeutic interventions in eight patients. J Clin Immunol 7:389–399

McMahon JN, Lieberman JE, Gordon-Smith EC et al (1981) Hereditary haemolytic anaemia due to red cell pyrimidine 5 '-nucleotidase deficiency in two Irish families with a note on the benefit of splenectomy. Clin Lab Haematol 3(1):27–34

Messenger LJ, Zalkin H (1979) Glutamine phosphoribosylpyrophosphateamidotransferase from *Escherichia coli*. Purification and properties. J Biol Chem 254(9):3382–3392

Morisaki T, Gross M, Morisaki H et al (1992) Molecular basis of AMP deaminase deficiency in skeletal muscle. Proc Natl Acad Sci 89(14):6457–6461

Mousson de Camaret B, Taanman JW, Padet S et al (2007) Kinetic properties of mutant deoxyguanosine kinase in a case of reversible hepatic mtDNA depletion. Biochem J 402(2):377–385

Ng SB, Buckingham KJ, Lee C et al (2010) Exome sequencing identifies the cause of a mendelian disorder. Nat Genet 42(1):30–35

Nishino I, Spinazzola A, Papadimitriou A et al (2000) Mitochondrial neurogastrointestinalencephalomyopathy: an autosomal recessive disorder due to thymidine phosphorylase mutations. Ann Neurol: Official J Am Neurolog Assoc Child Neurol Soc 47(6):792–800

Nyhan WL, James JA, Teberg AJ et al (1969) A new disorder of purine metabolism with behavioral manifestations. J Pediatr 74(1):20–27

Ohdoi C, Nyhan WL, Kuhara T (2003) Chemical diagnosis of Lesch–Nyhan syndrome using gas chromatography-mass spectrometry detection. J Chromatogr B 792(1):123-130

Page T, Bakay B, Nissinen E et al (1981) Hypoxanthine-guanine phosphoribosyltransferase variants: correlation of clinical phenotype with enzyme activity. J Inherit Metab Dis 4:203–206

Page T, Yu A, Fontanesi J, Nyhan WL (1997) Developmental disorder associated with increased cellular nucleotidase activity. Proc Natl Acad Sci USA 94(21):11601–11606

Paglia DE, Valentine WN (1980) Hereditary and acquired defects in the pyrimidine nucleotidase of human erythrocytes. Curr Top Hematol 3:75–109

Patten B (1982) Beneficial effect of D-ribose in patient with myoadenylate deaminase deficiency. Lancet 319(8280):1071

Peters GJ, Carrey EA, Šebesta I (2006) Purine and pyrimidine metabolism: a firm basis for a transformed society. Nucleos Nucleot Nucleic Acids 25(9–11):971–974

Pralong E, Pollo C, Coubes P et al (2005) Electrophysiological characteristics of limbic and motor globus pallidus internus (GPI) neurons in two cases of Lesch–Nyhan syndrome. Neurophysiol Clin/Clin Neurophysiol 35(5–6):168-173

Puig JG, Torres RJ, Mateos FA et al (2001) The spectrum of hypoxanthine-guanine phosphoribosyltransferase (HPRT) deficiency: clinical experience based on 22 patients from 18 Spanish families. Medicine 80(2):102–112

Reiter S, Simmonds HA, Zöllner N et al (1990) Demonstration of a combined deficiency of xanthine oxidase and aldehyde oxidase in xanthinuric patients not forming oxipurinol. Clin Chim Acta 187(3):221–234

Roessler BJ, Nosal JM, Smith PR et al (1993) Human X-linked phosphoribosylpyrophosphate synthetase superactivity is associated with distinct point mutations in the PRPS1 gene. J Biol Chem 268(35):26476–26481

Saada A, Shaag A, Mandel H et al (2001) Mutant mitochondrial thymidine kinase in mitochondrial DNA depletion myopathy. Nat Genet 29(3):342–344

Salerno C, Celli M, Finocchiaro R et al (1998) Effect of D-ribose administration to a patient with inherited deficit of adenylosuccinase. In: Purine and pyrimidine metabolism in man IX. Springer US, Boston, pp 177–180

Schendel FJ, Mueller E, Stubbe J et al (1989) Formylglycinamide ribonucleotide synthetase from Escherichia coli: cloning, sequencing, overproduction, isolation, and characterization. Biochemistry 28(6):2459–2471

Schrimsher JL, Schendel FJ, Stubbe J et al (1986) Purification and characterization of aminoimidazole ribonucleotide synthetase from Escherichia coli. Biochemistry 25(15):4366–4371

Schwartzberg LS (2006) Neutropenia: etiology and pathogenesis. Clin Cornerstone 8:S5–S11

Seegmiller JE, Rosenbloom FM, Kelley WN (1967) Enzyme defect associated with a sex-linked human neurological disorder and excessive purine synthesis. Science 155(3770):1682–1684

Simmonds HA, Van Gennip AH (2003) Purine and pyrimidine disorders. In: Physician's guide to the laboratory diagnosis of metabolic diseases. Springer, Berlin/Heidelberg, pp 445–466

Simmonds HA, Webster DR, Wilson J et al (1984) Evidence of a new syndrome involving hereditary uric acid overproduction, neurological complications and deafness. In: Purine metabolism in Man-IV: Part A: Clinical and therapeutic aspects; regulatory mechanisms. Springer US, Boston, pp 97–102

Smith JL (1998) Glutamine PRPP amidotransferase: snapshots of an enzyme in action. Curr Opin Struct Biol 8(6):686–694

Smith LH Jr (1973) Pyrimidine metabolism in man. N Engl J Med 288(15):764–771

Stone RL, Aimi J, Barshop BA et al (1992) A mutation in adenylosuccinate lyase associated with mental retardation and autistic features. Nat Genet 1(1):59–63

Sumi S, Marinaki AM, Arenas M et al (2002) Genetic basis of inosine triphosphate pyrophosphohydrolase deficiency. Hum Genet 111:360–367
Taira M, Ishijima S, Kita K et al (1987) Nucleotide and deduced amino acid sequences of two distinct cDNAs for rat phosphoribosylpyrophosphate synthetase. J Biol Chem 262(31):14867–14870
Valentine WN, Fink K, Paglia DE et al (1974) Hereditary hemolytic anemia with human erythrocyte pyrimidine 5′-nucleotidase deficiency. J Clin Invest 54(4):866–879
Van Gennip AH, Abeling NGGM, Vreken P et al (1997) Inborn errors of pyrimidine degradation: clinical, biochemical and molecular aspects. J Inherit Metab Dis 20:203–213
Van Kuilenburg AB, Meinsma R, Zonnenberg BA et al (2003) Dihydropyrimidinase deficiency and severe 5-fluorouracil toxicity. Clin Cancer Res 9(12):4363–4367
Van Kuilenburg AB, Meinsma R, Beke E et al (2004) β-Ureidopropionase deficiency: an inborn error of pyrimidine degradation associated with neurological abnormalities. Hum Mol Genet 13(22):2793–2801
Van Werkhoven MA, Duley JA, McGown I et al (2013) Early diagnosis of adenylosuccinate lyase deficiency using a high-throughput screening method and a trial of oral S-adenosyl-l-methionine as a treatment method. Dev Med Child Neurol 55(11):1060–1064
Vlassov VV, Pyshnyi DV, Vorobjev PE (2009) Nucleic acids: structures, functions, and applications. In: Handbook of nucleic acids purification
Wadman SK, Duran M, Beemer FA et al (1983) Absence of hepatic molybdenum cofactor: an inborn error of metabolism leading to a combined deficiency of sulphite oxidase and xanthine dehydrogenase. J Inher Metab Dis 6:78–83
Wadman SK, Beemer FA, De Bree PK et al (1984) New defects of pyrimidine metabolism. Purine Metabol Man-IV: Part A: Clin Therap Aspects Regul Mech:109–114
Waldman B, Ansquer JC, Sullivan DR et al (2018) Effect of fenofibrate on uric acid and gout in type 2 diabetes: a post-hoc analysis of the randomised, controlled FIELD study. Lancet Diabetes Endocrinol 6(4):310–318
Weinshilboum R (1989) Thiol S-methyltransferases, I: Biochemistry. Sulfur-containing drugs and related organic compounds, Damani LA Ellis Horwood, Chichester, 2(Part A), 121ą142
Weinshilboum R, Wang L (2006) Pharmacogenomics: bench to bedside. Focus 3(3):739–441
Wennerstrand P, Dametto P, Hennig J et al (2012) Structural characteristics determine the cause of the low enzyme activity of two thiopurine S-methyltransferase allelic variants: a biophysical characterization of TPMT* 2 and TPMT* 5. Biochemistry 51(30):5912–5920
Westheimer FH (1981) Monomeric metaphosphates. Chem Rev 81(4):313–326
Zalkin H (1993) The amidotransferases. Adv Enzymol Relat Areas Mol Biol 66:203–309
Zöllner N, Reiter S, Gross M et al (1986) Myoadenylate deaminase deficiency: successful symptomatic therapy by high-dose oral administration of ribose. Klin Wochenschr 64:1281–1290

Vitamins and Their Impact on Human Health

6

Neeraj Tiwari, Neha Mehrotra, Prabhanshu Tripathi, and Sapna Sharma

Abstract

Vitamins are essential organic micronutrients that play significant role as coenzymes or enzymes in many vital processes such as normal cell function, growth, and development. These micronutrients are an integral part of human health which needs to be supplied by the diet in smaller amounts for the maintenance of the metabolic functions. The human body is unable to synthesize vitamins, either at all or even if produced it is not in adequate amounts. As such, they must be obtained through the diet. A total number of 13 vitamins such as vitamins A, C, D, E, K, and the B complex vitamins have been identified till date and many of them have been investigated for their impact on human health. The deficiency or deregulation of vitamins adversely affects the metabolism, thus implicating human health. There are several reports suggesting their therapeutic role in various diseases by participating in different biochemical pathways of the human body. In addition, they have prime therapeutic role in several diseases such as IBD, cardiovascular diseases, cancer, and neurodegenerative disorders, etc. Recently, research investigators have shown keen interest in exploring the relationship between vitamins and human health. The present chapter provides an understanding of vitamins, their structural chemistry, biological functions, and its impact on human body.

N. Tiwari · S. Sharma (✉)
Faculty of Biosciences, Institute of Biosciences and Technology, Shri RamSwaroop Memorial University, Lucknow-Deva Road, Barabanki, Uttar Pradesh, India

N. Mehrotra · P. Tripathi
FEST Division, CSIR-Indian Institute of Toxicology Research, Lucknow, Uttar Pradesh, India

R. L. Singh et al. (eds.), *Clinical Applications of Biomolecules in Disease Diagnosis*, https://doi.org/10.1007/978-981-97-4723-8_6

Keywords

Vitamin · Human health · Antioxidative activity · Anti-inflammatory potential · Micronutrients · Enzymes

6.1 Introduction

Vitamins are essential micronutrients that are required in tiny quantities for various metabolic processes and the normal functioning of human body (Akram et al. 2020). They are organic in nature and cannot be synthesized within the body and usually obtained from daily balanced diet. They are organic compounds that cannot be synthesized within the body or obtained through our daily diet or supplements (Mahmood 2014). However, there are some expectations such as "Vitamin D" which can be formed under the skin upon exposure to ultraviolet radiation (Engelsen 2010; Holick 2018). The vitamins were discovered in the early twentieth century when scientists first observed diseases like scurvy, which are caused by nutritional deficiencies of vitamin C (Semba 2012). In 1906, Sir Frederick Hopkins, a British Biochemist, demonstrated that certain foods contained accessory factors, which were distinct from proteins, carbohydrates, minerals, etc. (McDowell 2013; Combs Jr and McClung 2016). In the same year, an American biochemist named Elmer McCollum started investigating the causes of rickets, a disease that causes softening and weakening of bones in children. McCollum discovered that certain foods contained a distinctive substance that was essential for preventing rickets, and he called this substance as "fat-soluble D." He also discovered another essential nutrient, "fat-soluble B," which was later identified as vitamin D (Holick 2023). Later, the term "vitamine" was coined by a Polish-American biochemist named Casimir Funk. He combined "vita," with "amine," meaning amine of life referring to the chemical structure, to form the term "vitamine" (Semba 2013). Funk believed that amines were vital for life. However, later it was found that not all vitamins were amines or contained nitrogen, specifically vitamin A (Piro et al. 2010; Behr et al. 2020). In 1920, the final "e" was dropped from the word "vitamines," and the term "vitamins" was adopted (Maltz 2013). The identification and discovery of vitamins revolutionized the field of nutrition, and many scientists began studying the effects of vitamins on the body (Eggersdorfer et al. 2012).

There are total 13 known vitamins, which are classified into two categories: fat-soluble vitamins (A, D, E, and K) which are preserved in the human body's tissues for longer periods of time and another one is water-soluble vitamins (B-complex and C) which are easily absorbed and not stored in the body (Lykstad and Sharma 2019). The B-complex vitamins include vitamin B1 (thiamine), vitamin B2 (riboflavin), vitamin B3 (niacin), vitamin B6 (pyridoxine), vitamin B5 (pantothenic acid), vitamin B7 (biotin), vitamin B9 (folate), and vitaminB12 (cobalamin) (Schellack et al. 2019). Each vitamin has a specific role to play in maintaining human health, and a deficiency or excess of any vitamin can have adverse effects on the body (Singh et al. 2014; Godswill et al. 2020). In general, all vitamins do not contribute for any energy generation or structural purpose in the body. However, these vitamins essentially act as cofactors for several enzymes (Dewangan and Bhatia 2023).

6.2 Structure and Physio-chemical Properties

6.2.1 Vitamin A

Retinoids are compounds that possess the same qualitative biological activity for example, retinol which are derived from its parent compound, a monocyclic consists of five carbon–carbon double bonds and additional functional group located at the end of the acyclic portion (Fig. 6.1) (Maurya et al. 2022). These compounds are collectively known as Vitamin A, and are referred to as retinoids due to their resemblance to retinol (Carazo et al. 2021). There are three active forms of retinoids namely retinol, retinal, and retinoic acid (Polcz and Barbul 2019). Vitamin A shows high solubility in organic solvents such as oils, fats, and lipids, but soluble in alcohols like ethanol, and in solubility in water (Ravisankar et al. 2015). Vitamin A is important for healthy vision and maintaining healthy skin, teeth, and bones (Gilbert 2013). Vitamin A can be found in animal products such as liver, fish, and eggs, as well as in plant sources such as sweet potatoes, carrots, and spinach (Gurmu et al. 2014).

6.2.2 Vitamin D

Vitamin D comprises of group of sterols, the first category is ergocalciferol, which are commonly known as vitamin D_2 and another ischolecalciferol, which are popularly known as vitamin D_3. Both of these forms are soluble in fat/fat solvents (Jäpelt and Jakobsen 2013). Vitamin D_2 and D_3 differ structurally only in the C-17 side chain, in which vitamin D_2 has a double bond and an additional methyl group (Fig. 6.2) (Bikle 2014). Since vitamin D_2 and D_3 molecules are biologically inactive, these are converted into dihydroxylated active forms via hydroxylation at the similar carbon atoms. In liver, the vitamin D gets metabolized into 25-hydroxy vitamin D which is the principal circulating form of vitamin D in blood. On reaching

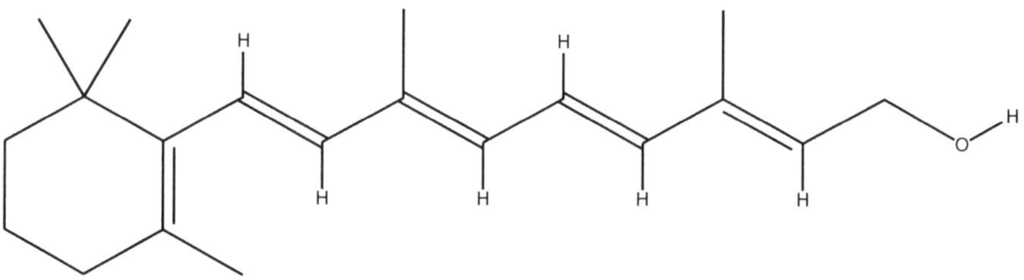

Fig. 6.1 Structure of Vitamin A

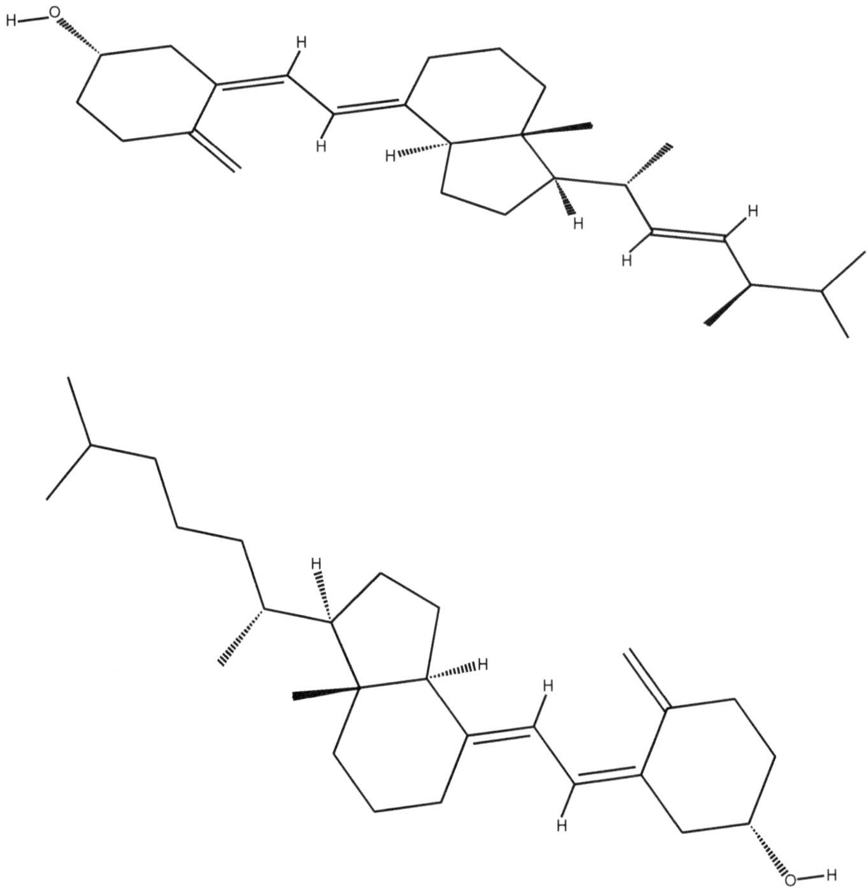

Fig. 6.2 Structure of Vitamin D_2 and Vitamin D_3

kidney, this 25(OH)D gets further metabolized to new intermediate compound 1, 25-dihydroxy vitamin D (1,25$(OH)_2$D) which is the principal hormonal form of vitamin D (Ross et al. 2011; Bikle 2014). Vitamin D is important for bone health and the absorption of calcium. Vitamin D can be obtained through exposure to sunlight, fortified foods such as milk and cereal, and fatty fish such as salmon and tuna (LoPiccolo and Lim 2010; McCabe et al. 2012; Spiro and Buttriss 2014).

6.2.3 Vitamin E

The existence of vitamin E in nature was first noted by Evans and Bishop in1923 (Bell 1987; Wolf 2005). Vitamin E denotes tocopherols and tocotrienols which are a group of structurally similar compounds. In total, eight compounds have been isolated which are four tocopherols popularly known as α, β, γ, and

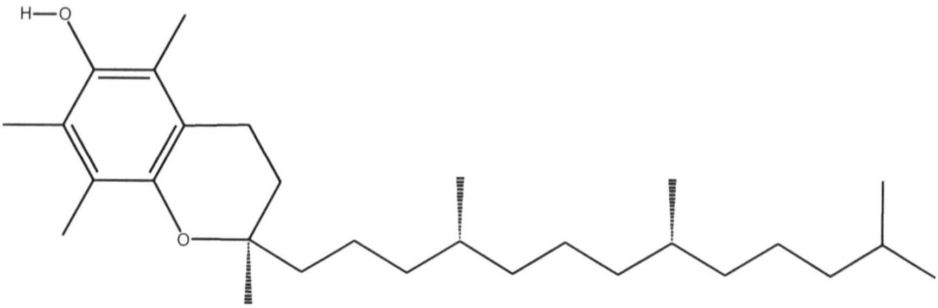

Fig. 6.3 Structure of Vitamin E

δ-tocopherols have been recognized (Wolf 2005; Azzi 2019). Although, they are structurally similar but differ only in methyl group position which is attached to aromatic ring.

Among them, β and γ-tocopherols structure consists of two methyl groups at different positions whereas, α-tocopherols have three methyl groups and δ-tocopherols have one methyl group (Fig. 6.3) (Falk and Munné-Bosch 2010). The tocopherols are oils possessing light yellow color when kept at room temperature. Being insoluble in water but they can easily be dissolved in vegetable oils and nonpolar solvents (Shahidi and De Camargo 2016). Since all tocopherols comprise of hydroquinone and phenolic hydrogen at the C-6 position of the chromanol nucleus. These can take an unpaired electron inside the resonance formation of the ring that transforms into a semi stable chromanoxyl radical before eventually converting into tocopheryl quinone which acts as an "antioxidant." However, the antioxidant capacity of vitamin E gets vanished in the presence of several oxidants or oxidizing agents (Azzi and Stocker 2000; Lampi 2011). Vitamin E is an antioxidant that protects cells from damage and helps to maintain a healthy immune system. Sources of vitamin E include nuts, seeds, vegetable oils, and leafy green vegetables (Rizvi et al. 2014).

6.2.4 Vitamin K

Vitamin K, a fat-soluble vitamin, includes all compounds which are derivatives of napthoquinone that plays an important role in various biological functions such as blood clotting (Shearer and Newman 2014; Ravisankar et al. 2015). It exists in two forms namely vitamin K_1 and vitamin K_2. In 1934, Dam and Schonheyder first realized the presence of Vitamin and later in 1939 Dam, Karrer, and co-workers isolated pure vitamin K_1 (phylloquinone) while Doisy and her co-workers isolated pure vitamin K_2 (menaquinone) (Halder et al. 2019). Phylloquinone is found in the thylakoid membranes of the chloroplasts and is an integral component of electron transport system (Gu et al. 2021).There are studies which claimed that Vitamin K_2 is synthesized exclusively by bacteria (Zhang et al. 2021). The primary structural difference between Vitamin K_1 and K_2 is in their side chain. Vitamin K_1 (phylloquinone) has a

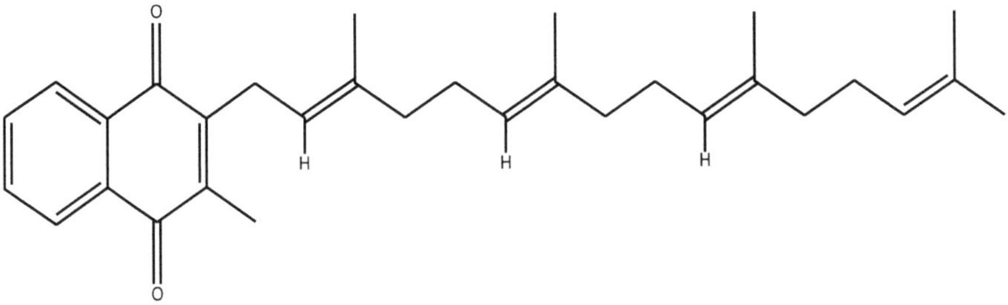

Fig. 6.4 Structure of Vitamin $K_1$and K_2

phytol chain whereas Vitamin K_2 (menaquinone) has a collection of isoprene units in its side chain (Fig. 6.4) (Bus and Szterk 2021). Vitamin K_1 is mainly found in green leafy vegetables, while Vitamin K_2 is present in fermented foods or animal sources such as cheese, egg yolk, and meat (Zhou et al. 2022). Vitamin K plays a role in blood clotting and bone health. It can be found in green leafy vegetables, such as spinach and kale, broccoli, and Brussels sprouts (Vermeer 2012; Sinbad et al. 2019).

6.2.5 Vitamin C

Vitamin C is a water-soluble vitamin that has a chemical name Ascorbic acid. At room temperature, it is a white or slightly yellowish crystalline powder that has no odor (Abbas et al. 2012). It falls under the category of water-soluble vitamins, as a result it is highly soluble in polar solvents like water and ethanol but has minimal solubility in non-polar solvents like chloroform and benzene (Bellows et al. 2012; El-Ishaq and Obirinakem 2015). The first reported isolation of Vitamin C was performed by Zilva in between 1917 and 1927, which was later followed by Szent Gyorgyi's discovery of ascorbic acid presence in oranges, adrenal glands, and cabbage as "hexuronic acid" (Carpenter 2012; Chambial et al. 2013). Later, the vitamin was crystallized from lemon juice and recognized as the same as Szent Gyorgyi's hexuronic acid by Glen King in 1932 (Paciolla et al. 2019). Finally, Haworth and Hirst determined the chemical structure of Vitamin C and also accomplished its synthesis (Paciolla et al. 2019). The molecule of Vitamin C consists of a cyclic hexose (cyclohexane) molecule with four hydroxyl (-OH) groups and a ketone group (>C=O) on the carbon atoms (Fig. 6.5). Ascorbic acid has two enantiomers L and D forms, out of which only the L-Ascorbic Acid is biologically active and important for human health (Olabisi 2005). Vitamin C is an antioxidant that plays a role in maintaining healthy skin, bones, and teeth. It is also important for maintaining a healthy immune system (Chambial et al. 2013). Foods rich in vitamin C include citrus fruits, strawberries, papaya, bell peppers, and kiwi (Yang et al. 2009).

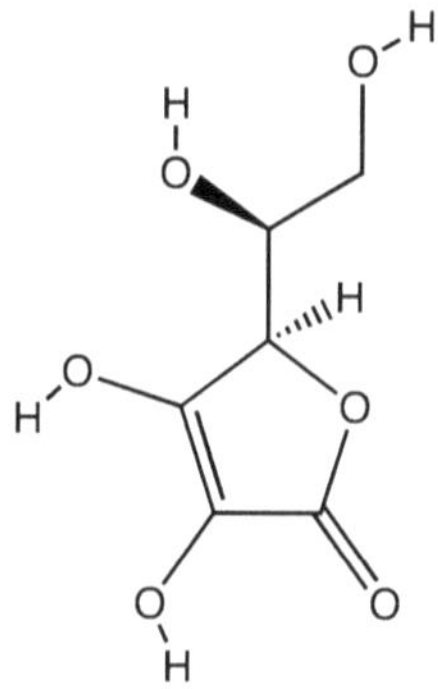

Fig. 6.5 Structure of Vitamin C

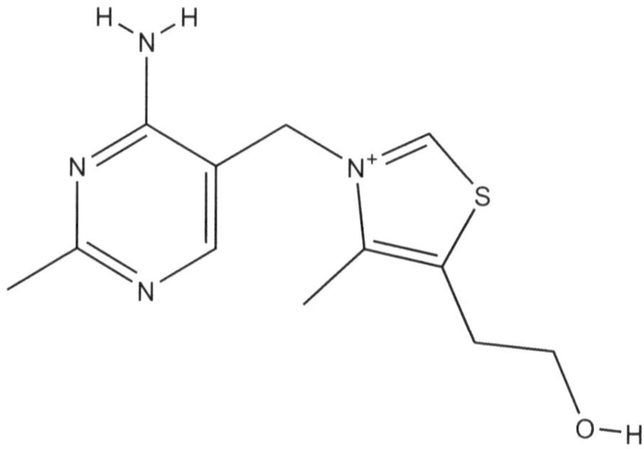

Fig. 6.6 Structure of Vitamin B1

6.2.6 Vitamin B1

Vitamin B1, also known as Thiamin, is an essential dietary water-soluble vitamin (Martel et al. 2018; Cornell and Arita 2021). The structure of thiamin is composed of pyrimidine and thiazole moieties that are substituted and connected through a methylene bridge (Fig. 6.6). It appears as a white, crystalline powder with a slight odor and is highly soluble in water, slightly soluble in ethanol and insoluble in ether (Schnellbaecher et al. 2019; Yoshii et al. 2019). Thiamin is sensitive to heat, light, and effected by change in pH. It can be decomposed by the presence of certain alkaline compounds used in food preparation such as baking soda and by unstable oxidizing agents (Hrubša et al. 2022). Since free thiamine is unstable due to the presence of quaternary nitrogen in polar compounds like water. It is cleaved into thiol form, as a result the hydrochloride and mononitrate forms are commercially used in the market for other preparations (Voelker et al. 2018). The vitamin B family plays crucial roles in several body functions, including the metabolism of enzymes, hormones, and neurotransmitters. They also help form red blood cells and aid in the maintenance of a healthy nervous system (Hanna et al. 2022). Among B complex vitamins, vitamin B1 can be found in whole grains, pork, beans, and nuts (Ramsey and Muskin 2013).

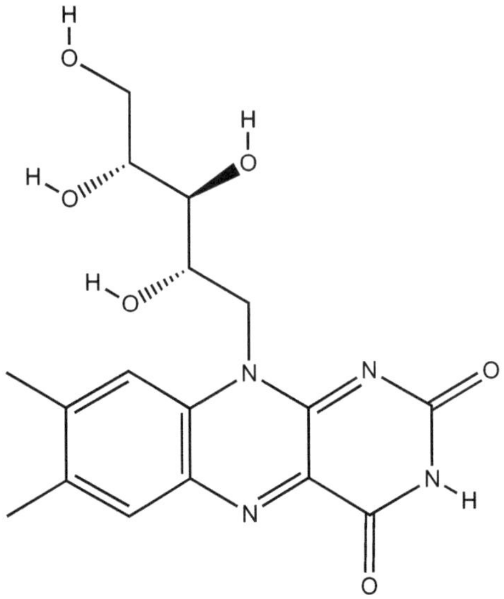

Fig. 6.7 Structure of Vitamin B2 (Riboflavin)

6.2.7 Vitamin B2

Vitamin B2, popularly known as Riboflavin, is a yellow to orange solid or powder with no odor or taste (Al-Shammary et al. 1990). It has a distinct yellow-green fluorescence under UV light which can be useful in its detection. It is soluble in water, sparingly soluble in ethanol and insoluble in ether (Abdullah 2016). The riboflavin gets phosphorylated into two metabolically active forms namely flavin mononucleotide (FMN) and flavin adenine dinucleotide (FAD) (Mosegaard et al. 2020). FMN is composed of riboflavin, one phosphate group, and one ribose sugar while FAD is composed of riboflavin, two phosphate groups, and an adenine nucleotide (Fig. 6.7). As a result, FAD has one extra phosphate group than FMN; therefore, FAD carries two negative charges in solution while FMN carries only one negative charge (Pinto and Rivlin 2013; Liu et al. 2020). These physio-chemical properties play an important role in Riboflavin's various biological functions, including its role as a cofactor for various enzymes involved in energy metabolism, antioxidant defense, and recycling of vitamins like glutathione and vitamin E (Parra et al. 2018). Vitamin B2 can be obtained from dairy products, whole grains, and leafy green vegetables (Buehler 2011).

6.2.8 Vitamin B3 (Niacin)

Vitamin B3, also referred to as Niacin or Nicotinic Acid, is the generic descriptor for pyridine 3-carboxylic acid derivatives (Ndez-Lainez et al. n.d.; Chand and Savitri 2016). It is a white or pale yellow crystalline powder that can appear pinkish-brown when exposed to light. Niacin exists as niacinamide or nicotinamide in its

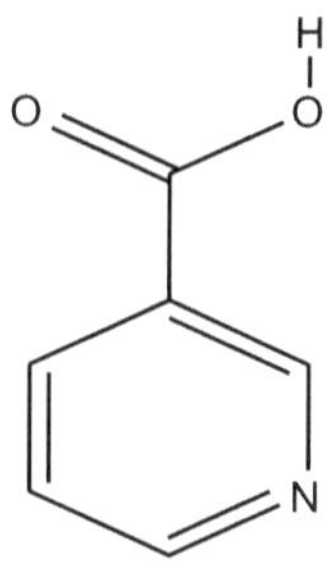

Fig. 6.8 Structure of Vitamin B3 (Niacin)

amide form. This amide group results from the interaction of amino group and carboxylic acid (Fig. 6.8). An alteration into activated forms, such as nicotinamide adenine dinucleotide (NAD^+) or nicotinamide adenine dinucleotide phosphate ($NADP^+$), occurs while niacin is converted to these essential forms. Both NAD^+ and $NADP^+$ act as cofactors in many enzymatic reactions (Chand and Savitri 2016; Bains et al. 2018). Vitamin B3 can be found in meat, fish, poultry, and whole grains (Peechakara and Gupta 2023).

6.2.9 Vitamin B5 (Pantothenic Acid)

Vitamin B5, also known as pantothenic acid, is a water-soluble vitamin that is essential for the human body to function properly (Said 2015). It plays a crucial role in energy metabolism, and is also involved in synthesizing and metabolizing fats, carbohydrates, and proteins (Tardy et al. 2020; Wan et al. 2022). Vitamin B5 is stable in acid and neutral conditions but it is relatively unstable under alkaline conditions. It can also withstand high cooking temperatures, and its content is not reduced even after boiling or baking (Schnellbaecher et al. 2019). In addition, Vitamin B5 has no color, taste, or odor, which makes it easy to incorporate into food and supplements. The structure of pantothenic acid includes a pantoic acid unit which consists of a carboxylic acid group (-COOH) and a five-carbon unit and a β-alanine unit (Fig. 6.9). The pantothenic acid and β-alanine units combine to form pantothenic acid's coenzyme A molecule, essential for energy metabolism within the cells (Sanvictores and Chauhan 2023). The major sources of pantothenic acid can be found in meat, poultry, fish, and whole grains (Chawla and Kvarnberg 2014).

6.2.10 Vitamin B6 (Pyridoxine)

Vitamin B6 also known as pyridoxine is a water-soluble vitamin, which is heat and light sensitive (Cornell and Arita 2021). Pyridoxine is a relatively reactive molecule that can undergo various chemical reactions, such as oxidation and reduction reactions (Parra et al. 2018). The vitamin was first identified as a growth factor in rats in 1934 by Paul Gyorgy, a Hungarian biochemist. Gyorgy named the vitamin "pyridoxine" because of its similarity in chemical structure to pyridine (Fig. 6.10)

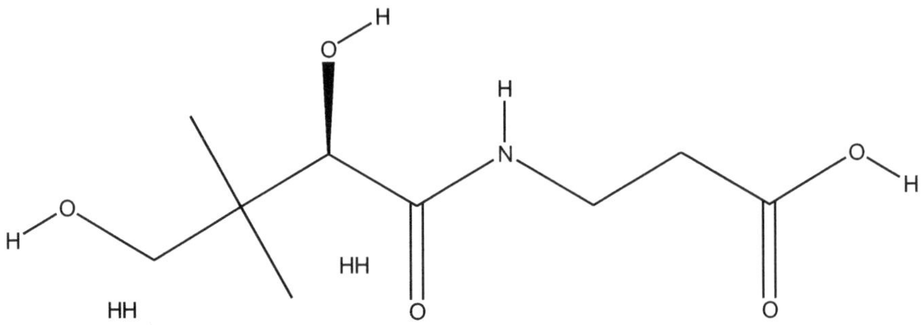

Fig. 6.9 Structure of Vitamin B5 (pantothenic acid)

Fig. 6.10 Structure of Vitamin B6 (Pyridoxine)

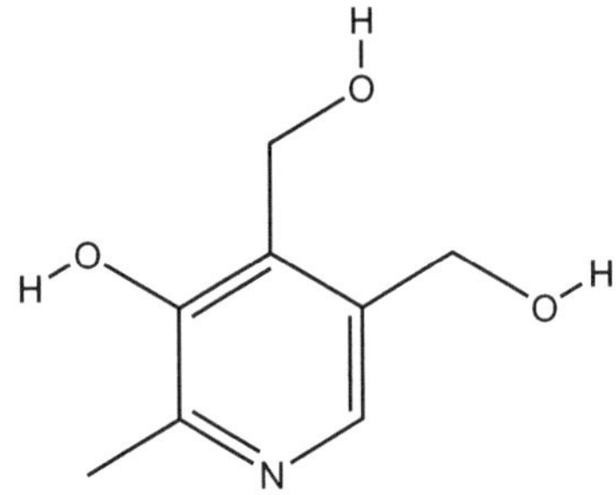

(Fitzpatrick 2011; O'Sullivan 2012). Later in 1938, Samuel Lepkovsky isolated vitamin B6 and established its chemical structure. He also discovered that pyridoxine had three active forms, namely pyridoxal, pyridoxine, and pyridoxamine. All these forms found to act as coenzymes, thus helping enzymes to perform necessary reactions within cells (Mooney et al. 2009; Rosenberg 2012). The pyridoxine and biotin can be obtained from fish, poultry, beans, nuts, bananas and egg yolks, liver, and nuts.

6.2.11 Vitamin B7 (Biotin)

Vitamin B7, popularly known as biotin, is a water-soluble vitamin that plays a crucial role in the metabolism of carbohydrates, fats, and proteins (Pathan et al. 2023). It is also essential for the synthesis of fatty acids, glucose, and some amino acids (Saleem and Soos 2023). It was discovered as Vitamin H in 1931 by Paul Gyorgy, further which led to the isolation and identification of biotin's chemical structure (Mironenko and Eliseeva 2019). The structure of biotin consists of a tetrahydrothiophene ring fused with an animidazolidone ring, which has a pentanoic acid side chain (Fig. 6.11). The molecule has three carbon atoms that are asymmetric. It is a white crystalline powder with no distinctive odor (Alban 2011).

Fig. 6.11 Structure of Vitamin B7 (Biotin)

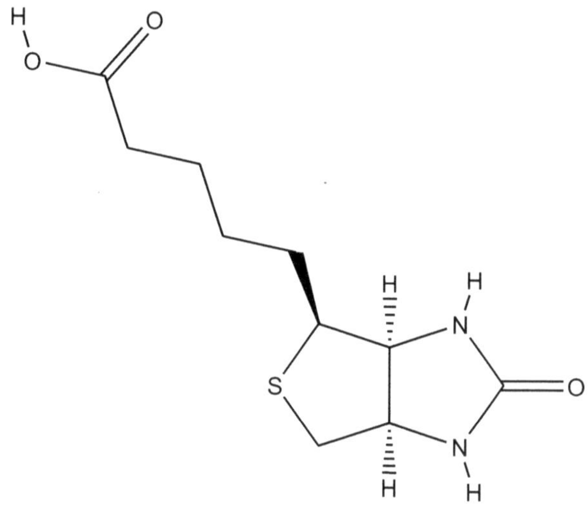

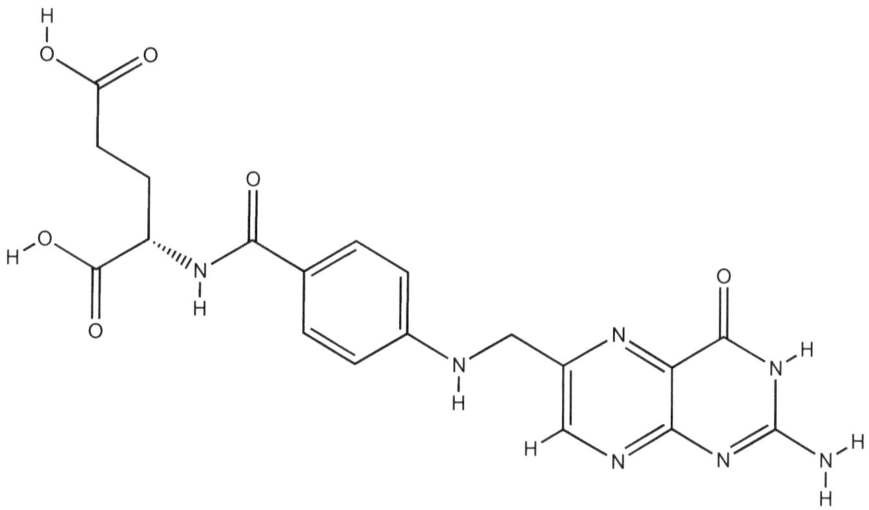

Fig. 6.12 Structure of Vitamin B9 (folate)

6.2.12 Vitamin B9 (Folate)

The structure of vitamin B9, also known as folate, comprises of a pteridine ring fused with a para-aminobenzoic acid known as PABA moiety along with glutamic acid residue (Chan et al. 2013). The pteridine ring contains two nitrogen atoms and one oxygen atom, while the PABA moiety contains an amino group and a carboxyl group (Fig. 6.12) (Tjong et al. 2022). Vitamin B9, or folate, is a water-soluble vitamin which is stable in acidic conditions, but is sensitive to heat, oxidation, and light. Folate is highly polar and can form chelates with metal ions such as zinc and copper (Premjit et al. 2023).

6.2.13 Vitamin B12 (Cobalamin)

Vitamin B12, also known as cobalamin, is a water-soluble vitamin containing a cobalt ion in its center, which is responsible for its biological activity (Osman et al. 2021). The cobalt ion is coordinated to four nitrogen atoms from the corrin ring and one nitrogen atom from a dimethylbenzimidazole group (Fig. 6.13) (Wdowik and Gryko 2022). The corrin ring of vitamin B12 contains four pyrrole rings, which are

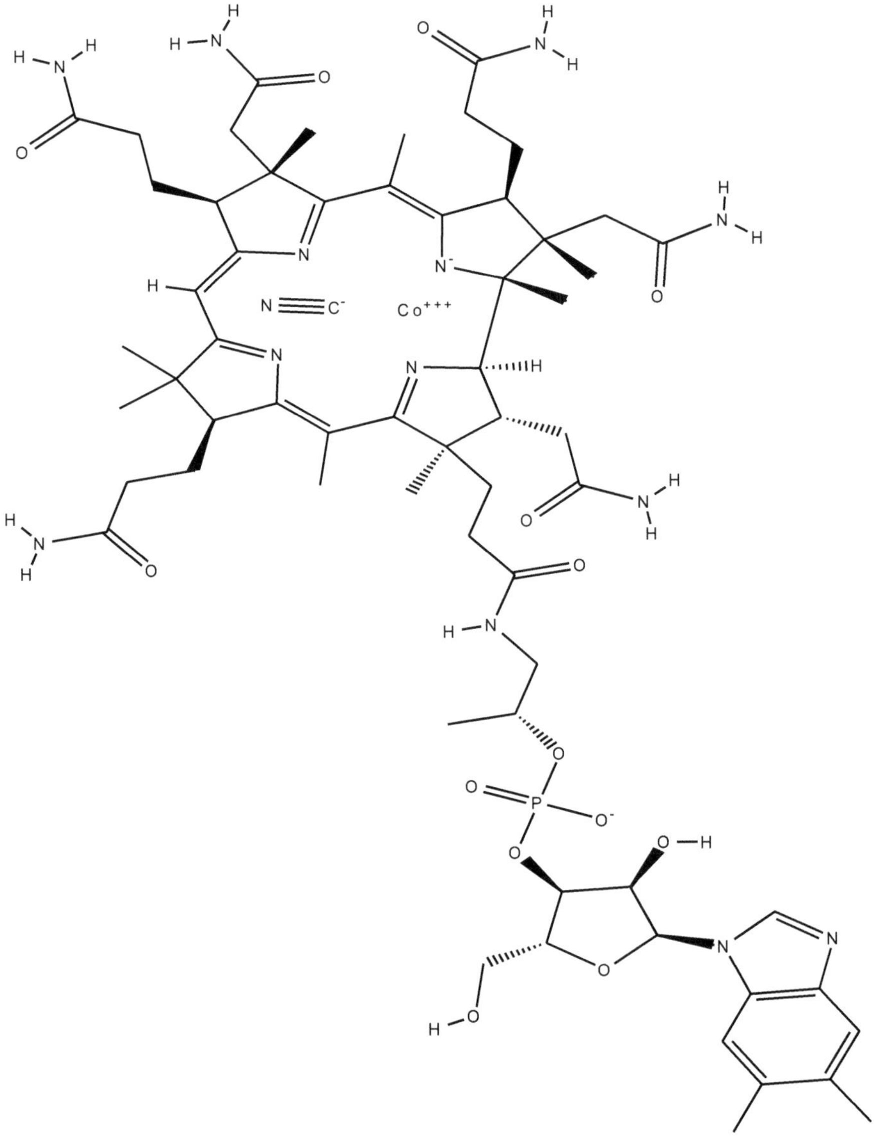

Fig. 6.13 Structure of Vitamin B12 (cobalamin)

connected by methylene bridges (Glew et al. 2006). Vitamin B12 is unique among the B vitamins in that it requires intrinsic factor, a protein produced by the stomach, for absorption in the small intestine (Marcus 2013). The folate and cobalamin could be obtained from dark green leafy vegetables, beans, fortified grains and meat, poultry, fish, and dairy products (Strain et al. 2017).

6.3 Dietary Sources and Its Bioavailability

Bioavailability of vitamins refers to the extent to which vitamins are absorbed, their availability for utilization in blood circulation within the body after their consumption (Sauberlich 1985). It could vary with age, gender, and physiological conditions. It could also be affected by certain drugs, caffeine products, smoking & alcohol habits, intake of phytates, etc. Therefore, it is important to consume a balanced diet that is rich in vitamins (variety of fruits and vegetables, whole grains, dairy products, etc.) to ensure a balanced intake of vitamins (Krebs 2001). This balanced intake is essential for maintaining optimal health and preventing vitamin deficiencies (Table 6.1).

Table 6.1 Details of different forms of vitamins, their dietary sources, recommended dietary intake, and list of diseases due to their deficiency

S.No.	Vitamin	Dietary source	Recommended dietary intake (RDI)	References
1	Vitamin A	Leafy green vegetables, tomatoes, beef liver, fish oils, milk, eggs	700–900 mcg	Carazo et al. (2021)
2	Vitamin D	Cod liver oil, salmon, tuna fish, dairy and plant milks fortified with vitamin D, beef liver, egg yolk	15–20 mcg	Dominguez et al. (2021)
3	Vitamin E	Sunflower, safflower, and soybean oil, almonds, peanuts, peanut butter, beet greens, collard greens, spinach, pumpkin, red bell pepper	15 mg	Niki and Abe (2019)
4	Vitamin K	Green leafy vegetables, soybean and canola oil, fermented soybeans, meat, cheese, eggs	90–120 mcg	Mladěnka et al. (2022)
5	Vitamin C	Citrus fruits (oranges, kiwi, lemon etc.), bell peppers, strawberries, tomatoes, cruciferous vegetables	75–90 mg	Doseděl et al. (2021)
6	Vitamin B1	Pork, fish, beans, lentils, green peas, enriched cereals, sunflower seeds, yogurt	1.1–1.2 mg	Mikkelsen and Apostolopoulos (2019)
7	Vitamin B2	Milk, yogurt, cheese, eggs, beef liver, chicken breast, salmon, almonds, spinach	1.1–1.3 mg	Mikkelsen and Apostolopoulos (2019)
8	Vitamin B3	Red meat, poultry, fish, nuts, seeds, legumes	14–16 mg	Mikkelsen and Apostolopoulos (2019)

(continued)

Table 6.1 (continued)

S.No.	Vitamin	Dietary source	Recommended dietary intake (RDI)	References
9	Vitamin B5	Liver, kidney, beef, mushrooms, avocados, nuts, seeds, milk, yogurt	5 mg	Hrubša et al. (2022)
10	Vitamin B6	Poultry, beef liver, tuna, salmon, chickpeas	1.3–1.7 mg	Cornell and Arita (2021)
11	Vitamin B7	Pork, beef liver, salmon, avocados, nuts, seeds	30 mcg	Suter (2020)
12	Vitamin B9	Dark green leafy, beans, fresh fruits, whole grains, liver, eggs	400 mcg	Mironenko and Eliseeva (2019)
13	Vitamin B12	Rd meat, fish, eggs, dairy products (milk, cheese, and yogurt) enriched soy milk	2.4 mcg	Butola et al. (2020)

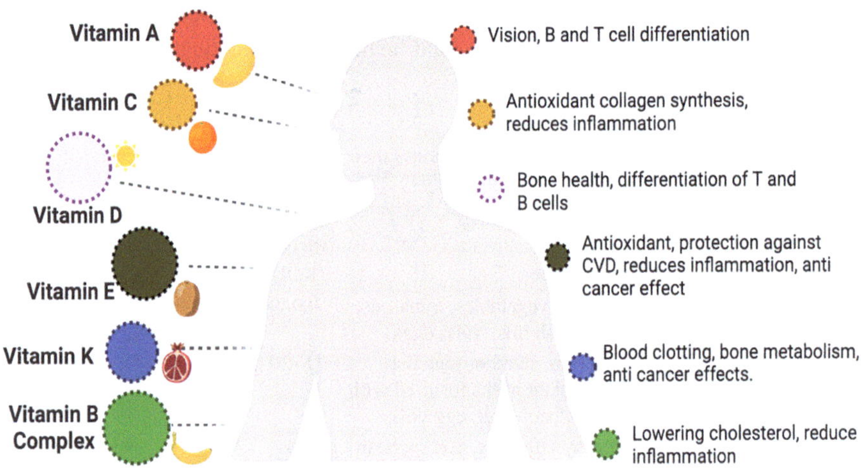

Fig. 6.14 Representation of vitamins from dietary sources and their role in human physiology

6.4 Vitamins Influences Human Health

In the above sections, we have discussed various types of vitamins, their structures, their natural sources. In this section, we will learn about their importance in human health. Vitamins possess many properties (Fig. 6.14) which have therapeutic role in disease progression (Table 6.2).

Vitamin A, a fat-soluble vitamin, plays a crucial role in several body functions, including vision, immune function, cell proliferation and differentiation (Aslam et al. 2017; Huang et al. 2018). Vitamin A is important for maintaining nourishing eyes and vision. It is an element of rhodopsin, a retina protein that helps us clear vision in low light environment. Its deficiency can cause night blindness and vision

Table 6.2 Common symptoms, prevention strategies, and histopathological/pathophysiological features associated with vitamins

	Vitamin A	Vitamin C	Vitamin D	Vitamin E	Vitamin K	Vitamin B complex
Disease & symptoms	Night blindness, high childhood, morbidity & mortality, diminished ability to fight infections, low bone, development	Scurvy, capillary hemorrhage, muscle degeneration, gingival bleeding, anorexia, anemia	Loss of bone density, leading to osteoporosis, and bone fractures, rickets in children, osteomalacia in adults	Retinopathy, skeletal, myopathy, ataxia, decreased immune function	Significant bleeding, osteoporosis, increased, cardiovascular disease	Dementia, anemia, nausea, dermatitis, fatigue, constipation, mouth ulcers
Prevention strategies	Consumption of adequate amounts of Vitamin A or carotenoid precursors	Taking vitamin C supplements and eating a diet rich in Vitamin C & consumption of fruits & vegetables such as citrus, berries, kiwi, green and red peppers	An adequate diet containing vitamin D and sunlight exposure	Consumption of vegetable oils like wheat germ, sunflower oils. Consumption of nuts & seeds such as peanuts, almonds & sunflower seeds	Diet rich in yoghurt, fermented foods, green beans, avocados, vegetable oils	Consumption of dairy products, eggs and meat (liver from beef, chicken) & sea food, like salmon. Legumes & leafy greens
Histopathology/ Pathophysiology	Bitotspots: conjunctival xerosis, corneal xerosis	Compromised collagen synthesis, thin osteoid matrix, dilated hair follicles	Decline in intestinal absorption & phosphorus absorption leading to hypocalcemia and osteomalacia	Hemolytic anemia, cerebellar ataxia & cognitive difficulties	Increased arterial calcification, excessive bleeding, decreased levels of Vit-K-dependent clotting factors clotting factors	Degeneration of peripheral nerves, altered iron absorption, inflammation, capillary dilation

problems (Sajovic et al. 2022). Vitamin A is also maintaining a healthy immune system. It helps in regulation of growth and differentiation of T cells and B cells and participates in defense system and the production of antibodies. Vitamin A also participates in maintaining healthy skin. It regulates the growth and differentiation of dermal cells and plays a role in the production of sebum, an oily substance that helps keep the skin moisturized and helps in protection against environmental antigens (Mora et al. 2008; Huang et al. 2018). This vitamin is essential for reproductive health in both men and women and participates in the development of sperm and eggs. Interestingly, this is also important for maintaining the health of the placenta during pregnancy (Clagett-Dame and Knutson 2011). Vitamin A has been shown to have anti-cancer effects in some studies. It does this by regulating cell growth and differentiation, and by promoting apoptosis, or programmed cell death, in cancer cells (Doldo et al. 2015; Takahashi et al. 2022). Vitamin A deficiency is linked with an increased risk of infectious diseases, such as measles and diarrhea. This is because vitamin A plays a crucial role in maintaining a healthy immune system. A balanced diet provides all the necessary vitamin A; few individuals may require supplements to meet their nutritional needs on the daily basis, whereas other can nourish themselves through healthy diet. However, it is always important to consult a physician/healthcare professional before taking any supplements, as excessive intake of vitamin A can lead to adverse effects or toxicity on individual health (Olson et al. 2022).

Vitamin C, popularly known as ascorbic acid, acting as a crucial role in maintaining human health and preventing various diseases (Singh et al. 2013; Chambial et al. 2013; Singh and Singh 2018). Scurvy is a disorder caused due to lack of vitamin C. The typical symptoms are fatigue, gum disease, and skin-related problems (Maxfield et al. 2023). Vitamin C is very essential element for collagen synthesis, which is principal factor for regulating healthy skin, gums, and other connective tissues in the human body (Iqbal et al. 2004; Pullar et al. 2017). There are several studies which show that vitamin C enhances the immune system and reduces inflammation. It also exhibits antioxidant properties that protect cells from cytotoxic effects due to generation of free radicals ions (Vishwakarma et al. 2013; Carr and Maggini 2017). This can prevent cancer-associated mechanism by balancing the increased oxidative stress-induced inflammation. Vitamin C is known to reduce the risk of cardiovascular disease by improving blood vessel function and reducing inflammation. It may also help lower cardiovascular-related biomarkers such as blood pressure and cholesterol levels (Chambial et al. 2013; Moser and Chun 2016).

Vitamin C also plays an important role in bone health. It promotes collagen synthesis and bone mineralization. There are reports indicating that vitamin C deficiency can lead to osteoporosis, a condition characterized by weak and brittle bones (Chin and Ima-Nirwana 2018; Brzezińska et al. 2020). Vitamin C can interfere in age-related macular degeneration (AMD), which is a leading cause of blindness in old adults. Reports have shown that vitamin C can protect retina from oxidative stress and inflammation (Yadav et al. 2011). Therefore, vitamin C plays a crucial role in maintaining human health and preventing several diseases. Although, a balanced diet can provide all the necessary vitamin C needed.

Vitamin D, a fat-soluble vitamin popularly known as calciferol, is essential for various bodily functions, including immune function, bone health, and cell growth/differentiation (Prietl et al. 2013). Vitamin D plays an elementary role in maintaining bone health. It helps in absorption of calcium and phosphorus from the diet, which is necessary for bone growth and remodeling (Laird et al. 2010). A deficiency of vitamin D can lead to medical conditions such as osteoporosis and frequent fractures (De Martinis et al. 2021).

Vitamin D also regulates immune system. There are several reports in the growth and differentiation of immune cells, such as T cells and B cells (Aranow 2011). It also acts as co-factors in various metabolic reactions which synthesize of several intermediate products or peptides that fight off infections (Gombart 2009). The lack of vitamin D has been linked to an increased risk of severe autoimmune diseases, like multiple sclerosis and rheumatoid arthritis (Kostoglou-Athanassiou et al. 2012; Sintzel et al. 2018). A deficiency of vitamin D has been linked to an increased risk of heart disease and stroke (Judd and Tangpricha 2009). There are several reports which claimed that Vitamin D has been shown to have anti-cancer effects (Fleet et al. 2012). Vitamin D is classified as a prohormone which, upon stimulation, helps in releasing an essential steroid hormone known as calcitriol (Ross et al. 2011). Research has revealed a close association between calcitriol and a reduced risk of cancer (Trump 2018). Vitamin D has been shown to improve insulin sensitivity and glucose tolerance in some studies (Talaei et al. 2013). A deficiency of vitamin D has been linked to an increased risk of type 2 diabetes. Many studies have looked at the role of vitamin D in diabetes and have shown an association between low levels of vitamin D and increased risk for type 2 diabetes and its complications (Lips et al. 2017).

Similarly, Vitamin D has also shown to have a protective effect against respiratory infections, such as the flu and pneumonia. It does this by boosting the immune system and reducing inflammation in the respiratory tract (Taha et al. 2021). A deficiency of vitamin D has been linked to an increased risk of respiratory infections and inflammation (Hughes and Norton 2009). Therefore, there are so many findings which supports that Vitamin D plays a crucial role in maintaining human health and preventing various diseases. While a balanced diet and exposure to sunlight can provide all the necessary vitamin D, although some people may require supplements to meet their needs for healthy and complete participation for proper functioning of human body.

Vitamin E, a fat-soluble vitamin acts as a potent antioxidant agent for the human body (Rizvi et al. 2014). It protects cells from damage caused by free radicals, which are unstable molecules known as oxidant that can cause oxidative stress and contribute to the development of various diseases (Packer 1991). Vitamin E has been shown to have a protective effect against cardiovascular disease. There are findings which conclude that it reduces inflammation, reducing blood pressure and improves endothelial function (Vardi et al. 2013). Vitamin E may prevent the oxidation of LDL cholesterol, which is a major risk factor for heart disease (Sesso et al. 2008). Vitamin E has been found to display anti-cancer effects in some studies. It may also enhance the immune system's ability to fight off cancer cells (Yu et al.

2009). Vitamin E has also been shown to have a protective effect against Alzheimer's disease (Lloret et al. 2019). Vitamin E has known to improve insulin sensitivity and glucose tolerance according to several research studies (Asbaghi et al. 2023). It also modulates oxidative stress-induced secondary complications, such as neuropathy and retinopathy (Pazdro and Burgess 2010). Vitamin E has protective effect against skin damage caused due to long exposure of UV radiation (Keen and Hassan 2016). Altogether, vitamin E plays a crucial role in maintaining human health and preventing various diseases.

Vitamin K, a fat-soluble vitamin, plays a critical role in blood clotting mechanism and bone metabolism (Akbari and Rasouli-Ghahroudi 2018). It is also involved in various other biological processes, including cell growth and proliferation, inflammation, and immune function (Namazi et al. 2019). Vitamin K is essential to produce several clotting factors that help prevent bleeding during trauma (Turcotte et al. 2020). Deficiency of vitamin K can lead to bleeding disorders, such as hemophilia and von Willebrand disease. On the other hand, excessive intake of vitamin K can interfere with the effectiveness of anticoagulant medications, such as warfarin (Park et al. 2015). Vitamin K plays a major role in bone metabolism by activating a protein called osteocalcin, which helps bind calcium to the bone matrix (Maresz 2015). Research studies have shown that vitamin K supplementation can improve bone density and reduce the risk of fractures in postmenopausal condition with osteoporosis (Elshaikh et al. 2020). Vitamin K has a protective effect against cardiovascular diseases by preventing the calcification of arteries (Shioi et al. 2020). This occurs when calcium deposits accumulate in the arterial walls, leading to stiffening and narrowing of the vessels. Vitamin K activates a protein called matrix Gla protein, which helps prevent calcification by binding to calcium ions (Roumeliotis et al. 2019; Wei et al. 2019). Vitamin K has been shown to have anti-cancer effects in some studies. It does this by inhibiting cancer cell growth and inducing apoptosis (programmed cell death). Vitamin K may also enhance the immune system's ability to fight off cancer cells (Dasari et al. 2017; Dasari et al. 2018). Studies have shown that Vitamin K may improve insulin sensitivity and glucose metabolism in people with type 2 diabetes. It does this by activating a protein called insulin-like growth factor-1, which helps regulate glucose uptake and utilization (Ho et al. 2020).

B complex vitamins are a group of eight water-soluble vitamins that play essential roles in various biological processes in the human body. These vitamins include thiamine (B1), riboflavin (B2), niacin (B3), pantothenic acid (B5), pyridoxine (B6), biotin (B7), folic acid (B9), and cobalamin (B12) (Lykstad and Sharma 2019). B complex vitamins, particularly niacin (B3), can help lower cholesterol levels and reduce the risk of heart disease. Niacin can also improve blood flow and reduce inflammation in the arteries (Ruparelia et al. 2011). Several B complex vitamins, including folate (B9) and cobalamin (B12), are essential for the production of red blood cells (Koury and Ponka 2004). Deficiency of these vitamins can lead to anemia, a condition characterized by low levels of hemoglobin in the blood. Folate (B9) is crucial for fetal development during pregnancy. Deficiency of folate can lead to neural tube defects, such as spina bifida, in the developing fetus (Greenberg et al. 2011; Imbard et al. 2013). Several B complex vitamins, including thiamine (B1), riboflavin (B2), niacin (B3), pyridoxine (B6), and cobalamin (B12), play a role in the production of neurotransmitters, such as serotonin and dopamine, which are involved in mood regulation. Deficiency of these vitamins can lead to depression and anxiety (Calderón-Ospina and Nava-Mesa 2020). B complex vitamins, particularly folate (B9) and cobalamin (B12), are essential for brain function and cognitive health. Deficiency of these vitamins can lead to cognitive decline and dementia. B complex vitamins, particularly biotin (B7) and pantothenic acid (B5), are essential for skin health. In conclusion, B complex vitamins play essential roles in various biological processes in the human body and can impact human health in numerous ways. While a balanced diet can provide all the necessary B complex vitamins, some people may require supplements to meet their needs. However, it is important to consult a healthcare professional before taking any supplements, as excessive intake of certain B complex vitamins can have adverse effects on health.

6.5 Summary

Vitamins are vital for good health; they are required in much smaller amounts than macro-nutrients, like carbohydrates, proteins, and fats. All the vitamins are essential and important for many daily bodily functions, such as cell reproduction and growth, but most importantly for the processing of energy in cells. Mostly all human beings can meet their daily vitamin needs by eating a range of plants or animals derived food forms.

References

Abbas S, Da Wei C, Hayat K et al (2012) Ascorbic acid: microencapsulation techniques and trends—a review. Food Rev Intl 28(4):343–374

Abdullah R (2016) Determination of water soluble vitamins (B group: B1, B2, and B6) by RP-HPLC. Chem Mater 8:1–10

Akbari S, Rasouli-Ghahroudi AA (2018) Vitamin K and bone metabolism: a review of the latest evidence in preclinical studies. Biomed Res Int 2018:4629383

Akram M, Munir N, Daniyal M et al (2020) Vitamins and minerals: types, sources and their functions. In: Functional foods and nutraceuticals: bioactive components, formulations and innovations, pp 149–172

Alban C (2011) Biotin (vitamin B8) synthesis in plants. Adv Bot Res 59:39–66

Al-Shammary FJ, Zubair MU, Mian MS et al (1990) Analytical profile of riboflavin. Analytical profiles of drug substances. Academic Press, pp 429–476

Aranow C (2011) Vitamin D and the immune system. J Investig Med 59(6):881–886

Asbaghi O, Nazarian B, Yousefi M et al (2023) Effect of vitamin E intake on glycemic control and insulin resistance in diabetic patients: an updated systematic review and meta-analysis of randomized controlled trials. Nutr J 22(1):10

Aslam MF, Majeed S, Aslam S, Irfan JA (2017) Vitamins: key role players in boosting up immune response-a mini review. Vitam Miner 6:153

Azzi A (2019) Tocopherols, tocotrienols and tocomonoenols: many similar molecules but only one vitamin E. Redox Biol 26:101259

Azzi A, Stocker A (2000) Vitamin E: non-antioxidant roles. Prog Lipid Res 39(3):231–255

Bains P, Kaur M, Kaur J et al (2018) Nicotinamide: mechanism of action and indications in dermatology. Indian J Dermatol Venereol Leprol 84:234
Behr A, Seidensticker T, Behr, A et al (2020) Vital amines-vitamins. Chem Rene: An Intro 295–307
Bell EF (1987) History of vitamin E in infant nutrition. Am J Clin Nutr 46(1):183–186
Bellows L, Moore R, Anderson J et al (2012) Water-soluble vitamins: B-complex and vitamin C. Food and nutrition series. Health 9.312
Bikle DD (2014) Vitamin D metabolism, mechanism of action, and clinical applications. Chem Biol 21(3):319–329
Brzezińska O, Łukasik Z, Makowska J et al (2020) Role of vitamin C in osteoporosis development and treatment—a literature review. Nutrients 12(8):2394
Buehler BA (2011) Vitamin B2: riboflavin. J Evid Based Comp Altern Med 16(2):88–90
Bus K, Szterk A (2021) Relationship between structure and biological activity of various vitamin K forms. Food Secur 10(12):3136
Butola LK, Kute PK, Anjankar A et al (2020) Vitamin B12-do you know everything. J Evol Med Dent Sci 9(42):3139–3134
Calderón-Ospina CA, Nava-Mesa MO (2020) B Vitamins in the nervous system: current knowledge of the biochemical modes of action and synergies of thiamine, pyridoxine, and cobalamin. CNS Neurosci Ther 26(1):5–13
Carazo A, Macáková K, Matoušová K et al (2021) Vitamin A update: forms, sources, kinetics, detection, function, deficiency, therapeutic use and toxicity. Nutrients 13(5):1703
Carpenter KJ (2012) The discovery of vitamin C. Ann Nutr Metab 61(3):259–264
Carr AC, Maggini S (2017) Vitamin C and immune function. Nutrients 9(11):1211
Chambial S, Dwivedi S, Shukla KK et al (2013) Vitamin C in disease prevention and cure: an overview. Indian J Clin Biochem 28:314–328
Chan YM, Bailey R, O'Connor DL (2013) Folate Adv Nutr 4(1):123
Chand T, Savitri B (2016) Vitamin B3, niacin. In: Industrial biotechnology of vitamins, biopigments, and antioxidants, pp 41–65
Chawla J, Kvarnberg D (2014) Hydrosoluble vitamins. Handb Clin Neurol 120:891–914
Chin KY, Ima-Nirwana S (2018) Vitamin C and bone health: evidence from cell, animal and human studies. Curr Drug Targets 19(5):439–450
Clagett-Dame M, Knutson D (2011) Vitamin A in reproduction and development. Nutrients 3(4):385–428
Combs GF Jr, McClung JP (2016) The vitamins: fundamental aspects in nutrition and health. Academic Press
Cornell L, Arita K (2021) Water soluble vitamins: B1, B2, B3, and B6. Geriatric gastroenterology. Springer, pp 569–596
Dasari S, Ali SM, Zheng G et al (2017) Vitamin K and its analogs: potential avenues for prostate cancer management. Oncotarget 8(34):57782
Dasari S, Samy AL, Kajdacsy-Balla A et al (2018) Vitamin K2, a menaquinone present in dairy products targets castration-resistant prostate cancer cell-line by activating apoptosis signaling. Food Chem Toxicol 115:218–227
De Martinis M, Allegra A, Sirufo MM et al (2021) Vitamin D deficiency, osteoporosis and effect on autoimmune diseases and hematopoiesis: a review. Int J Mol Sci 22(16):8855
Dewangan S, Bhatia AK (2023) Vitamins and metabolites. In: Handbook of biomolecules. Elsevier, pp 119–131
Doldo E, Costanza G, Agostinelli S, Tarquini C, Ferlosio A, Arcuri G, Passeri D, Scioli MG, Orlandi A (2015) Vitamin a, cancer treatment and prevention: the new role of cellular retinol binding proteins. Biomed Res Int 2015:624627
Dominguez LJ, Farruggia M, Veronese N et al (2021) Vitamin D sources, metabolism, and deficiency: available compounds and guidelines for its treatment. Meta 11(4):255
Doseděl M, Jirkovský E, Macáková K et al (2021) Vitamin C—sources, physiological role, kinetics, deficiency, use, toxicity, and determination. Nutrients 13(2):615
Eggersdorfer M, Laudert D, Letinois U et al (2012) One hundred years of vitamins—a success story of the natural sciences. Angew Chem Int Ed 51(52):12960–12990

El-Ishaq A, Obirinakem S (2015) Effect of temperature and storage on vitamin C content in fruits juice. IJCBS 1(2):17–21
Elshaikh AO, Shah L, Mathew CJ et al (2020) Influence of vitamin K on bone mineral density and osteoporosis. Cureus 12(10)
Engelsen O (2010) The relationship between ultraviolet radiation exposure and vitamin D status. Nutrient 2(5):482–495
Falk J, Munné-Bosch S (2010) Tocochromanol functions in plants: antioxidation and beyond. J Exp Bot 61(6):1549–1566
Fitzpatrick TB (2011) Vitamin B6 in plants: more than meets the eye. In: Advances in botanical research. Elsevier, pp 1–38
Fleet JC, Desmet M, Johnson R et al (2012) Vitamin D and cancer: a review of molecular mechanisms. Biochem J 441(1):61–76
Gilbert C (2013) What is vitamin A and why do we need it? Community Eye Health 26(84):65
Glew RH, McCarthy DM, Vanderjagt DJ (2006) Vitamin B12 deficiency: is it underestimated in pregnant women? Acta Obstet Gynecol Scand 85(2):241–242
Godswill AG, Somtochukwu IV, Ikechukwu AO et al (2020) Health benefits of micronutrients (vitamins and minerals) and their associated deficiency diseases: a systematic review. Int J Food Sci 3(1):1–32
Gombart AF (2009) The vitamin D–antimicrobial peptide pathway and its role in protection against infection. Future Microbiol 4(9):1151–1165
Greenberg JA, Bell SJ, Guan Y et al (2011) Folic acid supplementation and pregnancy: more than just neural tube defect prevention. Rev Obstet Gynecol 4(2):52
Gu X, Chen IG, Harding SA et al (2021) Plasma membrane phylloquinone biosynthesis in nonphotosynthetic parasitic plants. Plant Physiol 185(4):1443–1456
Gurmu F, Hussein S, Laing M (2014) The potential of orange-fleshed sweet potato to prevent vitamin A deficiency in Africa. Int J Vitam Nutr Res 84(1–2):65–78
Halder M, Petsophonsakul P, Akbulut AC et al (2019) Vitamin K: double bonds beyond coagulation insights into differences between vitamin K1 and K2 in health and disease. Int J Mol Sci 20(4):896
Hanna M, Jaqua E, Nguyen V et al (2022) Vitamins: functions and uses in medicine. Perm J 26:89–97
Ho HJ, Komai M, Shirakawa H (2020) Beneficial effects of vitamin K status on glycemic regulation and diabetes mellitus: a mini-review. Nutrients 12(8):2485
Holick MF (2018) Photobiology of vitamin D. In: Vitamin D (fourth edition) volume 1: biochemistry, physiology and diagnostics. Elsevier, pp 45–55
Holick MF (2023) The one-hundred-year anniversary of the discovery of the sunshine vitamin D3: historical, personal experience and evidence-based perspectives. Nutrients 15(3):593
Hrubša M, Siatka T, Nejmanová I et al (2022) Biological properties of vitamins of the B-complex, part 1: vitamins B1, B2, B3, and B5. Nutrients 14(3):484
Huang Z, Liu Y, Qi G et al (2018) Role of vitamin A in the immune system. J Clin Med 7(9):258
Hughes DA, Norton R (2009) Vitamin D and respiratory health. CEI 158(1):20–25
Imbard A, Benoist JF, Blom HJ (2013) Neural tube defects, folic acid and methylation. IJERPH 10(9):4352–4389
Iqbal K, Khan A, Khattak M (2004) Biological significance of ascorbic acid (vitamin C) in human health-a review. Pakistan J Nutr 3(1):5–13
Jäpelt RB, Jakobsen J (2013) Vitamin D in plants: a review of occurrence, analysis, and biosynthesis front. Plant Sci 4:50324
Judd SE, Tangpricha V (2009) Vitamin D deficiency and risk for cardiovascular disease. AJMS 338(1):40–44
Keen MA, Hassan I (2016) Vitamin E in dermatology. Indian Dermatol Online J 7(4):311–315
Kostoglou-Athanassiou I, Athanassiou P, Lyraki A et al (2012) Vitamin D and rheumatoid arthritis. Ther Adv Endocrinol Metab 3(6):181–187
Koury MJ, Ponka P (2004) New insights into erythropoiesis: the roles of folate, vitamin B12, and iron. Annu Rev Nutr 24:105–131

Krebs NF (2001) Bioavailability of dietary supplements and impact of physiologic state: infants, children and adolescents. J Nutr 131(4):1351S–1354S
Laird E, Ward M, McSorley E et al (2010) Vitamin D and bone health. Pot Mech Nutr 2(7):693–724
Lampi AM (2011) Analysis of tocopherols and tocotrienols by HPLC. Selected topics in the analysis of lipids
Lips P, Eekhoff M, van Schoor N et al (2017) Vitamin D and type 2 diabetes. J Steroid Biochem Mol Biol 173:280–285
Liu S, Hu W, Wang Z et al (2020) Production of riboflavin and related cofactors by biotechnological processes. Microb Cell Factories 19(1):31
Lloret A, Esteve D, Monllor P et al (2019) The effectiveness of vitamin E treatment in Alzheimer's disease. Int J Mol Sci 20(4):879
LoPiccolo MC, Lim HW (2010) Vitamin D in health and disease. Photodermatol Photoimmunol Photomed 26(5):224–229
Lykstad J, Sharma S (2019) Biochemistry, water soluble vitamins. In: StatPearls [Internet]. StatPearls Publishing, Treasure Island. Available from: https://www.ncbi.nlm.nih.gov/books/NBK538510/
Mahmood L (2014) The metabolic processes of folic acid and vitamin B12 deficiency. J Health Res (In Developing Countries) 1(1):5–9
Maltz A (2013) Casimer Funk, nonconformist nomenclature, and networks surrounding the discovery of Vitamins1. J Nutr 143(7):1013–1020
Marcus JB (2013) Culinary nutrition: the science and practice of healthy cooking. Academic Press
Maresz K (2015) Proper calcium use: vitamin K2 as a promoter of bone and cardiovascular health. Integ Med J Clin 14(1):34
Martel JL, Kerndt CC, Doshi H et al (2018) Vitamin B1 (thiamine). In: StatPearls [Internet]. StatPearls Publishing, Treasure Island. Available from: https://www.ncbi.nlm.nih.gov/books/NBK482360/
Maurya VK, Shakya A, Bashir K et al (2022) Vitamin A fortification: recent advances in encapsulation technologies. Compr Rev Food Sci Food Saf 21(3):2772–2819
Maxfield L, Daley SF, Crane JS (2023) Vitamin C deficiency. In: StatPearls [Internet]. StatPearls Publishing, Treasure Island. 2024 Jan-. Available from: https://www.ncbi.nlm.nih.gov/books/NBK493187/
McCabe MP, Smyth MP, Richardson DR (2012) Current concept review: vitamin D and stress fractures. Foot Ankle Int 33(6):526–533
McDowell L (2013) Vitamin history, the early years. First Edition Design Pub
Mikkelsen K, Apostolopoulos V (2019) Vitamin B1, B2, B3, B5, and B6 and the immune system. Nutr Immun:115–125
Mironenko A, Eliseeva T (2019) Vitamin B9–description, benefits, effects on the body and best sources. J Hum Nutr Diet 4(10):88–100
Mladěnka P, Macáková K, Kujovská Krčmová L et al (2022) Vitamin K–sources, physiological role, kinetics, deficiency, detection, therapeutic use, and toxicity. Nutr Rev 80(4):677–698
Mooney S, Leuendorf JE, Hendrickson C, Hellmann H (2009) Vitamin B6: a long known compound of surprising complexity. Molecules 14(1):329–351
Mora JR, Iwata M, Von Andrian UH (2008) Vitamin effects on the immune system: vitamins A and D take centre stage. Nat Rev Immunol 8(9):685–698
Mosegaard S, Dipace G, Bross P et al (2020) Riboflavin deficiency—implications for general human health and inborn errors of metabolism. Int J Mol Sci 21(11):3847
Moser MA, Chun OK (2016) Vitamin C and heart health: a review based on findings from epidemiologic studies. Int J Mol Sci 17(8):1328
Namazi N, Larijani B, Azadbakht L (2019) Vitamin K and the immune system. Nutr Immun:75–79
Ndez-Lainez CF, Lezc IIG, Fernandez-Mejia CR (n.d.) The chemistry and biochemistry of niacin (B3)
Niki E, Abe K (2019) In: Niki E (ed) Vitamin E: chemistry and nutritional benefits, vol 2019. The Royal Society of Chemistry, pp 1–11

O'Sullivan T (2012) Exploring B vitamins and their impact on cognitive function and mood from conception to early adulthood. Nutrition and mental performance: a lifespan. Perspective 53
Olabisi AO (2005) The chemistry of L-ascorbic acid derivatives in the asymmetric synthesis of C2-and C3-substituted aldono-γ-lactones. Wichita State University
Olson JM, Ameer MA, Goyal A (2022) Vitamin A toxicity. In: StatPearls [Internet]. StatPearls Publishing
Osman D, Cooke A, Young TR et al (2021) The requirement for cobalt in vitamin B12: a paradigm for protein metalation. Biochim Biophys Acta Mol 1868(1):118896
Paciolla C, Fortunato S, Dipierro N et al (2019) Vitamin C in plants: from functions to biofortification. Antioxidants 8(11):519
Packer L (1991) Protective role of vitamin E in biological systems. Am J Clin Nutr 53(4):1050S–1055S
Park JN, Lee JS, Noh MY et al (2015) Association between usual vitamin K intake and anticoagulation in patients under warfarin therapy. Clin Nutr Res 4(4):235
Parra M, Stahl S, Hellmann H (2018) Vitamin B6 and its role in cell metabolism and physiology. Cells 7(7):84
Pathan AS, Jain PG, Mahajan AB et al (2023) Beneficial effects of water-soluble vitamins in nutrition and health promotion. Vit Nutraceut: Rec Adv Appl 26:235–251
Pazdro R, Burgess JR (2010) The role of vitamin E and oxidative stress in diabetes complications. Mech Ageing Dev 131(4):276–286
Peechakara BV, Gupta M (2023) Vitamin B3. In: StatPearls. StatPearls Publishing, Treasure Island
Pinto J, Rivlin R (2013) Riboflavin (vitamin B2). Handbook Vitam 5:191–266
Piro A, Tagarelli G, Lagonia P et al (2010) Casimir Funk: his discovery of the vitamins and their deficiency disorders. Ann Nutr Metab 57(2):85–88
Polcz ME, Barbul A (2019) The role of vitamin A in wound healing. Nutr Clin Pract 34(5):695–700
Premjit Y, Pandey S, Mitra J (2023) Recent trends in folic acid (vitamin B9) encapsulation, controlled release, and mathematical modelling. Food Rev Int 39(8):5528–5562
Prietl B, Treiber G, Pieber TR et al (2013) Vitamin D and immune function. Nutrients 5(7):2502–2521
Pullar JM, Carr AC, Vissers M (2017) The roles of vitamin C in skin health. Nutrients 9(8):866
Ramsey D, Muskin PR (2013) Vitamin deficiencies and mental health: how are they linked. Curr Psychiatr Ther 12(1):37–43
Ravisankar P, Reddy AA, Nagalakshmi B et al (2015) The comprehensive review on fat soluble vitamins. IOSR J Pharm 5(11):12–280
Rizvi S, Raza ST, Ahmed F et al (2014) The role of vitamin E in human health and some diseases. Sultan Qaboos Univ Med J 14(2):e157
Rosenberg IH (2012) A history of the isolation and identification of vitamin B6. Ann Nutr Metab 61(3):236–238
Ross AC, Taylor CL, Yaktine AL et al (2011) Overview of vitamin D. Dietary reference intakes for calcium and vitamin D. National Academies Press (US)
Roumeliotis S, Dounousi E, Eleftheriadis T et al (2019) Association of the inactive circulating matrix Gla protein with vitamin K intake, calcification, mortality, and cardiovascular disease: a review. Int J Mol Sci 20(3):628
Ruparelia N, Digby JE, Choudhury RP (2011) Effects of niacin on atherosclerosis and vascular function. Curr Opin Cardiol 26(1):66–70
Said HM (2015) Water-soluble vitamins. Nutr Prim Care Provid 111:30–37
Sajovic J, Meglič A, Glavač D et al (2022) The role of vitamin A in retinal diseases. Int J Mol Sci 23(3):1014
Saleem F, Soos MP (2023) Biotin deficiency. In: StatPearls [Internet]. StatPearls Publishing, Treasure Island. Jan-. Available from: https://www.ncbi.nlm.nih.gov/books/NBK547751/
Sanvictores T, Chauhan S (2023) Vitamin B5 (pantothenic acid). In: StatPearls [Internet]. StatPearls Publishing, Treasure Island
Sauberlich HE (1985) Bioavailability of vitamins. Prog Food Nutr Sci 9(1–2):1–33

Schellack G, Harirari P, Schellack N (2019) Vitamin B-complex deficiency, supplementation and management SA. Pharm J 86(3):23–29
Schnellbaecher A, Binder D, Bellmaine S et al (2019) Vitamins in cell culture media: stability and stabilization strategies. Biotechnol Bioeng 116(6):1537–1555
Semba RD (2012) The discovery of the vitamins. Int J Vitam Nutr Res 82(5):310–315
Semba RD (2013) The vitamin A story. Ophthalmologica 230(1):1–54
Sesso HD, Buring JE, Christen WG et al (2008) Vitamins E and C in the prevention of cardiovascular disease in men: the Physicians' Health Study II randomized controlled trial. JAMA 300(18):2123–2133
Shahidi F, De Camargo AC (2016) Tocopherols and tocotrienols in common and emerging dietary sources: occurrence, applications, and health benefits. Int J Mol Sci 17(10):1745
Shearer MJ, Newman P (2014) Recent trends in the metabolism and cell biology of vitamin K with special reference to vitamin K cycling and MK-4 biosynthesis. J Lipid Res 55(3):345–362
Shioi A, Morioka T, Shoji T et al (2020) The inhibitory roles of vitamin K in progression of vascular calcification. Nutrients 12(2):583
Sinbad OO, Folorunsho AA, Olabisi OL et al (2019) Vitamins as antioxidants. Int J Food Sci Nutr 2(3):214–235
Singh P, Singh RL (2018) Quantification of phytochemicals imparting antioxidant activities in commonly used vegetables. Int J Appl Sci Biotech 6(2):107–112
Singh P, Vishwakarma SP, Singh RL (2013) Evaluation of antioxidant, oxidative DNA damage protective and antimicrobial activities of *Foeniculum vulgare* plant. J Med Plant Res 7(35):2551–2563
Singh RL, Vishwakarma SP, Singh P (2014) Vitamins and minerals: roles and plant sources. In: Prakash D (ed) Phytochemicals of nutraceutical importance and their role in human health. CABI-UK, pp 248–265
Sintzel MB, Rametta M, Reder AT (2018) Vitamin D and multiple sclerosis: a comprehensive review. J Neurol:59–85
Spiro A, Buttriss J (2014) Vitamin D: an overview of vitamin D status and intake in Europe. Nutr Bull 39(4):322–350
Strain JJ, Hughes C, Pentieva K et al (2017) The B-vitamins. Sustain Nutr Chang World:185–203
Suter PM (2020) The B-vitamins. In: Essential and toxic trace elements and vitamins in human health. Elsevier, pp 217–239
Taha R, Abureesh S, Alghamdi S et al (2021) The relationship between vitamin D and infections including COVID-19: any hopes? Int J Gen Med:3849–3870
Takahashi N, Saito D, Hasegawa S et al (2022) Vitamin A in health care: suppression of growth and induction of differentiation in cancer cells by vitamin A and its derivatives and their mechanisms of action. Pharmacol Ther 230:107942
Talaei A, Mohamadi M, Adgi Z (2013) The effect of vitamin D on insulin resistance in patients with type 2 diabetes. Diabetol Metab Syndr 5:1–5
Tardy AL, Pouteau E, Marquez D et al (2020) Vitamins and minerals for energy, fatigue and cognition: a narrative review of the biochemical and clinical evidence. Nutrients 12(1):228
Tjong E, Dimri M, Mohiuddin SS (2022) Biochemistry, tetrahydrofolate. In: StatPearls [Internet]. StatPearls Publishing
Trump DL (2018) Calcitriol and cancer therapy: A missed opportunity. Bone Rep 9:110–119
Turcotte JJ, Stone AH, Gilmor RJ et al (2020) The effect of neuraxial anesthesia on postoperative outcomes in total joint arthroplasty with rapid recovery protocols. J Arthroplast 35(4):950–954
Vardi M, Levy NS, Levy AP (2013) Vitamin E in the prevention of cardiovascular disease: the importance of proper patient selection. J Lipid Res 54(9):2307–2314
Vermeer CV (2012) Vitamin K: the effect on health beyond coagulation–an overview. Food Nutr Res 56(1):5329
Vishwakarma SP, Singh P, Singh U, Singh RL (2013) Antioxidant activities of some tuberous plants leaves. Int J Pharm Sci Rev Res 20(1):28–33
Voelker AL, Miller J, Running CA et al (2018) Chemical stability and reaction kinetics of two thiamine salts (thiamine mononitrate and thiamine chloride hydrochloride) in solution. Food Res Int 112:443–456

Wan Z, Zheng J, Zhu Z et al (2022) Intermediate role of gut microbiota in vitamin B nutrition and its influences on human health. Front Nutr 9:1031502
Wdowik T, Gryko D (2022) C–C bond forming reactions enabled by vitamin B12 – opportunities and challenges. ACS Catal 12(11):6517–6531
Wei FF, Trenson S, Verhamme P, Vermeer C, Staessen JA (2019) Vitamin K–dependent matrix Gla protein as multifaceted protector of vascular and tissue integrity. Hypertension 73(6):1160–1169
Wolf G (2005) The discovery of the antioxidant function of vitamin E: the contribution of Henry A. Mattill. J Nutr Mar 135(3):363–366
Yadav CSU, Kalariya NM, Ramana KV (2011) Emerging role of antioxidants in the protection of uveitis complications. Curr Med Chem 18(6):931–942
Yang J, Liu J, Parry J (2009) Vitamin C: daily requirements, dietary sources and adverse effects. In: Handbook of vitamin C research. Nova Science Publishers, New York, pp 237–260
Yoshii K, Hosomi K, Sawane K et al (2019) Metabolism of dietary and microbial vitamin B family in the regulation of host immunity. Front Nutr 6:48
Yu W, Jia L, Park SK et al (2009) Anticancer actions of natural and synthetic vitamin E forms: RRR-α-tocopherol blocks the anticancer actions of γ-tocopherol. Mol Nutr Food Res 53(12):1573–1581
Zhang Z, Liu L, Liu C et al (2021) New aspects of microbial vitamin K2 production by expanding the product spectrum. Microb Cell Factories 20:1–12
Zhou S, Mehta BM, Feeney EL (2022) A narrative review of vitamin K forms in cheese and their potential role in cardiovascular disease. Int J Dairy Technol 75(4):726–737

Electrolytes and Acid-Base Disorders

7

Shelly Sehgal, Sachin Gupta, and Manish Kumar Mishra

Abstract

An array of chemical components is required by the human body for the reactions governing daily metabolic activities. Since, most of these reactions are interconnected, the concentration of the participating chemicals needs to be in ideal balance because any deviation from this may have adverse effects on various organs as well as organ systems. Subsequent to water, the electrolytes are highly significant in carrying out the chemical reactions of life. Apart from maintaining the osmotic balance and conducting electrical impulses across the cell membranes, they also aid in the release of vital hormones. This chapter discusses the ways in which the body maintains these critical balances. Several hormones like angiotensin, ADH (antidiuretic hormone), PTH (parathyroid hormone), and aldosterone regulate the electrolyte balance in the body along with hypothalamus and kidneys. The pH of bodily fluids such as blood is maintained between 7 and 7.5 by the acids and bases present in phosphates, bicarbonate ions, cellular proteins, plasma, hemoglobin, etc., in conjunction with electrolytes. Any deviation from the maintenance and regulation of acid-base and electrolyte homeostasis may lead to disorders like metabolic acidosis, alkalosis, ketosis with a major effect on kidneys and renal system. Along with the basic concepts, this chapter will provide insights into the current advances in the pathology, clinical presentation, and treatment of the disorders associated with the imbalances of these ions.

S. Sehgal (✉) · M. K. Mishra
Centre for Molecular Biology, Central University of Jammu,
Jammu, Jammu and Kashmir, India
e-mail: shelly.molb@cujammu.ac.in

S. Gupta
Department of ENT, SMGS Hospital Jammu, Jammu, Jammu and Kashmir, India

R. L. Singh et al. (eds.), *Clinical Applications of Biomolecules in Disease Diagnosis*, https://doi.org/10.1007/978-981-97-4723-8_7

Keywords

Electrolytes · Buffers · Osmolality · Acidosis · Alkalosis

7.1 Introduction to Electrolytes and Acid-Base Homeostasis

Most of the cellular functions are carried out in fluid medium present in the human body (blood, plasma, urine, etc.) which is accurately balanced, and several mechanisms maintain the optimum pH required to achieve equilibrium (Lord 1999). The molecules which control the cell charge neutrality may be dysregulated/dysfunctional, because of high or low levels due to several stress conditions. If the balance of electrolytes, ions, buffers, and acid-bases is not well regulated and exceeds or decreases than the normal level, serious life-threatening conditions can arise (Lord 1999; Batlle et al. 2012). Major organs like kidneys and lungs play an essential role in the regulation of these molecules inside the body by the mechanism of excretion and reabsorption. Along with this, plasma osmolarity also plays a role in regulating the balance of ions and turbidity.

7.2 Electrolytes

Electrolytes are molecules that possess a net positive or negative charge when dissolved in a liquid medium (Grimm et al. 2021). Electrolytes are responsible for the maintenance of electrical neutrality in cells or intercellular electrical neutrality, nerves, and muscle action potential. Fluctuations in levels of Na, P, Ca, HCO_3^-, Mg, Cl, and K pose hindrances in the normal functioning of the body leading to electrolyte disorders (Elliott 2017). Bicarbonate is responsible for managing systemic acid-base balance and is regulated within the arterial pH range of 7.36–7.44 and that of intercellular is 7.2 approximately (Romero and Rossano 2019). Imbalance within this range can be detrimental to cellular processes viz. restriction in enzymatic functions ultimately causing cell death. Different homeostatic mechanism contributes to keeping check on fluctuating pH.

1. Chemical acid-base buffer systems
2. Respiration and its control by the respiratory center
3. The kidneys

Blood buffer system provides defense/escape against changes in systemic pH.

7.2.1 Role of HCO_3^-/CO_2 Buffer in Acid-Base Homeostasis

Quantitatively, bicarbonate system is the largest buffer system of the blood. It takes few seconds to minutes to regulate the pH change. For homeostasis, both HCO_3^-/

CO_2 buffer regulates the acid/alkali loads quantitatively and independently regulate HCO_3^- & PCO_2 by kidneys and lungs (Occhipinti and Boron 2019). Arterial CO_2 is managed by alveolar ventilation. Upon its production in peripheral tissues as carbonic acid, it gets released as H^+ & HCO_3^- (Silva et al. 2009). The reaction is being catalyzed by a metalloenzyme present in the blood. The reaction followed is in accordance with the Le Chatelier's principle, which implies that to neutralize the effects of changes on system at equilibrium, the same tends to adjust the direction of reaction of either forward or backward as required to be at equilibrium (Norwich 2010).

$$CO_2 + H_2O \rightarrow H_2CO_3 \leftrightarrow H^+ + HCO_3^-$$

Renal regulation of acid-base balance is achieved by Predominant reabsorption & absorption and excretion of HCO_3^- leads to regulation of plasma HCO_3^- by this renal acid-base handling. In response to high/low CO_2 production, Alveolar production is decreased/increased to maintain PCO_2 and constant pH that is any increase in the plasma CO_2 and decrease in pH stimulates ventilation and it gets increased (Occhipinti and Boron 2019). Thereby bringing PCO_2 at normalcy unlike in renal control of plasma HCO_3^- where kidney achieves the balance by completely reabsorbing of all the filtered HCO_3^- and generation of HCO_3^- as the replacement of consumed acids. However, HCO_3^- production depends on net acid excretion into urine. To ensure homeostasis, add on of acid and loss of alkali are equivalent and equilibrium is achieved by shifting or favoring the above reaction in forward direction or backward as is the case. Mismanagement of acid and base may have clinical implications related to metabolic acidosis and respiratory disorders (Quintanilla 1976; Romero and Rossano 2019).

7.2.2 Major Ions Present in the Human Body

There is a strong interconnection between ions, water, and health. Our body fluid is rich in electrolytes, which may be defined as the chemicals produce charged ions when dissolved in water (Lord 1999; Schwaller 2010; Grimm et al. 2021). The major ions present in the body are sodium, potassium, chloride, nitrogen, phosphate, sulfate, calcium which perform various functions such as ion exchange, muscle movement, maintains body homeostasis, etc. (Kilner et al. 2023).

7.2.3 Role of Major Ions in the Human Body

The ions hence generated, facilitate the course of electrical signals throughout the body. Numerous ions, also known as electrolytes, are found throughout the body and serve a range of purposes (Table 7.1). Enzyme protein structures are stabilized with the aid of other ions. Others still help endocrine glands release hormones. Certain ions help electrical impulses in neurons and muscles travel over their cell membranes. Every ion present in plasma plays a role in maintaining the osmotic

Table 7.1 Function and effects of various ions on the human body

S.No.	Ions	Functions	If taken in less amount	If it is in high amount
1	Na^+	Acid – base balance Regulation of osmotic pressure Transmission of nerve signals	Muscle cramps Hyponatremia (less than 135 mmol/L) Seizures Nausea	Hypernatremia (more than 145 mmol/L) Hypertension Nausea
2	K^+	Promotes membrane potential Nerve – impulse generation Contraction of muscles	Hypokalemia (less than 3.6 mmol/L) Apathy Irregular heartbeat	Hyperkalemia (more than 5.5 mmol/L) Cardiac arrhythmia Cardiac failure
3	Mg^{2+}	Activation of enzymes Promotes protein synthesis Regulation of blood sugar level	Muscle weakness Muscle cramps Hypomagnesemia (less than 1.46 mg/dL)	Diarrhea Nausea Hypermagnesemia (more than 2.68 mg/dL)
4	Ca^{2+}	Blood clotting Aids in nerve and muscle functioning Muscle contraction	Osteoporosis Brittle bones Hypocalcemia (less than 8.8 mg/dL)	Cardiac arrhythmia Hypercalcemia (more than 10.7 mg/dL)
5	Cl^-	Enables the secretion of stomach acid Regulation of osmotic pressure	Convulsions Hypochloremia (less than 96 mmol/L) Respiratory acidosis	Hypertension Hyperchloremia (more than 106 mmol/L)

Table 7.2 Dietary sources of essential ions

S.No.	Ions	Dietary sources
1	Na^+	Pickled foods, cheese
2	Cl^+	Table salt
3	K^+	Fruits, vegetables
4	Mg^{2+}	Seeds, nuts
5	Ca^{2+}	Dairy products, green leafy vegetables

equilibrium, which regulates the flow of fluid among the cells and their surroundings. Fruits and vegetables are rich source of minerals (Singh et al. 2014). Dietary sources of essential ions are summarized in Table 7.2.

7.2.4 Electrolytes as Osmotic Regulators

Electrolytes are crucial in regulating the osmotic pressure within the cells which sustains the functioning of muscle and nerve cells. Any deviation in the levels of electrolytes severely affects the cellular functioning and may cause detrimental impacts on health. The five key electrolytes necessary for normal body functions are Calcium (Ca), Chloride (Cl), Potassium (K), Magnesium (Mg), and Sodium (Na)

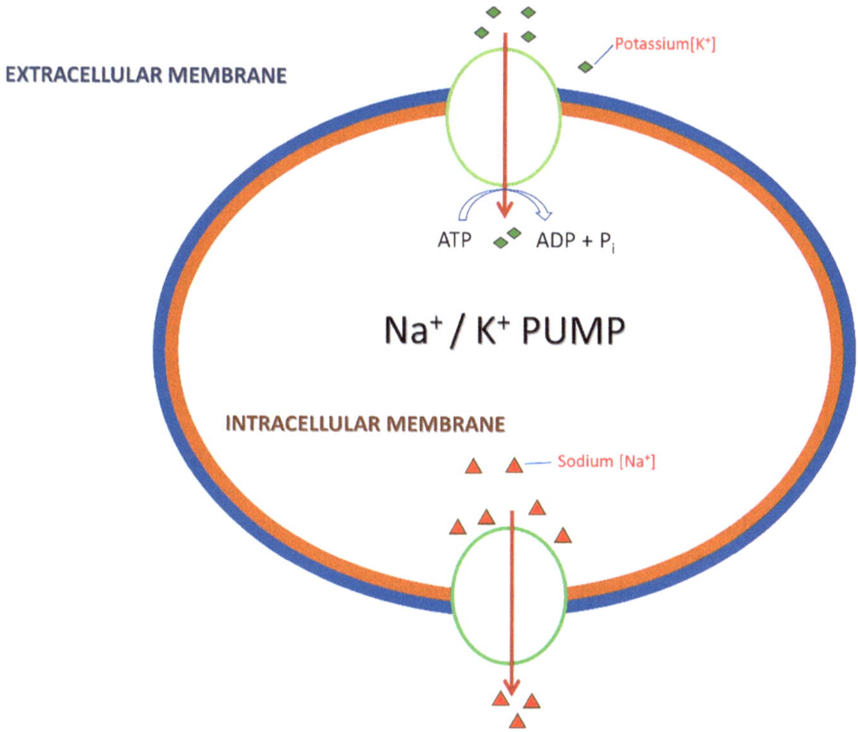

Fig 7.1 Graphical depiction of sodium-potassium pump. ATP is used for intake of potassium and sodium is fluxed out from the cell to maintain ionic balance both in intracellular and extracellular environment

(Schwaller 2010). These form electrolytes when they are dissolved in a fluid via metabolic processes which results in positively or negatively charged ions (Fig 7.1).

Basic functions of ions present in the body are

- Balancing of fluids
- Maintenance of blood's proper pH
- Carry electrical signals from nerves to muscles

7.2.4.1 Role of Electrolytes

These five ions support the function of neuronal excitability, the release of hormones, membrane permeability, bodily fluid buffering, and fluid transport between divisions.

Sodium: An essential extracellular cation for preserving osmotic pressure, cellular excitability, and fluid balance is sodium. It actively contributes to blood pressure management and aids in nerve impulse transmission. One important cellular process that maintains the correct distribution of potassium and sodium across cell membranes and contributes to the resting potential of the membrane is the sodium-potassium pump.

Potassium: Potassium, which is primarily found inside cells, is essential for biological functions like muscular contraction and nerve signal transmission. The gradient in potassium concentration has a significant impact on the membrane potential at rest of cells. Its critical significance in preserving regular physiological function is highlighted by the fact that potassium deficiency can result in heart arrhythmias and muscular weakness.

Magnesium: Magnesium, also known as the "forgotten cation," is essential to more than 300 enzymatic processes in the body. It is necessary for DNA replication, ATP synthesis, and the proper operation of muscles and nerves. Magnesium is essential for maintaining general health because a lack can cause cramping in the muscles, neurological problems, and anomalies in the cardiovascular system.

Calcium: As a multipurpose ion, calcium is important for the development of bones and teeth, blood coagulation, contraction of muscles, and the transfer of nerve impulses. Hormones including calcitonin and parathyroid hormone (PTH) are essential in controlling calcium levels. Enzyme activity, cellular signaling, and structural integrity depend on the precise equilibrium between the mineral calcium in the extracellular and intracellular compartments.

Chloride: Chloride, the principal anion in extracellular fluid, helps to keep osmotic pressure and electrolyte balance. It is an essential part of the stomach's hydrochloric acid, which helps indigestion. Chloride and sodium work together to regulate the distribution of water and the acid-base balance.

Together, these ions support the preservation of equilibrium, a constantly changing equilibrium essential to the healthy operation of human beings. Ion concentrations are regulated by complex regulatory mechanisms that maintain homeostasis in the face of internal and external fluctuations.

7.2.4.2 Regulatory Mechanisms

Ion concentrations are mostly controlled by the endocrine system. Adrenal gland-produced hormones like aldosterone encourage the kidneys to reabsorb sodium, which affects blood pressure and the body's overall sodium balance. The kidneys' ability to reabsorb water is controlled by antidiuretic hormone (ADH), which has an impact on blood ion concentration.

The glands of the parathyroid gland, which release parathyroid hormone (PTH) in reaction to low calcium in the blood levels, strictly control calcium homeostasis. PTH increases the absorption of calcium in the intestines encourages the absorption of calcium in the kidneys and stimulates the release of calcium from bones. The thyroid gland calcitonin promotes calcium deposition in the bones, which counteracts PTH. An essential physiological process known as the sodium-potassium pump actively moves potassium into cells against concentration gradients and sodium out of cells. This mechanism uses ATP as energy and keeps the electrochemical gradient that is required for cell excitability and communication in place.

7.3 Buffers

7.3.1 Introduction to Buffers

There are different types of buffers that are actively involved in various metabolic activities, depending on the pH range that needs to be maintained and the specific chemical and biological requirements of the system (Shah and Mandiga 2022). Some common biological buffer systems include the phosphate buffer, the bicarbonate buffer, and the amino acid buffer (Hoenig and Zeidel 2014; Alzeer and Alzeer 2023).

7.3.2 Role of Buffers

In the human body, buffers play a crucial role in maintaining pH balance. Buffers are substances that resist changes in the acidity or alkalinity of a solution. In biological systems, the pH level needs to be tightly regulated for optimal functioning of enzymes and other biochemical processes. Buffers help prevent drastic pH changes by accepting or donating protons (H+ ions), stabilizing the internal environment of cells and tissues. This is essential for processes like metabolism and cellular function. Examples of biological buffers include bicarbonate ions in the blood and proteins within cells (Gehlbach and Schimdt 2004).

7.3.2.1 Maintaining Homeostasis

The carbonic acid–bicarbonate buffer system is the most important buffer for maintaining the pH of homeostasis of blood (Silva et al. 2009; Occhipinti and Boron 2019). The bicarbonate buffer system is an acid-base homeostatic mechanism involving the balance of carbonic acid, bicarbonate ion, and carbon dioxide in order to maintain pH in the blood and organs or the tissue of the blood to support proper metabolic function.

Bicarbonate buffer in the blood have a pH of 7.4. In this buffer, the pH is balanced by the presence of both weak acid and its conjugate base so that any excess acid or base introduced to the system is neutralized. An excess of acids in the blood is called acidosis, while an excess of bases is called alkalosis, any deviation or change in blood pH can affect organ function, sometimes cause death (Alzeer and Alzeer 2023). Bicarbonate ions (HCO_3^-) play a significant role in maintaining the pH balance in the human body. They act as a crucial buffer system, particularly in the blood.

7.3.2.1.1 Blood pH Regulation

Bicarbonate ions help to regulate the pH of the blood by acting as a buffering agent. They can accept or donate protons (H+ ions) to maintain the blood's pH within a narrow and optimal range, typically around 7.35–7.45.

7.3.2.1.2 Acid-Base Balance

In the carbonic acid-bicarbonate buffer system, bicarbonate ions combine with hydrogen ions to form carbonic acid (H_2CO_3), which can then dissociate into water and carbon dioxide. This reversible reaction help resist sudden changes in blood pH, neutralizing excess acids or bases.

7.3.2.1.3 Respiratory Regulation

The bicarbonate ion is linked to the respiratory system. When there is an increase in acidity (more H+ ions) in the blood, the respiratory system can increase the elimination of carbon dioxide (CO_2), reducing the concentration of carbonic acid and helping restore pH balance. Overall, bicarbonate ions play key roles in maintaining the acid-base equilibrium, which is critical for various biochemical processes and the proper functioning of enzymes in the body (Ellison et al. 1958).

7.3.2.2 Regulation of Enzymatic Functions

Enzymes are very specific in nature and function optimally at the right temperature and right pH. In order for enzyme to function optimally, cells secrete acids and bases to maintain the correct pH levels. If the pH of blood goes out of range, the enzyme stops working and this in turn affects a lot of biological processes in human body, leading to various diseases. Buffers in the human body help regulate enzymatic functions by maintaining the pH within a narrow range. Enzymes functions optimally at specific pH levels and buffers resist changes in pH by either accepting or donating protons. This stability ensures that enzymatic reactions occur efficient, supporting various physiological processes essential for the body's proper functioning.

Buffers in the body, such as bicarbonate and phosphate ions, play a crucial role in regulating enzymatic functions by maintaining a stable pH environment. Enzymes often have an optimal pH at which they function most effectively. Buffers help resist changes in pH by either accepting or releasing protons (H+ ions) (Stangherlin et al. 2021). For example, if there is an increase in acidity (more H+ ions), the buffer system absorbs excess protons, preventing a significant drop in pH. Conversely, if there is an increase in alkalinity (fewer H+ ions), the buffer releases protons to maintain a balanced pH. This pH stability is vital for the proper functioning of enzymes. Enzymes are sensitive to changes in pH, and deviations from their optimal pH impeding denaturation or reduced activity. Buffers act as a protective mechanism that ensures that enzymatic reactions occur well within the appropriate pH range for optimal efficiency in various physiological processes.

7.4 Maintenance of Plasma Osmolality

7.4.1 Plasma and Plasma Osmolality

Plasma is basically the liquid part of the blood in which clotting is prevented by adding an anti-coagulant. Plasma is obtained by performing centrifugation of whole blood in which an anti-coagulant is present. It is a light-yellow liquid that is clear in appearance with a density of 1.025 g/ml (Rowlands et al. 2022). Plasma osmolality may be defined as the test used to measure the concentration of different solutes present in the plasma (Lord 1999). It tells us about the water balance.

7.4.2 Composition of Plasma

Plasma is composed of four constituents namely water, inorganic substances, organic substances, and gases.

7.4.2.1 Water

90% of the plasma is water having electrolytes which are the charged minerals and control the amount of fluid in the blood (Mathew et al. 2023).

7.4.2.2 Inorganic Substances

0.9% of plasma contains inorganic substances which consist of cations [Na^+, H^+, Ca^{2+}, Mg^{2+}] and anions [Cl^-, HCO_3^-, PO_4^{3-}, SO_4^{2-}].

- Na^+ – has a role in maintaining extracellular fluid volume and membrane potential (Kilner et al. 2023).
- Ca^{2+} – muscle contraction, clotting of blood, hormone secretion, imparting nerve impulse (Kilner et al. 2023).
- Mg^{2+} – performs the function of release of neurotransmitters, muscle contractions, and metabolism of adenosine triphosphate (Kilner et al. 2023).
- PO_4^{3-} – involved in metabolic pathways, an important part of ATP (Kilner et al. 2023).

7.4.2.3 Organic Substances

9.0% of organic substances are there in plasma and these are further present in three different forms which are:

- Plasma protein (albumin, globulin, fibrinogen, prothrombin). Albumin is the white protein portion, which is small and present in large numbers. Globulins are globular proteins larger in size and less in number, prothrombin cleaves to form thrombin which enzymatically converts fibrinogen into fibrin which clots blood. Plasma lipids and plasma proteins combine to form lipoproteins.
- Plasma lipids which include cholesterol, triglycerides, and phospholipids.
- The last organic substance includes miscellaneous compounds such as glucose, amino acids, vitamins, nitrogenous waste, etc.

7.4.2.4 Gases

The other constituent of plasma is gases like O_2, CO_2, etc.

7.4.3 Measurement of Osmolality

Osmolality is a clinical test that is performed to measure the concentration of all the chemical particles that are present in the body fluids such as blood, urine, and stool. These particles can be glucose, amino acids, electrolytes (Na, Cl, K), and urea. In other words, osmolality is the total number of osmoles/solutes present in 1 kg water (Lord 1999). It is very important to maintain the osmolality of the fluid concentrations. Slight changes in the concentrations can be hazardous. The imbalance in the fluid concentrations can be due to dehydration, heart disease, kidney disease, and diabetes. A device named Osmometer is used to measure the osmolality in the labs.

Based on the source of fluid used to measure the fluid concentration, osmolality is of three types: -

i. *Blood osmolality* – To calculate blood osmolality, plasma or serum are used which are the liquid portion of the blood.
ii. *Urine osmolality* – It is performed to know the reason behind increased or decreased urination.
iii. *Stool osmolality* – It is used in the identification of chronic diarrhea.

Osmolality can be measured as osmolality or calculated osmolality. Measured osmolality is determined using an osmometer in the labs. Calculated osmolality is determined using a formula:

$$1.86[\mathrm{Na}] + [\mathrm{Glucose}]/18 + [\mathrm{BUN}]/2.8 + 9$$

Another important aspect is the gaps present in the osmolality which are known as the osmolar gaps. It is the value that comes in between measured osmolality and calculated osmolality. There are four types of gaps – Serum anion, Urine anion, Serum osmolar, and stool osmolar gap. The normal range of plasma osmolality is 275–290 m OSmol/kg (Lord 1999). There are osmoreceptors present in the hypothalamus which basically senses the plasma osmolality. So slight change in the plasma osmolality which may be even 1% can activate the osmoreceptors which in turn stimulates the thirst. Plasma osmolality can only be maintained by the regulation of water intake and excretion of urine. When the plasma osmolality increases, osmoreceptors sense that and further instruct the neurons in the hypothalamus region to secrete ADH (Antidiuretic hormone). ADH that is circulating binds with the V2 receptors of the collecting duct in the kidney which leads to water channel insertions, i.e., the Renal collecting duct containing Aquaporin-2 which leads to the increased permeability of water. Then in the hypertonic renal medulla, the water is reabsorbed. This leads to an increase in urine concentration and decreased water excretion. Urine contains solutes such as urea, and sodium which are also excreted along with urine. So basically, plasma osmolality is linked with ADH secretion. A 1% change in plasma osmolality can lead to the secretion of ADH, so secretion of ADH is regulated. With low plasma osmolality, secretion stops which leads to the excretion of diluted urine, i.e., Na+-free urine is excreted out which corrects the plasma osmolality. Water intake is regulated by the osmoreceptors in the hypothalamus which stimulates the thirst with the increase in the serum osmolality. This shows that change in plasma osmolality is linked to both urine osmolality and serum osmolality. The human body also prevents the loss of water by the process of oxidation from the skin, lungs, urine, and gastrointestinal tract. Adequate intravascular volume needs to be maintained for the controlled osmolality. An increase in this leads to the stimulation of both thirst and ADH secretion irrespective of the plasma osmolality. For thirst to be initiated, there needs to be a moderate decrease in the volume but plasma osmolality should have only a 1–2% change.

Maintenance of plasma osmolality is done by the posterior pituitary and renal organ system and Renin Angiotensin Aldosterone System (RAAS). RAAS maintains the extracellular volume, sodium concentration, and blood pressure by secreting the hormone "RENIN" which circulates to the adrenal glands.

7.4.4 Mechanisms to Maintain Plasma Osmolality

There are various systems involved in the maintenance of plasma osmolality as mentioned below.

7.4.4.1 Posterior Pituitary/Renal Systems

- Hyperosmolar state – When plasma is having less fluid, the body becomes dehydrated leading to increased plasma osmolality making the cell to shrink. When neurons experience this shrinkage as they act as osmoreceptors, they undergo depolarization through transient receptor potential vanilloid (TRVP) cation channels which increases the cell's charge. This action leads to the release of antidiuretic hormone (ADH). ADH works under the influence of V2 receptors increasing the intracellular cAMP creating an aquaporin channel due to which the filtrate water can be reabsorbed, decreasing the plasma osmolality, and thus maintaining the homeostasis (Venugopal et al. 2017; Danziger and Zeidel 2015; Shah and Mandiga 2022).
- Hypo-osmolar state – Release of ADH is reduced due to the hyperpolarization which commands the kidneys to excrete a higher amount of urine, thus again increasing the plasma osmolality (Venugopal et al. 2017; Shah and Mandiga 2022).

7.4.4.2 Renin Angiotensin Aldosterone System

The primary function of macula densa cells in the distal convoluted tubule is to analyze NaCl concentration. Two situations can happen:

- Decreased NaCl concentration in the filtrate – This decrease is sensed by macula densa which dilates the afferent renal arterioles increasing the hydrostatic pressure which further increases the NaCl concentration (Hoenig and Zeidel 2014). Renin is released by the juxtaglomerular cells which receive a signal by prostaglandins secreted by macula densa. Renin converts angiotensinogen into Angiotensin 1 and further into Angiotensin 2.

7.4.4.3 Functions Angiotensin 2

i. The posterior pituitary receives a signal to release ADH (Mecawi et al. 2015).
ii. Blood pressure is increased by vascular myocyte contraction leading to higher hydrostatic pressure and filtrate production and NaCl concentration is filtered.
iii. Increases sympathetic activity.
iv. Reabsorption of tubular NaCl and excretion of potassium and water is increased which overall increases the plasma NaCl concentration and thus increasing plasma osmolality.
v. Increased NaCl concentration in the filtrate – Prostaglandins secretion is reduced which blocks the RAAS pathway. The mechanism of how it works is mentioned earlier.

7.5 Renal Regulation of Acid-Base Balance

In order to function properly, our body needs the right acid and base balance of compounds (Imenez Silva and Mohebbi 2022). In other words, it refers to the balance between intake production and elimination of H^+ ions (Lee Hamm et al. 2015).

7.5.1 Major Organs Involved in the Acid-Base Balance Regulation

7.5.1.1 Lungs

The foremost organ involved in maintaining the acid-base balance is lungs. Lungs are crucial for excretion of CO_2 (12,000–13,000 mmol/day), which aids in regulation (Occhipinti and Boron 2019). Blood pH is maintained by releasing CO_2 from the lungs.CO_2 is constantly produced by cells which passes into the blood; this carbon dioxide is then transported by the blood which is further exhaled by the lungs. The pH of the blood decreases as CO_2 accumulates in the blood. In other words, lungs ensure carbonic acid removal and the decreased blood CO_2 results in the decreased blood H_2CO_3 increasing the pH, hence restoring the acid-base hemostasis. Regulation of CO_2 and carbonic acid levels is done via pulmonary capillaries of blood.

7.5.1.2 Kidneys

Another essential role in maintaining the acid-base balance is performed by the pair of kidneys and non-volatile acids are excreted via them (amount involved 70–100 mmol/much smaller) (Tannen 1983; Hoenig and Zeidel 2014). The bicarbonate that is filtered can be reabsorbed through the kidneys. Against the non-volatile acids, the concentration of plasma bicarbonate should be fortified in opposition to renal loss, as it is the chief extracellular buffer. In order to first understand the physiology of renal acid-base, the relationship between the roles of the kidney to the lungs for the maintenance of the systemic balance of acid-base must be understood. Dietary food contains carbohydrates and fats in the majority and to metabolize them completely insulin and O_2 are required which yields water and carbon dioxide. With the lung's normal functionality, the produced CO_2 (20 mol/day) is excreted maintaining the systemic acid-base balance. Depending on the specific amino acids, there is a production of either alkali or acids on the metabolic activity of these amino acids, but dietary proteins produce net acids on metabolic activity often called "non-volatile acids" which are usually buffered rapidly. The lungs help in the excretion of the carbon dioxide produced, whereas kidneys excrete the acidic salt (Na^+) along with NH_4^+. HCO_3^- is lost during the process of titrating the non-volatile acids, hence more HCO_3^- is required produced via the process of excretion of NH_4^+ which returns to the blood to replace the lost bicarbonate (Koeppen 2009). pH levels are affected by CO_2 and HCO_3^- together leading to the distinction between volatile and non-volatile acid. If the production of CO_2 is more than the excretion, an increase in [H^+] would be abruptly seen, thus, leading to acidosis but if the vice versa happens, a decrease in [H^+] concentration leads to alkalosis.

7.5.1.2.1 Role of Kidneys

The bicarbonate reabsorption is filtered (4000–5000 mmol/day). Non-volatile acids and excretion of anionic and H^+ ions associated are with it (1 mmol per kilogram per day) (Fig 7.2).

Two main tubular mechanisms are involved: Proximal: helps in NH_3 production and bicarbonate reabsorption (approx-85%). Carbonic anhydrase catalyzes the reaction via two routes, namely (Na^+-H^+) antiporter and the second one which act as

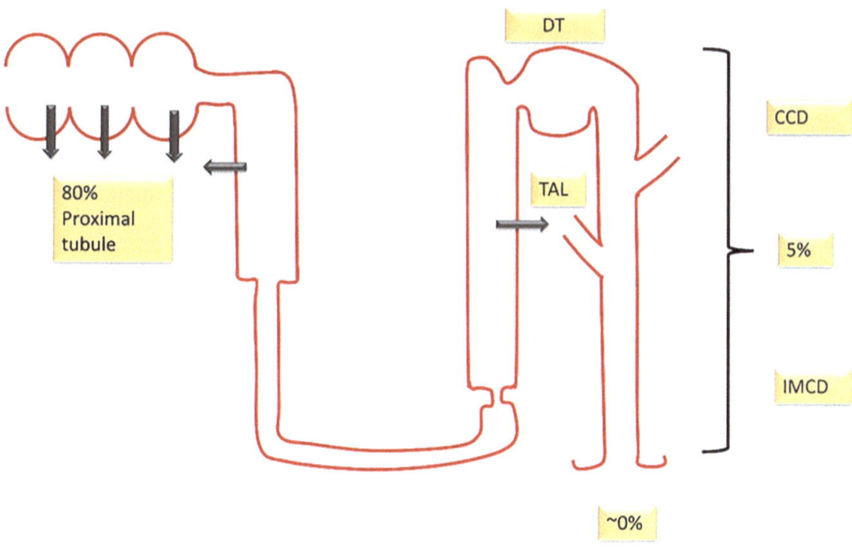

Fig. 7.2 Reabsorption of bicarbonates along the nephron; DT Distal tubule, CCD Cortical collecting duct, IMCD inner medullary collecting duct, TAL thick ascending limb

a proton pump, called H^+- ATPase are involved which helps the hydrogen ion to leave the PTC (proximal tubule cell) and its lumen.

Reabsorption of bicarbonate can be controlled on the basis of four main factors-

- Luminar HCO_3^- conc.
- Arterial PCO_2
- Luminar flow rate
- Angiotensin 2

It is pertinent to mention that if any of the above factors has even the slightest increase in the value, it will affect directly affect the reabsorption of HCO_3^-.

Also, an increase in the hormone called parathyroid will lead to an increase in cAMP and a decrease in bicarbonate reabsorption. Within PTCs, NH_4 is majorly produced via glutamine with the help of an enzyme called glutaminase which acts on it. Metabolized glutamate produces Alpha–ketoglutarate which leads to the production of ammonium. NH_3 is entirely present in the acid form NH_4^+, i.e., its pKa value is 9.2 (very high). If the pH of the urine is low, then the concentration of NH_3 that is needed to be excreted would be high. Also, if acidosis is present, then it would further lead in the increment in ammonium concentration.

Distal- It contains intercalated cells which reabsorb the left amount of bicarbonate. Here, the excreted quantity of non-volatile acids is low, i.e., with low capacity and a high gradient system.

It includes three processes:

- TA genesis
- Luminal flow is introduced with NH_3 (via a process called medullary cycling-regulates the conc. of urine)
- Reabsorption of remaining bicarbonate (PTC absorbs about 85%, loop of henle-10% and distal about 5% of the filtered bicarbonate. Also, on further removal of water from Loop of Henle, there is a decrease in vol. of the filtrate.) Metabolic alkalosis can be maintained by the reduction in GFR. With the reduction in GFR is directly proportional to the filtered load of bicarbonate.

7.6 Metabolic Acidosis and Alkalosis

7.6.1 Metabolic Acidosis

When body fluids like blood, urine, sweat, etc., have elevated levels of acids present, it is known as acidosis (Seifter 2014). It may also occur when the kidneys cannot remove the acid in the form of urine. A plasma bicarbonate concentration of less than 20 mmol/L is indicative of an acid-base imbalance, as defined by the Henderson-Hasselbalch method that defines metabolic acidosis (Jung et al. 2019). "Acidemia" is the word used to describe the lowered pH caused by this imbalance; the condition is usually considered serious when the pH reaches or drops below 7.20 (Jung et al. 2019).

Intensive care unit or emergency room patients frequently develop metabolic acidosis. Different plasma and urine tests are available to doctors to help them diagnose metabolic acidosis, define its underlying causes, and direct patient care (Kimmoun et al. 2015). In particular, respiratory (leading to increased ventilatory demand) and cardiovascular (causing arterial vasodilation, reductions in cardiac inotropism and cardiac output, and ventricular arrhythmia) complications may be brought on by acute metabolic acidosis, which can coexist with a variety of illnesses (Kraut and Madias 2010, 2012; Kimmoun et al. 2015). Since there is little human clinical data available, the effects of acute metabolic acidosis on these organ dysfunctions are primarily deduced from studies conducted on animals or in vitro.

For metabolic acidosis to be managed, the underlying acid-base disturbance must be addressed. For example, electrolyte imbalance correction and appropriate fluid resuscitation are required in cases of sepsis and diabetic ketoacidosis. Other treatment options could include antibiotics, dialysis, bicarbonate supplementation in some cases, and antidotes for acidosis poisoning.

The various types of metabolic acidosis are:

- Diabetic acidosis occurs due to the buildup of substances called ketone bodies (acidic in Nature) during uncontrolled diabetes (Wallace and Matthews 2004). Irregular release of insulin may cause the release of counter-regulatory hormones like catecholamine, cortisol, glucagon, and growth hormone which results in hyperglycemia due to the enhancement of gluconeogenesis and glycogenolysis.
- The loss of excessive sodium bicarbonate from the body during severe diarrhea may lead to hyperchloremic acidosis.
- Acidosis can also be caused by various kidney-related disorders like uremia, distal renal tubular acidosis, etc.
- In lactic acidosis, the blood lactate levels are increased by overproduction of lactic acid in muscle tissues indicating unbalanced glucose metabolism (Spiegelberg et al. 2022). Very high accumulation of L-lactic acid leads to a state of severe acidemia and cardiovascular collapse.

7.6.2 Metabolic Alkalosis

A normal human body pH function at around 7.35–7.45. An increase in pH beyond this range is known as alkalosis. Thus, metabolic alkalosis is a condition in which the body's pH is greater than 7.45 secondary to some metabolic process. Bicarbonate and carbon dioxide are the main components of the pH buffer system in the human body (Khanna and Kurtzman 2001; Gillion et al. 2019). Bicarbonate is basic while carbon dioxide is acidic. Thus, an increase in bicarbonate concentration and a reduction in carbon dioxide concentration make blood more basic in nature. Metabolic alkalosis is due increase in serum bicarbonate levels that are regulated via the renal system. The causes of metabolic alkalosis can be caused due to gastrointestinal loss of hydrogen ions, intracellular shift of hydrogen ions, excessive renal hydrogen ion loss, retention or addition of bicarbonate ions or volume contraction around a constant amount of extracellular bicarbonate called contraction alkalosis. Usually, metabolic alkalosis indicates an overabundance of HCO_3^-. There are two possible origins of this excess HCO_3^-: external, endogenous, or a combination of both (Brinkman and Sharma 2023). Exogenous sources include salts of precursors like lactate, acetate, or citrate, which completely oxidize to produce HCO_3^-, or salts of Na^+ or K^+, HCO_3^-. The two main endogenous sources that can produce appreciable amounts of HCO_3^- are the stomach and the kidneys. The body must eliminate H^+ in order for the process of net endogenous HCO_3^- production to occur (Aronson and Giebisch 2011). The production of HCO_3^- occurs when HCl is secreted into the stomach lumen; however, external loss of HCl, usually through vomiting and/or suction, is necessary for the net accumulation of HCO_3^- in the extracellular fluid (ECF) (Tinawi 2021).

7.7 Electrolyte Disorders

7.7.1 Understanding Electrolyte Disorders

The minerals that carry electric charge when dissolved in water in the human body are called electrolytes (Chenevey 1987). As described previously, sodium, potassium, calcium, chlorine, and magnesium constitute the most commonly found electrolytes in the human body (Elliott 2017). These ions have an important role in physiological processes such as essential for keeping your nervous system, muscle contraction, and internal environment balanced. In sustaining homeostasis of the body also, these electrolytes play a major role. Sudden change in the electrolyte's concentrations can lead to various disorders which can be quite fatal for one's health (Reddi 2023). The changed concentration is due to the issues in the mechanisms that maintain electrolyte balance (Timerga et al. 2020). Nowadays on a large scale, the cases of electrolyte disorders are expanding thus leading to severe illness and even death (Timerga et al. 2020). Numerous factors contribute to this imbalance, and the major one being sweating which leads to loss of Na^+, renal loss, vomiting leading to gastrointestinal loss (Hawkins 2003).

7.7.2 Factors Causing Electrolyte Imbalance

The term "electrolyte imbalance" refers to a change in the normal levels of these ions in the body (Allison 2004). This imbalance can be caused by various factors, including:

i. *Dehydration*: When you lose fluids through activities like sweating, vomiting, or diarrhea, it can disrupt the balance of electrolytes, particularly sodium and potassium (Quintanilla 1976; Lee Hamm et al. 2015).

ii. *Kidney Problems*: The major role of the kidney is to regulation of electrolyte levels. Kidney disorders can lead to imbalances. (The electrolyte homeostasis is maintained by the kidney.)
iii. *Diet*: Consuming an excessive amount of certain foods or drinks (e.g., high-sodium or low-potassium diets) can impact electrolyte balance (Demigné et al. 2004).
iv. *Health Conditions*: Various medical conditions, such as diabetes, heart problems, and hormonal disorders, can disrupt electrolyte levels (Smiley et al. 2011).
v. *Excessive Exercise*: Prolonged or intense physical activity can lead to electrolyte imbalances due to excessive sweating (Rowlands et al. 2022).
vi. *Some Medications*: Medications, e.g., steroids, laxatives, diuretics, and some chemotherapy drugs can also lead to electrolyte disorders.
vii. *Nutritional and water-intake*: Lack of proper nutrition and intake of more water can also lead to electrolyte disorders.
viii. *Genetic conditions*: Electrolyte disorder can also be caused in some cases where people may have some inherited genetic conditions that disturb the body's capability to maintain appropriate electrolyte balance.

Electrolyte imbalances can have bad effects on our body, including:

i. Functioning of muscle: Some of the crucial electrolytes that play important roles in the contraction of muscle such as magnesium, potassium, and calcium, the imbalance of these electrolytes can lead to muscle cramps, weakness, spasms, and twitching, irregular heart rhythms (arrhythmia), and fatigue (Balcı et al. 2013).
ii. Irregular heart rhythms (arrhythmias): The electrical activity of the heart is disturbed due to electrolyte imbalance which may lead to life-threatening conditions.
iii. Weakness: Phosphate and calcium are important electrolytes for bone health and imbalance of these electrolytes leads to the weakening of bones and some bone disease (Lee and Cho 2015).
iv. Functioning of the nervous system: In the nervous system, various electrolytes such as calcium and sodium act as helping hand in transmitting signals. The improper concentration of these electrolytes leads to some symptoms such as numbness, confusion, seizures, etc. (Jacoby 2020).
v. Irregular Blood pressure: The imbalance of electrolytes such as chloride and sodium leads to high blood pressure or low blood pressure.
vi. Acid-base balance: Bicarbonate is one of the crucial electrolytes that help in balancing the pH in the body. Alkalosis or acidosis is the condition caused by imbalance of these electrolytes.

It is important to maintain a proper balance of electrolytes through a balanced diet, hydration, and monitoring if you have specific medical conditions that may affect your electrolyte levels (Dhatariya and Matfin 2018).

7.7.3 Common Electrolyte Disorders

7.7.3.1 Hyponatremia

This occurs when blood sodium levels are too low, below 135 mmol/L. It can lead to symptoms like confusion, seizures, and muscle cramps. Treatment involves addressing the underlying cause, such as fluid restriction or administration of hypertonic saline (Hoorn and Zietse 2017).

7.7.3.2 Hypernatremia

This condition arises due to an increased level of sodium in blood, beyond 145 mmol/L. The symptoms presented by restlessness, difficulty in sleeping, and rapid breathing (tachypnoea). Treatment includes correcting fluid deficits either orally or intravenously, and addressing the underlying cause (AL-Nattah et al. 2023).

7.7.3.3 Hypokalemia

When potassium levels in serum falls below 3.6 mmol/L, the condition is termed as hypokalemia. This is often caused by gastrointestinal disorders, due to which the body loses high amounts of potassium muscle twitching, fatigue, weakness, and hypokalemic paralysis (Viera and Wouk 2015). Treatment involves oral or intravenous potassium replacement, along with addressing the underlying cause (Kardalas et al. 2018).

7.7.3.4 Hyperkalemia

High levels of potassium in serum, above 5.5 mmol/L can arise hyperkalemic conditions. Cardiac arrythmias is a common symptom of hyperkalemic condition (Youn and McDonough 2009; Viera and Wouk 2015). Treatment may include dietary restrictions, medications to shift potassium into cells, and in severe cases, dialysis (Palmer et al. 2021).

7.7.3.5 Hypocalcemia and Hypercalcemia

Abnormal levels of calcium like more than normal range or less than normal range which is required by the body can impact bone health and neuromuscular function (Stewart 1983). Treatment for hypocalcemia involves oral or intravenous calcium supplementation, and for hypercalcemia includes hydration, loop diuretics, bisphosphonates, or calcitonin depending on the severity and underlying cause (Cooper and Gittoes 2008; Minisola et al. 2015).

7.8 Conclusions

Acid-base homeostasis is important for the proper functioning of various enzymes and metabolites inside the human body. Electrolytes such as Na, K, Cl, Ca, Mg, etc., play major roles in regulating acid-base homeostasis in the human body. The role of carbonic acid (HCO_3^-/CO_2) buffer is also instrumental in maintaining the constant pH of bodily fluids. Kidney is the primary organ that maintain a pH in the range of

7–7.5, which is optimal for various enzymes. Any dysregulation in the pH of blood or other body fluids affects the normal functioning of organs or organ systems. Diseases such as acidosis and alkalosis are caused due to accumulation of acids such as lactic acid or bicarbonate ions in the blood, respectively. Dietary habits and lifestyle are important factors which influence the homeostasis. Various diseases such as hyponatremia, hypokalemia, hypocalcemia, and hypercalcemia are some common types of electrolyte disorders.

References

Allison S (2004) Fluid, electrolytes and nutrition. Clin Med 4:573

AL-Nattah M, Abdullah A, Alkhateeb N et al (2023) Navigating a complex presentation: management of hypernatremic dehydration, acute kidney injury, hyperkalemia, and metabolic acidosis in a patient with down syndrome: a case report. Cureus 15

Alzeer J, Alzeer J (2023) The role of buffers in establishing a balance of homeostasis and maintaining health. Am J Med Chem 2023:1–6

Aronson PS, Giebisch G (2011) Effects of pH on potassium: new explanations for old observations. J Am Soc Nephrol 22:1981–1989

Balcı AK, Koksal O, Kose A et al (2013) General characteristics of patients with electrolyte imbalance admitted to emergency department. World J Emerg Med 4:113

Batlle D, Chen S, Haque SK (2012) Physiologic principles in the clinical evaluation of electrolyte, water, and acid-base disorders. In: Seldin and Giebisch's the kidney: physiology and pathophysiology, pp 2477–2511

Brinkman JE, Sharma S (2023) Physiology, metabolic alkalosis. In: StatPearls [Internet]. StatPearls Publishing, Treasure Island. Available from: https://www.ncbi.nlm.nih.gov/books/NBK482291/

Chenevey B (1987) Overview of fluids and electrolytes. Nurs Clin North Am 22:749–759

Cooper MS, Gittoes NJL (2008) Diagnosis and management of hypocalcaemia. BMJ: Br Med J 336:1298

Danziger J, Zeidel ML (2015) Osmotic homeostasis. Clin J Am Soc Nephrol 10:852–862

de Mecawi A, Ruginsk SG, Elias LLK et al (2015) Neuroendocrine regulation of hydromineral homeostasis. Compr Physiol 5:1465–1516

Demigné C, Sabboh H, Rémésy C, Meneton P (2004) Protective effects of high dietary potassium: nutritional and metabolic aspects. J Nutr 134:2903–2906

Dhatariya K, Matfin G (2018) Severe hyperglycemia, diabetic ketoacidosis, and hyperglycemic hyperosmolar state. Endocr Metab Med Emer:531–547

Elliott TL (2017) Electrolytes: potassium disorders. FP Essent 459:21–28

Ellison G, Straumfjord JV, Hummel JP (1958) Buffer capacities of human blood and plasma. Clin Chem 4:452–461

Gehlbach BK, Schimdt GA (2004) Bench-to-bedside review: treating acid–base abnormalities in the intensive care unit – the role of buffers. Crit Care 8:259

Gillion V, Jadoul M, Devuyst O, Pochet JM (2019) The patient with metabolic alkalosis. Acta Clin Belg 74:34–40

Grimm LM, Humann-Ziehank E, Zinne N et al (2021) Analysis of pH and electrolytes in blood and ruminal fluid, including kidney function tests, in sheep undergoing long-term surgical procedures. Acta Vet Scand 63:43

Hawkins RC (2003) Gender and age as risk factors for hypokalemia and hyperkalemia in a multiethnic Asian population. Clinica Chimica Acta 331:171–172

Hoenig MP, Zeidel ML (2014) Homeostasis, the milieu inte'rieur, and the wisdom of the nephron. Clin J Am Soc Nephrol 9:1272–1281

Hoorn EJ, Zietse R (2017) Diagnosis and treatment of hyponatremia: compilation of the guidelines. J Am Soc Nephrol 28:1340

Imenez Silva PH, Mohebbi N (2022) Kidney metabolism and acid-base control: back to the basics. Pflugers Arch 474:919–934

Jacoby N (2020) Electrolyte disorders and the nervous system. Continuum (Minneap Minn) 26:632–658

Jung B, Martinez M, Claessens YE et al (2019) Diagnosis and management of metabolic acidosis: guidelines from a French expert panel. Ann Intensive Care 9:92

Kardalas E, Paschou SA, Anagnostis P et al (2018) Hypokalemia: a clinical update. Endocr Connect 7:R135

Khanna A, Kurtzman NA (2001) Metabolic alkalosis. Respir Care 46:354–365

Kilner JA, Druce J, Ishihara T (2023) Electrolytes. High-temperature solid oxide fuel cells for the 21st century: fundamentals. Des Appl:85–132

Kimmoun A, Novy E, Auchet T et al (2015) Hemodynamic consequences of severe lactic acidosis in shock states: from bench to bedside. Crit Care 19(1):175

Koeppen BM (2009) The kidney and acid-base regulation. Adv Physiol Educ 33:275–281

Kraut JA, Madias NE (2010) Metabolic acidosis: pathophysiology, diagnosis and management. Na Rev Nephrol 6(5):274–285

Kraut JA, Madias NE (2012) Treatment of acute metabolic acidosis: a pathophysiologic approach. Nat Rev Nephrol 8:589–601

Lee AW, Cho SS (2015) Association between phosphorus intake and bone health in the NHANES population. Nutr J 14:28

Lee Hamm L, Nakhoul N, Hering-Smith KS (2015) Acid-base homeostasis. Clin J Am Soc Nephrol 10:2232–2242

Lord RCC (1999) Osmosis, osmometry, and osmoregulation. Postgrad Med J 75:67–73

Mathew J, Sankar P, Varacallo M (2023) Physiology, blood plasma. In: StatPearls [Internet]. StatPearls Publishing. Treasure Island. Available from: https://www.ncbi.nlm.nih.gov/books/NBK531504/

Minisola S, Pepe J, Piemonte S, Cipriani C (2015) The diagnosis and management of hypercalcaemia. BMJ 350:h2723

Norwich KH (2010) Le Chatelier's principle in sensation and perception: fractal-like enfolding at different scales. Front Physiol 1:1–7

Occhipinti R, Boron WF (2019) Role of carbonic anhydrases and inhibitors in acid–base physiology: insights from mathematical modeling. Int J Mol Sci 20(15):384120

Palmer BF, Carrero JJ, Clegg DJ et al (2021) Clinical management of hyperkalemia. Mayo Clin Proc 96:744–762

Quintanilla AP (1976) Acute acid base disorders. II. Specific disturbances. Postgrad Med 60:75–83

Reddi AS (2023) Electrolyte and acid-base disorders: clinical evaluation and management 1–729.

Romero MF, Rossano AJ (2019) Acid-Base Basics. Semin Nephrol 39:316–327. https://doi.org/10.1016/J.SEMNEPHROL.2019.04.002

Rowlands DS, Kopetschny BH, Badenhorst CE (2022) The hydrating effects of hypertonic, isotonic and hypotonic sports drinks and waters on central hydration during continuous exercise: a systematic meta-analysis and perspective. Sports Med 52:349–375

Schwaller B (2010) Cytosolic Ca^{2+} buffers. Cold Spring Harb Perspect Biol 12:a035543

Seifter JL (2014) Integration of acid–base and electrolyte disorders. New Engl J Med 371:1821–1831

Shah MM, Mandiga P (2022) Physiology, plasma osmolality and oncotic pressure. In: StatPearls [Internet]. StatPearls Publishing, Treasure Island. Available from: https://www.ncbi.nlm.nih.gov/books/NBK544365/

Silva AS, Yunes JA, Gillies RJ, Gatenby RA (2009) The potential role of systemic buffers in reducing intratumoral extracellular pH and acid-mediated invasion. Cancer Res 69:2677–2684

Singh RL, Vishwakarma SP, Singh P (2014) Vitamins and minerals: roles and plant sources. In: Prakash D (ed) Phytochemicals of nutraceutical importance and their role in human health. CABI-UK, pp 248–265

Smiley D, Chandra P, Umpierrez GE (2011) Update on diagnosis, pathogenesis and management of ketosis-prone Type 2 diabetes mellitus. Diabetes Manage 1:589–600

Spiegelberg J, Lederer AK, Claus S et al (2022) Severe hyperlactatemia in unselected surgical patients: retrospective analysis of prognostic outcome factors. BMC Surg 22(1):312

Stangherlin A, Watson JL, Wong DCS et al (2021) Compensatory ion transport buffers daily protein rhythms to regulate osmotic balance and cellular physiology. Nat Commun 12:6035

Stewart PA (1983) Modern quantitative acid-base chemistry. Can J Physiol Pharmacol 61:1444–1461

Tannen RL (1983) Ammonia and acid-base homeostasis. Med Clin North Am 67:781–798

Timerga A, Kelta E, Kenenisa C et al (2020) Serum electrolytes disorder and its associated factors among adults admitted with metabolic syndrome in Jimma Medical Center, South West Ethiopia: facility based crossectional study. PLoS One 15(11):e0241486

Tinawi M (2021) Pathophysiology, evaluation, and management of metabolic alkalosis. Cureus 13(1):e12841

Venugopal V, Carone V, Kenchaiah M (2017) Hyperosmolar hyperglycaemic state. Acute medicine – a practical guide to the management of medical emergencies, 5th edn, pp 499–500

Viera AJ, Wouk N (2015) Potassium disorders: hypokalemia and hyperkalemia. Am Fam Physician 92:487–495

Wallace TM, Matthews DR (2004) Recent advances in the monitoring and management of diabetic ketoacidosis. QJM 97:773–780

Youn JH, McDonough AA (2009) Recent advances in understanding integrative control of potassium homeostasis. Annu Rev Physiol 71:381–401

Metabolism of Macro-elements (Calcium, Magnesium, Sodium, Potassium, Chloride and Phosphorus) and Associated Disorders

8

Shivmurat Yadav, Jyoti Yadav, Sandeep Kumar, and Pankaj Singh

Abstract

The human body requires a variety of minerals to maintain its intricate daily homeostasis. These minerals participate in a broad spectrum of biological processes, intricately involved in cell signaling, nerve function, muscle operation, bone integrity, fluid equilibrium, and metabolic regulation. Macro-elements, typically requiring over 100 mg/day, play a pivotal role in ensuring optimal physiological processes. Despite stringent regulatory mechanisms, occasional imbalances may lead to a variety of disorders and many serious health consequences. This chapter scrutinizes the roles of six major macro minerals – calcium (Ca), phosphorus (P), magnesium (Mg), sodium (Na), potassium (K), and chlorine (Cl) in human health. Imbalances in calcium affect bones, muscles, nerves, kidneys, and heart rhythms. Phosphorus disorders impact bones, soft tissues, and kidney function. Magnesium disorders contribute to conditions like hypertension, insulin resistance, headaches, and mood disorders. Sodium deficiency causes nausea and seizures, while excess leads to hypertension, edema, and heart issues. Potassium imbalance influences muscles, nerves, and cardiovascular health. Chloride disorders result in complications like vomiting, diarrhea, metabolic alkalosis, and cystic fibrosis. This comprehensive exploration covers various dimensions of these minerals, including functions, metabolism, deficiency risks, and potential consequences of excess intake, associated disorders, and their regulation.

S. Yadav · J. Yadav · S. Kumar
Department of Medicine, University of Oklahoma Health Sciences Center, Oklahoma City, OK, USA

P. Singh (✉)
Department of Biotechnology, Dr. Rammanohar Lohia Avadh University, Ayodhya, India
e-mail: pankaj@rmlau.ac.in

R. L. Singh et al. (eds.), *Clinical Applications of Biomolecules in Disease Diagnosis*, https://doi.org/10.1007/978-981-97-4723-8_8

Keywords
Macro-element · Metabolic disorders · Calcium · Phosphorus · Magnesium · Sodium · Potassium · Chloride

8.1 Introduction

Maintaining daily homeostasis in the human body requires a variety of minerals crucial for vital biological functions, such as cell signaling, nerve function, muscle operation, bone integrity, fluid balance, and metabolic regulation. These minerals are categorized into micro-minerals and macro-minerals based on their required daily intake. Macro-elements, requiring over 100 mg/day, play a pivotal role in ensuring optimal physiological processes. Imbalances of macro-elements including calcium, phosphorus, magnesium, sodium, potassium, and chloride can lead to serious health disorders. Proper daily intake is crucial for normal physiological function.

Calcium is essential for bone and teeth formation, muscle contraction, and blood clotting, regulated by hormones like parathyroid hormone (PTH) and vitamin D3. Phosphate is integral to cellular metabolism, playing a central role in energy transfer, nucleic acid synthesis, and cellular signaling. In addition to PTH and vitamin D3 hormones, Fibroblast growth factor-23 (FGF23) also regulates phosphate levels. Magnesium is crucial for over 300 enzymatic reactions, impacting energy metabolism, muscle function, and nerve transmission. Kidney and intestinal regulation prevent deficiencies or excesses, avoiding metabolic disorders of magnesium affecting cardiovascular and neurological health. Potassium, vital for cellular function, nerve transmission, and muscle contraction, is regulated by kidneys, with dietary intake as a factor. Its imbalances can lead to metabolic disorders affecting the heart, muscles, and nerves.

Sodium, crucial for fluid balance, nerve function, and cell membrane potential, is regulated by kidneys influenced by aldosterone. Imbalances can result in neurological manifestations and symptoms like tachypnea. Chloride, linked to sodium and regulated by the kidneys, contributes to fluid and electrolyte balance, with imbalances affecting gastrointestinal and water retention conditions. Studies show that antioxidant-rich foods prevent micro and macro element induced metabolic-related diseases (Singh et al. 2014; Singh and Singh 2018). Macro-element metabolism is influenced by factors like genetics, hormonal imbalances, kidney function, and diet. This chapter covers the importance, regulation, and main disorders or consequences associated with macro-elements (Fig. 8.1). Recommended dietary intakes of macronutrients are summarized in Table 8.1.

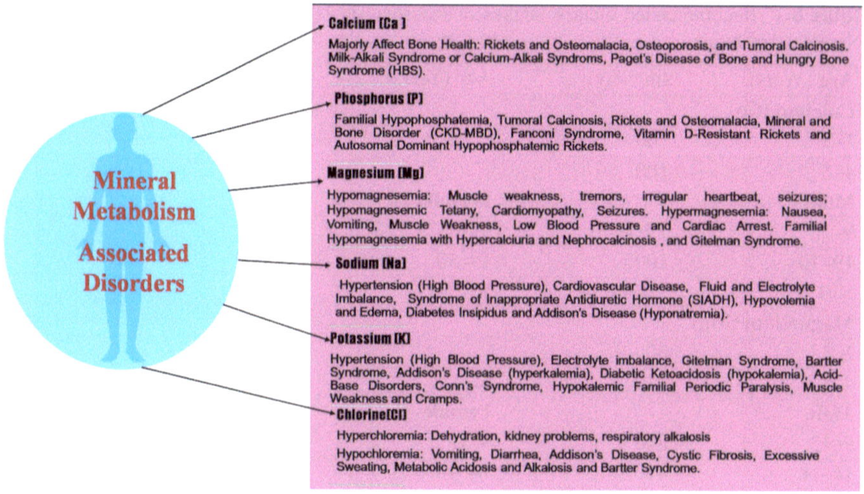

Fig. 8.1 Adequate dietary supply of non-synthesizing macro-elements is crucial, as disturbance in these minerals can lead to various serious metabolic disorders and diseases

8.2 Metabolic Regulation, Functions, and Disorders of Calcium Metabolism

Calcium is essential for bone and teeth formation, muscle contraction, blood clotting, cell signaling, enzyme activation, neurotransmitter release, and hormone secretion, and plays a crucial role in diverse physiological functions and metabolic pathways in the human body. Metabolism of calcium is tightly regulated to maintain a balance between absorption, excretion, and storage of calcium. Parathyroid hormone (PTH), crucial in calcium regulation, directly affects the kidney and bone through PTH1R receptors on tubular cells, osteoblasts, and osteocytes. It also indirectly impacts the gastrointestinal tract (Fermor and Skerry 1995; Datta and Abou-Samra 2009). In the kidney, PTH enhances calcium reabsorption and vitamin D secretion, enhancing calcium absorption in the gut. PTH regulates bone remodeling by directly stimulating osteogenesis and bone formation and indirectly promoting osteoclast resorption (Kousteni and Bilezikian 2008). Impaired PTH release leads to pathological conditions like hyperparathyroidism or hypoparathyroidism, characterized by hypercalcemia or hypocalcemia, respectively.

8.2.1 Metabolic Disorders of Calcium

The total serum calcium concentration typically ranges between 8.5 and 10.5 mg/dL, and the normal range for ionized calcium is 4.65–5.25 mg/dL (Goyal et al. 2023). Deviations above or below these normal ranges result in hypercalcemia or

Table 8.1 Recommended dietary intakes of macronutrients by the Food and Nutrition Board, National Academies

Age (Years)	(mg/day)	Age (Years)	(mg/day)
Calcium (Ca)			
1–3	700		
4–8	100		
Male		**Female**	
9–18	1,300	9–18	1300*
19–70	1000	19–50	1000*
>70	1200	>50	1200*
Magnesium (Mg)			
1–3	80		
4–8	130		
Male		**Female**	
9–13	240	9–13	240
14–18	410	14–18	360^{n}, 360^{L}, and 400^{P}
19–30	400	19–30	310^{n}, 310^{L}, and 350^{P}
>30	420	>30	320^{n}, 320^{L}, and 360^{P}
Phosphorus (P)			
1–3	460		
4–8	500		
Male		**Female**	
9–18	1250	9–18	1250*
>18	700	>18	700*
Potassium (K)			
1–3	2000		
4–8	2300		
Male		**Female**	
9–13	2500	9–13	2300
14–18	3000	14–18	2300^{n}, 2600^{P}, and 2500^{L}
>18	3400	>18	2600^{n}, 2900^{P}, and 2800^{L}
Sodium (Na)			
1–3	800		
4–8	1000		
Male		**Female**	
9–13	1200	9–13	1200*
>14	1500	>14	1500*
Chloride (Cl)			
1–3	1500		
4–8	1900		
Male		**Female**	
9–50	2300	9–50	2300*
50–70	2000	50–70	2000
>70	1800	>70	1800

*For all types of women (pregnant/lactating and neither pregnant nor lactating), *P* for pregnant women, *L* for lactating women, and *n* for those who are not pregnant or lactating hypocalcemia, and most calcium disorders are associated with one of these conditions.

Many diseases and conditions are linked to calcium metabolic disorders are:

8.2.1.1 Hypocalcemia

Hypocalcemia, characterized by low blood calcium levels, has three primary causes. First, PTH-related issues as it is crucial for calcium balance (Zivin et al. 2001). Second, renal failure, often resulting from kidney diseases, diminishes vitamin D activation, affecting both calcium absorption and renal calcium processes. Nutritional deficiencies, namely, vitamin D deficiency and calcium deficiency cause low level of calcium in blood (Zivin et al. 2001). Unlike hyperparathyroidism, chronic hypocalcemia stems primarily from hypoparathyroidism, caused by factors like parathyroidectomy, radiation damage, or congenital conditions. Pseudohypoparathyroidism (PsHP), another condition resistant to PTH, also manifests as hypocalcemia. Additionally, hypocalcemia can arise from vitamin D deficiency due to inadequate UV exposure or kidney dysfunction (Cooper and Gittoes 2008).

8.2.1.2 Hypercalcemia

Hypercalcemia, characterized by elevated blood calcium levels, can result from various causes, including primary hyperparathyroidism, malignancy, certain medications, hyperthyroidism, granulomatous diseases, excessive calcium intake, and adrenal gland disorders.

8.2.1.3 Malignancy and Hyperparathyroidism

Hypercalcemia is a condition seen in about 1 in 500 adults, primarily caused by malignancy or hyperparathyroidism (Bilezikian and Silverberg 2004; Stewart 2005; Waters 2009). In cancer patients, hypercalcemia is often due to bone invasion (Local Osteolytic Hypercalcemia; LOH), excessive parathyroid hormone-related protein (PTHrP) release (Humoral Hypercalcemia of Malignancy; HHM), or increased 1,25(OH)2D synthesis (Insogna and Broadus 1987). Bone absorption increases due to osteoclast-activating factors in LOH, while HHM results from excessive PTHrP secretion from malignant cells (Guise et al. 1996; Roodman 2004). PTHrP binds to PTH receptors, promoting bone absorption and renal calcium reabsorption (Stewart et al. 1982; Burtis et al. 1988). Certain lymphoid tumors, like Hodgkin's lymphoma, over release 1,25(OH)2D, thereby intensifying both intestinal calcium absorption and renal reabsorption processes (Seymour et al. 1994).

8.2.1.4 Hyperparathyroidism

The parathyroid glands, located near the thyroid in the neck, produce parathyroid hormone (PTH) to regulate calcium levels. Hyperparathyroidism, marked by excessive PTH production, impacts bones significantly as it leads to increased calcium release from bones into the bloodstream. It has two main categories: primary hyperparathyroidism (PHPT), often due to adenomas, and secondary hyperparathyroidism (SHPT), caused by hypocalcemia-driven hyperplasia. Unlike PHPT, where the problem is with the parathyroid glands themselves, SHPT is often a result of an underlying health issue affecting calcium balance. This type is usually a response to conditions that lower calcium levels in the body, such as chronic kidney disease or dietary calcium deficiency. In SHPT, the parathyroid glands become overactive as they try to compensate for low calcium levels by producing more PTH. Elevated PTH levels during hyperparathyroidism stimulate 1,25(OH)2D synthesis, increasing renal calcium reabsorption, osteoclastic activity, and intestinal calcium absorption, ultimately leading to hypercalcemia.

8.2.1.5 Calcium-Sensing Receptor (CaSR) Associated Disorders

The mutation in CaSR is associated with three major disorders: Familial Hypocalciuric Hypercalcemia (FHH), Neonatal Severe Hyperparathyroidism (NSHPT) and Autosomal Dominant Hypocalcemia (ADH).

8.2.1.5.1 Familial Hypocalciuric Hypercalcemia (FHH)

It stems from mutated CaSR, and sarcoidosis, a granulomatous disorder with elevated 1, 25(OH)2D levels, also contribute to hypercalcemia (Carroll and Schade 2003; Iannuzzi et al. 2007). FHH is characterized by elevated blood calcium levels and reduced urinary calcium excretion due to decreased CaSR sensitivity to calcium levels. This results in a diminished response to changes in serum calcium, leading to inadequate suppression of parathyroid hormone (PTH) secretion (Pearce et al. 1996; Christensen et al. 2008).

8.2.1.5.2 Neonatal Severe Hyperparathyroidism (NSHPT)

NSHPT in newborns is due to nonfunctional or absent of CaSR. This disrupts the normal PTH regulation, resulting in excessive PTH production and high blood calcium levels (Gorvin 2019).

8.2.1.5.3 Autosomal Dominant Hypocalcemia (ADH)

ADH leads to low blood calcium levels because the receptors are more sensitive toward calcium-sensing receptor (CaSR). This causes even slightly low calcium levels to trigger an excessive reduction in PTH secretion, resulting in hypocalcemia. This contrasts with FHH, where reduced receptor sensitivity causes insufficient suppression of PTH and leads to hypercalcemia (Pearce et al. 1996).

8.2.1.6 Nutritional Rickets and Osteomalacia

These are conditions characterized by deficiencies in essential nutrients, particularly vitamin D, calcium, and phosphorus. Rickets primarily affects children and is marked by the softening and weakening of the bones, leading to deformities. Osteomalacia, on the other hand, occurs in adults and is characterized by the softening of the bones due to poor mineralization. Vitamin D plays a crucial role in maintaining proper bone health by aiding in the absorption of calcium and phosphorus in the intestine. Inadequate intake of vitamin D, along with insufficient calcium and phosphorus in the diet, can lead to these conditions.

8.2.1.7 Osteoporosis

Osteoporosis weakens the bones, making them porous and brittle due to decreased bone mineral density (BMD) and quality, which elevates the risk of fractures, especially in the hip, spine, and wrist. It progresses symptom-free until a fracture manifests, often termed the "silent disease." Adequate calcium intake is associated with higher BMD. Low BMD is a key characteristic of osteoporosis, indicating a reduction in the amount of mineralized bone tissue.

8.2.1.8 Tumor-Induced Osteomalacia (TIO)

TIO is a rare paraneoplastic syndrome characterized by osteomalacia, where bones become soft and weak due to poor mineralization (Florenzano et al. 2017). Unlike nutritional causes, TIO is linked to a specific tumor called a phosphaturic mesenchymal tumor, which produces excess fibroblast growth factor 23 (FGF23). Elevated FGF23 leads to increased urinary phosphate excretion, impaired kidney phosphate absorption, and low blood phosphate levels, ultimately causing osteomalacia. Treatment involves surgical tumor removal and supplementation with phosphate and vitamin D to restore normal bone mineralization.

8.2.1.9 Tumoral Calcinosis

Tumoral calcinosis (TC) is a rare syndrome marked by calcium salt deposition in peri-articular soft tissues, typically presenting in childhood or adolescence as painless masses around joints, potentially causing limitations in joint function, particularly with larger masses (Fathi and Sakr 2014).

8.2.1.10 Milk-Alkali Syndrome or Calcium-Alkali Syndrome

In this syndrome, hypercalcemia occurs due to excessive intake of calcium, alkali-rich foods or supplements, antacids, and vitamin (Beall et al. 2006; Patel et al. 2013).

8.2.1.11 Paget's Disease of Bone

Paget's disease is a chronic disorder affecting bone remodeling, increasing bone turnover, and releasing calcium into the bloodstream. Hyperparathyroidism, often linked to hypersensitive osteoclasts, is the leading cause of hypercalcemia in these patients, and adenectomy (parathyroid gland removal) can alleviate both the biochemical and clinical symptoms of Paget's disease (Green and Altman 2009).

8.2.1.12 Neonatal Hypocalcemia

Neonatal hypocalcemia is linked to disruptions in calcium metabolism, with insufficient calcium transfer from the mother during pregnancy posing a risk (Vuralli 2019). Premature infants are particularly vulnerable as their underdeveloped parathyroid glands may hinder calcium regulation, exacerbated by potential calcium insufficiency in the later stages of pregnancy (Vuralli 2019). Maternal factors like diabetes or certain medications can influence calcium metabolism, contributing to neonatal hypocalcemia. Additionally, impaired parathyroid gland function and vitamin D deficiency play crucial roles, impacting calcium regulation in newborns (Elsary et al. 2018).

8.2.1.13 Hypoparathyroidism

Hypoparathyroidism is a rare endocrine disorder characterized by abnormally low or absence of production of PTH, leading to reduced serum calcium and elevated serum phosphorus levels (Abate and Clarke 2016). The primary cause of hypoparathyroidism often stems from injury or unintentional removal of the parathyroid glands during thyroid surgery (Abate and Clarke 2016).

8.2.1.14 Thyrotoxicosis

Thyrotoxicosis, commonly induced by conditions such as Graves' disease or toxic nodular goiter (Ozkaya et al. 2015), as well as in cases of acromegaly (Manroa et al. 2014), pheochromocytoma (Heath and Edis 1979), and adrenal insufficiency (Katahira et al. 2004) which causes hypercalcemia. In these conditions, hypercalcemia is often a secondary effect of the underlying disease, and treatment focuses on managing the primary condition to restore normal calcium levels.

8.2.1.15 Hungry Bone Syndrome (HBS)

HBS typically arises in post-operative patients undergoing parathyroidectomy or thyroidectomy, and it may also occur in individuals with osteoblastic bone metastasis (Witteveen et al. 2013; Jain and Reilly 2017; Karunakaran et al. 2018; Sakai et al. 2020). This syndrome is characterized by a rapid and profound drop in serum calcium levels following surgery. During hyperparathyroidism, the body adapts to high levels of PTH by increasing bone resorption and releasing calcium into the bloodstream. When the hyper functioning parathyroid gland is removed, this adaptive mechanism continues, but there is no longer an excess of PTH to stimulate bone resorption. As a result, calcium is rapidly deposited into the bones, leading to a condition known as hungry bone syndrome. During HBS severe hypocalcemia, hypomagnesemia, and hypophosphatemia may occur (Witteveen et al. 2013).

8.3 Metabolic Regulation, Functions, and Disorders of Phosphate Metabolism

Phosphate is integral to cellular metabolism, serving essential functions in energy transfer, nucleic acid synthesis, and cellular signaling. As a key component of ATP, phosphate plays a central role in energy storage and transfer within cells. It is also a crucial component of nucleic acids, contributing to the backbone of DNA and RNA. Involvement of phosphate in phosphorylation and dephosphorylation events regulates the activity of enzymes and proteins, influencing cellular processes such as signal transduction and gene expression. Additionally, phosphate contributes to bone and teeth formation, acts as a buffer for pH regulation, and participates in the homeostasis of cellular phosphate levels. In addition to regulating calcium, vitamin D3, PTH, and calcitonin also control serum phosphate levels. Fibroblast Growth Factor 23 (FGF23), primarily secreted by osteoblasts and osteocytes, is another hormone that regulates phosphate levels in the kidneys by inhibiting phosphate reabsorption.

8.3.1 Metabolic Disorders of Phosphate

Disruptions in regulatory mechanisms may cause hypophosphatemia or hyperphosphatemia, affecting bone health, cellular function, and metabolic stability. Common causes of phosphate imbalance include inadequate dietary intake, gastrointestinal disorders, and renal excretion dysfunction. Numerous diseases and conditions are associated with disruptions in phosphate balance are:

8.3.1.1 Hypophosphatemia

Hypophosphatemia is defined as a serum phosphate level below 2.5 mg/dL, caused by inadequate intake, reduced intestinal absorption, excessive urinary excretion, or intracellular phosphate shift. Renal phosphate wasting, attributed to genetic or acquired renal disorders, may result from vitamin D deficiency, hyperparathyroidism, oncogenic osteomalacia, or Fanconi syndrome (Assadi 2010). Genetic disorders, like X-linked hypophosphatemic rickets, typically manifest in infancy. Most patients with hypophosphatemia are asymptomatic, but severe cases can lead to skeletal muscle weakness, myocardial dysfunction, rhabdomyolysis, and altered mental status.

8.3.1.2 Hyperphosphatemia

Elevated level of phosphate in the blood is known as hyperphosphatemia which arises due to increased intestinal absorption, reduced renal excretion in kidney disease, or heightened tissue release of phosphorus into extracellular fluids. Notably, apart from kidney disease-related hypocalcemia, stemming from impaired vitamin D synthesis, hyperphosphatemia can also manifest in chronic kidney disease (CKD) due to phosphorus excretion failure. This condition contributes to calcium deposits, leading to anomalies in bone mineralization and blood vessel calcification (Hamano et al. 2017). In 2006, Kidney Disease Improving Global Outcomes (KDIGO) reported that hyperphosphatemia is a hallmark disorder and designated as chronic kidney disease-mineral and bone disorder (CKD-MBD) (Khwaja 2010; Bover et al. 2016; Ketteler et al. 2017). Hypoparathyroidism-linked hyperphosphatemia occurs when low PTH levels, typically responsible for phosphate excretion, lead to elevated blood phosphate levels. Similarly, conditions like tumor lysis syndrome, rhabdomyolysis, hemolysis, hyperthermia, and acute leukemia can trigger heightened phosphorus release from tissues (Voelkl et al. 2019).

8.3.1.3 Hypoparathyroidism

In hypoparathyroidism, the elevated phosphate levels result from inadequate renal excretion due to the absence of PTH (Zavatta et al. 2021). In a normal physiological state, PTH stimulates the kidneys to excrete excess phosphate while retaining calcium. However, when PTH is deficient, as is the case in hypoparathyroidism, this regulatory mechanism is disrupted.

8.3.1.4 Familial Hypophosphatemia (X-linked Hypophosphatemia)

X-linked hypophosphatemia (XLH) is an X-linked dominant condition resulting from mutations in the phosphate regulating gene with homology to endopeptidases (PHEX) gene. This gene encodes a cell-surface-bound enzyme known as phosphate-regulating neutral endopeptidase PHEX. The enzyme is primarily expressed in osteoblasts, osteocytes, and dental tissues, including odontoblasts and cementoblasts. Animal studies indicate that Phex dysfunction boosts fibroblast growth factor 23 (FGF23) secretions, mainly from osteocytes (Haffner et al. 2019). This mechanism explains the hallmark characteristics of the disease, including renal phosphate wasting, resulting in hypophosphatemia, decreased active vitamin D synthesis, and the manifestation of features such as rickets, osteomalacia, odontomalacia, and disproportionate short stature (Haffner et al. 2019).

8.3.1.5 Oncogenic Osteomalacia

Oncogenic osteomalacia, alternatively referred to as tumor-induced osteomalacia, is an uncommon paraneoplastic syndrome resulting from mesenchymal tumors that produce FGF-23 (Dey et al. 2017). Distinctive biochemical indicators associated with this condition include hypophosphatemia and reduced levels of 1,25-dihydroxy vitamin D (Dey et al. 2017).

8.3.1.6 Hyperphosphatemic Familial Tumoral Calcinosis (HFTC)

HFTC is a rare disorder caused by disruptions in FGF23-mediated phosphate regulation, resulting in either FGF23 deficiency or resistance. This leads to hyperphosphatemia and ectopic calcifications in patients (Boyce et al. 2020). Ectopic calcifications can cause pain, swelling, and discomfort in joints, impacting joint mobility.

8.3.1.7 Fanconi Syndrome

Fanconi syndrome is characterized by proximal tubule dysfunction, resulting in excessive urinary excretion of amino acids, glucose, phosphate, bicarbonate, uric acid, and other reabsorbed solutes. Severe cases can lead to acidosis, dehydration, electrolyte imbalances, rickets, osteomalacia, and growth failure (Foreman 2019).

8.3.1.8 Tertiary Hyperparathyroidism

Tertiary hyperparathyroidism is characterized by persistent, autonomous over activity of the parathyroid glands, often stemming from long-standing SHPT, notably associated with chronic kidney disease (CKD) (Palumbo et al. 2021). CKD induces metabolic disturbances, including hyperphosphatemia, calcitriol deficiency, and hypocalcemia, prompting increased parathyroid hormone (PTH) secretion. Hyperphosphatemia directly stimulates the parathyroid gland, causing nodular hyperplasia and elevated PTH levels. Even after correcting CKD through renal transplant, the hypertrophied parathyroid tissue persists, leading to continued excessive PTH secretion, despite normal or elevated serum calcium levels (Palumbo et al. 2021).

8.3.1.9 Hereditary Hypophosphatemic Rickets with Hypercalciuria (HHRH)

HHRH is an extremely rare autosomal recessive disorder, estimated to affect 1 in 250,000 individuals (Bergwitz and Miyamoto 2019). Mutations in the Solute Carrier Family 34 Member 3 (SLC34A3) gene, which encodes the sodium-dependent phosphate transporter (NaPi-IIc), are associated with this condition. NaPi-IIc dysfunction in the proximal tubules of the nephron leads to a phosphate wasting disorder. Patients diagnosed with HHRH, who present with Genu Valgum deformity, may experience pain resolution when treated with phosphorus supplementation and undergo surgical correction of the valgum deformity (Colazo et al. 2020).

8.3.1.10 Vitamin D-Resistant Rickets

Vitamin D-resistant rickets is a rare group of disorders marked by insufficient response to vitamin D treatment, stemming from disruptions in vitamin D and/or phosphate metabolism. The main form is X-linked hypophosphatemic vitamin D-resistant rickets, and other variations include oncogenic hypophosphatemic osteomalacia, autosomal dominant hypophosphatemic rickets, hereditary hypophosphatemic rickets with hypercalciuria, and pseudo-vitamin D-deficient rickets types I and II (Grzanka and Kucharz 2004).

8.3.1.11 Autosomal Dominant Hypophosphatemic Rickets (ADHR)

Autosomal dominant hypophosphatemic rickets (ADHR) is an exceptionally rare genetic form of rickets resulting from mutations in the fibroblast growth factor 23 gene (Mameli et al. 2021). It is distinguished by hypophosphatemia, primarily due to isolated renal phosphate wasting.

8.4 Metabolic Regulation and Functions of Magnesium

Magnesium, an essential intracellular cation, plays a vital role in ATP metabolism, muscle function, neurological processes, and neurotransmitter release. It aids in calcium re-uptake by the sarcoplasmic reticulum's calcium-activated ATPase during muscle contraction. The kidney plays a central role in regulating Mg metabolism by adjusting the excretion in the urine, specifically in thick ascending limb (TAL) and distal collecting tubule (DCT) (Tinawi 2020). It ensures that normomagnesemia is maintained until the glomerular filtration rate (GFR) falls below 10–30 mL/min. Hypomagnesemia, with serum levels below 1.46 mg/dL, can result from conditions like alcohol use disorder, gastrointestinal issues, excessive renal loss, or certain medications (e.g., omeprazole). Manifestations include ventricular arrhythmias, such as torsades de pointes.

8.4.1 Metabolic Disorders of Magnesium

Metabolic disorders related to magnesium can arise from abnormalities in absorption, distribution, or excretion of magnesium.

8.4.1.1 Hypomagnesemia

Normal serum Mg concentration is 1.7–2.6 mg/dL. Hypomagnesemia, defined as serum Mg < 1.7 mg/dL, results from inadequate dietary intake, malabsorption, kidney disorders, or excessive loss through urine or the gastrointestinal tract. Symptoms include muscle cramps, weakness, and irregular heartbeat, with significant manifestations observed when serum Mg falls below 1.2 mg/dL (Pham et al. 2014).

8.4.1.2 Hypermagnesemia

Hypermagnesemia, defined as serum Mg > 2.6 mg/dL, is the opposite of hypomagnesemia and is relatively rare. It is often associated with kidney dysfunction. Symptoms may include nausea, vomiting, weakness, and, in severe cases, respiratory and cardiac arrest. Significant manifestations occur when serum Mg exceeds 4.8 mg/dL (Pham et al. 2014).

8.4.1.3 Renal Magnesium Wasting

Renal magnesium wasting is characterized by urinary excretion exceeding 1 mmol/day in conjunction with hypomagnesemia. Congenital forms, as seen in syndromes like Bartter's and Gitelman's, are associated with hypercalciuria and hypocalciuria, respectively, potentially arising from defects in the thick ascending limb of Henle's loop and the distal convoluted tubule (Sutton and Domrongkitchaiporn 1993). These conditions can result in electrolyte imbalance.

8.4.1.4 Familial Hypomagnesemia with Hypercalciuria and Nephrocalcinosis (FHHNC)

FHHNC is a rare autosomal recessive disorder caused by mutations in genes encoding tight junction proteins, specifically claudin-16 and claudin-19 (Margabandhu and Doshi 2019). This condition involves renal magnesium and calcium wasting, leading to the onset of nephrocalcinosis and renal stones in early childhood.

8.4.1.5 Primary Aldosteronism

Primary aldosteronism, known for causing secondary hypertension with common features of hypertension, hypokalemia, and metabolic alkalosis, also influences calcium, magnesium, and phosphate metabolism. In a case of severe primary aldosteronism linked to adrenal adenoma, the patient exhibited prolonged hypocalcemia, hypomagnesemia, and hypophosphatemia, along with uncontrolled hypertension and hypokalemia (Toh et al. 2017).

8.5 Metabolic Regulation, Functions, and Disorders of Sodium Metabolism

Sodium, a crucial extracellular electrolyte, plays a vital role in maintaining fluid volume, nerve function, blood pressure, and cell membrane potential. It undergoes active transport with potassium across cell membranes. Kidneys regulate sodium, with the proximal tubule handling most of the reabsorption. In the distal convoluted tubule, sodium reabsorption occurs through sodium-chloride symporters, influenced by aldosterone – a hormone produced by the adrenal glands.

Hyponatremia, the most common electrolyte disorder, is diagnosed when serum sodium falls below 135 mmol/L, leading to neurological manifestations like headaches, confusion, and nausea. Hypernatremia, with serum sodium exceeding 145 mmol/L, presents symptoms such as tachypnea (Adrogué and Madias 2000). Serum sodium levels are intricately controlled by water homeostasis through the coordination of thirst, antidiuretic hormone (ADH), and the renin-angiotensin-aldosterone system. In normal osmoregulation, thirst and vasopressin secretion are inhibited when plasma sodium falls below 135 mmol and rising linearly with increasing sodium levels (Balanescu et al. 2011). Vasopressin binds to V2 receptors in renal collecting ducts, causing aquaporin insertion into the luminal membrane, enabling water reabsorption (Bichet 2012).

8.5.1 Metabolic Disorders of Sodium

Sodium is a crucial electrolyte in the body, and imbalances in sodium levels can lead to various health issues. Here are some conditions associated with sodium:

8.5.1.1 Hyponatremia

This is a condition characterized by low levels of sodium in the blood. It can result from excessive fluid intake, certain medications, kidney problems, or hormonal imbalances. Symptoms can include nausea, headache, confusion, and, in severe cases, seizures. Various factors lead to hyponatremia through different mechanism. Excessive fluid intake known as water intoxication, or dilutional hyponatremia, is rare and occurs when the kidneys are unable to excrete the surplus water due to excessive intake (Joo and Kim 2013). Endurance exercise, such as marathons may induce hyponatremia, as athletes may consume excessive water without adequately replacing lost sodium due to sweating (Hew-Butler et al. 2017).

Several medications, including diuretics (Mann 2008), antidepressants (Gartlehner et al. 2008; Holland and Bhogle 2013; Leth-Møller et al. 2016), antipsychotics (Ali and Bazzano 2018), and certain antiepileptic drugs (Lu and Wang 2017), can disrupt sodium balance and decrease the sodium level in blood. Additionally, several medical conditions contribute to hyponatremia. Heart failure, characterized by reduced cardiac output, activates arginine vasopressin (AVP), resulting in kidney water reabsorption and hyponatremia. Thirst and angiotensin II can exacerbate this condition by promoting water intake. Moreover, a low glomerular filtration rate (GFR) causes water retention, further diluting sodium and worsening hyponatremia (Rodriguez et al. 2019). Individuals with chronic kidney disease (CKD) have a high prevalence of both hyponatremia and hypernatremia (Arzhan et al. 2021). Hyponatremia is also well documented in patients with liver cirrhosis (John and Thuluvath 2015).

8.5.1.2 Hypernatremia

Hypernatremia occurs when there is an excessive amount of sodium in the blood, which can happen due to dehydration, diabetes insipidus, or inadequate water intake, leading to symptoms like extreme thirst, dry mucous membranes, and confusion. This condition can result from various causes, including dehydration, where insufficient fluid intake or excessive fluid loss plays a significant role, often due to factors like sweating, fever, vomiting, or diarrhea (Das et al. 2015). Dehydration, different from hypovolemia, in which both salt and water are lost (Mange et al. 1997). Hypernatremia increases plasma tonicity, stimulating thirst and the release of antidiuretic hormone (ADH) to minimize water loss and increase water intake (Adrogué and Madias 2000). In contrast to hyponatremia, where water moves into the cells, during hypernatremia, increased plasma tonicity usually pulls water out of the cells, resulting in a decrease in intracellular volume.

Hypernatremia can also result from the ingestion of table salt, a rare but potentially fatal condition, particularly when a high amount (>70 g) of salt is consumed (Moder and Hurley 1990). Other factors that contribute to elevated sodium levels include the administration of intravenous (IV) fluids or tube feedings with high sodium content, consistently consuming a diet excessively rich in sodium without adequate water intake, and various kidney disorders, such as acute kidney injury and chronic kidney disease, affecting sodium excretion (Arzhan et al. 2021). Infants and the elderly are more susceptible to hypernatremia due to their decreased ability to regulate fluid balance and sense thirst (Schlanger et al. 2010; Durrani et al. 2022). Furthermore, patients with primary aldosteronism may develop mild hypernatremia, this can often be corrected through medical or surgical management of the primary aldosteronism (Gregoire 1994). Treatment for hypernatremia depends on its underlying cause and severity, often involving the correction of fluid and sodium balance through oral or intravenous rehydration. Identifying and addressing the root cause is crucial to prevent recurrence, as severe hypernatremia can lead to symptoms such as confusion, muscle twitching, and seizures.

8.5.1.3 Hypovolemia and Edema

The two disorders of sodium balance are hypovolemia and edema. Hypovolemia refers to any condition in which the extracellular fluid (ECF) volume is reduced, often due to salt and water losses not being replaced, leading to symptoms like hypotension or shock, induced by vomiting, diarrhea, diuretic therapy, bleeding, or third-space sequestration. On the other hand, edema, including ascites, is a manifestation of excess sodium and an expanded ECF volume, often mediated by an increase in capillary hydraulic pressure.

8.5.1.4 Vomiting and Diarrhea

Vomiting and diarrhea are physiological responses often associated with various medical conditions. When these conditions are severe and prolonged, they can lead to significant disturbances in the body"s electrolyte balance, particularly involving sodium, and result in dehydration. Hypernatremia, hypokalemia, and metabolic acidosis have all been observed in cholera patients with severe volume loss (>3 L) due to diarrhea (Watten et al. 1959). Hyponatremia can occur in children with significant volume loss (Tayab and Hoq 2020). Hypokalemia occurs due to fecal K^+ loss with large volume diarrhea leading to total body K^+ depletion (Watten et al. 1959; Tayab and Hoq 2020).

8.5.1.5 Syndrome of Inappropriate Antidiuretic Hormone (SIADH)

Excessive release of antidiuretic hormone (ADH) can cause water retention and dilutional hyponatremia. SIADH is an overproduction disorder of antidiuretic hormone (ADH), leading to excessive water retention and ultimately causing hyponatremia with hypervolemia due to impaired water excretion (Jones 2018). Disturbances in plasma sodium levels subject cells to hypotonic or hypertonic conditions, affecting all cells in the body. However, the clinical consequences of hyponatremia and hypernatremia are predominantly neurological, and swift fluctuations in plasma sodium levels, whether too low or too high, can result in severe, enduring, and occasionally fatal brain damage (Sterns 2015).

8.5.1.6 Diabetes Insipidus

ADH regulates the sodium and water but dilute urine with high plasma sodium (>145 mmol), which may occur either due to deficient vasopressin secretion (neurogenic diabetes insipidus) or kidney insensitivity for vasopressin (nephrogenic diabetes insipidus) (Bichet 2012). Neurogenic diabetes insipidus may occur due to head injuries, brain tumors, infections, or other conditions that damage the hypothalamus or pituitary gland, which are responsible for producing and releasing vasopressin. In nephrogenic diabetes insipidus, the kidneys are unable to respond to the normal levels of vasopressin in the body. This condition can be inherited (genetic) or acquired due to factors like certain medications, chronic kidney disease, or electrolyte imbalances. In both neurogenic and nephrogenic diabetes insipidus conations, kidneys excrete large amounts of dilute urine, leading to excessive thirst and frequent urination. Diabetes insipidus (with total absence of vasopressin or no tubular response to vasopressin) generally does not cause hypernatremia, because thirst prompts replacement of urinary losses of water (Bourque 2008). However, hypernatremia or hyponatremia can occur with insufficient or excessive water intake (Adrogué and Madias 2000; Atsariyasing and Goldman 2014).

8.5.1.7 Addison's Disease

In Addison's disease, hyponatremia occurs due to a deficiency of aldosterone, a hormone produced by the adrenal glands and acts on the kidneys to promote the retention of sodium while excreting potassium. When aldosterone is deficient, the kidneys lose their ability to adequately retain sodium. Hyponatremia in Addison's disease arises due to aldosterone deficiency, causing the kidneys to lose sodium, and increased water retention triggered by elevated ADH release in response to a decrease in systemic blood pressure and cardiac output (Benner et al. 2019).

8.6 Metabolic Regulation, Functions, and Disorders of Potassium Metabolism

Potassium, an important intracellular ion, is essential for cell membrane potential, muscle (including heart) function, and nerve transmission. Kidneys regulate potassium levels through excretion adjustments based on dietary intake and hormonal signals. Aldosterone influences potassium balance, affecting secretion in distal tubules. The sodium–potassium pump maintains homeostasis by pumping out sodium and bringing potassium into cells. Kidneys filter potassium at the glomerulus and reabsorb it in the proximal convoluted tubule and thick ascending loop of Henle. Aldosterone facilitates potassium secretion in the distal convoluted tubule. Additional secretion occurs through channels and potassium chloride cotransporters at the apical tubular membrane. Potassium imbalances may lead to cardiac arrhythmias. Hypokalemia (serum potassium <3.6 mmol/L) causes weakness, fatigue, and muscle twitching. Hypokalemic paralysis, a generalized weakness, can be familial or sporadic. Hyperkalemia (serum potassium >5.5 mmol/L) may result in arrhythmias, muscle cramps, weakness, rhabdomyolysis, and myoglobinuria.

8.6.1 Metabolic Disorders of Potassium

Potassium is a vital electrolyte that plays a crucial role in maintaining proper cellular function, nerve transmission, and muscle contraction. Imbalances in potassium levels can lead to various health issues. Some disorders associated with potassium are:

8.6.1.1 Hypokalemia

Low levels of potassium in the blood or the use of certain medications like diuretics causes excessive loss of potassium through vomiting, diarrhea. Hypokalemia, a frequently encountered electrolyte imbalance in clinical settings, can arise from diminished potassium intake, potassium shift into cells intracellular, or both renal potassium excretion and extrarenal potassium loss due to diarrhea, vomiting, and excessive sweating (Unwin et al. 2011). Three key factors influencing renal potassium excretion include heightened mineralocorticoid receptor activation, increased sodium delivery in the distal regions, and elevated non-reabsorbable ions in distal nephrons (Unwin et al. 2011). In the distal nephron, potassium (K^+) is transported into cells through basolateral Na-K-ATPases and released into the luminal fluid via apical renal outer medullary potassium (ROMK) channels. Normally, at the physiological intracellular magnesium (Mg^{2+}) concentration of 1 mM, ROMK allows more K^+ ions to move into cells than out, as intracellular Mg^{2+} binds to ROMK, preventing K^+ efflux (Huang and Kuo 2007). However, magnesium deficiency disrupts this mechanism, potentially causing hypokalemia.

8.6.1.2 Hyperkalemia

Hyperkalemia occurs when there is an elevated level of potassium in the blood. It can result from kidney dysfunction, certain medications, high potassium intake, or conditions like Addison's disease. Symptoms can range from muscle weakness to potentially life-threatening cardiac arrhythmias.

8.6.1.3 Gitelman Syndrome

Gitelman syndrome, sometimes known as familial hypokalemia-hypomagnesemia, is a rare genetic disorder, impairs kidney electrolyte reabsorption, resulting in hypokalemic metabolic alkalosis, hypomagnesemia, and reduced urinary calcium excretion (Knoers and Levtchenko 2008).

8.6.1.4 Bartter Syndrome

Bartter syndrome is another inherited renal tubular disorder that arises from impaired salt reabsorption in the thick ascending limb of the loop of Henle, leading to salt loss, hypokalemia, and metabolic alkalosis (Da and Unha 2018). In metabolic alkalosis, as the blood becomes more alkaline as an adaptive response, body attempts to correct this imbalance by shifting potassium from the bloodstream into cells and increasing the concentration of hydrogen ions in the bloodstream, resulting in hypokalemia (Unwin et al. 2011).

8.6.1.5 Medication Induced Disorder

Various medications, including antibiotics, diuretics, mineralocorticoids, and glucocorticoids, can decrease the potassium level (Veltri and Mason 2015). Literature showed that antioxidant containing food has the ability to reduce the impact of medication and it protects the liver which is the main site of toxicant metabolization (Singh et al. 2012; Vishwakarma et al. 2013). Diuretics are often prescribed for hypertension to reduce blood pressure by increasing sodium and water excretion through urine, with certain types like loop and thiazide diuretics potentially causing hypokalemia due to potassium loss (Wile 2012; Veltri and Mason 2015). Also, hypokalemia occurs during distal renal tubular acidosis due to impaired H^+-ATPase pump, leading to increased potassium secretion in the cortical collecting duct (Simpson and Schwartz 2001). Hypokalemia is reported in 15.8% of 5,100 patients with hypertension, whereas 7.3% were hyperkalemic and the prevalence of primary aldosteronism in those with hypokalemia was 28.1% which increases with decreasing potassium concentrations (Burrello et al. 2020). Also, administration of intravenous fluids (Dextrose) without adequate potassium supplementation can cause hypokalemia and hyponatremia (Armon et al. 2008).

8.6.1.6 Kidney Disease

Patients with chronic kidney disease (CKD) often exhibit hyperkalemia, characterized by elevated serum potassium levels (Watanabe 2020). This occurs due to reduced renal ion excretion and the usage of medications aimed at slowing CKD progression or managing related conditions like diabetes mellitus and heart failure.

8.6.1.7 Addison's Disease

In Addison's disease, hyperkalemia is primarily caused by hypoaldosteronism, a condition where the adrenal glands produce insufficient aldosterone. Aldosterone plays a crucial role in regulating potassium levels in the body. In the absence of an adequate amount of aldosterone, the kidneys lose their ability to effectively excrete potassium into the urine (DeFronzo 1980). Aldosterone normally stimulates the renal tubules to reabsorb sodium and excrete potassium. However, with hypoaldosteronism, this regulatory mechanism is compromised, leading to decreased potassium excretion. As a result, there is an accumulation of potassium in the bloodstream, contributing to hyperkalemia.

8.6.1.8 Diabetic Ketoacidosis

In diabetic ketoacidosis (DKA), hypokalemia results from osmotic diuresis induced by increased urine production due to substances like glucose. DKA, triggered by insulin deficiency, prompts the use of alternative energy sources, leading to ketone production and metabolic acidosis. The kidneys respond by boosting urine output to eliminate excess acids and glucose, causing potassium loss. DKA commonly induces hypokalemia through osmotic diuresis, resulting in a total body potassium deficiency ranging from 3 to 6 mEq/kg (Kitabchi et al. 2004; Wolfsdorf et al. 2014). Administering insulin to these patients without addressing potassium levels first can exacerbate the decrease in extracellular potassium, potentially causing cardiac dysrhythmias and death. Therefore, initiating potassium therapy along with careful fluid management is essential before starting insulin to prevent cardiac complications from hypokalemia (Gupta and El-Wiher 2019).

8.6.1.9 Acid-Base Disorders

Clinical scenarios often demonstrate a close connection between acid-base and potassium imbalances. Acidosis reduces K^+ secretion and increases reabsorption in the collecting duct, resulting in hypokalemia, while alkalosis has opposing effects (Lee Hamm et al. 2013). Additionally, potassium disorders impact acid-base homeostasis, with potassium depletion increasing H^+ secretion, ammoniagenesis, and H-K-ATPase activity. Conversely, hyperkalemia decreases ammoniagenesis and $NH4^+$ transport in the thick ascending limb (Lee Hamm et al. 2013). In summary, disruptions in potassium and acid-base homeostasis are mechanistically linked and hold clinical significance (Lee Hamm et al. 2013). Moreover, renal tubular acidosis (RTA) refers to a group of disorders characterized by compromise of renal acid-base regulation, causing impaired urinary acidification, net acid retention, and hyperchloremic metabolic acidosis (Yaxley and Pirrone 2016).

8.6.1.10 Conn's Syndrome (Primary Hyperaldosteronism)

Conn's syndrome, also known as primary hyperaldosteronism, is a medical condition characterized by excessive secretion of aldosterone, leading to increased retention of sodium and water, while causing the excretion of potassium (Balhana et al. 2022). The elevated aldosterone levels result in hypertension (high blood pressure) and low potassium levels in the blood (hypokalemia). Conn's syndrome can be caused by adrenal tumors (usually aldosterone-producing adenomas) or, less commonly, by adrenal hyperplasia.

8.6.1.11 Hereditary Periodic Paralysis

Hypokalemic familial periodic paralysis, a rare autosomal dominant channelopathy, manifests as muscle weakness and a concurrent decrease in blood potassium levels. It is linked to a genetic disorder involving the gene responsible for the dihydropyridine-sensitive calcium channel (Robinson et al. 2000).

8.6.1.12 Hyperthyroid Periodic Paralysis

Hyperthyroid periodic paralysis (HPP) is a rare but potentially lethal manifestation of hyperthyroidism, typically affecting young Asian males aged 20–40, despite hyperthyroidism being more common in females. It is characterized by acute paralytic attacks and hypokalemia (Neki 2016)

8.7 Metabolic Regulation, Functions, and Disorders of Chloride Metabolism

Chloride, an extracellular anion, is closely linked to sodium and regulated by the kidneys in serum. After glomerular filtration, most chloride is reabsorbed in both proximal and distal tubules through a combination of active and passive transport mechanisms. The normal blood chloride concentrations in healthy person typically fall within the range of 96–106 mEq/L. Hyperchloremia, an elevated blood chloride concentration, may result from conditions involving gastrointestinal bicarbonate loss. Conversely, hypochloremia, a lower-than-normal chloride level, can occur in situations such as vomiting or excessive water retention, as seen in conditions like congestive heart failure. The intricate balance and regulation of chloride levels by the kidneys are crucial for maintaining overall electrolyte and fluid balance in the body.

8.7.1 Metabolic Disorders of Chloride

Chloride is an essential electrolyte in the body, and its levels are regulated within a narrow range to maintain proper physiological functions. While chloride imbalances themselves may not cause specific diseases, they are often associated with underlying conditions. Some disorder conditions and situations related to chloride levels are:

8.7.1.1 Hypochloremia

Hypochloremia, abnormally low levels of chloride in the blood, often associated with conditions such as vomiting, diarrhea, certain kidney disorders, or metabolic alkalosis. It is primarily caused by renal or extra-renal factors. Common renal causes include loop diuretic use and osmotic diuresis, while extra-renal factors involve insufficient oral intake, vomiting, nasogastric suction, diarrhea, skin disruption (burn), and excess water in conditions like congestive heart failure. Diagnosis relies on elevated urine chloride and sodium levels, indicative of renal hypochloremia (Akoum et al. 2021). Although hypochloremia lacks specific symptoms, clinical manifestations depend on the underlying cause, with physical findings reflecting hypovolemia and extracellular fluid volume contraction. Notably, individuals with renal or extra-renal chloride losses exhibit similar physical findings, and elevated urine chloride and sodium levels confirm renal involvement despite evidence of extracellular fluid volume contraction (Akoum et al. 2021).

8.7.1.2 Hyperchloremia

Hyperchloremia is characterized by an electrolyte imbalance, wherein the serum chloride anion concentration surpasses 107 mmol/L and causes of hyperchloremia encompass excessive chloride intake (such as in fluid therapy with high-chloride solutions), water loss (due to conditions like diabetes insipidus, diuretic use, polyuric phase of renal failure, hypermetabolism, or diarrhea), and increased rates of chloride reabsorption (Astapenko et al. 2020).

8.7.1.3 Cystic Fibrosis

Cystic fibrosis (CF) is a genetic disorder caused by mutations in the CF transmembrane conductance regulator (CFTR) gene, which encodes a protein regulating the movement of chloride and sodium ions across epithelial cell membranes. This condition is characterized by the buildup of thick mucus in organs, especially the lungs and digestive system, leading to respiratory insufficiency and systemic obstructions (Elborn 2016). Reduced mucociliary clearance and altered ion transport facilitate bacterial colonization in the respiratory tract, initiating chronic infection and inflammation, which can potentially result in airway destruction (Bell et al. 2015).

8.7.1.4 Dehydration

Dehydration leads to hyperchloremia, an increase in blood chloride levels due to water loss and reduced fluid volume. The body responds by conserving water, leading to elevated chloride levels alongside other electrolyte imbalances. The severity and duration of dehydration determine the clinical significance of hyperchloremia, affecting acid-base balance and potentially causing metabolic acidosis.

8.7.1.5 Diabetic Ketoacidosis (DKA)

Diabetic ketoacidosis (DKA) is a severe complication of diabetes, primarily affecting individuals with type 1 diabetes due to severe insulin deficiency. Without insulin, cells cannot utilize glucose for energy, leading to the breakdown of fats and the release of acidic ketones into the bloodstream, causing acidosis. Symptoms include thirst, frequent urination, abdominal pain, vomiting, and fruity-scented breath. DKA is characterized by high blood sugar, ketones in the urine, and metabolic acidosis. Additionally, it often manifests with decreased electrolyte levels, including chloride (Chiasson et al. 2003).

8.7.1.6 Renal Tubular Acidosis (RTA)

Renal tubular acidosis (RTA) is a kidney disorder marked by impaired urine acidification, causing acid accumulation and electrolyte imbalances. It involves a dysfunction in proton secretion by the kidney nephron, leading to bicarbonate wasting (proximal) or reduced acid secretion (distal), resulting in persistent non-anion gap (hyperchloremic) metabolic acidosis (Brunner et al. 2015). RTA has four types, each with distinct causes and features. Type 1 (distal) RTA is characterized by impaired acid secretion in distal tubules, Type 2 (proximal) RTA features a defect in bicarbonate reabsorption in proximal tubules, and Type 3 is a mixed form combining features of both Type 1 and Type 2 RTA. Type 4 RTA, or hyperkalemic RTA, is characterized by impaired potassium excretion, leading to elevated blood potassium levels. Unlike other RTA types, it primarily involves disturbances in potassium handling rather than defects in tubular acidification. Reduced aldosterone activity contributes to impaired potassium excretion, resulting in hyperkalemia. Type 4 RTA is often associated with conditions such as diabetic nephropathy and chronic kidney disease. Clinical manifestations include metabolic acidosis, electrolyte imbalances, and potential bone abnormalities.

8.7.1.7 Addison's Disease

In Addison's disease, insufficient aldosterone production disrupts electrolyte balance, potentially causing chloride imbalances. Aldosterone, responsible for sodium reabsorption and potassium excretion in the kidneys, regulates chloride levels. Reduced aldosterone leads to sodium loss, potassium and hydrogen ion accumulation, affecting chloride balance. Chloride imbalances in Addison's disease contribute to disruptions in the body's acid-base equilibrium, resulting in symptoms like dehydration, weakness, fatigue, and blood pressure changes. Additionally, chloride imbalances may be observed in conjunction with other electrolyte abnormalities.

8.7.1.8 Bartter Syndrome

Bartter syndrome, a rare genetic disorder impacting the kidneys, causes significant electrolyte imbalances, including sodium, potassium, and chloride. Specific dysfunction in the transport of electrolytes, such as sodium, potassium, and chloride, across the renal tubules, disrupts the normal balance, leading to symptoms such as hypokalemia, hypochloremic metabolic alkalosis, and renal salt wasting (Choi et al. 2023).

8.8 Conclusions

The intricate role of macro-element metabolism orchestrates vital physiological processes, maintaining cellular integrity, electrical excitability, osmotic balance, and bone health. These elements are not mere biological constituents but architects of our vitality. The precise balance of macro-elements reflects evolutionary adaptations that enable the body to navigate diverse environments. Disruptions to this equilibrium can result in a spectrum of disorders affecting cellular, tissue, and systemic levels. From the parathyroid glands orchestrating calcium homeostasis to the sodium–potassium pump maintaining neuronal membrane potential, the metabolic interplay of macro-elements involves multifaceted intricate biochemical processes. Understanding the pathophysiology of disorders linked to macro-element metabolism is crucial for unraveling mysteries associated with conditions such as bone disorders (rickets, osteomalacia, and osteoporosis), electrolyte imbalances, and renal disorders. Delving deeper into these disorders reveals not only the challenges they pose to health but also potential avenues for therapeutic interventions and preventive strategies. Moreover, the exploration of macro-element metabolism extends beyond clinical medicine, fostering groundbreaking research and technological innovations. Encouraging a holistic understanding of the intricate connections between diet, lifestyle, and overall well-being, the knowledge in this chapter promotes collaboration across disciplines. In the ever-evolving field of medical science, it serves as a foundational guide in the continuously growing area of research.

References

Abate EG, Clarke BL (2016) Review of hypoparathyroidism. Front Endocrinol 7:172

Adrogué HJ, Madias NE (2000) Hypernatremia. N Engl J Med 342(20):1493–1499

Akoum A, Radwan R, Hage E, JEL M, Ezzeddine S (2021) Severe symptomatic hypochloremia associated with rare signet ring cell carcinoma of the ampulla of vater: a case report. Cureus 13(11):e19492

Ali SN, Bazzano LA (2018) Hyponatremia in association with second-generation antipsychotics: a systematic review of case reports. Ochsner J 18(3):230–235

Armon K, Riordan A, Playfor S, Millman G, Khader A (2008) Hyponatraemia and hypokalaemia during intravenous fluid administration. Arch Dis Child 93(4):285–287

Arzhan S, Lew SQ, Ing TS, Tzamaloukas AH, Unruh ML (2021) Dysnatremias in chronic kidney disease: pathophysiology, manifestations, and treatment. Front Med 8

Assadi F (2010) Hypomagnesemia: an evidence-based approach to clinical cases. Iran J Kidney Dis 4(1):13–19

Astapenko D, Navratil P, Pouska J, Cerny V (2020) Clinical physiology aspects of chloremia in fluid therapy: a systematic review. Perioper Med 9:40

Atsariyasing W, Goldman MB (2014) A systematic review of the ability of urine concentration to distinguish antipsychotic – from psychosis-induced hyponatremia. Psychiatry Res 217(3):129–133

Balanescu S, Kopp P, Gaskill MB, Morgenthaler NG, Schindler C, Rutishauser J (2011) Correlation of plasma copeptin and vasopressin concentrations in Hypo-, Iso-, and hyperosmolar states. J Clin Endocrinol Metab 96(4):1046–1052

Balhana S, Pina H, Machete M, Aragüés M, Noronha C (2022) Conn's syndrome: an unusual cause of periodic paralysis. Cureus 14(5):e24880

Beall DP, Henslee HB, Webb HR, Scofield RH (2006) Milk-alkali syndrome: a historical review and description of the modern version of the syndrome. Am J Med Sci 331(5):233–242

Bell SC, De Boeck K, Amaral MD (2015) New pharmacological approaches for cystic fibrosis: promises, progress, pitfalls. Pharmacol Ther 145:19–34

Benner BJM, Alsma J, Feelders RA (2019) Hyponatraemia and hyperpigmentation in primary adrenal insufficiency. BMJ Case Rep 12(3)

Bergwitz C, Miyamoto KI (2019) Hereditary hypophosphatemic rickets with hypercalciuria: pathophysiology, clinical presentation, diagnosis and therapy. Pflug Arch Eur J Physiol 471(1):149–163

Bichet DG (2012) Physiopathology of hereditary polyuric states: a molecular view of renal function. Swiss Med Weekly 142

Bilezikian JP, Silverberg SJ (2004) Clinical practice. Asymptomatic primary hyperparathyroidism. N Engl J Med 350(17):1746–1751

Bourque CW (2008) Central mechanisms of osmosensation and systemic osmoregulation. Nat Rev Neurosci 9(7):519–531

Bover J, Ureña-Torres P, Górriz JL, Lloret MJ, da Silva I, Ruiz-García C, Chang P, Rodríguez M, Ballarín J (2016) Cardiovascular calcifications in chronic kidney disease: potential therapeutic implications. Nefrologia 36(6):597–608

Boyce AM, Lee AE, Roszko KL, Gafni RI (2020) Hyperphosphatemic tumoral calcinosis: pathogenesis, clinical presentation, and challenges in management. Front Endocrinol 11:534162

Brunner R, Drolz A, Scherzer TM, Staufer K, Fuhrmann V, Zauner C, Holzinger U, Schneeweiß B (2015) Renal tubular acidosis is highly prevalent in critically ill patients. Crit Care 19(1):1–8

Burrello J, Monticone S, Losano I, Cavaglià G, Buffolo F, Tetti M, Covella M, Rabbia F, Veglio F, Pasini B, Williams TA, Mulatero P (2020) Prevalence of hypokalemia and primary aldosteronism in 5100 patients referred to a tertiary hypertension unit. Hypertension 75(4):1025–1033

Burtis WJ, Wu TL, Insogna KL, Stewart AF (1988) Humoral hypercalcemia of malignancy. Ann Intern Med 108(3):454–457

Carroll MF, Schade DS (2003) A practical approach to hypercalcemia. Am Fam Physician 69(9):1959–1966

Chiasson JL, Aris-Jilwan N, Bélanger R, Bertrand S, Beauregard H, Ekoé JM, Fournier H, Havrankova J (2003) Diagnosis and treatment of diabetic ketoacidosis and the hyperglycemic hyperosmolar state. CMAJ 168(7):859–866

Choi N, Kim SH, Bae EH, Yang EM, Lee GH, Lee SH, Lee JH, Ahn YH, Cheong HIL, Kang HG, Hyun HS, Kim JH (2023) Long-term outcome of Bartter syndrome in 54 patients: a multicenter study in Korea. Front Med 10:1099840

Christensen SE, Nissen PH, Vestergaard P, Heickendorff L, Rejnmark L, Brixen K, Mosekilde L (2008) Plasma 25-hydroxyvitamin D, 1,25-dihydroxyvitamin D, and parathyroid hormone in familial hypocalciuric hypercalcemia and primary hyperparathyroidism. Eur J Endocrinol 159(6):719–727

Colazo JM, Reasoner SA, Holt G, Faugere MCM, Dahir KM (2020) Hereditary hypophosphatemic rickets with hypercalciuria (HHRH) presenting with genu valgum deformity: treatment with phosphate supplementation and surgical correction. Case Rep Endocrinol 9(2020):1047327

Cooper MS, Gittoes NJL (2008) Diagnosis and management of hypocalcaemia. BMJ (Clin Res Ed.) 336(7656):1298–1302

Da T, Unha S (2018) Bartter syndrome: causes, diagnosis, and treatment. Int J Nephrol Renovasc Dis:11–291

Das SK, Afroze F, Ahmed T, Faruque ASG, Sarker SA, Huq S, Islam MM, Shahrin L, Matin FB, Chisti MJ (2015) Extreme hypernatremic dehydration due to potential sodium intoxication: consequences and management for an infant with diarrhea at an urban intensive care unit in Bangladesh: a case report. J Med Case Rep 9:124

Datta NS, Abou-Samra AB (2009) PTH and PTHrP signaling in osteoblasts. Cell Signal 21(8):1245–1254

DeFronzo RA (1980) Hyperkalemia and hyporeninemic hypoaldosteronism. Kid Int 17(1):118–134
Dey B, Gochhait D, Subramanian H, Ponnusamy M (2017) Oncogenic osteomalacia: an approach to diagnosis with a case report. J Clin Diagn Res 11(4):ED05
Durrani NUR, Imam AA, Soni N (2022) Hypernatremia in newborns: a practical approach to management. Biomed Hub 7(2):55–69
Elborn JS (2016) Cystic fibrosis. Lancet (London, England) 388(10059):2519–2531
Elsary AY, Elgameel AA, Mohammed WS, Zaki OM, Taha SA (2018) Neonatal hypocalcemia and its relation to vitamin D and calcium supplementation. Saudi Med J 39(3):247
Fathi I, Sakr M (2014) Review of tumoral calcinosis: a rare clinico-pathological entity. World J Clin Cases 2(9):409
Fermor B, Skerry TM (1995) PTH/PTHrP receptor expression on osteoblasts and osteocytes but not resorbing bone surfaces in growing rats. J Bone Miner Res 10(12):1935–1943
Florenzano P, Gafni RI, Collins MT (2017) Tumor-induced osteomalacia. Bone Reports 7:90
Foreman JW (2019) Fanconi Syndrome. Pediatr Clin North Am 66(1):159–167
Gartlehner G, Thieda P, Hansen RA, Gaynes BN, DeVeaugh-Geiss A, Krebs EE, Lohr KN (2008) Comparative risk for harms of second-generation antidepressants: a systematic review and meta-analysis. Drug Safety 31(10):851–865
Gorvin CM (2019) Molecular and clinical insights from studies of calcium-sensing receptor mutations. J Mol Endocrinol 63(2):R1–R16
Goyal A, Anastasopoulou C, Ngu M, Singh S (2023) Hypocalcemia endocrinology and diabetes: a problem oriented approach, 2nd edn, pp 219–230
Green I, Altman A (2009) Paget's disease and hypercalcemia: coincidence or causal relationship. Harefuah 148(10):708–710
Gregoire JR (1994) Adjustment of the osmostat in primary aldosteronism. Mayo Clinic Proc 69(11):1108–1110
Grzanka K, Kucharz EJ (2004) Vitamin D-resistant rickets. Wiadomości Lekarskie (Warsaw, Poland:1960) 57(11–12):663–671
Guise TA, Yin JJ, Taylor SD, Kumagai Y, Dallas M, Boyce BF, Yoneda T, Mundy GR (1996) Evidence for a causal role of parathyroid hormone-related protein in the pathogenesis of human breast cancer-mediated osteolysis. J Clin Invest 98(7):1544–1549
Gupta A, El-Wiher N (2019) Therapeutic challenges in management of severe acidosis and profound hypokalemia in pediatric diabetic ketoacidosis. Glob Pediatr Health 6:1–4
Haffner D, Emma F, Eastwood DM, Duplan MB, Bacchetta J, Schnabel D, Wicart P, Bockenhauer D, Santos F, Levtchenko E, Harvengt P, Kirchhoff M, Di Rocco F, Chaussain C, Brandi ML, Savendahl L, Briot K, Kamenicky P, Rejnmark L, Linglart A (2019) Clinical practice recommendations for the diagnosis and management of X-linked hypophosphataemia. Nat Rev Nephrol 15(7):435–455
Hamano N, Komaba H, Fukagawa M (2017) Magnesium as a new player in CKD: too little is as bad as too much? Kid Int 92:1034–1036
Heath H, Edis AJ (1979) Pheochromocytoma associated with hypercalcemia and ectopic secretion of calcitonin. Ann Intern Med 91(2):208–210
Hew-Butler T, Loi V, Pani A, Rosner MH (2017) Exercise-associated hyponatremia: 2017 update. Front Med 4:21
Holland J, Bhogle M (2013) Sertraline and mirtazapine as geriatric antidepressants. Psychiatr Danub 25(Suppl 2):S286–S290
Huang CL, Kuo E (2007) Mechanism of hypokalemia in magnesium deficiency. J Am Soc Nephrol 18(10):2649–2652
Iannuzzi MC, Rybicki BA, Teirstein AS (2007) Sarcoidosis. N Engl J Med 357(1):2153–2165
Insogna KL, Broadus AE (1987) Hypercalcemia of malignancy. Ann Rev Med 38:241–256
Jain N, Reilly RF (2017) Hungry bone syndrome. Curr Opin Nephrol Hypertens 26(4):250–255
John S, Thuluvath PJ (2015) Hyponatremia in cirrhosis: pathophysiology and management. World J Gastroenterol 21(11):3197–3205
Jones DP (2018) Syndrome of inappropriate secretion of antidiuretic hormone and hyponatremia. Pediatr Rev 39(1):27–35

Joo MA, Kim EY (2013) Hyponatremia caused by excessive intake of water as a form of child abuse. Ann Pediatr Endocrinol Metab 18(2):95

Karunakaran P, Maharajan C, Ramalingam S, Rachmadugu SV (2018) Is hungry bone syndrome a cause of postoperative hypocalcemia after total thyroidectomy in thyrotoxicosis? A prospective study with bone mineral density correlation. Surgery 163(2):367–372

Katahira M, Yamada T, Kawai M (2004) A case of cushing syndrome with both secondary hypothyroidism and hypercalcemia due to postoperative adrenal insufficiency. Endocrine J 51(1):105–113

Ketteler M, Block GA, Evenepoel P, Fukagawa M, Herzog CA, McCann L, Moe SM, Shroff R, Tonelli MA, Toussaint ND, Vervloet MG, Leonard MB (2017) Executive summary of the 2017 KDIGO Chronic Kidney Disease-Mineral and Bone Disorder (CKD-MBD) Guideline Update: what's changed and why it matters. Kid Int 92(1):26–36

Khwaja A (2010) Chronic kidney disease-mineral and bone disorder KDIGO guidelines. Nephron Clin Pract 116(1):c25–c26

Kitabchi AE, Umpierrez GE, Murphy MB, Barrett EJ (2004) Hyperglycemic crises in diabetes. Diabetes Care 27(suppl_1):s94–s102

Knoers NV, Levtchenko EN (2008) Gitelman syndrome. Orphanet J Rare Dis 3:22

Kousteni S, Bilezikian JP (2008) The cell biology of parathyroid hormone in osteoblasts. Curr Osteoporos Rep 6(2):72–76

Lee Hamm L, Hering-Smith KS, Nakhoul NL (2013) Acid-Base and potassium homeostasis. Sem Nephrol 33(3):257–264

Leth-Møller KB, Hansen AH, Torstensson M, Andersen SE, Ødum L, Gislasson G, Torp-Pedersen C, Holm EA (2016) Antidepressants and the risk of hyponatremia: a Danish register-based population study. BMJ Open 6(5):e011200

Lu X, Wang X (2017) Hyponatremia induced by antiepileptic drugs in patients with epilepsy. Expert Opin Drug Saf 16(1):77–87

Mameli C, Sangiorgio A, Colombo V, Gambino M, Spaccini L, Cattaneo E, Zuccotti GV (2021) Autosomal dominant hypophosphatemic rickets: a case report and review of the literature. Int J Environ Res Public Health 18(16):8771

Mange K, Matsuura D, Cizman B, Soto H, Ziyadeh FN, Goldfarb S, Neilson EG (1997) Language guiding therapy: the case of dehydration versus volume depletion. Ann Int Med 127(9):848–853

Mann SJ (2008) The silent epidemic of thiazide-induced hyponatremia. J Clin Hypertens (Greenwich, Conn.) 10(6):477–484

Manroa P, Kannan S, Hatipoglu B, Licata A (2014) Hypercalcemia and acromegaly – clarifying the connections. A case report and review of the literature. Endocr Pract 20(5):e86–e90

Margabandhu S, Doshi M (2019) Familial Hypomagnesemia, Hypercalciuria and Nephrocalcinosis with Novel Mutation. Ind J Nephrol 29(1):57

Moder KG, Hurley DL (1990) Fatal hypernatremia from exogenous salt intake: report of a case and review of the literature. Mayo Clin Proc 65(12):1587–1594

Neki NS (2016) Hyperthyroid hypokalemic periodic paralysis. Pak J Med Sci 32(4):1051–1052

Ozkaya HM, Keskin FE, Haliloglu OA, Senel TE, Kadioglu P (2015) Life-threatening hypercalcemia due to graves' disease and concomitant adrenal failure: a case report and review of the literature. Case Rep Endocrinol 2015:684648

Palumbo VD, Damiano G, Messina M, Fazzotta S, Monte GLO, Monte AIL (2021) Tertiary hyperparathyroidism: a review. La Clinica Terapeutica 172(3):241–246

Patel AM, Adeseun GA, Goldfarb S (2013) Calcium-alkali syndrome in the modern era. Nutrients 5(12):4880–4893

Pearce SHS, Williamson C, Kifor O, Bai M, Coulthard MG, Davies M, Lewis-Barned N, McCredie D, Powell H, Kendall-Taylor P, Brown EM, Thakker RV (1996) A familial syndrome of hypocalcemia with hypercalciuria due to mutations in the calcium-sensing receptor. N Engl J Med 335(15):1115–1122

Pham PCT, Pham PAT, Pham SV, Pham PTT, Pham PMT, Pham PTT (2014) Hypomagnesemia: a clinical perspective. Int J Nephrol Renovasc Dis 7:219

Robinson JE, Morin VI, Douglas MJ, Wilson RD (2000) Familial hypokalemic periodic paralysis and Wolff-Parkinson-White syndrome in pregnancy. Can J Anaesth 47(2):160–164
Rodriguez M, Hernandez M, Cheungpasitporn W, Kashani KB, Riaz I, Rangaswami J, Herzog E, Guglin M, Krittanawong C (2019) Hyponatremia in heart failure: pathogenesis and management. Curr Cardiol Rev 15(4):252–261
Roodman GD (2004) Mechanisms of bone metastasis. N Engl J Med 350(16):1655–1664
Sakai K, Tomoda Y, Saito H, Tanaka K (2020) Hungry bone syndrome and osteoblastic bone metastasis from gastric cancer, QJM. Int J Med 113(12):903–904
Schlanger LE, Bailey JL, Sands JM (2010) Electrolytes in the aging. Adv Chronic Kidney Dis 17(4):308–319
Seymour JF, Gagel RF, Hagemeister FB, Dimopoulos MA, Cabanillas F (1994) Calcitriol production in hypercalcemic and normocalcemic patients with non-Hodgkin lymphoma. Ann Int Med 121(9):633–640
Simpson AM, Schwartz GJ (2001) Distal renal tubular acidosis with severe hypokalaemia, probably caused by colonic H^+-K^+-ATPase deficiency. Arch Dis Child 84:504–507
Singh P, Singh RL (2018) Quantification of phytochemicals imparting antioxidant activities in commonly used vegetables. Int J Appl Sci Biotech 6(2):107–112
Singh P, Vishwakarma SP, Singh U, Shukla M, Singh R, Singh RK, Singh RB, Wilson DW, Singh RL (2012) Quantification and evaluation of antioxidant activity of some bioactive phytochemicals in different medicinal plants. Open Nutr J 5:179–186
Singh RL, Vishwakarma SP, Singh P (2014) Vitamins and minerals: roles and plant sources. In: Prakash D (ed) Phytochemicals of Nutraceutical Importance and their role in human health. CABI-UK, pp 248–265
Sterns RH (2015) Disorders of plasma sodium – causes, consequences, and correction. N Engl J Med 372(1):55–65
Stewart AF (2005) Clinical practice. Hypercalcemia associated with cancer. N Engl J Med 352(4):373–379
Stewart AF, Vignery A, Silverglate A, Ravin ND, LiVolsi V, Broadus AE, Baron R (1982) Quantitative bone histomorphometry in humoral hypercalcemia of malignancy: uncoupling of bone cell activity. J Clin Endocrinol Metab 55(2):219–227
Sutton RAL, Domrongkitchaiporn S (1993) Abnormal renal magnesium handling. Miner Electrolyte Metab 19(4–5):232–240
Tayab MA, Hoq MA (2020) Acid-base and electrolyte disturbances in children presenting with acute watery diarrhoea in emergency observation and referral unit of Dhaka Shishu (children) hospital. Dhaka Shishu (Children) Hosp J 36(2):120–124
Tinawi M (2020) Disorders of magnesium metabolism: hypomagnesemia and hypermagnesemia. Arch Clin Biomed Res 4(3):205–220
Toh TH, Tong CV, Chong HC (2017) Primary aldosteronism-not just about potassium and blood pressure learning point for clinicians. QJM: An Int J Med 110(3):175–177
Unwin RJ, Luft FC, Shirley DG (2011) Pathophysiology and management of hypokalemia: a clinical perspective. Nat Rev Nephrol 7(2):75–84
Veltri KT, Mason C (2015) Medication-induced hypokalemia. Pharm Therap 40(3):185–190
Vishwakarma SP, Singh P, Singh U, Singh RL (2013) Antioxidant activities of some tuberous plant leaves. Int J Pharm Sci Rev Res 20(1):28–33
Voelkl J, Lang F, Eckardt KU, Amann K, Kuro-o M, Pasch A, Pieske B, Alesutan I (2019) Signaling pathways involved in vascular smooth muscle cell calcification during hyperphosphatemia. Cell Mol Life Sci 76(11):2077–2091
Vuralli D (2019) Clinical approach to hypocalcemia in newborn period and infancy: who should be treated? Int J Pediatr 2019:4318075
Watanabe R (2020) Hyperkalemia in chronic kidney disease. Rev Assoc Med Bras 66(1):31–36
Waters DM (2009) Hypercalcaemia. InnovAiT 2(12):698–701
Watten RH, Morgan FM, Songkhla YN, Vanikiati B, Phillips RA (1959) Water and electrolyte studies in cholera. J Clin Invest 38(11):1879–1889
Wile D (2012) Diuretics: a review. Ann Clin Biochem 49(5):419–431

Witteveen JE, van Thiel S, Romijn JA, Hamdy NA (2013) Hungry bone syndrome: still a challenge in the post-operative management of primary hyperparathyroidism: a systematic review of the literature. Eur J Endocrinol 168(3):R45–R53

Wolfsdorf JI, Allgrove J, Craig ME, Edge J, Glaser N, Jain V, Lee WW, Mungai LN, Rosenbloom AL, Sperling MA, Hanas R (2014) ISPAD Clinical Practice Consensus Guidelines 2014. Diabetic ketoacidosis and hyperglycemic hyperosmolar state. Pediatr Diabetes 15(20):154–179

Yaxley J, Pirrone C (2016) Review of the diagnostic evaluation of renal tubular acidosis. Ochsner J 16(4):525–530

Zavatta G, Altieri P, Vandi G, Vicennati V, Pagotto U, Vescini F (2021) Phosphate metabolism and pathophysiology in parathyroid disorders and endocrine tumors. Int J Mol Sci 22(23):12975

Zivin JR, Gooley T, Zager RA, Ryan MJ (2001) Hypocalcemia: a pervasive metabolic abnormality in the critically ill. Am J Kidney Dis 37(4):689–698

Metabolism of Essential Trace Elements and Associated Disorders

9

Saba Ubaid and Shivani Pandey

Abstract

Trace elements are inorganic substances that exist naturally and are necessary for human health at doses less than 100 mg daily. They mediate vital biochemical reactions, are crucial components of biological structures, and have a significant impact and function as major players in a number of activities required for life. When these elements are present in excess – that is, in greater quantities than what is necessary for biological functions – they can be hazardous to the health of the organism. As a result, it has been shown that variations from the ideal concentrations of trace elements can negatively impact biological functions and are linked to a number of deadly illnesses, including cancer. The focus of this chapter is to improve our understanding of the connection between trace elements and how diseases are affected by them. This chapter will also focus on the metabolism of essential trace elements, i.e., iron, iodine, zinc, fluoride, copper, manganese, selenium, magnesium, cobalt, and chromium and related diseases due to its alteration.

Keywords

Human health · Trace elements · Iodine · Zinc · Fluoride · Copper · Manganese

S. Ubaid · S. Pandey (✉)
Department of Biochemistry, King George's Medical University,
Lucknow, Uttar Pradesh, India
e-mail: shivanipandey@kgmcindia.edu

R. L. Singh et al. (eds.), *Clinical Applications of Biomolecules in Disease Diagnosis*, https://doi.org/10.1007/978-981-97-4723-8_9

9.1 Introduction

Chemical elements found at a concentration of <100 ppm is known as trace elements. It is a dietary mineral required for the development, optimal growth, and functioning of the organism (Bowen 1966). Because these trace elements directly affect the organism's metabolic and physiologic processes, their absence or excess can cause serious health problems, even death in extreme circumstances (Skalnaya and Skalny 2018). Certain trace elements regulate crucial biological functions by attaching themselves to molecules on the cell membrane's receptor site or by changing the membrane's structure to stop particular molecules from entering the cell (Prashanth et al. 2015). Trace elements play a crucial role in maintaining the stability of cellular structures at normal levels, but when they are lacking, they can activate other pathways and result in illnesses (Prashanth et al. 2015).

All trace elements are toxic if consumed at higher concentration for long periods. For some critical trace elements, there is a large difference between the optimal consumption and harmful intake to meet physiological needs, whereas for other elements, the difference is considerably smaller. The aim of this chapter is to elucidate the metabolism of essential trace elements: iron, iodine, zinc, fluoride, copper, manganese, selenium, magnesium, cobalt, and chromium, as well as deficiency and overload of these trace elements, which may complicate many diseased states. For example, iron deficiency can cause anemia, while its overload, known as hemochromatosis, can lead to organ damage, particularly affecting the heart, liver, and pancreas (Clark 2008; Halliday and Powell 2018). Iodine deficiency can lead to goiter, hypothyroidism, and cretinism, while intake of excess iodine leads to thyroid dysfunction (Alam Khan et al. 2002). Insufficient zinc can result in stunted growth, delayed sexual maturity, and weakened immune response. Conversely, excessive zinc intake can lead to copper deficiency, gastrointestinal (GI) disturbances, and impaired immune function (Bhowmik et al. 2010).

Fluoride is essential for dental health, while copper is involved in the formation of connective tissue, energy production, and the function of the nervous and immune systems (Bhattacharya et al. 2016). Deficiencies of copper can cause osteoporosis and neurologic disorders. Excessive copper intake, on the other hand, can result in Wilson's disease (Hermann 2019). Manganese is a cofactor for several enzymes involved in the metabolism of carbohydrates, amino acids, and cholesterol (Zabłocka-Słowińska and Grajeta 2012). Selenium is vital for selenoproteins, which have antioxidant properties and plays an important role in thyroid metabolism and supports immune system (Brown and Arthur 2001). Magnesium is known to involve in over 300 enzymatic reactions, including those for protein synthesis, energy generation, and the proper functioning of muscles and nerves. (Bhutto et al. 2005) (Table 9.1). Cobalt deficiency is rare but can lead to megaloblastic anemia and neurological disorders (Babior and Bunn 2001). Excessive cobalt intake, on the other hand, can lead to cobalt toxicity, resulting in thyroid dysfunction and cardiomyopathy (Maret 2019). Chromium is essential for glucose metabolism, as it enhances insulin function. A lack of chromium can cause problems with glucose tolerance and increase the risk of type 2 diabetes. Excessive chromium intake, however, can result in gastrointestinal

Table 9.1 Metabolic function of essential trace elements and their related disorders

Trace elements	Metabolic function	Deficiency	Excess	Genetic disorder
Iron (Fe)	Oxygen transport, deoxyribonucleic acid (DNA) synthesis, and electron transport (Abbaspour et al. 2014)	Anemia (Lopez et al. 2016)	Thalassemias, sideroblastic anemias, congenital dyserythropoietic anemias, myelodysplastic syndromes (Pietrangelo 2016)	Not observed
Iodine (I)	Precursor of thyroid hormones thyroxine and triiodothyronine, (Khurana 2006)	Goiter endemic cretinism	Hyperthyroidism or hypothyroidism (Laurberg et al. 2010)	Not observed
Copper (Cu)	Required for enzymes involved in aerobic metabolism (Harris 1995)	Not observed	Not observed	Menkes disease (Menkes et al. 1962); Wilson Disease (Cumings 1948); Occipital horn syndrome (OHS) (Tümer and Horn 1997)
Fluoride (F)	Preventive agent for dental caries (Ekstrand and Ehrnebo 1983)	Not observed	Dental fluorosis (Richards et al. 1986)	Not observed
Zinc (Zn)	Required for cell growth, cell division, the breakdown of carbohydrates, and wound healing, (Cousins 1985)	Dermatology issues, alopecia, Impact on growth malabsorption syndrome, steatorrhea, Crohn's disease, regional ileitis	Not observed	Acrodermatitis enteropathica
Manganese (Mn)	Required for normal growth and metabolism	Not observed	Not observed	Hypermanganesemia with dystonia 1 (HMNDYT1), CDG2N-SLC39A8 deficiency
Selenium (Se)	Biosynthesis of selenoproteins (Fan et al. 1988)	Keshan disease (cardiomyopathy) and Kashin–Beck disease (Combs and Combs 1986)	Selenosis (Tinggi 2003)	Not observed
Magnesium (Mg)	Plays an essential physiological role in many functions of the body	Hypomagnesemia (Classen et al. 2012)	Hypermagnesium (Girard et al. 2005)	Not observed

disturbances and renal impairment (Maret 2019). Understanding the metabolism of these essential trace elements and the disorders that can arise either from their deficiency or overload is crucial for maintaining human health.

9.2 Iron Metabolism

9.2.1 Systemic Iron Homeostasis

Humans effectively recycle and conserve iron, reflecting the limited amount of biologically accessible iron (Pantopoulos et al. 2012). The majority of iron found in hemoglobin (~2 g iron) is present in red blood cells, and it is recycled by reticuloendothelial macrophages during erythrophagocytosis. In addition to liver hepatocytes, these iron-recycling macrophages constitute an important source of iron storage. Lower levels of iron are present in all other cells for vital biological functions. The amount of iron in the bloodstream is quite low (2–4 mg), and it needs to be changed frequently to meet the daily needs of the body (~20–25 mg) for erythropoiesis and other processes. The duodenum absorbs about 1–2 mg of iron from food. Most of the circulating iron is protein-bound (attached to transferrin, TF) and the kidney has several ways for reclaiming iron, urinary excretion of iron appears to be quite low (Yang et al. 2003). Intake of dietary iron and its release from recycled hepatocytes and macrophages are the main ways to maintain systemic iron balance because loss of iron from the body is not regulated (Fig. 9.1).

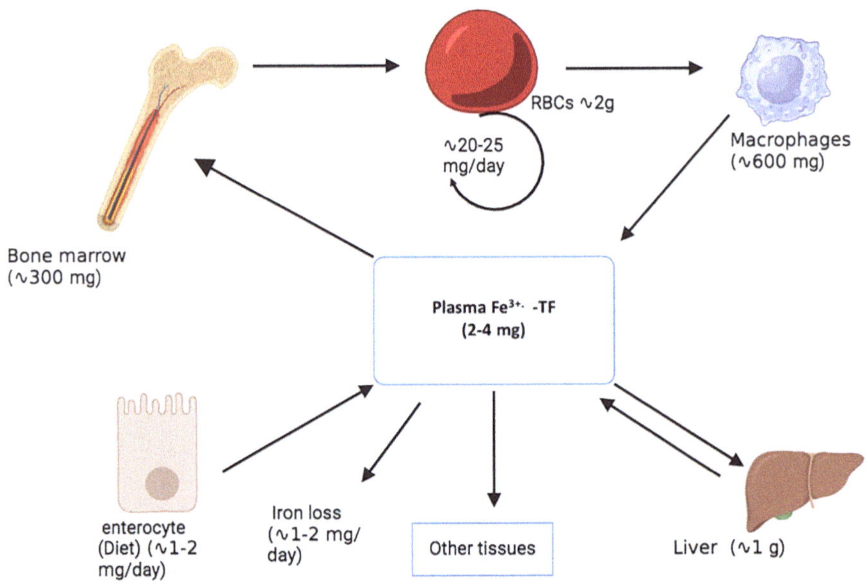

Fig. 9.1 Systemic iron homeostasis

9.2.2 Recycling and Absorption

Dietary iron is taken up in various states, including ferritin, heme, and inorganic forms. Inorganic dietary iron is primarily present in the oxidized form (Fe^{3+}), requiring reduction to the Fe^{2+} form before uptake by intestine. The reduction of Fe^{3+} to Fe^{2+} is facilitated by ferrireductases in the apical membrane of intestinal cells, such as duodenal cytochrome B (DCYTB), with the assistance of ascorbic acid (McKie et al. 2001). Fe^{2+} is then transported across the apical membrane by divalent metal transporter 1 (DMT1) (Fleming et al. 1997). The microenvironment created by Na^+/H^+ exchanger present on the brush border facilitate Fe^{2+} uptake. The exact mechanism of heme and ferritin uptake is unknown. Studies suggest that the inorganic form of iron is subsequently released in the enterocytes (Weintraub et al. 1968).

Iron absorbed by enterocytes can be either stored for future utilization, transported across the basolateral membrane for systemic distribution, or employed immediately for essential cellular metabolic functions. Similar to other cells, enterocytes store iron mostly as ferritin, which is a globular protein made up of 24 polypeptide chain encircling an iron core (Arosio et al. 2015). This makes it possible to keep iron in an inert state, which diminishes the production of ROS. Ferritin storage allows for the regulation of the amount of iron provided systemically by intestinal cell desquamation, which occurs every few days, as well as the more accurate distribution of iron to basolateral iron exporters (Vanoaica et al. 2010). Ferroportin, a basolateral iron exporter, is the mediator of systemic iron distribution (Donovan et al. 2000). While the precise molecular mechanisms underlying ferroportin's iron export process are not fully elucidated, current understanding suggests that iron is likely exported in its Fe^{2+} state. Subsequently, this Fe^{2+} is oxidized to Fe^{3+} by enzymes such as hephaestin and ceruloplasmin. Following this oxidation, the Fe^{3+} ions are bound to transferrin, which serves as the primary carrier for iron within the plasma (Drakesmith et al. 2015).

9.2.3 Circulating Iron

Normally, in the bloodstream, iron circulates in transferrin-bound form. Up to two iron molecules can be carried by transferrin. Transferrin transports iron to tissues so that the widely expressed transferrin receptor 1 (TFR1) may absorb it (Pantopoulos et al. 2012). Transferrin receptor 2 (TFR2), a homologue with more restricted expression, might act as a detector of iron levels bound to transferrin in various organs, such as red blood cells and the liver (Ganz 2013). Non-transferrin bound iron (NTBI) is absorbed through different pathways. For instance, ZIP14 plays a key role in hepatocyte NTBI uptake (Jenkitkasemwong et al. 2015). Ferritin levels in the blood can also be measured. Macrophages are the primary source of circulating ferritin, which is mostly correlated with body reserves of iron. However, other disorders such as inflammation, infection, liver disease, and cancer can also affect ferritin levels (Wood 2014).

9.2.4 Cellular Iron Homeostasis

Certain absorption mechanisms transport circulating iron to erythrocytes and other bodily cells (Pantopoulos et al. 2012). The acidic endosome releases iron, and the ferriductase STEAP3 reduces Fe^{3+} to Fe^{2+}. Nuclear receptor coactivator 4 (NCOA4), the cargo receptor, mediates autophagic turnover (ferritinophagy), which releases iron deposited in ferritin for export or consumption (Mancias et al. 2014). Other transporters, such as FLVCR1A for heme iron and ferroportin for elemental iron, as previously mentioned, are also involved in the export of iron from cells (Keel et al. 2008). Even while our understanding of the cellular iron trafficking pathways has advanced significantly, many of these channels remain poorly known.

9.2.5 Disorders

9.2.5.1 Iron Deficiency

Anemia is a well-known indicator of low iron levels. Iron deficiency mainly occurs during pregnancy, fast growth of children, increased blood loss, or insufficient dietary iron absorption (Lopez et al. 2016). CKD patients are more prone to iron deficiency due to several factors such as insufficient intake of nutrients, interference with iron absorption by certain medications such as phosphate binders and antacids (Zumbrennen-Bullough and Babitt 2014). Less frequently, mutations in genes related to erythroid iron intake or usage, duodenal iron absorption, iron mobilization from body reserves, or transferrin (TF) might result in iron deficiency anemia (Mims et al. 2005). Excess tissue iron is linked to iron deficiency anemia in many of these situations. Iron refractory iron deficiency anemia arises from mutations in the hepcidin regulatory gene, TMPRSS6, which prevents the liver from producing hepcidin (Finberg et al. 2008).

9.2.5.2 Iron Overload

Iron loading anemias (congenital dyserythropoietic anemias, thalassemias, myelodysplastic syndromes, sideroblastic anemias,), transfusional or other secondary types of iron excess, and genetic hemochromatosis are the main causes of systemic iron overload (Pietrangelo 2016). When iron levels in these disorders surpass transferrin's buffering ability, highly reactive forms of non-transferrin-binding iron (NTBI) are absorbed by the heart, liver, and endocrine glands. The excess iron in these organs starts oxidative damage and organ dysfunction, resulting in endocrinopathies such as cirrhosis, cardiomyopathy, diabetes mellitus, and others.

Mutations in the genes encoding hepcidin itself (HAMP) or in the genes hemojuvelin (HFE2, also known as HJV), transferrin receptor 2 (TFR2), or HFE are the primary triggers for hepcidin expression in reaction to iron levels and lead to hereditary hemochromatosis (Roetto et al. 2003). These genes' mutations affect hepatocytes' capacity to recognize rising iron levels and initiate signaling through the SMAD-BMP cascade, which in turn stimulates the transcription of hepcidin (Zumbrennen-Bullough and Babitt 2014). Absence of functionality a unique condition known as ferroportin disease is caused by mutations in ferroportin and is typified by the build-up of iron in macrophages (Pietrangelo 2016).

9.2.5.3 Iron and Liver Disease

In cases of iron overload, the liver is the principal organ that removes excess circulating NTBI from the body and serves as a major iron storage facility. As observed in hereditary hemochromatosis and β-thalassemia, high iron concentration can result liver damage, cirrhosis, and cancer when the liver's capacity for storing iron and producing antioxidants is surpassed (Pietrangelo 2016). Liver iron loading is also linked to a number of other chronic liver illnesses, such as nonalcoholic fatty liver disease, viral hepatitis, and alcoholic liver disease (Pietrangelo 2016). Hepcidin expression may occasionally be changed by the pathogenic causes linked to these illnesses and the liver damage itself, which can exacerbate iron excess (Pietrangelo 2016).

9.2.5.4 Iron and Diabetes Mellitus

The frequent side effects of iron overload conditions may cause diabetes mellitus including β-thalassemia and hemochromatosis, which was formerly known as "bronze diabetes" (McClain et al. 2006). Elevated body iron reserves, as determined by serum ferritin, and elevated dietary heme iron are linked to a higher risk of type 2 diabetes and other insulin resistance conditions, even in populations that appear healthy (Ford and Cogswell 1999). The fact that glycemic control is improved in individuals with reduced body iron through phlebotomy or iron chelation suggests that iron may have a causal role in the development of diabetes mellitus (Dymock et al. 1972). It is still unclear how iron contributes to the etiology of diabetes.

9.2.5.5 Iron and Cardiovascular Disease

Iron-overload cardiomyopathy has a complex etiology that involves oxidant-mediated damage, disruption of heart electrical activity, and fibrosis promotion (Pennell et al. 2013). Lack of iron has negative effects and it is linked to lower cardiovascular outcomes and more severe heart failure symptoms (Comín-Colet et al. 2013).

9.2.5.6 Iron and Neurodegenerative Diseases

The brain requires iron, for myelinogenesis, the synthesis of various neurotransmitters, including norepinephrine and dopamine, as well as to maintain the brain's high level of respiratory activity (Moos et al. 2007). It is commonly known that iron deficiency affects cognitive function and brain development (Beard 2003). Conversely, a number of neurodegenerative illnesses have been linked to iron excess or the pathological buildup in the brain (Singh et al. 2014). These diseases primarily affect the basal ganglia, one of the brain's iron-rich regions (Singh et al. 2014). The iron homeostasis genes ferritin light chain or ceruloplasmin mutations cause the most pronounced causal relationship in a subset of NBIA illnesses. These causes oxidative damage in astrocytes due to iron overload and maybe neuronal iron deficiency toxicity as well (Singh et al. 2014). Iron-mediated oxidative damage is caused by abnormalities in the ferritin light chain, which compromise the safe storage of iron in neuroferritinopathy (Singh et al. 2014). Friedreich's ataxia disease is believed to be caused by oxidative stress, mitochondrial malfunction, and may be due to the stimulation of sphingolipid production and other signaling pathways as a result of iron build-up (Singh et al. 2014). Additionally, iron build-up is linked to more prevalent neurological illnesses including Parkinson's and Alzheimer's disease (Singh et al. 2014)

9.2.5.7 Iron and Cancer

Patients with hereditary hemochromatosis and β-thalassemia are at heightened risk for hepatocellular carcinoma and potentially other malignancies due to iron overload (Fracanzani et al. 2001; Osborne et al. 2010). Two primary mechanisms are thought to be involved in the development of cancer when iron overload occurs (Manz et al. 2016). First, iron's pro-oxidant properties have the potential to harm DNA, which in turn encourages oncogenesis. Second, in order to sustain their fast pace of growth, cancer cells are more dependent on iron. These proteins include TFR1, which is expressed more frequently to increase iron uptake, ferroportin, which is expressed less frequently to cause less iron export and hepcidin, which is produced more frequently to autocrinely down regulate ferroportin expression, which reduces iron export. Low ferroportin tumor expression in breast cancer patients is linked to a decreased chance of survival and the spread of metastases. Low ferroportin tumor expression in breast cancer is linked to metastasis and decreased survival, whereas high ferroportin tumor expression along with low hepcidin expression indicates a good prognosis (Pinnix et al. 2010).

9.2.5.8 Iron and Kidney Disease

Numerous models of acute kidney injury (AKI) have been linked to the pathophysiology of iron and iron-induced reactive oxygen species (Baliga et al. 1993). Models of chronic renal illness have also shown kidney iron buildup (Nankivell et al. 1992). It is believed that oxidative damage is caused by an increase in iron concentration, which occurs when the kidney is subjected to ischemia or toxic insults. More recently, the pathophysiology of renal ischemia-reperfusion injury has been linked to ferroptosis (Linkermann et al. 2014). The protective benefits of iron chelator treatment in a variety of AKI animals (Paller et al. 1988). It has been shown that the iron metabolism proteins ferritin, hepcidin, and heme-oxygenase 1 (HO-1) all contribute to antioxidant defense, sequester iron, and modify iron trafficking, all of which protect against AKI (Zarjou et al. 2013; Nath 2006).

9.2.5.9 Iron and Bone Disorders

Patients with β-thalassemia and hemochromatosis are prone to osteoporosis (Guggenbuhl et al. 2005). By stimulating the synthesis of the hormone fibroblast growth factor 23, or FGF23, which is produced from bone and controls phosphorus and vitamin D balance, iron shortage may also have an effect on bone health (David et al. 2017). In healthy individuals, elevated cleavage levels coincide with elevated

FGF23 synthesis, resulting in negligible alterations to bioactive, intact FGF23 concentrations.

9.3 Iodine Metabolism

The majority of iodine is consumed through food, especially vegetables cultivated in iodine-rich soil; drinking water provides the remaining iodine needed (Dhaar and Robbani 2006). Several Asian cultures employ seaweeds like wakame, nori, or mekabu, which are abundant in iodine and may be used to make soups, salads, and sauces. Nature contains iodine in a variety of forms, including organic monoatomic iodine, inorganic diatomic iodine (also known as molecular iodine or I), and salts of sodium and potassium (iodides and iodates) (Patrick 2008). The thyroid gland is essential for the metabolism of iodine and this gland is lined by follicular cells which is filled by a clear viscous material called thyroglobulin (Pal 2021). The steps of iodine metabolism are as follows:

9.3.1 Iodine Trapping

The initial stage of iodine metabolism is called iodine entrapment. Iodide is taken up by an active transport system from the capillary and enters the gland's follicular cell to start the process. The sodium/iodine symporter protein (NIS), which is located in the basolateral membrane of the follicular cell, enables this against chemical and electrical gradients; the ATPase-dependent Na^{+}/K^{-} pump provides the energy needed for this activity (Khurana 2006).

9.3.2 Iodine Synthesis and Secretion

The second stage involves the synthesis and secretion of thyroglobulin. Synthesis initiates in the rough endoplasmic reticulum. This process occurs independently within the follicular cell. Subsequently, these units combine to form a dimer, after which additional carbohydrate components are added in Golgi apparatus. The entire thyroglobulin molecule comprises roughly 140 tyrosine residues, serving as a substrate for thyroid hormone production (Khurana 2006).

9.3.3 Oxidation of Iodine

Iodide oxidation is the third stage. Iodide in the follicular cell is transported by the sodium-independent iodide/chloride transporter pendrin. Then, (I) quickly oxidizes the iodide (I′) to iodine (Khurana 2006).

9.3.4 Iodination of the Tyrosine

Tyrosine undergoes iodination resulting in monoiodotyrosine (MIT), and diiodotyrosine (DIT). Following tyrosine iodination, a coupling process occurs where one MIT combines with one DIT to form triiodothyronine (T3) hormone, while two DIT produce thyroxine (T4) hormone. This reaction is facilitated by thyroid peroxidase (TPO) (Patrick 2008). The thyroid follicles store the thyroid hormones as colloid which is later used by the body (Khurana 2006).

The iodinated thyroglobulin-containing colloid is taken up by the epithelial cells through the process of endocytosis. This process is made possible by the presence of the TG receptor megalin on the apical membrane. The colloid now enters the cytoplasm as colloid droplets, which may travel in the direction of the basal membrane by the action of microtubules and microfilaments. The lysosomal vesicles, which carry proteolytic enzymes, fuse with the colloid droplets after that. The thyroglobulin molecule is broken down by the proteases, which releases MIT, DIT, T3, and T4 into the cytoplasm. The MIT and DIT are quickly deiodinated by the enzyme deiodinase, while T4 and T3 diffuse into the bloodstream via the basal surface. This process aids in the retrieval of tyrosine and iodide for recycling (Pal 2021; Khurana 2006). TSH, the hormone used by the pituitary gland to regulate thyroid output, operates through a feedback mechanism calibrated to the blood level of T4.

9.3.5 Secretion

Eighty micrograms of iodine per day are secreted by the thyroid gland as T3 and T4 hormones; an additional 40 μg are secreted each day and show up in extracellular fluid (ECF). The liver metabolizes T4 and T3, releasing 20 μg of iodine into the bile to be expelled in feces and 60 μg into ECF. Iodine excretes as 20 μg in stools and 480 micrograms in urine daily on average (Pal 2021). The thyroid gland typically maintains iodine content due to its highly efficient iodine-trapping mechanism (Detels et al. 2011). The thyroid to serum (T/S) ratio is used to determine how efficient iodide trapping is. Radiation-active iodide is used to measure the T/S [I]. The T/S ratio for iodide is controlled by thyroid stimulating hormone (TSH). Following a hypophysectomy, the TSH level declines, which lowers the T/S ratio. T/S ratio rises with high TSH levels, as in secondary (pituitary) hyperthyroidism (Porterfield 2001)

9.3.6 Disorders

9.3.6.1 Iodine Deficiency Disorders (IDD)

9.3.6.1.1 Goiter

There are four reasons to support the theory that iodine deficiency causes endemic goiter (WHO 2007). A strong correlation exists between the disease's occurrence in the population and low iodine content in food and water (Assey et al. 2006). A significant decrease in incidence occurs when iodine is added to the diet (Pearce et al. 2004). Iodine metabolism in endemic goiter patients fits the expected pattern of iodine deficiency and can be reversed by iodine repletion and iodine deficiency results in abnormalities in animals' thyroid glands that are comparable to those observed in humans (Hetzel and Pandav 1996; Haldimann et al. 2005). When iodine intake is carefully evaluated in a population with goiter, the results are almost always significantly below normal.

9.3.6.2 Endemic Cretinism

After surviving the crucial postnatal period, infants of iodine-deficient mothers may acquire endemic cretinism. Profound intellectual disability characterizes the neurological form of endemic cretinism, often presenting alongside cerebral diplegia and deaf-mutism (Detels et al. 2011). Children living in endemic areas have delayed mental and physical development, low IQs, and poor academic achievement. The decline in the quality of our human resources is a clear indication of the serious effects of iodine shortage on children's ability to learn and their overall quality of life (Gopalan 2000).

9.4 Copper Metabolism

9.4.1 Absorption

Animals ingest copper mostly through their digestive systems. The remainder of the dietary copper is eliminated through the feces, with just 20–60% of it being absorbed. Absorption of copper is influenced by several factors. For instance, eating phosphate, citrate, and animal protein increases the absorption of copper. Copper ions are more readily absorbed from salts than from oxides, such as copper gluconate, copper acetate, or copper sulfate (Turnlund et al. 1983). Dietary absorption of copper is inhibited by large intakes of phytate, simple sugars (sucrose and fructose), and elevated levels of dietary zinc and cadmium (Wapnir 1998). The metal is transferred to the liver, where it is bound by serum albumin, after passing through the basolateral membrane.

9.4.2 Transportation and Secretion

Copper is transported to the peripheral organs through the plasma that is bound to low-molecular-mass complexes, ceruloplasmin, or serum albumin (Harris 2000). Copper is found in three different pools in blood: one linked to proteins, one composed of non-exchangeable red blood cells, and one that is labile and contains low molecular mass complexes. In humans, ceruloplasmin firmly binds 80–90% of the plasma copper, with the remaining portion being linked to albumin and amino acids. The small intestine is the primary site of intestinal copper absorption in mammals

(Wapnir 1998). Ceruloplasmin is released into the bloodstream, where it provides extrahepatic organs with copper (Harris 2000). Together with albumin and other low-molecular-mass proteins, copper can also circulate in the body as complexes (Harris 2000). Three potential secretory pathways exist in mammals:

(a) A Cu-ATPase, which has been found in Wilson's disease patients (also known as Wilson's protein or ATP7B)
(b) Amultiorganic cation transporter (cMoat) (Elferink and Jansen 1994).
(c) Secretion by lysosomes (Gross et al. 1989)

The copper chaperone ATOX1 transports copper to ATP7B in the Golgi apparatus after it enters the cell via the copper transporter CTR1. Copper is integrated into a number of cuproenzymes in the Golgi apparatus, including ceruloplasmin. A vesicular compartment is where ATP7B redistributes when the cell's copper levels rise. After ATP7B relocalizes, copper is eliminated from the hepatocyte through the bile by an unidentified process that most likely involves COMMD1 (De Bie et al. 2005)

9.4.3 Disorders

9.4.3.1 Menkes Disease (MD)

Menkes originally described the rare hereditary illness known as Menkes disease in 1962 (Menkes et al. 1962). It is typified by aberrant copper metabolism (Turnlund et al. 1983). Mutations in ATP7A cause this condition, which is inherited in an X-linked recessive fashion. Mutations in the ATP7A gene affect the intestinal function of ATP7A, which lowers the amount of copper that enters the bloodstream and causes copper to build up in intestinal cells (Møller et al. 2009). A severe systemic copper deficit and reduced production of several key enzymes would result if copper cannot be delivered to the body's various organs, including the liver and brain (Shim and Harris 2003). For example, poor synapse function and axonal development may arise from disruption of dopamine β-hydroxylase-like monooxygenase function (Kaler et al. 1993). Neonates with inborn vascular abnormalities are defined by the disease of the tetralogy of Fallot (Szauter et al. 2005). Menkes disease patients frequently exhibit signs such as skin laxity, aneurysms, spontaneous fractures, and bone loss (Royce et al. 1980).

9.4.3.2 Wilson Disease

Wilson disease is a genetic mutation-related illness characterized by a pathological copper buildup (Litwin et al. 2019). Wilson disease is mainly caused by mutations in the ATP7B gene, which in turn causes the transmembrane copper-transporting ATPase to become inactive, obstructing the biliary tract's ability to excrete copper and ultimately upsetting the balance of copper in the body (Cumings 1948). Moreover, cirrhosis and liver fibrosis are caused by copper overload in hepatocytes

(Przybyłkowski et al. 2014). Extrahepatic toxicity results from the slow accumulation of copper released from liver cells in other organs (Walshe and Potter 1977).

Wilson disease patients may have copper concentrations that are 10–15 times greater than those of healthy people (Horoupian et al. 1988), indicating a substantial correlation between brain copper levels and the degree of neuropathology. Astrocytes first act as a buffer against the harmful effects of copper, while also promoting its proliferation. To improve metallothionein's capacity to store copper, its synthesis is subsequently increased. Elevated copper levels have the potential to cause astrocyte damage, malfunction of the blood-brain barrier, and various brain tissue diseases (Bertrand et al. 2001). In Wilson disease, there is a correlation between the degree of brain pathology and neurological disability and abnormalities in the morphology and function of the retina (Langwińska-Wośko et al. 2016).

9.4.3.3 Occipital Horn Syndrome (OHS)

Occipital horn syndrome (OHS), previously referred to as X-linked cutis laxa or Ehlers-Danlos syndrome type IX, is a rare disorder characterized by notable connective tissue irregularities, including hernias, bladder diverticula, joint laxity, and cutis laxa. It also manifests distinctive exostoses known as "occipital horns," which point downward and are situated in the tendinous insertions of the sternocleidomastoid and trapezius muscles. Patients may exhibit a range of intellectual disabilities, from mild to severe (Tümer and Horn 1997). OHS is considered to represent the less severe end of the phenotypic spectrum due to pathogenic mutations in ATP7A, which encodes a copper transporter, and its allelic association with Menkes disease (MD) (Horn and Tümer 2002). Dysfunction of cupro-enzymes, such as dopamine ß-hydroxylase, lysyl oxidase, cytochrome C oxidase, and tyrosinase, is implicated in various clinical features of both disorders (Byers et al. 1980).

9.5 Fluoride Metabolism

9.5.1 Absorption

Fluoride is predominantly absorbed in the stomach through the gastrointestinal tract without relying on specialized enzymatic systems (Whitford 1996). It travels over epithelia as hydrogen fluoride, or undissociated acid. The physiological activity of fluoride is explained by its diffusibility in hydrogen. The prevalence of the undissociated form of hydrogen fluoride increases at lower pH levels (<3.5), whereas the ionized form becomes more prevalent at higher pH levels (Whitford 1994). Recent studies indicate that the majority of fluoride absorption occurs in the small intestine and is independent of pH, with undissociated acid traversing the stomach. There is evidence supporting the existence of several carrier-mediated, pH gradient-dependent pathways for intestinal fluoride transport (Villa et al. 2009). There is evidence to support the existence of multiple carrier-mediated, pH gradient-dependent pathways for intestinal fluoride transport (Villa et al. 2009).

The stomach's acidity has a direct impact on how quickly fluoride is absorbed from it (Whitford and Pashley 1984). The solubility of the ingested fluoride molecule is one of the elements that affects the rate of absorption, though. Enhanced absorption rates are attained with more soluble fluoride compounds such as sodium fluoride (NaF) and hydrogen fluoride, facilitating faster absorption. Conversely, slower absorption rates are associated with less soluble fluoride compounds like calcium fluoride (CaF_2) and magnesium fluoride (MgF_2) (Ekstrand and Ehrnebo 1983).

9.5.2 Distribution

Plasma fluoride levels rise as soon as fluoride is absorbed (at 10 min), peaking at 60 min. It takes 11–15 h for basal levels to restore (Whitford and Pashley 1984). The fact that plasma contains two different types of fluoride is widely known. Ionic fluoride is the name given to one portion, whereas nonionic or bonded fluoride, which is made up of lipid-soluble organic fluoro-compounds, is the name given to the second fraction. Uncertainty surrounds the nonionic fraction's biological importance (Ekstrand and Ehrnebo 1983). Fluoride is quickly deposited in the skeleton or eliminated by the kidneys after it reaches plasma. Aside from age, other variables that affect fluoride skeletal absorption include bone remodeling and modeling activity (Dunipace et al. 1995). The amount of fluoride that remains in the bones decreases with an individual's age. When fluoride is absorbed for the first time in subjects who have never been exposed to it before, saturation is reached (Hodge et al. 1970). Fluoride can accumulate in the crystal structures, the adsorbed layers, or the bone matrix. Fluoride can be gradually eliminated after it has been incorporated and when bone saturation approaches. Prior findings following the withdrawal of fluoride from public water sources indicate that the half-life of fluoride loss in adults is 120 weeks, whereas it is only 70 weeks in youngsters (Likins et al. 1956).

9.5.3 Secretion

Saliva secretes fluoride, which is found in higher concentrations in plasma. While salivary fluoride levels typically range from 0.01 to 0.06 ppm in individuals exposed to fluoride, these levels play a critical role in fluoride's efficacy as a preventive agent against dental cavities (Ingram et al. 2005).

9.5.4 Excretion

A small amount of fluoride that is not retained in bone is expelled primarily through the kidneys and through feces. The renal clearance of fluoride from plasma is regulated by both pH and urine flow (Whitford and Pashley 1984). Age also has an impact on fractional urinary fluoride excretion, which is the percentage of ingested fluoride that is eliminated in the urine and exhibits unique nocturnal and diurnal rhythms (Ekstrand and Ehrnebo 1983). Similar to how pH and other factors affect fluoride absorption, fractional urine excretion affects both adults and children. Diets high in fruits and vegetables cause urine to have an alkaline pH, while diets high in protein cause urine to become more acidic. The fractional excretion of fluoride in the urine is subsequently altered by these variations in urine pH.

9.5.5 Disorders

9.5.5.1 Dental Fluorosis

Excessive fluoride consumption during tooth formation results in dental fluorosis, a hypomineralization of dental enamel (Richards et al. 1986). It has been proposed that in order for dental fluorosis to manifest, there must be an excessive concentration of fluoride in the enamel's developing environment at a key window of higher vulnerability (also known as the "window of susceptibility") for that particular surface (Evans and Darvell 1995). Dental fluorosis develops at the cellular level in response to fluoride concentrations in the extracellular fluid surrounding the growing dental enamel (Fejerskov et al. 1977).

9.6 Zinc Metabolism

The small intestine facilitates zinc absorption through a carrier-mediated mechanism (Cousins 1985). Normally, under physiological conditions, zinc uptake processes are not saturated. However, determining the precise fraction of absorbed zinc is complex due to zinc secretion into the gut. In fasting individuals, administration of zinc in aqueous solutions results in efficient absorption, typically around 60–70%, whereas absorption from solid diets is less efficient and varies based on zinc content and diet composition (WHO 2004). On average, approximately 33% of zinc is considered absorbed in humans (Cousins 1985).

Zinc absorption is influenced by concentration, increasing with higher dietary zinc levels up to a maximum rate. Moreover, the zinc status of individuals can affect absorption (WHO 2004). Zinc-deprived individuals absorb zinc more efficiently, whereas those on a high-zinc diet show reduced absorption efficiency (Krebs 2000). During digestion, zinc is released from food as free ions. These ions may then bind to naturally occurring ligands before being transported into the enterocytes in the duodenum and jejunum (Cousins 1985). Serum zinc represents only a small fraction of the body's total zinc, it undergoes rapid turnover to meet the needs of various tissues (Cousins 1985).

9.6.1 Zinc Transporters (ZnTs)

Zinc transporters are integral for maintaining cellular zinc homeostasis, with human cells containing a minimum of 10 ZnTs and 15 ZIP transporters, each with opposing functions (Cousins et al. 2006). The expression and cellular localization of ZnTs are meticulously regulated in response to fluctuations in zinc levels. Both ZnT and ZIP transporter families demonstrate distinct tissue-specific expression patterns and respond differently to variations in dietary zinc levels or imbalances. Furthermore, they exhibit varied responses to physiological stimuli mediated by hormones and cytokines (McMahon and Cousins 1998).

9.6.2 Role of Metallothionein

Metallothionein serves as a key player in zinc homeostasis, but its exact interaction with zinc transporters, especially ZnTs, is not fully understood. The stimulation of MT synthesis by factors like dietary zinc and inflammation suggests its involvement in cellular responses to changes in zinc levels and physiological conditions. Further research is needed to elucidate the intricate relationship between MT, zinc transporters, and zinc absorption regulation (Davis et al. 1998).

9.6.3 Homeostasis

For survival, the cellular zinc status must be maintained in a steady condition, or homeostasis. Zinc homeostasis is primarily maintained by changes in endogenous intestinal excretion and total zinc absorption (Hambidge and Krebs 2001). In both animals and humans, adjustments in total zinc absorption and endogenous intestinal excretion are the primary mechanisms for maintaining zinc homeostasis (Hambidge and Krebs 2001). Changes in plasma zinc turnover rates, variations in urinary zinc excretion, and maybe a redistribution of zinc from some tissues –like bone – to other tissues in order to maintain their function are some examples of these adaptations (King et al. 2000).

9.6.4 Excretion

Approximately half of the zinc eliminated from the body exits through the gastrointestinal tract. A significant portion of zinc is excreted via biliary and intestinal secretions, yet the majority is reabsorbed, playing a critical role in regulating zinc balance.

9.6.5 Disorders

9.6.5.1 Acrodermatitis Enteropathica (AE)

Acrodermatitis enteropathica is an autosomal recessive disorder that typically affects infants of Italian, Armenian, or Iranian descent (Prasad 2004). The disease usually manifests in the early months of life, shortly after weaning from breast feeding. Patients with AE experiencing severe zinc deficiency may present with various dermatological symptoms, including bullous pustular dermatitis on the extremities and around the oral, anal, and genital areas, as well as paronychia and alopecia. Ophthalmic manifestations may include blepharitis, conjunctivitis, photophobia, and corneal opacities. Neuropsychiatric symptoms might manifest as irritability, discomfort, and occasional cerebellar ataxia. Additionally, congenital malformations are commonly observed in fetuses and infants born to women with AE (Cavdar et al. 1980).

9.6.5.2 Total Parenteral Nutrition (TPN)

Zinc depletion may result with total parenteral nutrition (TPN), with patients who experience diarrhoea. This loss can lead to severe zinc deficiency, characterized by alopecia, dermatologic issues, neuropsychiatric symptoms, weight loss, and increased susceptibility to infections.

9.6.5.3 Penicillamine Therapy

Patients with Wilson's disease undergoing penicillamine therapy can experience severe zinc deficiency due to excessive zinc loss induced by the treatment.

9.6.5.4 Zinc and Growth

Zinc deficiency significantly impacts growth, being the primary limiting factor in experimental animals (Nam et al. 2013). Zinc deficiency leads to a decrease in circulating insulin-like growth factor 1 (IGF-1) concentration, impacting cell cycle regulation and cell division. Zinc plays a pivotal role in activating IGF-1 receptors and enhancing the activity of deoxythymidine kinase (TK), which is essential for growth and DNA synthesis (Aisen et al. 1978).

9.6.5.5 GI Disorders

Numerous gastrointestinal conditions have been linked to a mild zinc deficiency. These consist of steatorrhea, regional ileitis, Crohn's disease, and malabsorption syndrome. MacMahon et al. (1968) were the first to document a zinc deficit in a patient with steatorrhea. Since then, awareness of the issue of zinc deficiency in patients with malabsorption diseases has become widespread among the medical community.

9.7 Manganese (Mn) Metabolism

9.7.1 Absorption

The most typical method of absorbing Manganese (Mn) is oral exposure consuming Mn-rich veggies and water. Supplements, nuts, and baby formula are the main food sources for Mn. In the gut, ingested magnesium is quickly absorbed and enters cells either by active transport or passive diffusion. Similar to other divalent cations like iron and calcium, magnesium is transported in human intestinal cells using a biphasic pattern and a saturable process (Leblondel and Allain 1999). After reaching a steady-state condition, it takes approximately an hour to activate the cellular components, mostly transporters, and then gradually accelerate Mn absorption (Leblondel and Allain 1999). The primary method of exposure for occupational Mn poisoning is inhalation of airborne particles. Another way to be exposed to Mn is by intravenous delivery of drugs containing high quantities of the metal.

9.7.2 Distribution and Regulation

The main organ that controls body Mn levels through endogenous Mn losses in the gut is the liver (Davis et al. 1993). On the cell membrane, the liver cells express a variety of Mn transporters that control Mn influx, such as citrate transporters (Osborne et al. 2010), ZIP14 (Nam et al. 2013), DMT1 (Nam et al. 2013) and transferrin/transferrin receptor (Tf/TfR) (Aisen et al. 1978). According to Rahil-Khazen et al. (2002), bone is another tissue that has a significant buildup of Mn, with a normal concentration of 1 mg/kg. There has been less research on the distribution of Mn in the kidney and pancreas. Neurons are particularly vulnerable to Mn poisoning, perhaps because of their extended life span and high energy requirements.

9.7.3 Elimination

The majority of excess Mn is converted by the liver to bile and excreted in the feces (Schroeder et al. 1966). The liver is said to be the main source of endogenous gastrointestinal losses of Mn; hence, it plays a crucial part in this process (Davis et al. 1993). Intoxication with Mn is more likely in those with hepatic issues. Apart from the major hepatobiliary elimination by feces, there have also been reports of Mn excretion through urine (Klaassen 1976), milk (Stastny et al. 1984), and sweat (Omokhodion and Howard 1994), albeit in very little amounts. However, in some situations, the percentage of elimination via various methods may vary.

9.7.4 Disorders

9.7.4.1 HMNDYT1-SLC30A10 Deficiency

Hypermanganesemia with dystonia 1 (HMNDYT1), caused by bi-allelic mutations in SLC30A10, was the first inherited manganese transporter defect to be identified (Quadri et al. 2012). More than 30 patients have been reported to date. This condition is characterized by systemic manganese accumulation, leading to a distinct syndrome featuring hypermanganesemia, dystonia, polycythemia, chronic liver disease, and depletion of iron stores.

9.7.4.2 CDG2N-SLC39A8 Deficiency

In 2015, the first inherited disorder linked to manganese (Mn) deficiency was identified. This disorder is caused by bi-allelic mutations in SLC39A8, a manganese uptake transporter (Boycott et al. 2015).

9.8 Magnesium Metabolism

9.8.1 Absorption and Transport

The amount of magnesium consumed is determined by the magnesium content of food and drink. Magnesium is abundant in green leafy vegetables, cereals, grains, legumes, and nuts (Fawcett et al. 1999). Typically, magnesium intake correlates directly with energy intake, except in cases where the primary source of energy is derived from refined sugars or alcohol. Food processing or refining can deplete magnesium content by as much as 85%. Additionally, cooking, especially boiling magnesium-rich foods can result in significant magnesium loss. Therefore, food processing and cooking practices may explain the seemingly high prevalence of low magnesium intake in many populations. The fractional intestinal absorption of magnesium is inversely related to intake, with 65% absorption at low intake and 11% at high intake. Most absorption occurs in the ileum and colon (Kayne and Lee 1993). Magnesium absorption is mainly passive under normal intake but shows saturation at low levels. Though mechanisms are unclear, parathyroid hormone (PTH) may regulate it, with controversial involvement of vitamin D.

The kidney is known to play an important role in magnesium homeostasis (Alan 2001). Roughly 15–20% of filtered magnesium is reabsorbed in the proximal tubular segments, 65–75% in the thick ascending limb of the loop of Henle (TALH), and the remainder in the distal segments (Alan 2001). Plasma magnesium concentration significantly affects urinary magnesium excretion. Hypermagnesemia leads to increased magnesium excretion due to higher filtered load and reduced reabsorption in the thick ascending limb of the loop of Henle (TALH).

9.8.2 Excretion

Decreased dietary magnesium intake can lead to a fractional excretion drop to below 0.5% of the filtered load, primarily attributed to heightened reabsorption in the thick ascending limb of the loop of henle (TALH). The precise mechanism driving this adaptation remains incompletely understood (Noronha and Matuschak 2009). Additionally, an elevation in the filtered load of magnesium, without concurrent alterations in plasma concentration, can also affect tubular handling of magnesium.

9.8.3 Disorders

9.8.3.1 Hypomagnesium

Hypomagnesemia can arise from a multitude of factors, including dietary deficiency (such as malnutrition or reliance on total parenteral nutrition), redistribution (as seen in hungry bone syndrome), gastrointestinal (GI) disorders, or renal wasting (Classen et al. 2012). Chronic use of proton pump inhibitors (PPIs) is a significant cause of hypomagnesemia (Calò et al. 2000). This loss of magnesium occurs due to decreased intestinal absorption, often attributed to the down regulation of TRPM6 channels (Al Alawi et al. 2018). Genetic disorders affecting magnesium levels may originate in the kidney, either in the distal convoluted tubule (DCT) or the thick ascending limb (TAL).

9.8.3.2 Hypermagnesium

Hypermagnesemia is less common than hypomagnesemia and typically manifests in patients with advanced chronic kidney disease (CKD) when the glomerular filtration rate (GFR) drops below 10–30 mL/min. This condition may coincide with the consumption of magnesium salts present in certain laxatives (such as magnesium citrate or magnesium hydroxide) or antacids

9.9 Cobalt Metabolism

Cobalt can be significantly absorbed through the lungs following inhalation, mainly in the form of metallic cobalt or cobalt oxide present in dust and welding fumes, often mixed with other metals. Minute particles can disseminate systemically through the lymphatic and vascular system, releasing soluble cobalt ions. After a single dose administration of cobalt (Co^{2+}) in humans, cobalt concentration in blood and serum initially peaks but then rapidly decreases. This decline is attributed to tissue uptake, primarily in the liver and kidney, as well as urinary and fecal excretion. By 24 h post-administration, cobalt levels in blood and serum reach a low level (Smith et al. 1972). Cobalt primarily accumulates in the liver, kidney, pancreas, and heart. Over time following cobalt administration, the relative content in the skeleton and skeletal muscle increases.

9.10 Chromium Metabolism

9.10.1 Absorption

Chromium (Cr) enters the body through various routes, including the skin, respiratory system, and digestive tract. Its metabolism is significantly affected by the entry route, as well as its oxidation state and legend nature. In the diet, chromium (Cr) exists in both inorganic and organic complexes. Elemental chromium from cooking or stainless-steel equipment is unlikely to be absorbed and has minimal to no nutritional value (Wallach 1985). The most common ionic forms are Cr^{3+} and Cr^{6+}. Hexavalent chromium, obtained through diet or industrial exposure, is rapidly reduced to Cr^{3+} in tissues and fluids, which is the biologically active form. The absorption of chromium (Cr) in the intestinal mucosa involves several mechanisms that are not yet fully understood. The exact mechanism responsible for intestinal Cr absorption remains uncertain. Some studies suggest that Cr may enter the body through passive diffusion, while others propose the involvement of transport proteins located in the intestinal mucosa. Results from various studies showed some conflict in the findings. Some suggest that trivalent Cr absorption is unsaturable, indicating passive diffusion, while others demonstrate a saturation effect with increasing concentrations, suggesting involvement of transport proteins (Mertz et al. 1965). In vitro studies using isolated intestines from deficient rats have shown a decrease in the percentage of absorbed trivalent Cr with increasing concentrations in the medium, indicating saturation. This suggests that transport proteins are involved in Cr absorption. Moreover, cellular energy does not appear to be required for Cr transport, leading to the conclusion that Cr^{3+} is absorbed through facilitated transport, involving transport proteins but not oxidative metabolism.

9.10.2 Blood Transport

Chromium (Cr) binds to transferrin, a protein known to have two binding sites, A and B, with varying affinities for iron depending on pH (Hopkins and Schwarz 1964). Previous data indicate that Cr exclusively binds to site B, leading to antagonism with Fe^{3+} at high iron concentrations, such as those seen in hemochromatosis. Recent research confirms that Cr^{3+} has a high affinity for plasma transferrin (Sayato et al. 1980). Additionally, a low-molecular-weight Cr binding substance (LMCr) facilitates significant transfer of Cr to transferrin (Yamamoto et al. 1984). Transferrin primarily binds newly absorbed Cr, with albumin taking over when transferrin is saturated. High concentrations of Cr may nonspecifically bind to other plasma proteins, including γ- and α-globulins and lipoproteins (Hopkins and Schwarz 1964). However, the blood level of Cr does not accurately reflect Cr's nutritional status as it is not in equilibrium with body stores.

9.10.3 Excretion

Excretion of Cr is primarily via urine, with smaller amounts in the sweat, hair, and bile (Ducros 1992). After oral absorption, fecal elimination is the major route due to limited absorption, while after injection, kidney eliminates approximately 80% of the dose (Stotz 1969). The mechanisms of Cr metabolism are unclear. Initial studies revealed that around 40% of plasma Cr III could be filtered, with 60–95% reabsorbed by renal tubules (Donaldson and Rennert 1981). Stable Cr complexes like Cr-EDTA are used to assess glomerular filtration rate. Differences in serum Cr distribution exist based on administration route. Recent research identified a low-molecular-weight substance binding trivalent Cr chemically in urine, indicating a binding mechanism rather than renal reabsorption (Table 9.1).

9.11 Conclusions

Trace elements play a vital role in maintaining proper bodily function, as they are essential components of biological structures and are involved in crucial biochemical reactions. However, imbalances in trace element levels can have detrimental effects on health, with both deficiency and excess leading to serious complications, including fatal diseases such as cancer. While considerable progress has been made in understanding the role of trace elements in disease, there are significant gaps in our knowledge that need to be addressed. Future research needs to be focused on elucidating the precise mechanisms by which trace elements influence biological processes and how these processes are dysregulated in disease states. Specifically, there is a need for further investigation into the relationship between trace elements and diseases, including cancers, to develop more effective prevention and management strategies. Additionally, understanding the determinants of trace element levels in the body and how they are influenced by factors such as diet, environment, and genetics is crucial. By addressing these gaps in research and advancing our understanding of trace elements and their role in health and disease, we can ultimately improve outcomes for individuals affected by trace element imbalances and associated conditions. This will pave the way for the development of targeted interventions and personalized treatment approaches to optimize health and well-being.

References

Abbaspour N, Hurrell R, Kelishadi R (2014) Review on iron and its importance for human health. J Res Med Sci 19(2):164

Aisen P, Leibman A, Zweier J (1978) Stoichiometric and site characteristics of the binding of iron to human transferrin. J Biol Chem 253(6):1930–1937

Al Alawi AM, Majoni SW, Falhammar H (2018) Magnesium and human health: perspectives and research directions. Int J Endocrinol 2018

Alam Khan V, Khan MA, Akhtar S (2002) Thyroid disorders, etiology and prevalence. J Med Sci 2(2):89–94

Alan SL (2001) Evolving concepts in epithelial magnesium transport. Curr Opin Nephrol Hypertens 10(5):649–653

Arosio P, Carmona F, Gozzelino R et al (2015) The importance of eukaryotic ferritins in iron handling and cytoprotection. Biochem J 472(1):1–15

Assey VD, Greiner T, Mzee RK et al (2006) Iodine deficiency persists in the Zanzibar Islands of Tanzania. Food Nutr Bull 27(4):292–299

Babior BM, Bunn HF (2001) Megaloblastic anemias. Harrisons Prin Int Med 1:674–680

Baliga R, Ueda N, Shah SV (1993) Increase in bleomycin-detectable iron in ischaemia/reperfusion injury to rat kidneys. Biochem J 291(3):901–905

Beard J (2003) Iron deficiency alters brain development and functioning. J Nutr 133(5): 1468S–1472S

Bertrand E, Lewandowska E, Szpak GM et al (2001) Neuropathological analysis of pathological forms of astroglia in Wilson's disease. Folia Neuropathol 39(2):73–79

Bhattacharya PT, Misra SR, Hussain M (2016) Nutritional aspects of essential trace elements in oral health and disease: an extensive review. Scientifica

Bhowmik D, Chiranjib K, Kumar S (2010) A potential medicinal importance of zinc in human health and chronic. Int J Pharm 1(1):05–11
Bhutto A, Mastoi A, Memon SA et al (2005) Magnesium and its essential role in health. Jlumhs 4(1):33–35
Bowen HJM (1966) Trace elements in biochemistry
Boycott KM, Beaulieu CL, Kernohan KD et al (2015) Autosomal-recessive intellectual disability with cerebellar atrophy syndrome caused by mutation of the manganese and zinc transporter gene SLC39A8. Am J Hum Genet 97(6):886–893
Brown KM, Arthur JR (2001) Selenium, selenoproteins and human health: a review. Public Health Nutr 4(2b):593–599
Byers PH, Siegel RC, Holbrook KA et al (1980) X-linked cutis laxa: defective cross-link formation in collagen due to decreased lysyl oxidase activity. N Engl J Med 303(2):61–65
Calò L, Punzi L, Semplicini A (2000) Hypomagnesemia and chondrocalcinosis in Bartter's Gitelman's syndrome: review of the pathogenetic mechanisms. Am J Nephrol 20(5):347
Cavdar AO, Babacan E, Arcasoy A et al (1980) Effect of nutrition on serum zinc concentration during pregnancy in Turkish women. Am J Clin 33(3):542–544
Clark SF (2008) Iron deficiency anemia. Nutrition in clinical practice 23(2):128–141
Classen HG, Gröber U, Kisters K (2012) Drug-induced magnesium deficiency. Med Monatsschr Pharm 35(8):274–280
Combs GJ, Combs SB (1986) The role of selenium in nutrition, pp xii+-532pp
Comín-Colet J, Enjuanes C, Gonzalez G et al (2013) Iron deficiency is a key determinant of health related quality of life in patients with chronic heart failure regardless of anaemia status. Eur J Heart Fail 15(10):1164–1172
Cousins RJ (1985) Absorption, transport, and hepatic metabolism of copper and zinc: special reference to metallothionein and ceruloplasmin. Physiol Rev 65(2):238–309
Cousins RJ, Liuzzi JP, Lichten LA (2006) Mammalian zinc transport, trafficking, and signals. J Biol Chem 281(34):24085–24089
Cumings JN (1948) The copper and iron content of brain and liver in the normal and in hepato-lenticular degeneration. Brain 71(4):410–415
David V, Francis C, Babitt JL (2017) Ironing out the cross talk between FGF23 and inflammation. Am J Physiol Renal Physiol 312(1):F1–F8
Davis CD, Zech L, Greger JL (1993) Manganese metabolism in rats: an improved methodology for assessing gut endogenous losses. Proc Soc Exp Biol Med 202(1):103–108
Davis SR, McMahon RJ, Cousins RJ (1998) Metallothionein knockout and transgenic mice exhibit altered intestinal processing of zinc with uniform zinc-dependent zinc transporter-1 expression. J Nutr 128(5):825–831
De Bie P, Van De Sluis B, Klomp L et al (2005) The many faces of the copper metabolism protein MURR1/COMMD1. J Hered 96(7):803–811
Detels R, Beaglehole R, Lansang MA et al (2011) Oxford textbook of public health. Oxford University Press
Dhaar GM, Robbani I (2006) Foundations of community medicine. Elsevier
Donaldson DL, Rennert OM (1981) Chromium (III) metabolism by the kidney. Ann Clin Lab Sci 11(5):377–385
Donovan A, Brownlie A, Zhou YI et al (2000) Positional cloning of zebrafish ferroportin1 identifies a conserved vertebrate iron exporter. Nature 403(6771):776–781
Drakesmith H, Nemeth E, Ganz T (2015) Ironing out ferroportin. Cell Metab 22(5):777–787
Ducros V (1992) Chromium metabolism: a literature review. Biol Trace Elem Res 32:65–77
Dunipace AJ, Brizendine EJ, Zhang W et al (1995) Effect of aging on animal response to chronic fluoride exposure. J Dent Res 74(1):358–368
Dymock IW, Cassar J, Pyke DA et al (1972) Observations on the pathogenesis, complications and treatment of diabetes in 115 cases of haemochromatosis. Am J Med 52(2):203–210

Ekstrand J, Ehrnebo M (1983) The relationship between plasma fluoride, urinary excretion rate and urine fluoride concentration in man. J Occup Med:745–748
Elferink RPO, Jansen PL (1994) The role of the canalicular multispecific organic anion transporter in the disposal of endo- and xenobiotics. PharmacolTher 64(1):77–97
Evans RW, Darvell BW (1995) Refining the estimate of the critical period for susceptibility to enamel fluorosis in human maxillary central incisors. J Public Health Dent 55(4):238–249
Fan AM, Book SA, Neutra RR et al (1988) Selenium and human health implications in California's San Joaquin Valley. J Toxicol Environ Health, Part A Curr Issues 23(4):539–559
Fawcett WJ, Haxby EJ, Male DA (1999) Magnesium: physiology and pharmacology. Br J Anaesth 83(2):302–320
Fejerskov O, Thylstrup A, Larser MJ (1977) Clinical and structural features and possible pathogenic mechanisms of dental fluorosis. Eur J Oral Sci 85(7):510–534
Finberg KE, Heeney MM, Campagna DR et al (2008) Mutations in TMPRSS6 cause iron-refractory iron deficiency anemia (IRIDA). Nat Genet 40(5):569–571
Fleming MD, Trenor CC III, Su MA et al (1997) Microcytic anaemia mice have a mutation in Nramp2, a candidate iron transporter gene. Nat genet 16(4):383–386
Ford ES, Cogswell ME (1999) Diabetes and serum ferritin concentration among US adults. Diabetes Care 22(12):1978–1983
Fracanzani AL, Conte D, Fraquelli M et al (2001) Increased cancer risk in a cohort of 230 patients with hereditary hemochromatosis in comparison to matched control patients with non–iron related chronic liver disease. Hepatology 33(3):647–651
Ganz T (2013) Systemic iron homeostasis. Physiol Rev 93(4):1721–1741
Girard B, Beucher G, Muris C et al (2005) Magnesium sulphate and severe preeclampsia: its use in current practice. J Gynecol Obstet Biol Reprod 34(1 Pt 1):17–22
Gopalan C (2000) Iodation of common salt for control of IDD-not the time to backtrack. NFI Bull 21(3):1–2
Gross JB, Myers BM, Kost LJ et al (1989) Biliary copper excretion by hepatocyte lysosomes in the rat. Major excretory pathway in experimental copper overload. J Clin Invest 83(1):30–39
Guggenbuhl P, Deugnier Y, Boisdet JF et al (2005) Bone mineral density in men with genetic hemochromatosis and HFE gene mutation. Osteoporos Int 16:1809–1814
Haldimann M, Alt A, Blanc A et al (2005) Iodine content of food groups. J Food Compos Anal 18(6):461–471
Halliday JW, Powell LW (2018) Hemochromatosis and other diseases associated with iron overload. In: Iron and human disease. CRC Press, pp 131–160
Hambidge M, Krebs NF (2001) Interrelationships of key variables of human zinc homeostasis: relevance to dietary zinc requirements. Annu Rev Nutr 21(1):429–452
Harris ED (1995) Role of ligands in the translocation of metals. In: Handbook of metal-ligand interactions in biological fluids. Marcel Dekker, Inc., New York, pp 71–88
Harris ED (2000) Cellular copper transport and metabolism. Annu Rev Nutr 20(1):291–310
Hermann W (2019) Classification and differential diagnosis of Wilson's disease. Ann Transl Med 7(Suppl 2)
Hetzel BS, Pandav CS (1996) SOS for a billion. The conquest of iodine deficiency disorders, 2nd edn. Oxford University Press, Delhi, pp 1–466
Hodge HC, Smith FA, Gedalia I (1970) Excretion of fluorides, Monograph series, vol 59. World Health Organization, pp 141–161
Hopkins LL Jr, Schwarz K (1964) Chromium (III) binding to serum proteins, specifically siderophilin. Biochim. Biophys (BBA)-Gen Subj 90(3):484–491
Horn N, Tümer Z (2002) Menkes disease and the occipital horn syndrome. Connect Tiss Herit Disord: Mol Genetic Med Aspects:651–685
Horoupian DS, Sternlieb I, Scheinberg IH (1988) Neuropathological findings in penicillamine-treated patients with Wilson's disease. Clin Neuropathol 7(2):62–67
Ingram GS, Agalamanyi EA, Higham SM (2005) Caries and fluoride processes. J Dent 33(3):187–191

Jenkitkasemwong S, Wang CY, Coffey R, Zhang W, Chan A, Biel T et al (2015) SLC39A14 is required for the development of hepatocellular iron overload in murine models of hereditary hemochromatosis. Cell Metab 22(1):138–150
Kaler SG, Ghal WA, Berry SA et al (1993) Predictive value of plasma catecholamine levels in neonatal detection of Menkes disease. J Inherit Metab Dis 16:907–908
Kayne LH, Lee DB (1993) Intestinal magnesium absorption. Miner Electrolyte Metab 19(4-5):210–217
Keel SB, Doty RT, Yang Z et al (2008) A heme export protein is required for red blood cell differentiation and iron homeostasis. Science 319(5864):825–828
Khurana I (2006) Endocrinal system, Textbook of medical physiology. Reed Elsevier India, pp 710–715
King JC, Shames DM, Woodhouse LR (2000) Zinc homeostasis in humans. J Nutr 130(5):1360S–1366S
Klaassen CD (1976) Biliary excretion of metals. Drug Metab Rev 5(2):165–196
Krebs NF (2000) Overview of zinc absorption and excretion in the human gastrointestinal tract. J Nutr 130(5):1374S–1377S
Langwińska-Wośko E, Litwin T, Szulborski K et al (2016) Optical coherence tomography and electrophysiology of retinal and visual pathways in Wilson's disease. Metab Brain Dis 31:405–415
Laurberg P, Cerqueira C, Ovesen L et al (2010) Iodine intake as a determinant of thyroid disorders in populations. Best Pract Res Clin Endocrinol Metab 24(1):13–27
Leblondel G, Allain P (1999) Manganese transport by Caco-2 cells. Biol Trace Elem Res 67:13–28
Likins RC, McClure FJ, Steere AC (1956) Urinary excretion of fluoride following defluoridation of a water supply. Public Health Rep 71(3):217
Linkermann A, Skouta R, Himmerkus N et al (2014) Synchronized renal tubular cell death involves ferroptosis. Proc Natl Acad Sci 111(47):16836–16841
Litwin T, Dzieżyc K, Członkowska A (2019) Wilson disease – treatment perspectives. Ann Transl Med 7(Suppl 2)
Lopez A, Cacoub P, Macdougall IC et al (2016) Iron deficiency anaemia. Lancet 387(10021):907–916
MacMahon RA, Parker MLM, McKinnon MC (1968) Zinc treatment in malabsorption. Med J Aust 2:210–212
Mancias JD, Wang X, Gygi SP et al (2014) Quantitative proteomics identifies NCOA4 as the cargo receptor mediating ferritinophagy. Nature 509(7498):105–109
Manz DH, Blanchette NL, Paul BT et al (2016) Iron and cancer: recent insights. Ann N Y Acad Sci 1368(1):149–161
Maret W (2019) Chromium supplementation in human health, metabolic syndrome, and diabetes. Met Ions Life Sci 19:231–251
McClain DA, Abraham D, Rogers J et al (2006) High prevalence of abnormal glucose homeostasis secondary to decreased insulin secretion in individuals with hereditary haemochromatosis. Diabetologia 49:1661–1669
McKie AT, Barrow D, Latunde-Dada GO et al (2001) An iron-regulated ferric reductase associated with the absorption of dietary iron. Science 291(5509):1755–1759
McMahon RJ, Cousins RJ (1998) Regulation of the zinc transporter ZnT-1 by dietary zinc. Proc Natl Acad Sci 95(9):4841–4846
Menkes JH, Alter M, Steigleder GK et al (1962) A sex-linked recessive disorder with retardation of growth, peculiar hair, and focal cerebral and cerebellar degeneration. Pediatrics 29(5):764–779
Mertz W, Roginski EE, Reba RC (1965) Biological activity and fate of trace quantities of intravenous chromium (III) in the rat. Am J Physiol-Legacy Cont 209(3):489–494
Mims MP, Guan Y, Pospisilova D et al (2005) Identification of a human mutation of DMT1 in a patient with microcytic anemia and iron overload. Blood 105(3):1337–1342
Møller LB, Mogensen M, Horn N (2009) Molecular diagnosis of Menkes disease: genotype–phenotype correlation. Biochimie 91(10):1273–1277
Moos T, Nielsen TR, Skjørringe T et al (2007) Iron trafficking inside the brain. J Neurochem 103(5):1730–1740

Nam H, Wang CY, Zhang L et al (2013) ZIP14 and DMT1 in the liver, pancreas, and heart are differentially regulated by iron deficiency and overload: implications for tissue iron uptake in iron-related disorders. Haematologica 98(7):1049
Nankivell BJ, Boadle RA, Harris DC (1992) Iron accumulation in human chronic renal disease. Am J Kidney Dis 20(6):580–584
Nath KA (2006) Heme oxygenase-1: a provenance for cytoprotective pathways in the kidney and other tissues. Kidney Int 70(3):432–443
Noronha JL, Matuschak GM (2009) Magnesium in critical illness: metabolism, assessment, and treatment. Appl Physiol Intens Care Med:201–213
Omokhodion FO, Howard JM (1994) Trace elements in the sweat of acclimatized persons. Clin Chim Acta 231(1):23–28
Osborne NJ, Gurrin LC, Allen KJ et al (2010) HFE C282Y homozygotes are at increased risk of breast and colorectal cancer. Hepatology 51(4):1311–1318
Pal GK (2021) Textbook of medical physiology, 4th edn. Elsevier Health Sciences
Paller MS, Hedlund BE, Sikora JJ et al (1988) Role of iron in postischemic renal injury in the rat. Kidney Int 34(4):474–480
Pantopoulos K, Porwal SK, Tartakoff A et al (2012) Mechanisms of mammalian iron homeostasis. Biochem 51(29):5705–5724
Patrick L (2008) Iodine: deficiency and therapeutic considerations. Altern Med Rev 13(2)
Pearce EN, Pino S, He X et al (2004) Sources of dietary iodine: bread, cows' milk, and infant formula in the Boston area. J Clin Endocrinol Metab 89(7):3421–3424
Pennell DJ, Udelson JE, Arai AE et al (2013) American Heart Association Committee on Heart Failure and Transplantation of the Council on Clinical Cardiology and Council on Cardiovascular Radiology and Imaging. Cardiovascular function and treatment in β-thalassemia major: a consensus statement from the American Heart Association. Circulation 128(3):281–308
Pietrangelo A (2016) Iron and the liver. Liver Int 36:116–123
Pinnix ZK, Miller LD, Wang WD et al (2010) Ferroportin and iron regulation in breast cancer progression and prognosis. Sci Transl Med 2(43):43ra56
Porterfield SP (2001) Endocrine physiology, Thyroid gland; The Mosby physiology monograph series, 2nd edn
Prasad AS (2004) Zinc deficiency: its characterization and treatment. Met Ions Biol Syst 41:103–138
Prashanth L, Kattapagari KK, Chitturi RT et al (2015) A review on role of essential trace elements in health and disease. J Dr YSR Univ Health Sci 4(2):75–85
Przybyłkowski A, Gromadzka G, Chabik G et al (2014) Liver cirrhosis in patients newly diagnosed with neurological phenotype of Wilson's disease. FunctNeurol 29(1):23
Quadri M, Federico A, Zhao T, et al (2012) Mutations in SLC30A10 cause parkinsonism and dystonia with hypermanganesemia, polycythemia, and chronic liver disease. Am J Hum Genet 90(3):467–477
Rahil-Khazen R, Bolann BJ, Myking A et al (2002) Multi-element analysis of trace element levels in human autopsy tissues by using inductively coupled atomic emission spectrometry technique (ICP-AES). J Trace Elem Med Biol 16(1):15–25
Richards A, Kragstrup J, Josephsen K et al (1986) Dental fluorosis developed in post-secretory enamel. J Dent Res 65(12):1406–1409
Roetto A, Papanikolaou G, Politou M et al (2003) Mutant antimicrobial peptide hepcidin is associated with severe juvenile hemochromatosis. Nat Genet 33(1):21–22
Royce PM, Camakaris J, Danks DM (1980) Reduced lysyl oxidase activity in skin fibroblasts from patients with Menkes' syndrome. Biochem J 192(2):579–586
Sayato Y, Nakamuro K, Matsui S et al (1980) Metabolic fate of chromium compounds. I. Comparative behavior of chromium in rat administered with $Na_2{}^{51}CrO_4$ and $^{51}CrCl_3$. J Pharmacobio-Dyn 3(1):17–23
Schroeder HA, Balassa JJ, Tipton IH (1966) Essential trace metals in man: manganese: a study in homeostasis. J Chronic Dis 19(5):545–571

Shim H, Harris ZL (2003) Genetic defects in copper metabolism. J Nutr 133(5):1527S–1531S
Singh N, Haldar S, Tripathi AK et al (2014) Brain iron homeostasis: from molecular mechanisms to clinical significance and therapeutic opportunities. Antioxid Redox Signal 20(8):1324–1363
Skalnaya MG, Skalny AV (2018) Essential trace elements in human health: a physician's view, vol 224. Publishing House of Tomsk State University, Tomsk, pp 1–222
Smith T, Edmonds CJ, Barnaby CF (1972) Absorption and retention of cobalt in man by whole-body counting. Health Phys 22(4):359–367
Stastny D, Vogel RS, Picciano MF (1984) Manganese intake and serum manganese concentration of human milk-fed and formula-fed infants. Am J Clin Nutr 39(6):872–878
Stotz EH (1969) Comprehensive biochemistry. Elsevier
Szauter KM, Cao T, Boyd CD et al (2005) Lysyl oxidase in development, aging and pathologies of the skin. PatholBiol 53(7):448–456
Tinggi U (2003) Essentiality and toxicity of selenium and its status in Australia: a review. Toxicol Lett 137(1-2):103–110
Tümer Z, Horn N (1997) Menkes disease: recent advances and new aspects. J Med Genet 34(4):265
Turnlund JR, Swanson CA, King JC (1983) Copper absorption and retention in pregnant women fed diets based on animal and plant proteins. J Nutr 113(11):2346–2352
Vanoaica L, Darshan D, Richman L et al (2010) Intestinal ferritin H is required for an accurate control of iron absorption. Cell Metab 12(3):273–282
Villa A, Cabezas L, Anabalon M et al (2009) The fractional urinary fluoride excretion of adults consuming naturally and artificially fluoridated water and the influence of water hardness: a randomized trial. Commun Dent Health 26(3):132–137
Wallach S (1985) Clinical and biochemical aspects of chromium deficiency. J Am Coll Nutr 4(1):107–120
Walshe JM, Potter G (1977) The pattern of the whole body distribution of radioactive copper (67Cu, 64Cu) in Wilson's disease and various control groups. QJM: An Int J Med 46(4):445–462
Wapnir RA (1998) Copper absorption and bioavailability. Am J Clin Nutr 67(5):1054S–1060S
Weintraub LR, Weinstein MB, Huser HJ et al (1968) Absorption of hemoglobin iron: the role of a heme-splitting substance in the intestinal mucosa. J Clin Invest 47(3):531–539
Whitford GM (1994) Effects of plasma fluoride and dietary calcium concentrations on GI absorption and secretion of fluoride in the rat. Calcif Tissue Int 54:421–425
Whitford GM (1996) The metabolism and toxicity of fluoride. Monog Oral Sci 16:1–153
Whitford GM, Pashley DH (1984) Fluoride absorption: the influence of gastric acidity. Calcif Tissue int 36:302–307
WHO F (2004) WHO expert consultation on human vitamin and mineral requirements. Vitamin and mineral requirements in human nutrition. World Health Organization, pp 45–58
WHO U (2007) International Council for the Control of Iodine Deficiency Disorders. Assessment of iodine deficiency disorders and monitoring their elimination: a guide for programme managers, 3rd edn. World Health Organization, Geneva
Wood JC (2014) Guidelines for quantifying iron overload. Hematol Am Soc Hematol Educ Prog 2014(1):210–215
Yamamoto A, Wada O, Ono T (1984) Distribution and chromium-binding capacity of a low-molecular-weight, chromium-binding substance in mice. J Inorg Biochem 22(2):91–102
Yang J, Mori K, Li JY (2003) Iron, lipocalin, and kidney epithelia. Am J Physiol Renal Physiol 285(1):F9–F18
Zabłocka-Słowińska K, Grajeta H (2012) The role of manganese in etiopathogenesis and prevention of selected diseases. Adv Hyg Exp Med 66:549–553
Zarjou A, Bolisetty S, Joseph R et al (2013) Proximal tubule H-ferritin mediates iron trafficking in acute kidney injury. J Clin Invest 123(10):4423–4434
Zumbrennen-Bullough K, Babitt JL (2014) The iron cycle in chronic kidney disease (CKD): from genetics and experimental models to CKD patients. Nephrol Dial Transplant 29(2):263–273

Enzymes and Their Clinical Applications 10

Pradeep Kumar Singh, Manikant Tripathi, Prachi Garg, Ram Lakhan Singh, Neelam Pathak, Shikha Shukla, and Pankaj Singh

Abstract

Enzymes are proteinaceous molecules which catalyze the chemical reactions in biological systems. Enzymes are found in almost all tissues and fluids of the body. They are responsible for maintaining the animal's homeostasis and body's all significant life processes depend on its enzyme activity. Enzymes accelerate the rate of reactions but never participate in any kind of biochemical reactions. Clinical enzymology is the study of enzymes which deal with applications of enzymes for diagnosis and treatment of diseases. The measurement of enzyme concentration in serum, plasma, and other body fluids is used as an indicator of organ disfunctioning or development of disease in a particular organ. There are a number of enzymes which are used for diagnostic purposes, such as aspartate aminotransferase (AST) and alanine aminotransferase (ALT), are used in investigation of liver disease and creatine kinase (CK) for muscular disorder, muscle damages, and brain injury. Similarly, lactate dehydrogenase (LDH) is a marker of hepatocyte damage and muscle damage and hemolysis; alkaline phosphatase (ALP) for liver disease, osteosarcoma, pancreatic carcinoma, pancreatitis, hepatic necrosis, induction of drugs/hormones; acid phosphatase (ACP) for prostate cancer, bone diseases, lysosomal disorders; amylase (AMS) for diagnosis of

Pradeep Kumar Singh, Prachi Garg and Pankaj Singh contributed equally with all other contributors.

P. K. Singh · P. Garg · R. L. Singh · N. Pathak
Department of Biochemistry, Dr. Rammanohar Lohia Avadh University, Ayodhya, India

M. Tripathi · P. Singh (✉)
Department of Biotechnology, Dr. Rammanohar Lohia Avadh University, Ayodhya, India
e-mail: pankaj@rmlau.ac.in

S. Shukla
Department of Zoology, RRPG College, Amethi, India

R. L. Singh et al. (eds.), *Clinical Applications of Biomolecules in Disease Diagnosis*, https://doi.org/10.1007/978-981-97-4723-8_10

pancreatitis and renal failure. Lipase is used as a marker for extra pancreatic neoplasia, pancreatic acinar cell damage, and glucose-6-phosphate dehydrogenase (G-6-PD) is used for disorder in spleen thymus, and erythrocytes.

Keywords

Clinical enzymology · Organ disfunctioning · Aspartate-aminotransferase · Alanine aminotransferase · Lactate dehydrogenase

10.1 Introduction

Enzymes are the biocatalysts that accelerate biochemical reactions in living organisms. It can also be extracted from the cells and used in commercially imperative processes. The German physiologist Wilhelm Kuhne in 1878 used the term "enzymes" which is derived from Greek words *en* (mean "within") and zyme (mean "yeast") to recognize the ability of yeast to produce alcohol from sugars (Robinson 2015). Most enzymes are proteinaceous in nature except ribozymes that have catalytic activity and are made up of small groups of RNA molecules playing a significant role in gene expression (Tjhung et al. 2020). Despite that, biochemists also develop antibodies with the catalytic activity and these are called as abzymes (Robinson 2015). The catalytic activity of any enzymes principally depends upon the integrity of their native protein conformation, thus one or more enzymes gets impaired in many diseases due to change in conformation out of mutations (Chaturvedi et al. 2016). The enzymes are highly specific as they possess remarkable specificity to a demonstrated group and its substrate. They are globular proteins consisting of 100–2000 or more amino-acid residues arranged in one or more polypeptide chains and form a specific three-dimensional structure with small specific areas called active sites. There are two models proposed for describing the enzyme binding to its substrate, i.e. "Lock and key model" and "Induced fit model" out of which the "induced fit model" is more acceptable and also called "Hand in glove model."

Although a large number of enzymes consist exclusively of protein but many of them also contain non-protein components, called cofactors. They may be organic or inorganic in nature which is necessary for enzyme's catalytic activity. When an enzyme requires co-factors for its activity, then the inactive enzyme component is termed as "apoenzyme." The combination of apoenzymes and cofactors is collectively referred as "Holoenzyme" (Robinson 2015). As mentioned above, enzymes accelerate biochemical reactions at very low concentration but it does not change thermodynamics of any chemical reactions because there is absolutely not any changes in fundamental structure and energetics of any substrate and product at the time of reactions rather than the speed of reactions goes up. There are more than 1300 enzymes present in human cells that performed different chemical processes and breakdown the large biomolecules to smaller one during the metabolic reactions.

In recent years, uses of clinically important enzymes in diagnosis as well as therapeutics grew enormously because of development of advanced techniques and procedures for isolation, purification, and quantification of enzymes. Enzymes are not only used in investigation but also involved in the treatment of significant pathologies such as cancer, renal disorders, liver diseases, and cardiovascular diseases (Chen et al. 2018; Triplett et al. 2018). The enzymes that are used in diagnosis or clinical therapeutics should have a high degree of purity. Enzymes provide insights into disease processes by diagnosis, prognosis, and assessment of response therapy.

10.2 Properties of Enzymes

As mentioned above, enzymes are complex macromolecules that catalyze biochemical reactions in the cells but do not affect the nature of the final product. There are several specific properties of enzymes as mentioned below.

10.2.1 Catalytic Properties

The enzymes have an extra-ordinary catalytic power thus only a small concentration of it is enough to convert large quantities of substrate into product. Enzyme immobilization reduces the catalytic power of enzymes. Currently we are using hydrophilic hollow layered double hydroxide as encapsulation carriers to maintain flexibility and biocatalytic properties of enzymes (He et al. 2023).

10.2.2 Specificity

All the enzymes are very specific in their nature because a particular enzyme acts only on particular substrates. The specificity of an enzyme is its ability to distinguish the substrate of well-defined structure and composition (Bourlieu et al. 2020). Depending upon the particular substrate and type of reaction, the specificity of enzymes is categorized into different forms (Fig. 10.1).

All enzymes are stable under certain conditions, which make them useful in diagnostic tests that require long-term storage. However, enzymes are sensitive to changes in their environment such as temperature, heat, pH, substrate concentration and also their activity can be affected by denaturation (Yeoman et al. 2010). Enzyme stabilization is also very crucial for the enzyme-based biosensing applications and clinical settings (Chapman and Stenzel 2019).

10.2.3 Enzymes Regulation

All the enzymes are regulated by different types of mechanism like allosteric regulation, covalent modification, feedback regulations, and proteolytic cleavage that makes them useful targets for drug development (Punekar 2018). These are the significant properties of enzymes that make them helpful in diagnostic tests, enzyme therapy, biomarker discovery, and drug development.

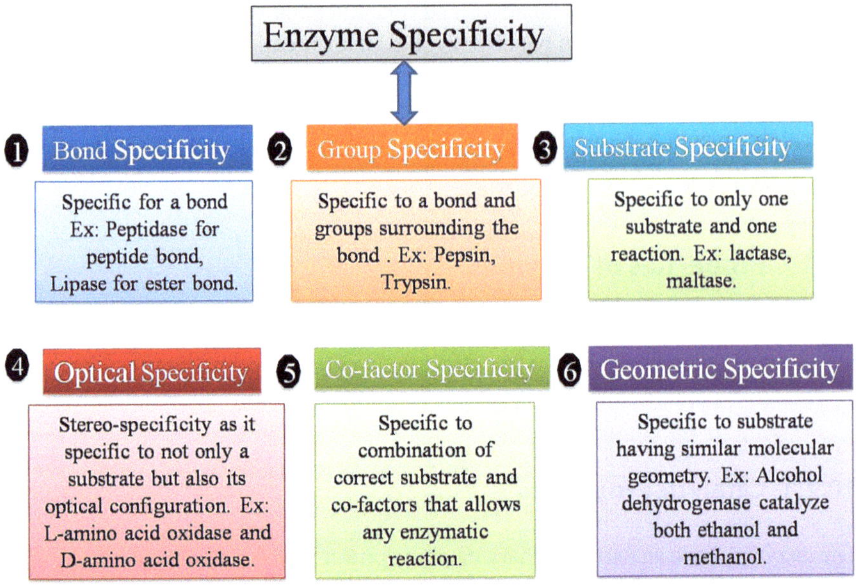

Fig. 10.1 Types of enzyme specificity

10.3 Enzymes in Clinical Diagnosis

10.3.1 Bone Diseases

Bone is a living tissue which is composed of calcium and collagen. Cortical bones are dense and form the periphery of the bones while trabecular bones form the inner layer of bones. Bones contribute to form a skeleton that protects the visceral organs and supports locomotion. Bones act as a pool for minerals such as phosphorus and calcium. Person having bone disease may affect its bone strength and flexibility. Diseases may arise due to environmental factors, genetics, diet, and infections. People with osteoporosis have more chances of distressing fractures. Primary osteoporosis is the most common form of osteoporosis. It is the result of deterioration of bone structure as a person's age increases (Seeman 2003). This deterioration can be managed and avoided by balanced nutrition and physical activity. Osteoporosis is very rare in the young population. In young adults, this disease is mainly caused by anorexia nervosa or glucocorticoid use (Khosla et al. 1994). There are many diseases and certain medications and toxic agents that lead to the occurrence of secondary osteoporosis (Stein and Shane 2003). A number of childhood diseases cause rickets, a condition in which deposition of calcium phosphate in growing bones is delayed. This delay is characterized by bowed legs in children. In adults, the corresponding disease is called osteomalacia. Both diseases are mainly prevented by administration of adequate levels of vitamin D (Chesney 2001; Pettifor 2002, 2003). People with renal disorders have more probability to gain complex bone disease known as renal osteodystrophy (Cunningham et al. 2004). Paget's disease of bone (Siris and Roodman 2003) is a progressive disorder that leads to remodeling of pelvis, legs, and spine or skull bones. A vast variety of genetic and developmental disorders leads to diseases of the skeleton. Among these the more common disorder is osteogenesis imperfecta (OI) (Whyte 2003; Rauch and Glorieux 2004).

10.3.1.1 Alkaline Phosphatase (ALP)

ALP is a hydrolase and functions best under alkaline pH, and it dephosphorylates the nucleotides and proteins. In children, the serum ALP levels are significantly greater than in adults and it correlates with the rate of bone growth. ALP value is slightly higher in men than in women, but at older age (more than 60 years) the enzyme value is equal or higher in women. ALP value enhances during pregnancy and puberty as well as after menopause (Corathers 2006). The increased levels of serum ALP indicate an increased rate of bone formation. Other pathological conditions that may lead to increased levels of ALP include rickets, hyperthyroidism, osteomalacia, and hyperparathyroidism. Low level of ALP is less common and is more likely to occur under genetic conditions or nutritional deficiency. A rare disorder, hypophosphatasia, is described by low levels of serum ALP causing abnormal metabolites phosphorylation and various skeletal abnormalities (Iqbal 1998).

10.3.1.2 Leukocyte Esterase

The presence of leukocyte esterase in synovial fluid is a really important addition to the physician's armamentarium for the diagnosis of joint infection (Parvizi et al. 2011).

10.3.2 Autoimmune Disorders

The immune system has the capacity of protecting hosts from microbial pathogens and disease. The autoimmune diseases occur after failure to discriminate self from non-self by breaching immunological tolerance. These diseases are very rare but, through rigorous epidemiological studies, it was found that 3–5% of the population are affected with autoimmune thyroid disease and type I diabetes (T1D). Although, there are nearly 100 different autoimmune diseases included under either organ specific such as Hashimoto's thyroiditis, autoimmune anemias, good pasture's syndrome, insulin-dependent diabetes mellitus, Graves' disease, Myasthenia gravis, and poststreptococcal glomerulonephritis or under systemic such as multiple sclerosis (MS), rheumatoid arthritis, and systemic lupus erythematosus (SLE) (Yu et al. 2014).

10.3.2.1 Lysozyme

Lysozyme cleaves (β1–4) glycosidic bond between the two types of sugar residue in the molecule, *N*-acetylmuramic acid (NAM) and *N*-acetylglucosamine (NAG), present in cell wall of bacteria. Tornsteinsdottir et al. (1999) demonstrated that lysozyme serum level increases in rheumatoid arthritis patients, which is considered as an indicator of macrophage/monocyte activity.

10.3.2.2 Gelatinase B

Gelatinase B is an enzyme that is secreted to regulate the cell-matrix composition. It is a member of the gelatinase subfamily of the matrix metalloproteinases (MMPs). Its main substrate is a denatured collagen (gelatin). Gelatinase B is mainly secreted by monocytes, keratinocytes, polymorphonuclear leukocytes, tissue macrophages, and by various malignant cells. Gelatinase B appears to be involved in several pathologic processes related to autoimmune diseases (Ram et al. 2006). Norga et al. (1996) has found that human gelatinase B is a marker enzyme for rheumatoid arthritis.

10.3.2.3 Tartrate-Resistant Acid Phosphatase (TRAP)

Mammalian TRAP is a 35 kDa glycosylated metalloenzyme. It is different from other mammalian acid phosphatases by its immunity to inhibition by tartrate and characteristic purple color. TRAP is released into circulation during bone resorption by osteoclasts. Elevated TRAP level is correlated with osteoporosis, rheumatoid arthritis, and metabolic bone disorders (Chao et al. 2010).

10.3.2.4 Cathepsin D

The high level of lysosomal cathepsin D can be used to know the onset of rheumatoid arthritis (Sohar et al. 2002). It has been reported that high value of cathepsin D is a clear indicator of collagen-induced arthritis in rats (Kumar et al. 2009).

10.3.3 Cancer

Cancer is a disease described by uncontrolled growth and spread of abnormal cells. It is one of the most fatal diseases in the world and the number of affected persons is more than all coronary heart diseases combined (Singh et al. 2022). The spread of cancerous cells is known as metastasis that may cause death. Cancer is caused by several external factors such as tobacco, radiation, chemicals, and infectious organisms as well as some internal factors such as hormones, inherited mutations, and random mutations. Many things augment the risk of cancer, including certain infections, dietary factors, lack of physical activity, obesity, and environmental pollutants (Anand et al. 2008). The enzymes which are commonly used as marker for cancer disease includes acid phosphatase (ACP), alanine transaminase (ALT), cathepsin D (CD), cyclooxygenase-2 (COX-2), lactate dehydrogenase (LDH), and thymidine kinase (TK).

Five different types of important ACPs are found in humans which differ in their amino acid homology, chromosomal origin, molecular weight, and sequence length (Bull et al. 2002). ACP level is widely used in the detection of prostate cancer and to monitor the prognosis (Kirschenbaum et al. 2011) whereas elevated level of ALT is an indicator of hepatocellular carcinoma (Ishiguro et al. 2009). Cathepsin D is widely used in diagnosing breast cancer. A number of reports showed that the level of pCD/CD can be used as a breast tumor marker (Schwartz 1992; Benes et al. 2008). A study reported by Taketo (1998) supports that expression of COX-2 can be used to know the progression of tumorigenesis. LDH is a valuable predictive marker in leukemia, lymphoma, and colon cancer based on LDH enzymatic activity (Schwartz 1992). Thymidine kinase (TK) plays an important role in the synthesis of DNA and increased levels of TK also indicate the occurrence of neoplasia at an early stage (He et al. 2010).

10.3.4 Diabetes (Type 2)

Diabetes mellitus (DM) is possibly one of the oldest known diseases of humans. Type 2 DM was first considered as a part of metabolic syndrome in 1988 (Patlak 2002). It is one of the fastest growing diseases due to defects in secretion and action of insulin hormone. It has been established that high oxidative stress is one of the important factors that cause diabetes in the whole world (Gupta et al. 2023). Type 2 DM (previously known as non-insulin dependent DM) is characterized by insulin resistance, hyperglycemia, and insulin deficiency (Maitra and Abbas 2005). Type 2 DM arises during the interactions among environmental, genetic, and behavioral risk factors (Chen et al. 2012). People suffering with type 2 DM are susceptible to a variety of short- and long-term problems. The increased morbidity and mortality of patients with type 2 DM arises because of the frequency of this type of diabetes. Ripsin et al. (2009) stated that the development of type 2 DM largely depends on a number of lifestyle and genetics factors. These factors include sedentary lifestyle, physical inactivity, cigarette smoking, and alcohol consumption (Hu et al. 2001). Obesity has been reported to be the reason for around 55% of cases of type 2 DM. Kocabay et al. (2011) have stated that besides the lifestyle and genetics risk factors, higher levels of ALP may also be an indicator of hepatic fibrosis in patients with type 2 diabetes and nonalcoholic steatohepatitis (NASH).

10.3.5 Gaucher's Disease (GD)

Gaucher's disease is a rare autosomal recessive heredity disorder which transfers from parents to children (Stirnemann et al. 2017). The result of this mutation causes a declined activity of glucocerebrosidase (also called glucosylceramidase or GCase or acid β-glucosidase,), which catalyzes the conversion of glucosylceramide (GlcCer) into ceramide and glucose. The deficiency of glucocerebrosidase activity causes accumulation of glucosylceramide in macrophages, which induces the conversion of macrophages into Gaucher cells. Glucosylceramide accumulation in the Gaucher cells is considered the first step towards neurological disorder leading to the vascular compression (Mikosch and Hughes 2010).

10.3.5.1 Symptoms of Gaucher Disease

The symptoms of Gaucher disease may vary from person to person. Major symptoms include enlarged spleen and liver, eye movement disorders, yellow spots in the eyes, low concentration of healthy red blood cells (anemia), extreme fatigue, bruising, lung problems, and seizures.

10.3.5.2 Enzymes Used in Diagnosis of Gaucher Disease

The main enzyme marker for Gaucher disease is chitotriosidase (Hollak et al. 1994). Chitotriosidase is an enzyme that is specifically produced and secreted by the pathological storage macrophages, known as Gaucher cells, which are characteristic of the disease (Aerts et al. 2005). Changes in plasma chitotriosidase levels can also reflect changes in clinical symptoms. In addition to chitotriosidase, another biomarker that can be used to monitor Gaucher disease is the chemokine CCL18/PARC, which is an inflammatory marker produced by T lymphocytes and monocytes (Giraldo et al. 2018). This biomarker can be used to monitor the disease in those patients who have been treated with enzyme replacement therapy (Giraldo et al. 2018). The enzyme test for Gaucher disease, called a beta-glucosidase leukocyte (BGL) test, is widely used and it is very easy standard blood test method for examination of deficiency of an enzyme called beta-glucosidase leukocyte (BGL) enzyme (Giuffrida et al. 2023). This test measures the activity of the enzyme beta-glucosidase in leukocytes (white blood cells).

10.3.6 Liver Diseases

Liver is the largest internal organ in the body and participates in homeostasis, filtration of particulates, protein synthesis, bioactivation, biotransformation, and detoxification of xenobiotics (Singh et al. 2023). Liver disease or hepatic disease is a medical condition that affects the liver's ability to function properly which includes detoxification of harmful products of metabolism, synthesis of clotting factors, metabolizing cholesterol, glucose, iron and controlling their levels, other proteins, fighting infections and illness, and excretion of bile (Sharma and Nagalli 2023). Liver disease when persisted for a long time, is termed as chronic liver disease. The most prevalent liver diseases include hepatitis (A, B, C, D and E), cirrhosis, and fatty liver disease. Liver disease is the major cause of death every year and approximately 29 million people suffer from chronic liver condition worldwide (Blachier et al. 2013). In liver cirrhosis, the liver forms fibrous tissue to kill hepatocytes which may be due to viral hepatitis, alcohol poisoning. or other toxic substances and can cause elevated liver enzymes.

The metabolic disorder such as diabetes affects many organs, including the liver, which is involved in the regulation of carbohydrate, lipid, and protein metabolism (Singh et al. 2016a, b). Elevated serum AST, ALT, and GGT are commonly observed in diabetes (Sheng et al. 2018). Liver function test provides an idea about the condition of the liver.

10.3.6.1 Symptoms of Liver Disease

Liver disease can have a wide range of symptoms like yellowing of the skin and eyes, abdominal pain, swelling in legs and ankles, itchy skin, pale stool color, loss of appetite, tiredness (fatigue), feeling general weakness (malaise), visible blood vessels that look like spiders (spider angiomas), memory loss, personality changes, or sleep disorders, and internal bleeding due to the bursting of enlarged veins in the esophagus, stomach, or intestines. These symptoms can be induced by other medical conditions as well and it is important to consult a clinician for an accurate diagnosis.

10.3.6.2 Enzymes Used in Diagnosis of Liver Diseases

A number of liver function parameters are available to test the proper function of the liver, (serum proteins; albumin and bilirubin (direct and indirect), ALT, AST, GGT, ALP, PT, and PTT) (Lala et al. 2023). These tests indicate overall condition of the liver and can be used to diagnose liver disease. Some common enzymes used for diagnosis of liver diseases include:

10.3.6.2.1 Alanine Transaminase (ALT)

Alanine transaminase (ALT) or serum glutamic-pyruvic transaminase (SGPT) is an enzyme found mostly in the liver, although it is also present in other body tissues. ALT is the most reliable clinical biomarker and widely used in liver dysfunction (Ozer et al. 2008). It is a liver damage marker enzyme found in higher concentration in the liver than the other organ and its function is to convert amino acids into intermediate metabolites of glycolysis and Kreb's cycle to provide the energy for the liver cells. When the liver is damaged due to any reasons, the level of ALT is elevated into the bloodstream. Some types of liver diseases cause high ALT levels before other symptoms of the disease appear, so an ALT blood test may help to diagnose certain liver diseases at an early stage. The normal range for ALT levels is 7–56 U/L (units per liter).

10.3.6.2.2 Aspartate Transaminase (AST)

Aspartate transaminase (AST), also known as SGOT (Serum glutamic oxaloacetic transaminase) and aspartate aminotransferase, exists in the liver, heart, brain, pancreas, kidneys, and other organs of the body (Huang et al. 2006). AST is an enzyme that is also involved in the metabolism of amino acids to produce the energy. AST is a pyridoxal phosphate-dependent transaminase enzyme that transfers the amino group from aspartate or glutamate to the corresponding keto acid. Like ALT, AST is released into the bloodstream when the liver is damaged (Amacher 1998). AST is similar to alanine transaminase (ALT), but ALT is more accurate than AST at detecting liver disease. AST is considered as less liver-specific marker enzyme than that of ALT because AST is also produced in other organs like the brain, myocardial cells, and skeletal muscles. Some clinicians use the ratio of AST to ALT to know liver health and to discriminate between extra hepatic and hepatic injury. The common reference range for AST blood test is 10–40 U/L.

10.3.6.2.3 Alkaline Phosphatase (ALP)

Alkaline phosphatase is present in different parts of the body, including the bone, liver, kidney, and digestive system. The higher concentration of ALP is released from liver and bone (>80%) in serum, and lower concentration from the intestine (Lowe et al. 2020). Elevated concentrations of ALP indicate liver injury. Abnormal levels of ALP in the blood can reflect damage to tissue or disruption of normal bodily processes. The ALP measurement in blood is commonly used to diagnose liver damage, bone disorders, and cancers (Lange et al. 1982; Karhade et al. 2019). However, this test alone cannot accurately diagnose the diseased condition, and hence other tests can be also used for correct diagnosis. ALP isoenzyme test is most common as a follow-up after finding abnormal levels of ALP on a prior test. The normal reference range of alkaline phosphatase in serum is 20–140 U/L.

10.3.6.2.4 Gamma-Glutamyl Transferase (GGT)

GGT is an enzyme found in the liver and other body parts such as kidney, bile duct, pancreas, gallbladder, spleen, and heart (Singh et al. 2014). Abnormal levels of GGT can indicate either damage to the liver's bile ducts or liver disease (Lala et al. 2023). It is a relatively non-specific liver marker enzyme and also showed increased levels in cardiac injury, muscular injury, and hepatic disease (Yang et al. 2014). In humans, an elevation in GGT is a clear marker of liver cirrhosis due to alcohol consumption.

10.3.7 Pancreatitis

Pancreatitis is a condition developed by inflammation of the pancreas, which can be characterized by severe progression and numerous complications (Vengadakrishnan and Koushik 2015; Liu et al. 2021). The pancreas is a large gland and releases digestive enzymes into our small intestine, and also releases proteinaceous hormone glucagon and insulin into the bloodstream. These hormones play a very important role in metabolism and help to maintain glucose concentration in blood. The two forms of pancreatitis have been detected.

10.3.7.1 Acute Pancreatitis

Acute pancreatitis is sudden inflammation in pancreas and its severity ranges from discomfort to life-threatening illness. In critical conditions, acute pancreatitis can cause bleeding, cysts development, tissue damage which leads to liver, kidney, heart, and lung dysfunctions. In chronic pancreatitis, pancreas gets damaged permanently from inflammation over many years. The causes of acute pancreatitis may be formation of gallstones (Matta et al. 2020), obstruction in duct (Gorelick and Lerch 2017), alcohol consumption (Yadav and Whitcomb 2010; Setiawan et al. 2016), drugs (Badalov et al. 2007; Wolfe et al. 2020), certain over medications (Vinklerova et al. 2010), and hypertriglyceridemia (Carr et al. 2016). Other reasons of acute pancreatitis include viral infections such as SARS-CoV-2, cytomegalovirus, mumps and coxsackie B virus, hypercalcemia, tumors, cardiac surgery, and insect bites (Mukherjee et al. 2020; Pandanaboyana et al. 2021).

10.3.7.2 Chronic Pancreatitis

Chronic pancreatitis is a syndrome involving long-lasting inflammation, fibrosis, and loss of acinar and islet cells that develop abdominal pain, malnutrition, and exocrine and endocrine insufficiency. It is supposed that a long period of acute pancreatitis develops into chronic pancreatitis. The factors responsible for chronic pancreatitis may include alcohol consumption for a long period of time (Pandol et al. 2010), hereditary (Muniraj et al. 2014), bile duct obstruction (Ghazale et al. 2008; Majumder and Chari 2016), hyperlipidemia and idiopathic i.e. unidentified reasons (Pham and Forsmark 2018), toxic-metabolic, autoimmune (immune system attack on the pancreas) (Pitchumoni and Chari 2013; Muniraj et al. 2014), recurrent and severe acute pancreatitis -associated chronic pancreatitis (Sankaran et al. 2015).

10.3.7.3 Symptoms of Pancreatitis

Symptoms of pancreatitis vary in acute and chronic pancreatitis. Acute pancreatitis symptoms include moderate to severe pain in the upper part of abdomen, higher heart rate, fever, swelling in the belly, nausea, and vomiting. Symptoms of chronic pancreatitis include constant pain in abdomen; diarrhea, upset stomach and vomiting, fatty and oily stools.

10.3.7.4 Enzymes Used in Diagnosis of Pancreatitis

Diagnosis of pancreatitis is based on blood tests to determine the level of serum lipase and amylase as well as imaging techniques. Measurements of serum and urinary enzymes such as amylase, lipase, trypsinogen, chymotrypsin, and elastase are used to diagnose pancreatitis.

10.3.7.4.1 Serum and Urinary Amylase

Pancreatic amylase is the only glycosidase enzyme secreted by pancreas as human pancreatic juice that hydrolyzes the α-1,4-glycosidic bond, has no effect on α-1,6-glycosidic bond (Whitcomb and Lowe 2007). In ingested food, it hydrolyses α-1,4-glycosidic bonds present in soluble starch, straight-chain starch, and glycogen. The human salivary glands also secrete alpha-amylase and can pass through the glomerulus and hence can be detected in urine. The use of the urine level of amylase is limited in practice because the diagnostic ability of urinary amylase is inferior to that of the serum level of amylase. Serum amylase has high values of sensitivity i.e., 85% and specificity 91% as compared to urinary amylase values, i.e. 83% and 88%, respectively (Baillie 1997). The level of urinary amylase is measured within the 24 h urine collection. Urinary amylase is also detected using urine test strips and can be considered useful beside blood tests for the diagnosis of patients with clinical pancreatitis (Braganza et al. 2011). The cut-off level of amylase is 1000 IU/L (Matull et al. 2006).

10.3.7.4.2 Lipase

In the detection of acute pancreatitis, one of the key blood biochemical parameters is serum lipase. Within a few hours of disease onset, the serum concentration of lipase increases gradually and reaches the peak level within 24 hours. Reports showed that lipase level generally stays elevated for up to two weeks; while the amylase level is elevated only up to five days (Greenberg et al. 2016). Detection of serum lipase test has a slightly higher sensitivity compared to the amylase test in the patients with acute pancreatitis (Gwozdz et al. 1990). The cut-off level of lipase in blood serum is 600 IU/L.

10.3.8 Myocardial Infarction (MI)

Myocardial infarction (MI) or heart attack is a condition in which blood flow to a part of the heart is blocked, causing damage to heart muscle. The blockage is usually caused by a buildup of plaque in the arteries, which can rupture and form a blood clot that blocks the artery. The blockage leads to myocardial ischemia, which is an imbalance between oxygen supply and demand that can sometimes lead to myocardial infarction. Treatment for a heart attack may include medications, such as aspirin and nitroglycerin, and medical procedures, such as angioplasty and stenting, to restore blood flow to the heart. Preventive measures to reduce the heart attack risk include maintaining a healthy lifestyle, such as eating a healthy diet, exercising regularly, no smoking, and managing stress. Prolonged myocardial ischemia activates cardiomyocyte death. Mitochondrial alterations are prominently involved in apoptosis and necrosis of cardiomyocytes in the infarcted heart.

10.3.8.1 Symptoms

The elevation in serum biomarkers, electrocardiographic findings, and patient's familial history helps to identify ischemic symptoms. The symptoms of a heart attack include chest pain or discomfort that may spread to the shoulders, neck, arms, or jaw, shortness of breath, sweating, nausea or vomiting, dizziness or fainting, and rapid or irregular pulse.

10.3.8.2 Enzymes Used in Diagnosis of Myocardial Infarction

Cardiac enzymes, i.e., creatine kinase (CK) and lactate dehydrogenase (LDH) are widely used as biomarkers for diagnosis of myocardial infarction (Basit and Huecker 2023).

10.3.8.2.1 Creatine Kinase (CK)

CK is a muscle enzyme that exists as isoenzymes, and is specific to myocardial cells. The CK level increases approximately 3–4 hours after MI and reaches peak level by 24 hours, remains elevated for 3–4 days, making it useful for detecting re-infarction in the window of 4–10 days after the initial insult. CK is a sensitive biomarker for diagnosis of acute MI as it has a relative specificity for cardiac tissue.

10.3.8.2.2 Lactate Dehydrogenase

Liver disease marker enzyme, lactate dehydrogenase (LDH), emerged as other alternative potential biomarker for detecting the myocardial infarction. In acute MI, LDH level starts increasing after 6–12 hours and reaches peaks within 24–72 hours. The level of LDH can also increase in many other conditions and hence cannot be considered as a specific marker for cardiac myocytes. The major limitation for the use of LDH as an enzyme biomarker is that it takes several hours to increase the level of LDH after onset of injury.

10.3.9 Renal Disorders

It includes a group of diseases that affect the functions of kidneys and their ability to filter waste and fluid from the blood. Some common types of kidney diseases include:

10.3.9.1 Chronic Kidney Disease (CKD)

Chronic Kidney Disease is characterized by a gradual loss of kidney's functions. It may be caused by factors such as diabetes, high blood pressure, and glomerulonephritis (Chen et al. 2019).

10.3.9.2 Polycystic Kidney Disease (PKD)

This is an autosomal dominant genetic disorder that causes cysts formation in kidneys and can cause kidney failure (Akbar and Bokhari 2023).

10.3.9.3 IgA Nephropathy

It is a condition when the immune system produces antibodies in the kidneys, triggering harmful inflammation that causes kidney damage. Currently, no treatment is available for IgA nephropathy, but treatments in IgA nephropathy can slow the damage to the kidneys (Rodrigues et al. 2017).

10.3.9.4 Interstitial Nephritis

Interstitial nephritis is caused by a reaction to a medicine and in this condition; the ability of the kidney goes lowers to purify blood and produces a large volume of urine. The effective treatment strategy of this disease is the discontinuing of medicine that causes interstitial nephritis (Sathick et al. 2013).

10.3.9.5 Lupus Nephritis

This is an autoimmune disease manifestation of systemic lupus erythematosus (SLE) that results in inflammation of the small blood vessels of kidney that filter wastes. It can cause kidney damage and lower the kidneys' ability to filter waste and fluid from the blood (Jaryal and Vikrant 2017).

10.3.9.6 Glomerulonephritis (Glomerular Disease)

Kidneys contain more than a million glomeruli, which filter and remove waste and fluid from blood. If glomeruli get damaged, this condition is known as glomerulonephritis. If not cured, it may lead to serious kidney problems, including kidney failure (AlYousef et al. 2020). Other kidney problems include kidney cysts and stone, acute kidney injury, and infections.

10.3.9.7 Symptoms of Renal Disorders

The symptoms of renal disorders can vary depending on the specific condition and the stage of the disease. Some common symptoms include: nausea and vomiting, muscle cramps, loss of appetite, swelling in the feet, ankles, or hands (edema), fatigue and weakness, over or less than usual urination, decreased mental sharpness, dry, itchy skin, high blood pressure (hypertension), chest pain, blood in the urine, difficulty in sleeping (insomnia), headaches, etc.

10.3.9.8 Enzymes Used in Diagnosis of Renal Disorders

The enzymes in urinary excretion are sensitive parameters and can be used as biomarkers for the detection of renal disorders. However, serum enzymes are more commonly used to assess the diagnosis of hepatobiliary disease. The enzymes used as biomarkers include lysosomal glycosidases and lysozyme (Hemalatha et al. 2013). Lysosomal glycosidases are widely used particularly in the early detection of diabetic nephropathy (Gatsing et al. 2006). However, accurate interpretation of serum concentrations of cardiac enzymes is extremely important since coronary artery disease is highly prevalent in patients with ESKD. In summary, while serum enzymes can be used to assess the diagnosis of hepatobiliary disease, urinary excretion of some enzymes can be used to detect early stages of renal dysfunction.

10.4 Enzyme as Therapeutics

For decades, the point of convergence is to fight human disorders by developing and identifying different therapeutic agents (Nash 2017). Current progress in biotechnology has expanded the potential of enzymes being used as biopharmaceuticals. The diminished activity of the enzyme is known to be the major cause of various disorders (Li 2018). High substrate specificity, minimum side effects, efficient catalysis are some of the significant advantages offered by the enzymes when used as therapeutics (Yari et al. 2017). Sources of therapeutic enzymes could be microbes, plants, or animals (Sabu 2003). However, the microbial enzyme sources are found to be the most efficient as it provides us the advantage of high production yield. Superior strains of microbes could be obtained through genetic manipulations using recombinant DNA techniques (Andexer et al. 2009). The promising potential of microbial enzymes as therapeutics is widely recognized by researchers nowadays (Taipa et al. 2019; Umamaheswari et al. 2016). Oncolytic enzymes such as asparaginase, diphtheria toxin (still in experimental stage), and neuraminidase are found to be extremely useful. Enzymes such as trypsin, papain, and collagenase could be used as debriding agents. Various enzymes have also been reported as thrombolytic factors which act as anti-inflammatory agents, combating infectious diseases, in enzyme replacement therapies (Kunamneni et al. 2018). In the recent research, Basit et al. (2020) found truncated human angiotensin converting enzyme 2 as a potential inhibitor of SARS-CoV-2 spike glycoprotein and potent COVID-19 therapeutic agent. It signifies the expanding potential of enzymes being used as an effective therapeutic agent.

Protein engineering and recent biotechnological advancements have paved ways for treatment of different diseases by studying and discovering the potential of enzymes as therapeutics. For therapeutics development, the most critical criteria are enzyme delivery inside the body. For binding to their possible substrates, the enzymes must be in their native state. The fragile structure of enzymes hinders their therapeutic development. Therefore, some complementary systems such as PEGylation, enzyme encapsulation, nano-carriers and many more have been developed to conquer complications of enzymes (Dean et al. 2017). Due to the high affinity and specificity of enzymes, therapeutic enzymes are more advantageous than known traditional therapeutic approaches (de la Fuente et al. 2021) (Table 10.1).

Recently, enzymes are emerging as promising therapeutic tools for different pathologies and also identified as promising drug targets. The enzyme application in pharmaceuticals is very vast and increasing interest in using enzymes for sustainable development from chemical products to enzymatic products (Mishra 2023). The enzyme technology is still in a progressive phase and expected to achieve impressive outcomes in the near future. The improvement in catalytic activity of enzymes and finding novel enzymes are the two main approaches to identify enzymes for useful applications. The pharmacogenetics and gene-based optimization of drug-metabolizing enzymes in pediatric medicine have been studied and their role in clinical applications is being explored (Baweja et al. 2016). Innovative biotechnology strategies are being developed for revolving enzyme drugs-related problems during treatments. Additionally, to improve the enzyme's characteristics, molecular modifications are being carried out; conjugation with biomolecules like antibodies (Arslan et al. 2021), DNA or metabolites are under study. Simultaneously, novel approaches of enzymes in therapy of cancer (Triplett et al. 2018), neurodegenerative disease (Liu et al. 2021), joint problems, and infections are being thoroughly investigated.

During the outbreak of covid-19, the enzyme therapeutics played an enormous role for the treatment of patients of SARS-CO-V2. Angiotensin-converting enzyme 2 (ACE2) having a binding affinity with SARS-CO-V2 led the development of human recombinant soluble ACE2 that not only helped in the virus entry prevention but also regulated downstream inflammation without antibody production impairment (Liu et al. 2021).

Table 10.1 Therapeutic enzyme used in the treatment of different diseases

Disease	Therapeutic enzyme	Reference
Alzheimer	Secretases	Strooper et al.(2010)
Celiac	Gluten degrading enzymes	Wei et al. (2020)
Glaucoma	Collagenases	Honkanen (2013)
Congenital sucrase-isomaltase deficiency (CSID)	Sacrosidase	Puntis and Zamvar (2015)
Acute intermittent porphyria (AIP)	Hydroxymethylbilane synthase, porphobilinogen deaminase	Fontanellas et al. (2016)

10.5 Factors Affecting Enzyme Activities

10.5.1 Concentration of Substrate

Recent studies showed that enzymatic hydrolysis is influenced by substrate concentration (Deng et al. 2018). Substrate concentration required for maximum velocity of enzymes is explained by the Michaelis–Menten model (Johnson and Goody 2011). The concentration of enzyme in the cell ranged between 0.01 and 10 μM and the concentration of substrate in the cell was between 1 and 100 μM. These are the usual concentration of enzyme and substrate in the cell. Vitolo (2020) found a rectangular hyperbola curve between substrate concentration and rate of reaction.

1. When substrate concentration is diminished, the rate of reaction decreases.
2. When substrate concentration is high, the rate of reaction increases.

Robinson (2015) reported that a lesser amount of substrate is required for enzyme saturation as it has a low km value. So even at lesser substrate concentration, maximum velocity is obtained. Therefore, they commonly mentioned km as estimation of affinity of enzyme for its substrate.

10.5.2 Effect of Temperature

All enzymes work for their function on the most favorable temperature at 37-40 °C (Shrief 2020) (Table 10.2). With increase in temperature, the rate of reaction increases but at higher temperature of more than 40 °C, protein gets denatured due to breaking of covalent bond or loss of interactive forces in 3-dimensional structure of protein. So the rate of reaction will decrease with increase in temperature. A bell-shaped curve is usually observed when enzyme activity is correlated with temperature (Peterson et al. 2007).

10.5.3 Effect of pH

Each enzyme has specific pH for its functions. Under favorable pH, every enzymes show high activity (Table 10.3). At higher or lower pH value than the optimum,

Table 10.2 Different types of enzyme and their optimum temperature range

Enzymes	Optimum temperature range (°C)
Amylase	37–40
Lipase	37–40
Pepsin	37–42
DNA polymerase	37
Restriction enzyme	37–40
Thermophilic enzyme	72–75
Catalase	37–40

Table 10.3 Enzymes and their optimum pH

Enzyme	Optimum pH
Lipase	8.0
Pepsin	1.5–1.6
Trypsin	7.8–8.7
Urease	7.0
Maltose	6.1–6.8
Catalase	7.0
Amylase (malt)	4.6–5.2
Amylase (pancreas)	6.7–7.0
Lipase(stomach)	4.0–5.0

most enzymes lose complete activity. pH affects the enzyme activity due to the proteinaceous nature of enzymes and charge on the amino acids change with change in pH.

10.5.4 Effect of Activator

In a study, Vitolo (2020) reported that activators promote enzymatic activity. It interacts with the enzyme molecule and performs catalysis. Enzymatic activators such as resveratrol which is a polyphenol, enhances certain biochemical functions like insulin sensitivity, inhibition of tumor formation and boosts endurance. Villalba and Alacain (2012) reported that Sirtuin activators can help in absolute stimulation of SIRT enzymes that can help in treatment of many diseases like obesity, formation of tumor, and decline in heart functions and other organ activity due to aging.

10.6 Advantages and Disadvantages of Use of Enzymes in Clinical Diagnosis

Enzymes are widely used in clinical diagnosis. Their specificity and affinity make them useful as diagnostic tools, reagents, and clinical therapeutics (Wang et al. 2022). Some advantages and disadvantages of clinical diagnosis of enzymes are reported below:

10.6.1 Advantages

- Enzymes are highly specific catalysts, facilitating appropriate reaction in subcellular compartments, cells, and tissues.
- Enzymes are catalytic and convert multiple target molecules to desired products making them specific and potent.
- Enzymes used in diagnostic tools and clinical therapeutic are prepared with high a degree of purity.
- Enzyme assays are helpful in the automation with clinical analyzers.
- Enzymes are mostly used in the therapies which have high affinity and specificity.

10.6.2 Disadvantages

- Difficulties in standardization and instability of enzymes can be a major disadvantage of enzyme assays (Nguyen and Kang 2019).
- Enzyme assays are an expensive and labor-intensive sophisticated technique and require high degree of purity.
- Enzymes assays require careful handling and storage to maintain their activity.

10.7 Future Prospectives and Challenges

Enzyme estimation in different biological samples have shown great potential in detecting and treating various pathologies. However, there are number of challenges towards enzymes used in clinical diagnosis and development of enzyme based therapies (de la Fuente et al. 2021; Silva et al. 2022) that include:

- Use of enzymes in clinical diagnosis procedures may be complex, tedious, and reaction conditions such as temperature and pH are narrowly controlled.
- For accurate and high reproducible results, substrate and reagents must be of high purity.
- Many enzymes degrade very rapidly within circulation and cannot be measured. For example, Isocitrate dehydrogenase, abundant in heart tissue, is so rapidly destroyed within the blood circulation that it cannot be measured during myocardial infractions.
- Short in vivo half-life limits their therapeutic efficacy because enzymes are rapidly cleared by the liver and kidney from the bloodstream.
- Lack of targeted action which leads to off-target effects and toxicity.
- Patient Immune system can synthesize antibodies that can neutralize the therapeutic effects of the enzyme.
- Undesired immunologic reactions and biodegradation limits the enzyme's efficacy.
- Instability of enzymes caused by the temperature, pH, and salinity of the body is also a challenging factor.
- Enzymes replacement therapy (ERT) is based on the periodic life-long intravenous administration of enzymes which is costly and can reach about $200,000/year.

To solve these problems, researchers are developing new drug delivery approaches like enzyme encapsulation in nanosystems to improve efficacy and safety in enzyme-based therapies. Enzymes inhibitors are also emerging as a key strategy in drug development and offering potential treatment for a wide range of diseases.

10.8 Conclusions

Enzymes have become an essential tool in clinical diagnosis as they are used as a markers and biosensor in detection/prediction of various kinds of diseases like diabetes, neurodegenerative diseases, jaundice, myocardial infraction, pancreatitis, etc. Even in treatment strategy, enzyme therapy is an emerging technology in a wide range of pathologies such as metabolic disorders, cancer, fibrosis, SARS-Co-V2 infections, and others. Enzyme encapsulation approaches, such as liposomes, membrane vesicles, nanoparticles, and erythrocytes improve in vivo half-life, tissue specificity, and reduce the immunogenicity of the enzymes. It is becoming the major choice for medical diagnostics because of its high specificity. Targeted enzyme modification technology, such as PEG conjugation improves functional bioavailability and reduces immunogenicity. The patient management to preserve efficacy and safety of therapy significantly improves by monitoring patients' immune response. In this context, microarray technology is a promising tool to improve anti-enzyme immune responses in patients treated with ERT. Though there are a number of research reports available on diagnostic uses of enzymes in medicine and healthcare from the last few decades, but still more research are needed to be focused on novel enzymes and technologies to reach safe, stable, and reliable approaches in clinical applications.

References

Aerts JMFG, Hollak CEM, van Breemen M, Maas M, Groener JEM, Boot RG (2005) Identification and use of biomarkers in Gaucher disease and other lysosomal storage diseases. Acta Paediatr Suppl 94(447):43–46

Akbar S, Bokhari SRA (2023) Polycystic kidney disease. In: StatPearls [Internet]. StatPearls Publishing, Treasure Island. https://www.ncbi.nlm.nih.gov/books/NBK532934/

AlYousef A, AlSahow A, AlHelal B et al (2020) Glomerulonephritis histopathological pattern change. BMC Nephrol 21:186. https://doi.org/10.1186/s12882-020-01836-3

Amacher DE (1998) Serum transaminase elevations as indicators of hepatic injury following the administration of drugs. Regul Toxicol Pharmacol 27:119–130

Anand P, Kunnumakkara AB, Sundaram C, Harikumar KB, Tharakan ST, Lai OS, Sung B, Aggarwal BB (2008) Cancer is a preventable disease that requires major lifestyle changes. Pharm Res 25(9):2097–2116

Andexer JN, Langermann JV, Kragl U, Pohl M (2009) How to overcome limitations in biotechnological process- examples from hydroxy nitrile lyase applications. Trends Biotechnol 27(10):599–607

Arslan FB, Ozturk K, Calis S (2021) Antibody-mediated drug delivery. Int J Pharm 596:120268. https://doi.org/10.1016/j.ijpharm.2021.120268

Badalov N, Baradarian R, Iswara K, Li J, Steinberg W, Tenner S (2007) Drug-induced acute pancreatitis: an evidence-based review. Clin Gastroenterol Hepatol 5:648–661

Baillie J (1997) Rapid measurement of urinary trypsinogen-2 as a screening test for acute pancreatitis. Gastrointest Endosc 46:385–386

Basit H, Huecker MR (2023) Myocardial infarction serum markers. In: StatPearls [Internet]. StatPearls Publishing, Treasure Island. https://www.ncbi.nlm.nih.gov/books/NBK532966/

Basit A, Ali T, Rehman SU (2020) Truncated human angiotensin converting enzymes 2; a potential inhibitor of SARS-CoV-2 spike glycoprotein and potent COVID-19 therapeutic agent. J Biomol Struct Dyn 39(10):1–10

Baweja M, Nain L, Kawarabayasi Y, Shukla P (2016) Current technological improvements in enzymes toward their biotechnological applications. Front Microbiol 7. https://doi.org/10.3389/fmicb.2016.00965

Benes P, Vetvicka V, Fusek M (2008) Cathepsin D—many functions of one aspartic protease. Crit Rev Oncol Hematol 68:12–28

Blachier M et al (2013) The burden of liver disease in Europe: a review of available epidemiological data. J Hepatol 58(3):593–608

Bourlieu C, Astruc T, Barbe S, Berrin JG, Bonnin E, Boutrou R, Hugouvieux V, Le Feunteun S, Paes G (2020) Enzymes to unravel bioproducts architecture. Biotech Advan 41:107546

Braganza JM, Lee SH, McCloy RF, McMahon MJ (2011) Chronic pancreatitis. Lancet 377:1184–1197

Bull H, Murray PG, Thomas D, Fraser AM, Nelson PN (2002) Acid phosphatases. Mol Pathol 55(65):65

Carr RA, Rejowski BJ, Cote GA, Pitt HA, Zyromski NJ (2016) Systematic review of hypertriglyceridemia-induced acute pancreatitis: a more virulent etiology. Pancreatology 16(4):469–476

Chao T, Yi-Ying W, Janckila AJ (2010) Tartrate-resistant acid phosphatase isoform 5b (TRACP 5b) as a serum maker for cancer with bone metastasis. Clin Chim Acta 411:1553–1564

Chapman R, Stenzel MH (2019) All wrapped up: stabilization of enzymes within single enzyme nanoparticles. J Am Chem Soc 141(7):2754–2769

Chaturvedi S, Singh AK, Keshari AK, Maity S, Sarkar S, Saha S (2016) Human metabolic enzymes deficiency: a genetic mutation based approach. Scientifica (Cairo). 9828672 2016:1

Chen L, Magliano D, Zimmet P (2012) The worldwide epidemiology of type 2 diabetes mellitus—present and future perspectives. Nat Rev Endocrinol 8:228–236

Chen H, McGowan EM, Ren N, Lal S, Nassif N, Shad-Kaneez F, Qu X, Lin Y (2018) Nattokinase: a promising alternative in prevention and treatment of cardiovascular diseases. Biomark Insights 13:117727191878513

Chen TK, Knicely DH, Grams ME (2019) Chronic kidney disease diagnosis and management: a review. JAMA 322(13):1294–1304

Chesney RW (2001) Vitamin D deficiency and rickets. Rev EndocrMetabDisord 2(2):145–151

Corathers SD (2006) The alkaline phosphatase level: nuances of a familiar test. Pediatr Rev 27:382

Cunningham J, Sprague SM, Cannata-Andia J, Coco M, Cohen-Solal M, Fitzpatrick L, Goltzmann D, Lafage-Proust MH, Leonard M, Ott S, Rodriguez M et al (2004) Osteoporosis in chronic kidney disease. Am J Kidney Dis 43(3):566–571

Dean SN, Turner KB, Medintz IL, Walper SA (2017) Targeting and delivery of therapeutic enzymes. Ther Deliv 8:577–595

de la Fuente M, Lombardero L, Gómez-González A, Solari C, Angulo-Barturen I, Acera A, Vecino E, Astigarraga E, Barreda-Gómez G (2021) Enzyme Therapy: Current challenges and future perspectives. Int J Mol Sci 22(17):9181

Deng Y, Butre IC, Wiereng AP (2018) Influence of substrate concentration on the extent of protein enzymatic hydrolysis. Int Dairy J 86:39–48

Fontanellas A, Avila MA, Berraondo P (2016) Emerging therapies for acute intermittent porphyria. Expert Rev Mol Med 18:e17

Gatsing D, Garba IB, Adoga GI (2006) The use of lysosomal enzymuria in the early detection and monitoring of the progression of diabetic nephropathy. Indian J Clin Biochem 21:42–48

Ghazale A, Chari ST, Zhang L et al (2008) Immunoglobulin G4-associated cholangitis: clinical profile and response to therapy. Gastroenterology 134(3):706–715

Giraldo P, Lopez de Frutos L, Cebolla JJ (2018) Biomarker combination is necessary for the assessment of Gaucher disease. Ann Transl Med 6(Suppl 1):S81

Giuffrida G, Markovic U, Condorelli A et al (2023) Glucosylsphingosine (Lyso-Gb1) as a reliable biomarker in Gaucher disease: a narrative review. Orphanet J Rare Dis 18:27

Gorelick FS, Lerch MM (2017) Do animal models of acute pancreatitis reproduce human disease? Cell Mol Gastroenterol Hepatol 4(2):251–262

Greenberg JA, Hsu J, Bawazeer M, Marshall J, Friedrich JO, Nathens A et al (2016) Clinical practice guideline: management of acute pancreatitis. Can J Surg 59:128–140

Gupta R, Shukla S, Gupta A, Singh RL, Singh P (2023) Estimation of phytochemical constituents and analysis of antioxidant activity in different parts of *Cassia fistula* plant. Int J Pharma Chem Ana 10(3):194–201

Gwozdz GP, Steinberg WM, Werner M, Henry JP, Pauley C (1990) Comparative evaluation of the diagnosis of acute pancreatitis based on serum and urine enzyme assays. Clin Chim Acta 187:243–254

He E, Xu XH, Guan H, Chen Y, Chen ZH, Pan ZL, Tang LL, Hu GZ, Li Y, Zhang M, Zhou J, Eriksson S, Fornander T, Skog S (2010) Thymidine kinase 1 is a potential marker for prognosis and monitoring the response to treatment of patients with breast, lung, and esophageal cancer and non-Hodgkin's lymphoma. Nucleotides Nucleic Acids 29(4–6):352–358

He W, Gan Y, Qi X, Wang H, Song H, Su P, Song J, Yang Y (2023) Enhancing enzyme activity using hydrophilic hollow layered double hydroxides as encapsulation carriers. ACS Appl Mater Interfaces 15:34513–34526

Hemalatha T, UmaMaheswari T, Krithiga G, Sankaranarayanan P, Puvanakrishnan R (2013) Enzymes in clinical medicine: an overview. Indian J Exp Biol 51:777–788

Hollak CE, van Weely S, van Oers MH et al (1994) Marked elevation of plasma chitotriosidase activity. A novel hallmark of Gaucher disease. J Clin Invest 93:1288–1292

Honkanen R (2013) Use of collagenase to treat glaucoma. US20150273028A1 Patent, Oct 24

Hu FB, Manson JE, Stampfer MJ, Colditz G, Liu S, Solomon CG et al (2001) Diet, lifestyle, and the risk of type 2 diabetes mellitus in women. N Engl J Med 345(11):790–797. https://doi.org/10.1056/NEJMoa010492

Huang XJ, Choi YK, Im HS, Yarimaga O, Yoon E, Kim HS (2006) Aspartate aminotransferase (AST/GOT) and alanine aminotransferase (ALT/GPT) detection techniques. Sensors (Basel) 6(7):756–782

Iqbal SJ (1998) Persistently raised serum acid phosphatase activity in a patient with hypophosphatasia: electrophoretic and molecular weight characterization as type 5. Clin Chim Acta 271(2):213–220

Ishiguro S, Inoue M, Tanaka Y, Mizokami M, Iwasaki M, Tsugane S (2009) Serum aminotransferase level and the risk of hepatocellular carcinoma: a population-based cohort study in Japan. Eur J Cancer Prev 18:26–32

Jaryal A, Vikrant S (2017) Current status of lupus nephritis. Indian J Med Res 45(2):167–178

Johnson KA, Goody RS (2011) The original Michaelis constant: translation of the 1913 Michaelis-Menton paper. J Biochem 50:8264–8269

Karhade AV, Thio QCBS, Kuverji M, Ogink PT, Ferrone ML, Schwab JH (2019) Prognostic value of serum alkaline phosphatase in spinal metastatic disease. Br J Cancer 120(6):640–646

Khosla S, Lufkin EG, Hodgson SF, Fitzpatrick LA, Melton LJ (1994) Epidemiology and clinical features of osteoporosis in young individuals. Bone 15(5):551–5

Kirschenbaum A, Liu XH, Yao S, Leiter A, Levine AC (2011) Prostatic acid phosphatase is expressed in human prostate cancer bone metastases and promotes osteoblast differentiation. Ann N Y Acad Sci 1237:64

Kocabay G, Telci A, Tutuncu Y, Tiryaki B, Ozel S, Cevikbas U, Okten A, Satman I (2011) Alkaline phosphatase: can it be considered as an indicator of liver fibrosis in non-alcoholic steatohepatitis with type 2 diabetes. Bratisl Lek Listy 112:626

Kumar DA, Ganesan G, Raju KVSN, Gangadharan AC, Manohar BM, Puvanakrishnan R (2009) Standardization of a model for collagen induced arthritis in rats. Biomedicine 29:109

Kunamneni A, Ogaugwu C, Goli D (2018) Chapter 15—Enzymes as therapeutic agents. In: Nunes CS, zaq Kumar V (eds) Enzymes in human and animal nutrition. Academic Press, pp 301–312

Lala V, Zubair M, Minter DA (2023) Liver function tests. In: StatPearls [Internet]. StatPearls Publishing, Treasure Island. https://www.ncbi.nlm.nih.gov/books/NBK482489/

Lange PH, Millan JL, Stigbrand T, Vessella RL, Ruoslahti E, Fishman WH (1982) Placental alkaline phosphatase as a tumor marker for seminoma. Cancer Res 42:3244–3247

Li M (2018) Enzyme replacement therapy: a review and its role in treating lysosomal storage diseases. Pediatr Ann 47:191–197

Liu W, Du JJ, Li ZH, Zhang XY, Zuo HD (2021) Liver injury associated with acute pancreatitis: the current status of clinical evaluation and involved mechanisms. World J Clin Cases 9(34):10418–10429

Lowe D, Sanvictores T, John S (2020) Alkaline phosphatase. In: StatPearls [Internet]. StatPearls Publishing, Treasure Island

Maitra A, Abbas AK (2005) Endocrine system. In: Kumar V, Fausto N, Abbas AK (eds) Robbins and cotran pathologic basis of disease, 7th edn. Saunders, Philadelphia, pp 1156–1226

Majumder S, Chari ST (2016) Chronic pancreatitis. Lancet 387(10031):1957–1966

Matta B, Gougol A, Gao X, Reddy N, Talukdar R, Kochhar R et al (2020) Worldwide variations in demographics, management, and outcomes of acute pancreatitis. Clin Gastroenterol Hepatol 18(7):1567–1575

Matull WR, Pereira SP, Donohue JWO (2006) Biochemical markers of acute pancreatitis. J Clin Pathol 59(4):340–344

Mikosch P, Hughes D (2010) An overview on bone manifestations in Gaucher disease. Wien Med Wochenschr 160:609–624

Mishra P (2023) Targeting enzymes as drug targets. Recent Advanc Fut Perspec 3(9):1–4

Mukherjee R, Smith A, Sutton R (2020) Covid-19-related pancreatic injury. Br J Surg 107(7):e190

Muniraj T, Aslanian HR, Farrell J et al (2014) Chronic pancreatitis, a comprehensive review and update. Part I: epidemiology, etiology, risk factors, genetics, pathophysiology, and clinical features. Dis Mon 60(12):530–550

Nash DB (2017) Trending BIO. Am Health Drug Benefits 10(7):344–345

Nguyen BT, Kang MJ (2019, 1977) Application of capillary electrophoresis with laser-induced fluorescence to immunoassays and enzyme assays. Molecules 24(10)

Norga K, Grillet B, Masure S, Paemen L, Opdenakker G (1996) Human gelatinase B, a marker enzyme in rheumatoid arthritis, is inhibited by D-penicillamine: anti-rheumatic activity by protease inhibition. Clin Rheumatol 15(1):31–34

Ozer J, Ratner M, Shaw M, Bailey W, Schomaker S (2008) The current state of serum biomarkers of hepatotoxicity. Toxicology 245(3):194–205

Pandanaboyana S, Moir J, Leeds JS, Oppong K, Kanwar A, Marzouk A et al (2021) SARS-CoV-2 infection in acute pancreatitis increases disease severity and 30-day mortality: COVID PAN collaborative study. Gut 70(6):1061–1069

Pandol SJ, Gorelick FS, Gerloff A et al (2010) Alcohol abuse, endoplasmic reticulum stress and pancreatitis. Dig Dis 28(6):776–782

Parvizi J, Jacovides C, Antoci V, Ghanem E (2011) Diagnosis of periprosthetic joint infection: the utility of a simple yet unappreciated enzyme. J Bone Joint Surg Am 93:2242–2248

Patlak M (2002) New weapons to combat an ancient disease: treating diabetes. FASEB J 16(14):1853. https://doi.org/10.1096/fj.02-0974bkt

Peterson ME, Daniel RM, Danson MJ, Eisenthal R (2007) The dependence of enzyme activity on temperature: determination and validation of parameter. Biochem J (Pt 2):331–337

Pettifor JM (2002) Rickets. Calcif Tissue Int 70(5):398–399

Pettifor JM (2003) Nutritional and drug-induced rickets and osteomalacia. In: Favus MJ (ed) Primer on the metabolic bone diseases and disorders of mineral metabolism. 5. American Society for Bone and Mineral Research, Washington, DC, pp 399–407

Pham A, Forsmark C (2018) Chronic pancreatitis: review and update of etiology, risk factors, and management. F1000Res 7:F1000. Faculty Rev-607, 7, 607

Pitchumoni CS, Chari S (2013) Ulcerative colitis and autoimmune pancreatitis. J Clin Gastroenterol 47(6):469

Punekar NS (2018) Regulation of enzyme activity. ENZYMES: Catalysis Kine. Mechan:461–492

Puntis JWL, Zamvar V (2015) Congenital sucrase- isomaltase deficiency: diagnostic challenges and response to enzyme replacement therapy. Arch Dis Child 100:869–871

Ram M, Sherer Y, Shoenfeld Y (2006) Matrix metalloproteinase-9 and autoimmune diseases.J. Clin Immunol 26:299–307

Rauch F, Glorieux FH (2004) Osteogenesis imperfecta. Lancet 363(9418):1377–1385

Ripsin CM, Kang H, Urban RJ (2009) Management of blood glucose in type 2 diabetes mellitus. Am Fam Physician 79(1):29–36

Robinson PK (2015) Enzymes: principles and biotechnological applications. Essays Biochem 59:1–41

Rodrigues JC, Haas M, Reich HN (2017) IgA nephropathy. Clin J Am Soc Nephrol 12(4):677–686

Sabu A (2003) Sources, properties and applications of microbial therapeutic enzymes. Indian J Biotechnol 2(3):334–341

Sankaran SJ, Xiao AY, Wu LM et al (2015) Frequency of progression from acute to chronic pancreatitis and risk factors: a meta-analysis. Gastroenterology 149(6):1490–1500

Sathick IJ, Zand L, Kamal AN, Norby SM, Garovic VD (2013) Acute interstitial nephritis: etiology, pathogenesis, diagnosis, treatment and prognosis. Nephrol Res Rev 5(1):13–20

Schwartz MK (1992) Enzymes as prognostic markers and therapeutic indicators in patients with cancer. Clin Chim Acta 206:77–82

Seeman E (2003) Invited review: pathogenesis of osteoporosis. J Appl Physiol 95(5):2142–2151

Setiawan VW, Pandol SJ, Porcel J, Wilkens LR, Le Marchand L, Pike MC et al (2016) Prospective study of alcohol drinking, smoking, and pancreatitis: the multiethnic cohort. Pancreas 45(6):819–825

Sharma A, Nagalli S (2023) Chronic liver disease. In: StatPearls [Internet]. StatPearls Publishing, Treasure Island. https://www.ncbi.nlm.nih.gov/books/NBK554597/

Sheng X, Che H, Ji Q, Yang F, Lv J, Wang Y et al (2018) The relationship between liver enzymes and insulin resistance in type 2 diabetes patients with nonalcoholic fatty liver disease. Horm Metab Res 50(05):397–402

Shrief E (2020) Factors affecting enzyme activity. Ain Shams University, Women faculty for Arts, Science and Education

Silva AKA, Sagné C, Gazeau F, Abasolo I (2022) Enzyme replacement therapy: current challenges and drug delivery prospects via extracellular vesicles. Rare Disea Orphan Drug J 1(3):13

Singh RL, Vishwakarma SP, Singh P (2014) Vitamins and minerals: roles and plant sources. In: Prakash D (ed) Phytochemicals of nutraceutical importance and their role in human health. CABI-UK, pp 248–265

Singh P, Singh RL, Kakkar P (2016a) Antioxidant, DNA damage protective and hepatoprotective activities of *Amorphophylluscampanulatus*. Int J Pharm Pharm Sci 8(3):330–338

Singh U, Shukla M, Singh P, Kakkar P, Singh RL (2016b) Role of *Vicia faba* fruit extract against cytotoxicity induced by acetaminophen in primary cultured rat hepatocytes. Int J Pharm Pharm Sci 8(8):71–78

Singh P, Singh RL, Pathak N, Singh PK, Tripathi M, Mondal S (2022) Phytochemistry and nutraceutical properties of *Carica papaya* (Linn.): a review. Diet Suppl Nutr 1(9):1–15

Singh U, Singh P, Singh PK, Shukla S, Singh RL (2023) Evaluation of antioxidant, oxidative DNA damage and hepatoprotective activities of ethanolic leaf extract of *Lagenaria siceraria* plant against acetaminophen induced hepatotoxiocity. Func Foods Health Dis 13(3):117–134

Siris ES, Roodman GD (2003) Paget's disease of bone. In: Favus MJ (ed) Primer on the metabolic bone diseases and disorders of mineral metabolism. 5. American Society for Bone and Mineral Research, Washington, DC, pp 495–506

Sohar N, Hammer H, Sohar I (2002) Lysosomal peptidases and glycosidases in rheumatoid arthritis. Biol Chem 383:865

Stein E, Shane E (2003) Secondary osteoporosis. Endocrinol Metab Clin N Am 32(1):115–134

Stirnemann J, Belmatoug N, Camou F, Serratrice C, Froissart R, Caillaud C, Levade T et al (2017) A review of gaucher disease pathophysiology, clinical presentation and treatments. Int J Mol Sci 18(2):441

Strooper BD, Vassar R, Golde T (2010) The secretases: enzymes with therapeutic potential in Alzheimer disease. Nat Rev Neurol 6(2):99–107

Taipa MA, Fernandes P, de Carvalho C (2019) Production and purification of therapeutic enzymes. Adv Exp Med Biol 1148:1–24

Taketo MM (1998) Cyclooxygenase-2 inhibitors in tumorigenesis (part I). J Natl Cancer Inst 90:1529–1536

Tjhung KF, Shokhirev MN, Horning DP, Joyce GF (2020) An RNA polymerase ribozyme that synthesizes its own ancestor. Proc N Acad Sci USA 117(6):2906–2913

Torsteinsdottir I, Hakansson L, Hallgren R, Gudbjornsson, Arvidson NG, Venge P (1999) Serum lysozyme: a potential marker of monocyte/macrophage activity in rheumatoid arthritis. Rheumatology 38:1249–1254

Triplett TA, Garrison KC, Marshall N, Donkor M, Blazeck J, Lamb C, Qerqez A et al (2018) Reversal of IDO-mediated cancer immune suppression by systemic kynurenine depletion with a therapeutic enzyme. Nat Biotechnol 36(8):758–764

Umamaheswari T, Hemalatha T, Sankaranarayanan P, Puvanakrishnan R (2016) Enzyme therapy: current perspectives. Indian Exp Biol 54:7–16

Vengadakrishnan K, Koushik AK (2015) A study of the clinical profile of acute pancreatitis and its correlation with severity indices. Int J Health Sci (Qassim) 9(4):410–417

Villalba JM, Alacain FJ (2012) Sirtuin activators and inhibitors. Biofactors 38(5):349–359

Vinklerova I, Prochazka M, Prochazka V, Urbanek K (2010) Incidence, severity, and etiology of drug-induced acute pancreatitis. Dig Dis Sci 55(10):2977–2981

Vitolo M (2020) Brief review on enzyme activity. World J Pharma Res 9(2):60–70

Wang L, Wang N, Zhang W, Cheng X, Yan Z, Shao G, Wang X, Wang R, Fu C (2022) Therapeutic peptides: current applications and future directions. Signal Transduc Targe Thera 7(1):1–27

Wei G, Helmerhorst EJ, Darwish G, Blumenkranz G, Schuppan D (2020) Gluten degrading enzymes for treatment of celiac disease. Nutrients 12:2095

Whitcomb DC, Lowe ME (2007) Human pancreatic digestive enzymes. Dig Dis Sci 52:1–17

Whyte MP (2003) Sclerosing bone disorders. In: Favus MJ (ed) Primer on the metabolic bone diseases and disorders of mineral metabolism. American Society for Bone and Mineral Research, Washington, DC, pp 449–466

Wolfe D, Kanji S, Yazdi F, Barbeau P, Rice D, Beck A, Butler C, Esmaeilisaraji L, Skidmore B, Moher D et al (2020) Drug induced pancreatitis: a systematic review of case reports to determine potential drug associations. PLoS One 15:e0231883

Yadav D, Whitcomb DC (2010) The role of alcohol and smoking in pancreatitis. Nat Rev Gastroenterol Hepatol 7(3):131–145

Yang X, Schnackenberg LK, Shi Q et al (2014) Hepatic toxicity biomarkers. In: Gupta RC (ed) Biomarkers in toxicology. Academic Press/Elsevier, pp 241–260

Yari M, Ghoshoon MB, Vakili B, Ghasemi Y (2017) Therapeutic enzymes: applications and approaches to pharmacological improvement. Curr Pharm Biotechnol 18:531–540

Yeoman CJ, Han Y, Dodd D, Schroeder CM, Mackie RI, Cann IKO (2010) Thermostable enzymes as biocatalysts in the biofuel industry. Adv Appl Micro 70:1–55

Yu C, Gershwin ME, Chang C (2014) Diagnostic criteria for systemic lupus erythematosus: a critical review. J Autoimmun 48–49:10–13

Hormonal Manifestations of Endocrine Diseases

11

Jyotsana Singh, Sarat Kumar Kottarath, Farhad Bano, Mohd Farhan, Amit Kumar Tripathi, and Yogesh Srivastava

Abstract

An essential part of the human body's homeostasis, the endocrine system keeps cells stable within certain ranges, including the regulation of electrolyte balance and temperature. Its smooth operation depends on the neurological and endocrine systems communicating with each other, with hormones as their main messengers. The endocrine system coordinates bodily reactions through chemical transmission via hormones, glands, and receptors. This dynamic system

Jyotsana Singh, Sarat Kumar Kottarath, Farhad Bano, Mohd Farhan, Amit Kumar Tripathi and Yogesh Srivastava contributed equally with all other contributors.

J. Singh (✉)
Department of Hematopoietic Biology and Malignancy, The University of Texas MD Anderson Cancer Center, Houston, TX, USA

S. K. Kottarath
Department of Experimental Therapeutics, The University of Texas MD Anderson Cancer Center, Houston, TX, USA

F. Bano
Special Centre for Molecular Medicine, Jawaharlal Nehru University, New Delhi, India

M. Farhan
Department of Integrative Biology and Pharmacology, McGovern Medical School University of Texas Health Science Center at Houston, Houston, TX, USA

A. K. Tripathi
Department of Microbiology, Immunology and Genetics, The University of North Texas Health Science Center, Fort Worth, TX, USA

Y. Srivastava
Department of Genetics, The University of Texas MD Anderson Cancer Center, Houston, TX, USA

R. L. Singh et al. (eds.), *Clinical Applications of Biomolecules in Disease Diagnosis*, https://doi.org/10.1007/978-981-97-4723-8_11

comprises the primary endocrine organs, whose main function is hormone secretion, as well as secondary endocrine organs better known for their roles in other organ systems. Hormones are secreted by glands or scattered cells and travel through the bloodstream to affect target cells that have the appropriate receptors, triggering a wide range of reactions. This complex system of tailored interactions is further enhanced by neurohormones, a fraction secreted by neuroendocrine cells that modulate neurotransmission, as demonstrated by oxytocin and vasopressin. This delicate balance can be upset by any disturbance in hormone secretion, whether it is an excess or a shortage, which can lead to several endocrine disorders. These diseases are caused by abnormalities in the signaling pathways that connect the various endocrine system glands. The neurological and endocrine systems are interdependent and essential in preserving homeostatic balance. This overview explores specific endocrine glands and clarifies the range of endocrine disorders that result from hormone dysregulation. Comprehending the subtleties of these glands' functioning and dysfunctions is essential to understanding the complexities of endocrine illnesses and creating successful treatment plans.

Keywords

Endocrine · Hormonal regulation · Hormonal imbalance · Disease manifestation

11.1 Introduction

The endocrine system plays a pivotal role in maintaining the homeostasis of the human body (Toto 1994). Cells require a stable environment to perform adequately, without any fluctuation beyond set limits, including temperature and electrolyte levels (Nagy and Malcomson 2022). For the body to respond appropriately to any change in internal or external stimuli, proper communication between the nervous and endocrine system is required. The endocrine system is also known as the "hormone system," since hormones are its primary mode of communication, and consists of a complex network of glands, hormones, and receptors.

Hormones are chemical messengers secreted by glands and released in the bloodstream and for distribution to distant sites, imparting slow but long-lasting responses (Fig. 11.1). These endocrine glands are derived from epithelial tissues and are broadly categorized as primary endocrine organs and secondary endocrine organs. The major function of primary endocrine organs is the secretion of hormones. The secondary endocrine organs have non-endocrine primary functions but also possess endocrine functions and include the kidney, heart, liver, stomach,

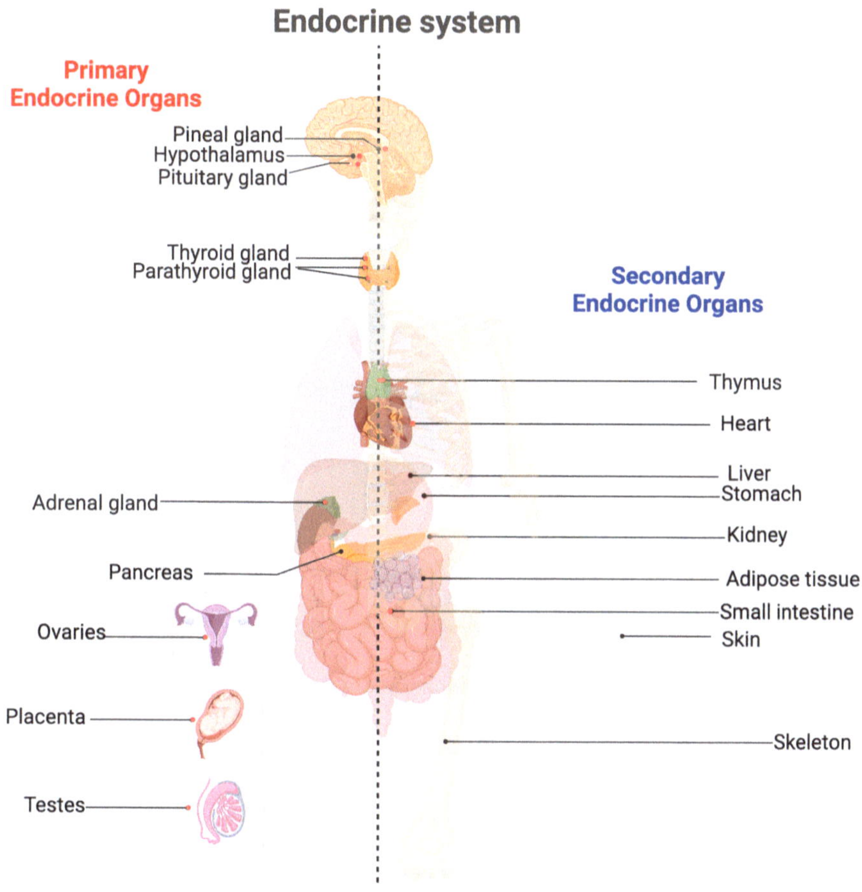

Fig. 11.1 The human endocrine system includes primary and secondary endocrine organs

small intestine, skin, and adipose tissue. Hormones are secreted by endocrine cells located within endocrine glands (or occasionally dispersed in other tissues), released into the interstitial fluid, and subsequently diffused in the blood circulatory system and transported to distant sites. Throughout the transport, only cells with specific receptors for the hormones, known as target cells, will respond (Stanfield 2013). For instance, testosterone is released by the testes and stimulates the development of male reproductive organs, sperm production, and protein metabolism. A subset of hormones categorized as neurohormones, such as vasopressin and oxytocin, are released by neurosecretory cells, also called neuroendocrine cells, in a process known as neurotransmission (Fig. 11.2) (Scharrer 1969). Indeed, the endocrine system is intricately interdependent with the nervous system in the regulation of homeostatic balance. Both excessive secretion and deficiency due to perturbation in release of hormones from endocrine glands will hamper this delicate balance, leading to the manifestation of various endocrine

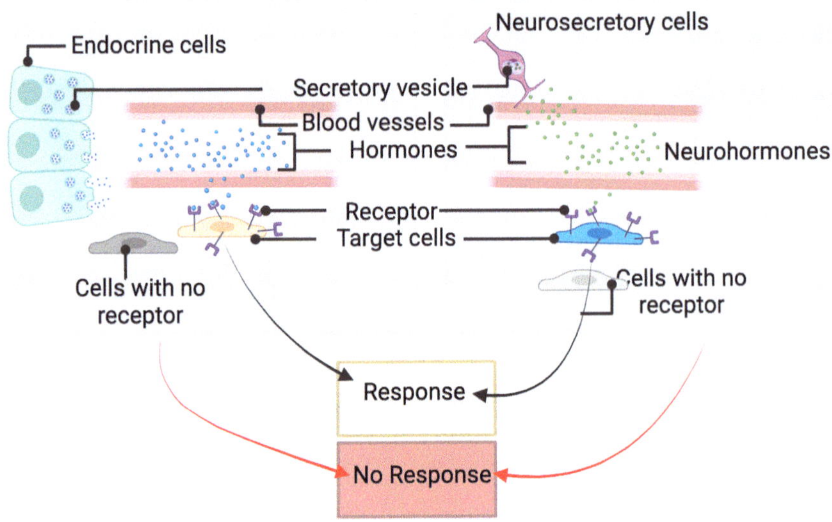

Fig. 11.2 Hormonal secretion from endocrine cells and neurosecretory cells. The hormones produce a response only when interacting with target cells with specific receptors

diseases (Yu 2014). We will discuss each endocrine gland and related endocrine diseases caused by hormonal imbalance.

11.2 Pineal Gland

The pineal gland is a small, pinecone-shaped endocrine gland located deep within the brain, in the epithalamus region. It plays a vital role in the regulation of various physiological processes, primarily through the secretion of the hormone melatonin. Melatonin is responsible for controlling the sleep–wake cycle and circadian rhythms (Favero et al. 2021). While the pineal gland is not commonly associated with the wide range of hormonal manifestations seen in endocrine diseases, a few notable conditions and hormonal imbalances are related to the pineal gland. Disorders involving the pineal gland are relatively rare, but they can have significant impacts on an individual's well-being and may require medical attention and intervention (Arendt et al. 2000).

11.2.1 Pineal Gland Calcification

Over time, the pineal gland may undergo calcification, in which calcium deposits accumulate within the gland. This calcification can affect pineal function and alter melatonin production. While pineal gland calcification is a normal part of aging, excessive or abnormal calcification may be associated with certain medical conditions (Tan et al. 2018).

11.2.2 Seasonal Affective Disorder

Seasonal affective disorder is a mood disorder characterized by depression that occurs during specific seasons, typically in the winter when there is reduced exposure to natural light. Reduced exposure to natural light can disrupt the pineal gland's melatonin production and circadian rhythms, potentially contributing to this disorder.

11.2.3 Jet Lag and Shift Work Sleep Disorder

Jet lag and shift work sleep disorder are conditions where changes in the sleep–wake cycle and circadian rhythms are disrupted due to altered light exposure. The pineal gland and its melatonin secretion play a role in adjusting to new time zones and work schedules. Disruptions can lead to hormonal manifestations such as fatigue, sleep disturbances, and mood swings. It's important to note that the pineal gland's primary function is the regulation of the sleep–wake cycle and circadian rhythms through melatonin production. While it may be indirectly linked to hormonal manifestations related to sleep and mood disorders, it is not as commonly associated with the wide range of hormonal disorders seen in other endocrine glands, like the thyroid, pancreas, or adrenal glands (Arendt et al. 2000).

11.2.4 Pineal Gland Tumors

Tumors that develop in the pineal gland can lead to various hormonal manifestations, such as changes in melatonin production. These tumors can be benign (pineocytomas) or malignant (pineoblastomas) (Favero et al. 2021). Pineocytomas are typically benign and originate from the pineal gland's cells. They tend to grow slowly and are less aggressive (Louis et al. 2007). On the other hand, pineoblastomas, which are less common, are malignant and can grow rapidly and invade nearby brain structures.

The symptoms of pineal gland tumors can vary depending on the size, location, and type of tumor. Common symptoms include headaches, nausea, vomiting, visual disturbances (such as double vision), balance and coordination problems, changes in personality or behavior, hormonal imbalances (including disruptions in melatonin production), and difficulties in sleeping and regulating the sleep–wake cycle (Osborn et al. 2012). In some cases, pineal gland tumors can also cause hydrocephalus, a condition where excess cerebrospinal fluid accumulates in the brain, leading to increased intracranial pressure and symptoms such as headaches, cognitive changes, and seizures. Diagnosis typically involves a combination of imaging studies, such as magnetic resonance imaging or computed tomography scans, to visualize the tumor and assess its size and location. A biopsy may be performed to determine the tumor's type and grade, particularly if it is malignant.

Treatment for pineal gland tumors often involves surgery to remove the tumor. The choice of surgical approach depends on factors such as tumor size and location.

Some tumors may be difficult to access due to their deep location in the brain. Radiation therapy and chemotherapy may also be used, especially for malignant tumors like pineoblastomas (Osborn et al. 2012). The goal of treatment is to remove or reduce the tumor and relieve any symptoms while preserving neurological function. The prognosis for individuals with pineal gland tumors varies based on the tumor's type, grade, size, and the extent of surgical removal. Benign pineocytomas generally have a better prognosis than malignant pineoblastomas. Pineal gland tumors are relatively rare, and their symptoms can be subtle or severe, making early diagnosis and intervention crucial. Medical management often involves a multidisciplinary approach, including neurosurgeons, oncologists, and neurologists, to provide the most appropriate treatment and supportive care for individuals with pineal gland tumors.

11.3 Thymus

The thymus is a specialized primary lymphoid organ located in the upper chest, behind the sternum (breastbone), and is primarily known for its role in the immune system's development and maturation of T lymphocytes (T cells). The thymus also produces several hormones, such as thymosin which is involved in immune tolerance, self-tolerance, and the maturation and function of T cells, which are critical for adaptive immunity (Dominari et al. 2020). While the thymus does not play as central a role in endocrine function as other glands like the pituitary or thyroid, it is still involved in some hormonal manifestations of endocrine diseases (Zdrojewicz et al. 2016).

11.3.1 Myasthenia Gravis

Myasthenia gravis is an autoimmune neuromuscular disorder that affects the neuromuscular junction, leading to muscle weakness and fatigue. While the thymus is not the direct cause of myasthenia gravis, it is often involved in the disease. Many individuals with myasthenia gravis have thymus abnormalities, such as thymic hyperplasia or thymic tumors. Surgical removal of the thymus (thymectomy) can be a treatment option for some individuals with myasthenia gravis, potentially leading to symptom improvement (Dominari et al. 2020).

11.3.2 Thymic Carcinomas and Thymomas

Thymic carcinomas and thymomas are tumors that originate in the thymus. These tumors can sometimes produce hormones that may lead to paraneoplastic syndromes, which are rare disorders that occur when cancerous tumors produce hormones or other substances that affect various organ systems. For example, thymic carcinomas can produce adrenocorticotropic hormone, which, in turn, increases cortisol levels and results in Cushing's syndrome. Treatment for thymic tumors and other thymic diseases often involves a combination of medical therapies, surgery, and supportive care, depending on the specific condition and its severity (Scorsetti et al. 2016).

11.4 Pituitary Gland and Hypothalamus

The pituitary gland and the hypothalamus, sometimes called "master glands," together regulate almost the entire physiology. The pituitary gland is a small, pea-sized gland encased in a region called the sella turcica at the base of the brain and is connected to the hypothalamus by the infundibulum or pituitary stalk. It regulates the function of the adrenal gland, ovaries, and testes and modulates vital physiological functions like metabolism, blood pressure, reproduction, growth, and development (Rongioletti and Smoller 2010).

The pituitary gland and hypothalamus closely coordinate to secrete, store, and release hormones. The pituitary gland has two parts with distinct functions: the anterior/front lobe (adenohypophysis) and posterior/back lobe (neurohypophysis). The anterior pituitary, developed during early gestation as Rathke's pouch, is capable (Lechan et al. 2019) of synthesizing and secreting its own hormones via specialized cell types, doing so under the influence of "releasing" hormones secreted by the hypothalamus. Meanwhile, the hormones of the posterior pituitary are synthesized in the cell bodies of the hypothalamus and stored in the posterior pituitary for release (Daughaday 1979; Davidovici et al. 2008; Rongioletti and Smoller 2010). The hormonal secretions of both lobes are under the control of the central nervous system: The posterior pituitary is directly regulated by a neurosecretory pathway via the anterior hypothalamus, and the anterior pituitary is controlled by the neuroendocrine system via the hypothalamic-hypophyseal portal system. The feedback loops between these organs and systems (positive and negative) are critical in the regulation of hypothalamic and anterior pituitary tropic hormones, including adrenocorticotropic hormone, thyrotropin-releasing hormone or thyrotropin, thyroid-stimulating hormone (TSH), corticotropin-releasing hormone, follicle-stimulating hormone (FSH), and luteinizing hormone (LH). Aberrant regulation of the pituitary leads to the manifestation of various diseases.

11.4.1 Kallmann Syndrome and Isolated/Idiopathic Hypogonadotropic Hypogonadism

Gonadotropin-releasing hormone (GnRH), a decapeptide (pGlu-His-Trp-Ser-Tyr--Gly-Leu-Arg-Pro-Gly·NH_2), is a critical regulator of human reproduction (Schally et al. 1971b). GnRH can stimulate the gonadotropins LH (luteinizing hormone) and FSH (follicle-stimulating hormone) from the pituitary gland (Schally et al. 1971a). Previously, this hormone was known as LH-releasing hormone (LH-RH) but has more recently been referred to as GnRH, reflecting its role in the regulation of both LH and FSH (Schally 2000).

Kallmann syndrome is a form of hypogonadotropic hypogonadism (decreased activity of gonads) caused by a deficiency of LH and FSH from the pituitary gland or hypothalamus, resulting in a lack of sexual development. In some cases, Kallmann syndrome is associated with isolated/idiopathic hypogonadotropic hypogonadism (IHH), which is due to a congenital defect (Table 11.1). In cases associated with IHH, Kallmann syndrome arises either from (1) impaired migration and homing of GnRH-secreting neurons or from (2) the absence of activation of GnRH-secreting neurons. In the first situation, IHH may be accompanied by anosmia, along with other Kallmann syndrome symptoms; however, with the lack of activation of GnRH-secreting neurons, IHH remains the only clinical manifestation. Males with Kallmann syndrome suffer from micropenis and undescended testes (cryptorchidism) and, during the onset of puberty, mostly lack the development of secondary sexual characteristics. In females, menstruation cycles and breast development are impaired. In both sexes, growth spurts are affected. Other associated symptoms are renal agenesis (missing kidney), cleft palate (hard and soft palate) and lipsor abnormality of the palate, hyposmia (impaired sense of smell, decreased smell sensation) or anosmia (lost smell), hearing impairment, and abnormality of eye and eye movement (Raivio et al. 2007).

11.4.2 Hypopituitarism

Hypopituitarism is a condition in which the pituitary gland does not produce sufficient hormones. This can lead to a range of hormonal imbalances, including deficiencies in growth hormone, TSH, adrenocorticotropic hormone, LH, FSH, and prolactin. Symptoms may include fatigue, weight loss or gain, sexual dysfunction, menstrual irregularities, low blood pressure, and problems with growth in children (Kim 2015).

11.4.3 Hyperpituitarism

Hyperpituitarism is the overproduction of hormones by the pituitary gland. This can result from a pituitary tumor or other causes. Common manifestations of hyperpituitarism include acromegaly (excess growth hormone), Cushing's disease (excess adrenocorticotropic hormone), and prolactinomas (excess prolactin). Symptoms can include excessive growth, weight gain, muscle weakness, and inappropriate lactation (Kopczak et al. 2017).

11.4.4 Pituitary Tumors

Pituitary tumors can be either benign (non-cancerous) or malignant (cancerous). Depending on the type and size of the tumor, they can compress surrounding brain structures and disrupt hormonal balance. Larger tumors can lead to visual

Table 11.1 Genetic mutations associated with Kallmann syndrome:

Gene	Abbreviation	Location	NCBI Gene ID	Function	Disease manifestation
CHD7	Chromodomain helicase DNA binding protein 7	8q12.2	55,636	Expressed in undifferentiated neuroepithelium and in mesenchyme of neural crest origin	Balasubramanian et al. (2014)
FGF8	Fibroblast growth factor 8	10q24	2253	Member of the fibroblast growth factor (FGF) family involved in organogenesis	Behre et al. (2010)
FGFR1	Fibroblast growth factor 1	8p11.2–p11.1	2260	Binds both acidic and basic fibroblast receptor growth factors	Trarbach et al. (2010)
GPR54 (*KISS1R*)	G protein-coupled receptor 54 19p13.3	19p13.3	84,634	Receptor for Kiss-1: stimulator of GnRH secretion	Moalla et al. (2019)
GNRH1	Gonadotropin-releasing hormone 1	8q21–p11.2	2796	Ligand of GnRH receptor	Behre et al. (2010)
GNRHR	Gonadotropin-releasing hormone receptor	4q21.2	2798	Receptor for the gonadotropin-releasing hormone	Beranova et al. (2001)
KAL1 (*ANOS1*)	Kallmann syndrome 1 sequence (anosmin 1)	Xp22.32	3730	Possible function in neural cell adhesion and axonal migration FGFR1	Behre et al. (2010)
KISS1	KiSS-1 metastasis suppressor (metastin)	1q32	3814	Ligand of GPR54: stimulator of GnRH secretion	Silveira et al. (2013)
PROK2	Prokineticin 2	3p13	60,675	Chemoattractant for neuronal precursor cells in the olfactory bulb	Dodé and Rondard (2013)
PROKR2	Prokineticin receptor 2	20p12.3	128,674	G protein–coupled receptor for prokineticins	Dodé and Rondard (2013)
TAC3	Tachykinin precursor 3 (neurokinin B)	12q13–q21	6866	Influences GnRH secretion	Gianetti et al. (2010)
TACR3	Tachykinin 3 (neurokinin B)	4q25	6870	Receptor for tachykinin 3 receptor	Gianetti et al. (2010)

disturbances, headaches, and hormonal imbalances and even affect the function of other nearby structures in the brain (Lake et al. 2013).

11.4.5 Diabetes Insipidus

Diabetes insipidus is a condition characterized by excessive thirst and excretion of large amounts of diluted urine. It can result from inadequate secretion of antidiuretic hormone by the pituitary gland (Christ-Crain et al. 2019).

11.4.6 Prolactinoma

A prolactinoma is a benign tumor of the pituitary gland that leads to an overproduction of prolactin, the hormone responsible for milk production in breastfeeding women. In addition to in appropriate lactation, it can cause menstrual irregularities and infertility in women and reduced libido and erectile dysfunction in men (Schlechte 2003).

11.4.7 Pituitary Apoplexy

Pituitary apoplexy is a rare but serious condition characterized by sudden hemorrhage or infarction (loss of blood supply) in the pituitary gland. It can lead to symptoms such as severe headaches, visual disturbances, and hormonal imbalances (Briet et al. 2015).

11.4.8 Therapy for Pituitary Gland Disorders

Treatment for pituitary gland disorders varies depending on the specific condition and may include medication, surgery, or radiation therapy. Some pituitary disorders may require lifelong hormone replacement therapy to restore hormonal balance. For example, individuals with hypopituitarism may need thyroid hormone, cortisol, or growth hormone replacement, among others. Timely diagnosis and appropriate treatment are essential for managing these endocrine diseases and mitigating their hormonal manifestations. The pituitary gland's central role in regulating other endocrine glands makes its proper function crucial for overall hormonal balance in the body (Orrego and Barkan 2000).

11.5 Thyroid Gland

In conjunction with the hypothalamus and pituitary gland, the thyroid gland completes the hypothalamus–pituitary–thyroid (HPT) axis. The coordinated functioning of the HPT axis is vital for the maintenance of physiological events, including

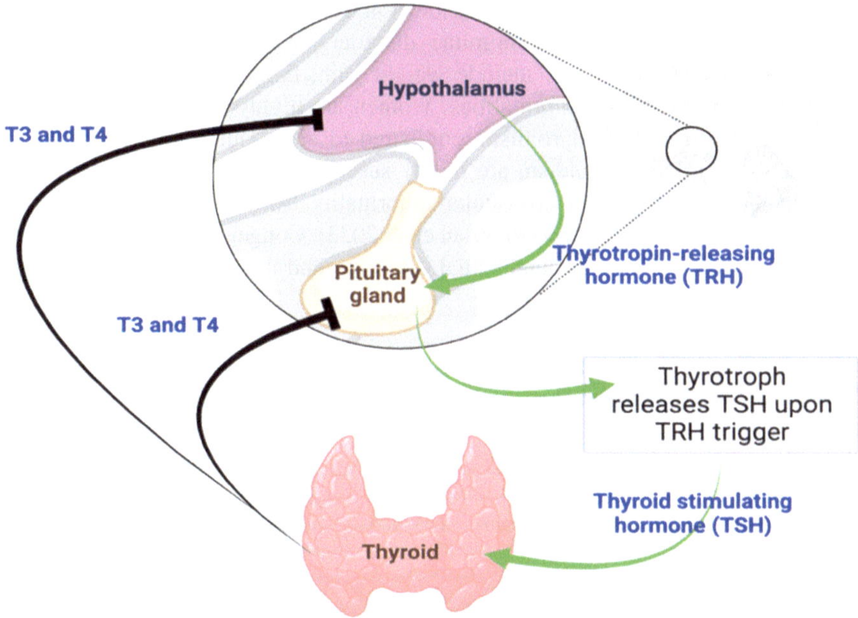

Fig. 11.3 Regulation of the hypothalamus–pituitary–thyroid axis. The hypothalamic paraventricular nucleus (PVN) in the hypothalamus releases thyrotropin-releasing hormone (TRH), which acts on pituitary thyrotropes to stimulate thyroid-stimulating hormone (TSH) synthesis. TSH acts on thyrocytes to stimulate all steps of thyroid hormone synthesis. The thyroid hormones T4 and T3 act on PVN neurons and on the thyrotropes to inhibit TRH and TSH synthesis and release. This feedback regulation is the main regulatory mechanism of thyroid function. The pituitary thyrotropes are also regulated by local factors through autocrine and paracrine pathways. Black arrows represent inhibition, and green arrows represent stimulation

metabolism, energy homeostasis, and reproduction. The thyroid gland regulates hormonal release through a feedback mechanism. The major hormones of the HPT axis are thyroid hormones triiodothyronine (T3) and tetra-iodothyronine (T4) as well as thyroid-stimulating hormone (TSH). The thyroid gland takes up dietary iodine and utilizes it to produce T3 and T4, which are released from the thyroid gland under the feedback regulation of TSH. Most of the T3 is produced from T4 by the action of the cellular 5′-deiodinase enzyme (DIO1 and DIO2) (Williams et al. 2008). The secretion of TSH through the pituitary is regulated by thyrotropin-releasing hormone secreted from the paraventricular neuron of the hypothalamus. TSH, in turn, stimulates the thyroid gland to produce and release thyroid hormones T3 and T4. Finally, the production of thyrotropin-releasing hormone is negatively regulated by thyroid hormones in a feedback loop (Fig. 11.3).

11.5.1 Hypothyroidism

Hypothyroidism manifests as reduced production of T3 and T4 and primarily occurs because of thyroid gland failure or insufficient stimulation of the thyroid by the hypothalamus or pituitary gland (Gaitonde et al. 2012). Moreover, hypothyroidism can be caused by certain autoimmune disorders, e.g., Hashimoto thyroiditis. Symptoms of hypothyroidism include weight gain, fatigue, abnormal cognition, infertility, and menstrual abnormalities. Women are highly susceptible to thyroid disorders, including hypothyroidism, compared to men (Dunn and Turner 2016). Individuals with hypothyroidism are highly susceptible to other health complications, including diabetes, cardiovascular abnormality, liver, kidney malfunctioning, and reproductive abnormalities (Rehman et al. 2023). Congenital hypothyroidism is one of the leading causes of intellectual disability and neurodevelopmental disorders in children (LaFranchi 2011).

11.5.2 Hyperthyroidism

Hyperthyroidism is manifested by the excess production of thyroid hormones and is most commonly due to the autoimmune condition Graves' disease. Other causes include thyroiditis, toxic adenomas, and toxic multinodular goiter (Reid and Wheeler 2005).

11.5.3 Treatment and Management

Hypothyroidism and hyperthyroidism are most commonly diagnosed via blood levels of hormones of the thyroid axis (T3, T4 and TSH) and thyroid auto antibodies (Reid and Wheeler 2005). Once confirmed, thyroid disorders are usually treated with medication. Thyroid hormone replacement therapy using T4 is the most reliable therapy used for hypothyroidism. In recent years, T3 and T4 combinatorial therapy has been tested for therapeutic potential in hypothyroidism but has not been proven to have benefits over traditional T4 monotherapy (Biondi and Wartofsky 2012), likely owing to the short half-life of T3. Toxic adenomas and toxic multinodular goiter associated with hyperthyroidism can be treated with the administration of radioactive iodine, anti-thyroid drugs (methimazole and propylthiouracil), or surgical thyroidectomy in severe cases (Reid and Wheeler 2005).

Lifestyle and dietary interventions can also be considered for thyroid disorders. A low metabolic rate in patients with hypothyroidism is one of the most common phenomena underlying the different symptoms, including fatigue, tiredness, poor cognition, and poor sleep quality. It is now proven that lifestyle changes, including exercise and a regulated diet, exert beneficial effects on hypothyroidism patients (Wu et al. 2021).

11.6 Adrenal Glands

The adrenal glands, also known as suprarenal glands, are located above the kidney (Fig. 11.1). Each gland has an outer cortex and inner medulla with distinct structures and functions, acting as two glands in one, much like the pituitary. The adrenal

Table 11.2 Major manifestations of adrenal disorders:

Disorder	Cause	Effect	Reference
Addison's disease (primary adrenal insufficiency)	Insufficient hormone production	Weakness, fatigue, loss of appetite, mood disorder and vertigo	Erichsen et al. (2009)
Adrenal nodules (asymptomatic or incidental)	Unknown	No major complications unless they remain unidentified, leading to excess hormone production or cancer	Glazer et al. (2021)
Pheochromocytoma	Cancer	High blood pressure, headache, flushing, heart palpitations	Tsirlin et al. (2014)
Primary adrenal Cushing's syndrome	Overproduction of cortisol	Unexplained weight gain, loss of fat from legs and arms, high blood pressure, elevated blood glucose levels, reddish-purple streaks or stretch marks on the skin	Buescher et al. (1992), Lake et al. (2013), Yu (2014); Lacroix et al. (2015)
Primary hyperaldosteronism (Conn's syndrome or primary aldosteronism)	Overproduction of steroid hormone aldosterone	High blood pressure and potassium deficiency	Rongioletti and Smoller (2010), Yu (2014)
Adrenocortical carcinoma	A very rare cancer	None, unless the adrenal glands are producing excessive hormones	Asa and Mete (2018), Edlow and Norwitz (2019a), Asa et al. (2023)
Isolated adrenal metastases	Metastasized from other parts of body to adrenal gland	Cancer	Paul et al. (2000)

cortex secretes adrenocorticosteroids, and the adrenal medulla secretes catecholamines (Stanfield 2013). These hormones help regulate metabolism, blood pressure, stress response, and the immune system, along with other functions. The major manifestations of adrenal gland disorders are outlined in Table 11.2.

11.7 Pancreas

The pancreas is a glandular organ devoid of bones, cartilage, or air-filled pockets; its name is aptly derived from the Greek phrase *pan kreas*, which means "all flesh." It is both an exocrine and an endocrine organ and is positioned in the upper abdominal cavity of the retroperitoneal region (Beger et al. 2023). In its exocrine function, the pancreas can secrete generous amounts of digestive enzymes synchronized by hormones and neuronal stimuli. The endocrine unit of the pancreas is composed of islets of Langerhans, which generate and release hormones that regulate blood sugar and additional glandular secretions (Karpińska and Czauderna 2022; Beger et al. 2023). These pancreatic islets mainly comprise four cell types, namely, α, β, δ, γ, and ε, which are unevenly dispersed within the pancreas and secrete different hormonal products (Table 11.3). The pancreatic islets are abundant in portal microcirculation in the order of cellular perfusion, thereby allowing the hormone-to-cell type interaction necessary for the regulation of subsequent hormones. This assists the regulation of α cells by β cells and the control of δ cell secretion by α cells (Dupont et al. 2022).

The primary endocrine activity of the pancreas is the management of glucose hemostasis and its exocrine function, contributing to 1–2% of the gland's function (Dupont et al. 2022; Bishop and Simo 2023). A variety of regulatory mechanisms including blood glucose levels and feedback lead to the secretion of insulin or glucagon. The functions of the other major hormones are described in Table 11.3.

In addition to the major hormones, vasoactive intestinal peptide, found in the gastrointestinal tract, stimulates insulin release and inhibits gastric secretions. It

Table 11.3 Endocrine cells of pancreatic islets, their distribution, and hormones released:

Cell type	Distribution in the human pancreas	Major hormone secreted	Function of hormone	Normal blood level of hormones
α	30–50%	Glucagon	Increases blood glucose level through glycogenolysis, and gluconeogenesis. Promotes oxidation of fat	183–367 pmol/L
β	50–60%	Insulin	Promotes glucose transport into cells. Inhibits glycogenolysis, breaks down fatty acids. Stimulates protein synthesis	174 pmol/L (fasting)
		Amylin	Suppresses glucagon secretion Limits food consumption by stimulating the satiety center of the brain. Delays gastric emptying. Decreases blood glucose	4 pmol/L (fasting) to 25 pmol/L (PP)
δ	~5%	Somatostatin	Inhibits secretion of hormones including insulin, growth hormone, glucagon, gastrin, thyroid-stimulating hormones, etc.	4.26–9.37 pmol/L
γ	<5%	Pancreatic polypeptide	Modulates digestion by inhibition of gastric emptying and biliary secretion	<837 pmol/L at 20–29 years and can increase up to 20 pg/mL every decade
ε	<1%	Ghrelin	Inhibits secretion of insulin Stimulates appetite, fat deposition, Releases growth hormone	166.0 ± 10.1 pmol/L

PP postprandial (Adopted from In't Veld and Marichal 2010; Da Silva Xavier 2018; Campbell and Newgard 2021; Bishop and Simo 2023).

also causes vasodilation and bronchodilation in the respiratory tract. Also, many other hormones, like pancreastatin, thyrotropin-releasing hormone, glicentin, gastrin-releasing factor, cholecystokinin, etc., have been described in islet cells (Dupont et al. 2022).

11.7.1 Clinical Significance of Insulin

Insulin is a peptide hormone made of 52 amino acids and is distributed as two peptide chains, each comprising 21 and 3 amino acid residues connected by a disulfide bond (Daghlas and Mohiuddin 2022). Insulin binds to insulin receptors on target cell membranes, leading to activation of phosphoinositide 3-kinase (PI3K), protein kinase B (Akt), and mitogen-activated protein kinase (MAPK) pathways (Zhang and Liu 2002; He et al. 2021). The major role of insulin is in glucose metabolism, which is carried out by insulin-mediated glucose uptake and glucose-stimulated insulin secretion. These pathways enable the uptake of glucose from skeletal muscle and adipose tissue and suppress glucose generation by hepatic cells. PI3K activation ultimately leads to extracellular expression of the GLUT-4 transporter on skeletal muscle, resulting in increased glucose entry for metabolism into glycogen (Vargas and Carrillo Sepulveda 2018). It has also been reported that insulin can inhibit NF-κB and the expression of pro-inflammatory cytokines, which helps reduce the inflammatory response (Chen et al. 2014; Sun et al. 2014). In the brain, insulin can suppress appetite and regulate cognitive function including memory (Scherer et al. 2021).

11.7.2 Diabetes Mellitus

Diabetes mellitus is a chronic medical condition characterized by high blood glucose levels (hyperglycemia) due to insufficient insulin production, reduced insulin function, or both. The condition that leads to insufficient insulin production is termed type 1 diabetes (T1DM) and is characterized by the destruction of β cells, generally with an autoimmune cause. On the other hand, type 2 diabetes (T2DM) is an imbalance of available insulin and insulin sensitivity, leading to impaired metabolism of glucose. T2DM arises from multivariable causes that develop due to obesity and aging. In addition, a type of diabetes diagnosed during pregnancy is termed gestational diabetes. The incidence of diabetes, T2DM, in particular, has doubled in the last two decades and remains a major health problem of this century (Soria 1998; Poretsky 2010).

11.7.3 Hyperinsulinemia

Hyperinsulinemia is a condition where insulin in the blood is higher than the normal level. This condition arises due to various factors including insulin resistance and excessive insulin production. Such a condition can result in hypoglycemia, cardiovascular disorders, inflammation, hyperlipidemia, and an increased risk of kidney damage (Thomas et al. 2019; Vaidya et al. 2023).

11.7.4 Glucagon

Glucagon is a 29-amino-acid peptide hormone secreted in response to hypoglycemia, exercise, and protein-rich meals (Gromada et al. 2007). It activates glucagon receptors, mostly expressed in the liver, pancreas, kidney, and gastrointestinal tract (Svoboda et al. 1994). Glucagon supports hepatic glucose production, the breakdown of fatty acids and amino acids, and appetite suppression and regulates heart rate. Dysregulated glucagon secretion leads to T1DM, T2DM, obesity, and hepatic steatosis (Svoboda et al. 1994).

11.8 Placenta and Pregnancy-Associated Endocrine Diseases

The placenta is the largest endocrine organ and is a temporary embryonic and, later, fetal organ necessary for both maternal and fetal health. Placental formation begins soon after implantation of a blastocyst and plays distinct yet coordinated roles such as providing immunity and facilitating nutrient, gas, and waste exchange between maternal and fetal circulation, along with acting as crucial endocrine organs during the entire gestation (Herrick and Bordoni 2023; Huppertz and Schleußner 2023). During pregnancy, the placental hormones influence the entire maternal endocrine system. The hormones secreted by the placenta can regulate the placental function locally in a paracrine and autocrine manner along with distant endocrine regulation. Placental hormones are either peptide hormones or steroid hormones. The major steroid hormones are glucocorticoids, progesterone, and estradiol. The peptide hormones include human chorionic gonadotropin, human placental lactogen), human placental growth hormone, leptin, corticotrophin-releasing hormone, adiponectin, activin A, and inhibin A (Stanfield 2013; Huppertz and Schleußner 2023). Owing to drastic hormonal profiles and fluctuations during pregnancy, endocrine homeostasis is critical for a successful pregnancy. While a complete description of the placenta and fetal development is beyond the scope of the book, here is an overview of endocrine manifestations of placenta-related diseases, which have become a major concern due to their rapidly increasing prevalence (Zhou and Acharya 2022).

11.8.1 Gestational Diabetes Mellitus.

With a global prevalence of 14.7% among pregnant individuals, gestational diabetes mellitus poses serious health consequences to both maternal and neonatal health (Saeedi et al. 2021; Mazumder et al. 2022). Increased insulin resistance and β-cell defects are underlying metabolic abnormalities for gestational diabetes mellitus, especially in women with preexisting obesity and high blood pressure (McIntyre et al. 2019).

Women with gestational diabetes mellitus are at higher risk of preeclampsia or eclampsia and hypertension or high blood pressure during gestation. These patients will also be at risk for T2DM, high blood pressure, and other cardiovascular diseases. Following birth, infants might suffer from hypoglycemia, macrosomia, breathing issues, and T2DM or insulin resistance (Najafi et al. 2019; Saeedi et al. 2021). The treatment course is mostly behavioral adjustments with a well-balanced diet, monitoring blood glucose level, exercise, and if needed, insulin injections (Stanfield 2013).

11.8.2 Pregnancy-Associated Endocrine Disorders

Pregnancy-related endocrine disorders are becoming more prevalent and can have adverse effects on the mother and fetus's health, hence affecting the overall outcome of successful pregnancy. Placental dysfunction is a major root cause of pregnancy-associated endocrine disorders (Zhou and Acharya 2022).

11.8.2.1 Preterm Labor and Birth

Recent studies have shown an association between gestational hormones and labor timing. Male fetuses with spontaneous preterm birth were observed with high estriol, progesterone, corticotropin-releasing hormone, total triiodothyronine, and free thyroxine. Similarly, female fetuses with high testosterone have an increased likelihood of intrauterine sexual dimorphism along with preterm birth (Cathey et al. 2021). Further, lower concentrations of adiponectin, leptin, insulin, and ghrelin were found in preterm infants (Han et al. 2021).

11.8.2.2 Pituitary Tumor in Pregnancy

Pituitary tumors or adenomas are common during pregnancy and are mostly caused by over- or under-secretion of pituitary hormones. Tumor size and location on and around the pituitary stalk impact the amount of hormonal secretion, and pregnancy itself alters the hormonal secretion and size of a pituitary tumor (Huang and Molitch 2019). These pituitary adenomas can affect fertility and pregnancy outcomes. The entire pituitary–pituitary tumor–pregnancy axis is very interesting but beyond the scope of this book (Table 11.4).

11.8.2.3 Overt Thyroid Dysfunction

Overt thyroid dysfunction, also called hyperthyroidism, poses an increased risk of adverse pregnancy outcomes with detrimental effects on fetal neurocognitive development (Table 11.5). It may cause maternal thyrotoxicosis (excess production of and exposure to thyroid hormone from any cause) and maternal hypothyroidism (inadequate thyroid hormone production).

Table 11.4 Other pregnancy-associated endocrine disorders:

Disorder	Etiology	Cause	References
Calcium disorder and parathyroid hormone	Primary hyperparathyroidism	Due to hypercalcemia and elevated parathyroid hormone	Bilezikian et al. (2014), Edlow and Norwitz (2019a), Calabria and Coggins (2024)
	Hypoparathyroidism	Due to incidental resection of the parathyroid glands at the time of thyroidectomy and hypocalcemia	Edlow and Norwitz (2019a)
Adrenal diseases	Adrenal insufficiency (Addison disease)	Caused by autoimmune destruction of the adrenal glands	Goldberg (1972), Edlow and Norwitz (2019a)
	Primary adrenal insufficiency	Due to deficiency of both cortisol and mineralocorticoid (mineralocorticoid production is typically preserved in secondary adrenal insufficiency)	Edlow and Norwitz (2019a)
	Secondary adrenal insufficiency	Caused by HPA-axis abnormalities that lead to ACTH deficiency and adrenocortical atrophy	Swartz et al. (1980), Edlow and Norwitz (2019a)
	Cushing syndrome	Exposure to excessive circulating cortisol (commonly due to pregnancy adrenal adenomas); could be ACTH-dependent or ACTH-independent	Buescher et al. (1992), Edlow and Norwitz (2019a)
	Addisonian crisis	State of acute adrenal insufficiency, rare during pregnancy but common in postpartum	Drucker et al. (1984)
	Congenital adrenal hyperplasia	Caused by increased fetal adrenal androgen production due to genetic disorders of steroidogenesis	Pang et al. (1988), New et al. (2001)
Pheochromocytoma (catecholamine-secreting tumors)		Caused by chromaffin cells (90% in the adrenal glands, 10% bilateral, 10% malignant)	Harper et al. (1989), Mannelli and Bemporad (2002), Edlow and Norwitz (2019b)
Ovarian endocrine tumors		Unknown	Hess et al. (1988), Katz et al. (1993), Edlow and Norwitz (2019b)

ACTH adrenocorticotropic hormone

Table 11.5 Etiology of overt thyroid dysfunction:

Thyroid dysfunction	Etiology	Cause	Reference
Maternal thyrotoxicosis	Graves' disease	Caused by circulating thyroid-stimulating immunoglobulin G autoantibodies, which can cross the placenta, leading to fetal thyroid dysfunction	Luton et al. (2005), Kempers et al. (2007), Laurberg et al. (2009)
	Thyroiditis	Due to hyperthyroidism and presence of large, palpable thyroid gland	Lazarus et al. (1997), Marqusee et al. (1997), Lazarus (2002), Stagnaro-Green (2012)
	Toxic multinodular goiter	Due to autonomously functioning thyrocytes, which can function and produce thyroid hormones without thyrotropin	Thomas and Croom (1987), Mandel et al. (1993), Edlow and Norwitz (2019b)
	Gestational trophoblastic neoplasia	Probably due to elevated human chorionic gonadotropin	Costa (2016)
	Gestational hyperthyroidism	Frequently associated with elevated human chorionic gonadotropin	Yu (2014), Huppertz and Schleußner (2023)
Thyrotoxic crisis or thyroid storm	–	Due to severe acute aggressive signs and symptoms of hyperthyroidism	Yu (2014), Huppertz and Schleußner (2023)
Maternal hypothyroidism: Primary hypothyroidism	Iodine deficiency	Low iodine	Huppertz and Schleußner (2023)
	Hashimoto thyroiditis (chronic autoimmune thyroiditis)	Caused by circulating antithyroglobulin and/or antimicrosomal autoantibodies	Huppertz and Schleußner (2023)
	Hypothyroid phase (silent/postpartum thyroiditis)	Self-limited subacute disorder involving hyperthyroidism followed by hypothyroidism and then recovery state	Lake et al. (2013), Huppertz and Schleußner (2023)

(continued)

Table 11.5 (continued)

Thyroid dysfunction	Etiology	Cause	Reference
Maternal hypothyroidism: Secondary hypothyroidism	Pituitary adenoma	Mostly are non-cancerous tumors in anterior pituitary	Orrego and Barkan (2000), Lake et al. (2013)
	Pituitary necrosis/hemorrhage	Due to blockage in blood flow or hemorrhage	Edlow and Norwitz (2019a)
	Lymphocytic hypophysitis	Neuroendocrine disorder due to autoimmune inflammation of pituitary gland	Rivera (2006), Christ-Crain et al. (2019), Edlow and Norwitz (2019a)
	Central nervous system sarcoidosis	Sarcoidosis of nervous system.	Pentland et al. (1985)
	Prior hypophysectomy	Hypophysis (pituitary gland)	Husami et al. (1977), Huang and Molitch (2019)
	Suprasellar/parasellar	–	Husami et al. (1977), Rodríguez-Porcel et al. (2013)
	Traumatic injury to pituitary/ hypothalamus	–	Kopczak et al. (2017), Christ-Crain et al. (2019)
Goiter	Endemic goiter	Due to iodine deficiency, enlarged thyroid glands	Schleußner (2023)
	Sporadic goiter (diffuse nontoxic goiter; multinodular goiter)	Enlarged thyroid glands with multiple areas of nodularity	Solis-Cohen and Steinbach (1939), Krassas et al. (2015)
	Diffuse toxic goiter (Graves' disease)	Autoimmune, enlarged thyroid glands	Botushanov et al. (2021)
	Thyroiditis (chronic autoimmune or Hashimoto disease; subacute; silent/ postpartum; suppurative)	Postpartum	Moleti et al. (2020)
	Diffuse malignant disease (lymphoma, anaplastic carcinoma)	–	Schlumberger et al. (2011), Edlow and Norwitz (2019a)
	Infiltrative diseases (Riedel thyroiditis, sarcoidosis, amyloidosis	–	Edlow and Norwitz (2019a)
Thyroid nodules	Benign	Benign abnormal growth in thyroid	Edlow and Norwitz (2019a)
	Malignant	Malignant abnormal growth in thyroid	Edlow and Norwitz (2019a)

Table 11.6 Etiology and disease manifestation of gonads:

Gonads	Endocrine manifestation of gonads	Etiology	Effects	Reference
Ovaries	Polycystic ovary syndrome	Could be (1) hereditary; (2) deficiencies in the cytochrome P450 pathway; (3) hyperinsulinemia	Female infertility, obesity, endometrial cancer, diabetes mellitus, impaired glucose tolerance, metabolic syndrome, cardiovascular risk, non-alcoholic fatty liver disease/non-alcoholic steatohepatitis	Behre et al. (2010), Unuane et al. (2011), Norman and Teede (2018), Huppertz and Schleußner (2023)
	Premature ovarian failure Primary ovarian insufficiency	Decline in normal ovarian function	Female infertility, irregular menstrual cycle, hot flashes, night sweats, etc.	Unuane et al. (2011)
Testes	Hypogonadism	Elevated gonadotropin levels with a low testosterone level	Affected fertility	Behre et al. (2010)
	Diabetes mellitus		Affected fertility	Behre et al. (2010)

11.9 Gonads

The major functions of gonads are regulation of steroid hormone production and gametogenesis. These steroid hormones are involved in sexual differentiation and maturation, fetal development, and growth. Ovaries (female gonads), located in the pelvis on each side of the uterus, are responsible for the production of ova and female hormones, estrogen, and progesterone (Table 11.6).

11.10 Secondary Endocrine Organs

Secondary endocrine organs are those that secrete hormones as a secondary function, along with their primary functions. The hormonal manifestations of conditions affecting secondary endocrine organs are listed Table 11.7.

Table 11.7 Endocrine manifestations of secondary endocrine organ dysfunction:

Secondary endocrine organs	Endocrine function	Disease manifestation	References
Heart	Atrial natriuretic peptide produced in stress response, reduces sodium reabsorption	No known direct manifestation; regulates hypertension, cardiovascular diseases, other heart diseases, renal failure, cirrhosis, hyper–/hypothyroidism	Vesely (2001), Lemaitre et al. (2022)
Thymus	Produces thymosin; involved in development and differentiation of T-lymphocytes	No known direct manifestation; enhances sex steroids, regulates various pituitary hormones like PRL, GH, LH, and FSH (Daneva et al. 1995), affects hypothalamus-pituitary axis (Daneva et al. 1995)	Daneva et al. (1995), Goldstein and Badamchian (2004)
Gastrointestinal tract	Stomach: gastrin produced in response to distention of stomach, regulates gastric acid secretion and proliferation of epithelial cells	Overstimulation leads to cancer known as gastrinoma, may affect gastrointestinal tract, pancreas, and liver	Koh and Chen (2000), Dockray (2004)
	Small intestine: secretin, cholecystokinin), glucose-dependent insulinotropic peptide (helps in digestion)	Endocrine manifestation still under investigation; dysregulation leads to various tumors, insulinoma, abnormal digestion, gut motility, appetite, and metabolism	Ahlman and Nilsson (2001), May and Kaestner (2010), Helander and Fändriks (2012)
Liver	Produces insulin-like growth factors including somatomedin (regulates growth), angiotensinogen (precursor to angiotensin and regulates blood pressure), thrombopoietin (stimulates thrombocyte production), and hepcidin (regulates iron hemostasis)	Endocrine manifestation still under investigation; liver damage can cause alteration in hypo- and hyperthyroidism, can regulate metabolic syndrome, induces hemostatic abnormalities, ovarian dysfunction, gonadal dysfunction, etc.	Burra (2013), Watt et al. (2019), Rhyu and Yu (2021)
Skin	Produces inactive form of 1,25-dihydroxyvitamin D_3, target of various hormones, and synthesizes various hormones	1. Low level of vitamin D3 is associated with increased risk if cancer, asthma, and multiple sclerosis. 2. Rickets in children; 3. Osteomalacia in adults	Davidovici et al. (2008), Rongioletti and Smoller (2010), Nagy and Malcomson (2022), Zouboulis (2009)

Kidney	Produces erythropoietin, which regulates blood calcium level and reabsorption of sodium and water Produces renin, responsible for regulation of blood pressure, ion channels, renal function, water retention Overall involved in vitamin D activation, metabolism of prostaglandins, insulin, gastric, parathyroid hormone, gluconeogenesis, thrombopoietin generation, urodilatin	Abnormal hemostasis	DeLuca (1975), Acharya and Olivero (2018)
Skeleton	Fibroblast growth factor 23 (FGF23)-regulates kidney function and vitamin D metabolism Osteocalcin stimulates pancreatic β cells and enhances insulin production, regulates glucose metabolism	Overexpression of FGF23:X-linked hypophosphatemic rickets and osteomalacia, autosomal dominant or recessive hypophosphatemic rickets and osteomalacia, tumor-induced rickets and osteomalacia hypophosphatemic rickets and osteomalacia associated with fibrous dysplasia with or without McCune–Albright syndrome. Impaired FGF23:Familial hyperphosphatemic tumoral calcinosis	DeLuca (1975), Acharya and Olivero (2018)
Adipose tissue	Produces and secretes several hormones that regulate lipid metabolism and storage, like leptin, adiponectin	Impaired adipo-renal axis, obesity	Zhu and Scherer (2018)

11.11 Additional Endocrine Cancers and the Role of Hormones in Cancer

Many cancers affecting the endocrine system are benign and are mostly considered inherited disorders. Others are malignant, leading to the invasion of other tissues and forming new tumors (Das and Dasari 2021; Rindi et al. 2022). It is well known that tumors of endocrine organs can themselves secrete hormones, disturbing the hormonal balance and causing serious health effects. Thyroid cancer is the seventh most diagnosed type of cancer. It is usually evident on physical examination or is visible and is an exception to other forms of endocrine tumors. These tumors differ based on the individual endocrine organ, anatomic location, secretory function, and the frequency of hereditary diseases (Wells 2016; Das and Dasari 2021; Rindi et al. 2022). Effective treatment strategies including surgery, chemotherapy, molecular targeted therapies, and radiation therapies are widely available for endocrine cancers (Debela et al. 2021).

Malignant cancers of neuroendocrine cells are called neuroendocrine tumors (NET). These affect mostly the gastrointestinal tract, lung, pancreas, breast, prostate thymus, and skin. NETs can produce and secrete bioactive molecules, including hormones leading to symptoms like flushing and diarrhea. Patients with advanced stages of NETs have a median survival period of 33 months, during which they suffer complications of abnormal hormone levels leading to fatal complications (Raphael et al. 2017; Patel and Galoian 2018). The treatment of choice for NETs is surgical removal of the affected gland in the early stages of tumor development. Furthermore, long-acting somatostatins, radionucleotides, chemotherapy, and immunotherapy can be used alone or in combination (Tsoli et al. 2019). Endocrine cancers also include gonadocorticoid-influenced cancer. Some well-known types include endometrial, ovarian, and breast cancer in females as well as prostate cancer in males. These are significantly affected by changes in levels of sex hormones like androgens, estrogens, and progestogens (Key 1995; Belfiore and Perks 2013; Emons 2022).

Estrogen and progesterone are crucial in the growth of female sex organs and pose a risk factor for the development of cancer. Increased levels of estrogen or estrogen receptor signaling, when kept unchecked, due to decreased levels of progesterone, can lead to hyperplasia or adenocarcinoma in the endometrium. Recent studies have also shown that insulin resistance combined with elevated levels of estrogen also plays a role in the proliferation of endometrial cancer (Rodriguez et al. 2019; Emons 2022). An imbalance of estrogen and progesterone plays a major role in breast and ovarian cancer as well. The risk of both endometrial and breast cancers increases with an earlier menarche and later menopause when the hormonal imbalance is at its maximum in females. Childbearing and breastfeeding have been positively correlated with a decreased risk of breast cancer (Group 2004; Yang and Wang 2019). In the case of ovarian cancers, gonadotropin-releasing hormones stimulate the secretion of luteinizing hormones and follicle-stimulating hormones, which act in its proliferation (Emons 2022).

Similarly, androgens, responsible for the normal development and functioning of the prostate, have been linked as a cause of prostate cancer and its progression. In

1966, Charles Brenton Huggins was awarded the Nobel Prize for the demonstration of androgen dependence as a potential cause of prostate cancer and point of intervention. Androgens like testosterone and dihydrotestosterone activate androgen receptors, which in turn stimulate genes responsible for the growth of prostate cells (Hååg et al. 2005; Girling et al. 2007; Massie et al. 2011). More recent studies indicate the role of estrogen in favor of prostate cancer development (Nelles et al. 2011).

Although the relationship between gonadotropins and cancers is not new, a large number of reports have now suggested a similar association with cancer for other hormones, like insulin and thyroid-stimulating hormones (Belfiore and Perks 2013). Evidence indicates an increased risk of cancer in the breast, endometrium, liver, colon, kidney, and pancreas due to obesity and T2DM. Various mechanistic reports point to the oncogenic potential of insulin receptors in malignant cells (Poloz and Stambolic 2015; Chiefari et al. 2021). More studies are needed to understand better the role of various hormones in cancer development, proliferation, and metastasis.

11.12 Summary

The endocrine system, a complex network of glands, plays a pivotal role in maintaining homeostasis within the human body. These glands secrete hormones, chemical messengers that regulate various physiological processes. Beyond their physiological functions, hormones are valuable tools in the realm of medical diagnosis. From thyroid hormones unraveling metabolic mysteries to hCG confirming pregnancy, hormonal assessments contribute significantly to accurate diagnoses. This biochemical approach not only reveals diagnostic nuances but also paves the way for tailored therapeutic strategies, aligning with the evolving landscape of precision medicine. The delicate balance of hormonal signaling is disrupted, and it can lead to a myriad of disorders. This hormonal imbalance arises from various factors including lifestyle choices, stress, inadequate nutrition, aging, and environmental factors to hormonal fluctuations. The effects of hormonal imbalance are far-reaching, impacting physical and mental well-being. An imbalance in certain hormones can lead to female- or male-centric disorders, while others can lead to disruption in sleep patterns, weight fluctuations, and compromised immune function. Mental health can also be affected, leading to mood swings, anxiety, and depression. Managing hormonal disruptions necessitates a multifaceted approach. Lifestyle modifications encompassing a nutrient-dense diet and regular exercise contribute to hormonal equilibrium. Medical interventions may include hormone-regulating medications like hormone replacement therapy (HRT) may be considered for severe deficiencies. Dietary supplements, such as vitamin D or omega-3 fatty acids, support hormonal health. Avoiding endocrine disruptors mitigates external influences. Regular monitoring through blood tests ensures precise adjustments in treatment plans for optimal hormonal balance. We have addressed herein a comprehensive finding related to the functionality of different endocrine glands, the clinical significance of hormones, and related disease manifestations. Understanding different endocrine systems and associated pathophysiological conditions is a primary requisite in recent times due to the high emergence of endocrine-related diseases. This will further help in disease management.

References

Acharya V, Olivero J (2018) The kidney as an endocrine organ. Methodist Debakey Cardiovasc J 14:305–307

Ahlman H, Nilsson O (2001) The gut as the largest endocrine organ in the body. Ann Oncol 12:S63–S68

Arendt J, Aulinas A, Feingold KR, Anawalt B, Blackman MR, Boyce A, Chrousos G, Corpas E, de Herder WW, Dhatariya K, Dungan K, Hofland J, Kalra S, Kaltsas G, Kapoor N, Koch C, Kopp P, Korbonits M, Kovacs CS, Kuohung W, Laferrère B, Levy M, McGee EA, McLach WD (2000) Physiology of the pineal gland and melatonin. In: Feingold KR, Anawalt B, Blackman MR, et al. (eds) Endotext. South Dartmouth (MA)

Asa SL, Mete O (2018) Endocrine pathology: past, present and future. Pathology 50:111–118

Asa SL, Erickson LA, Rindi G (2023) The spectrum of endocrine pathology. Endocr Pathol. https://doi.org/10.1007/s12022-023-09758-0

Balasubramanian R, Choi JH, Francescatto L et al (2014) Functionally compromised CHD7 alleles in patients with isolated GnRH deficiency. Proc Natl Acad Sci USA 111:17953–17958

Beger HG, Buchler MW, Hruban RH et al (2023) The pancreas: an integrated textbook of basic science, medicine, and surgery, 4th edn. Wiley

Behre HM, Nieschlag E, Partsch C-J, Wieacker P, Simoni M (2010) Diseases of the hypothalamus and the pituitary gland. In: Nieschlag E, Behre HM, Nieschlag S (eds) Andrology. Male reproductive health and dysfunction, 3rd edn. Springer, Berlin, pp 169–192

Belfiore A, Perks CM (2013) Grand challenges in cancer endocrinology: endocrine related cancers, an expanding concept. Front Endocrinol (Lausanne) 4:141

Beranova M, Oliveira LM, Bédécarrats GY et al (2001) Prevalence, phenotypic spectrum, and modes of inheritance of gonadotropin-releasing hormone receptor mutations in idiopathic hypogonadotropic hypogonadism. J Clin Endocrinol Metab 86:1580–1588

Bilezikian JP, Marcus R, Levine MA et al (2014) The parathyroids: basic and clinical concepts. Academic Press

Biondi B, Wartofsky L (2012) Combination treatment with T4 and T3: toward personalized replacement therapy in hypothyroidism? J Clin Endocrinol Metab 97:2256–2271

Bishop MA, Simo K (2023) Pancreatectomy. In: StatPearls [Internet]. StatPearls Publishing

Botushanov NP, Botushanov AN, Botushanova AD (2021) Graves' disease and pregnancy. Graves' Dis 205

Briet C, Salenave S, Bonneville JF et al (2015) Pituitary apoplexy. Endocr Rev 36:622–645

Buescher MA, McClamrock HD, Adashi EY (1992) Cushing syndrome in pregnancy. Obstet Gynecol 79:130–137

Burra P (2013) Liver abnormalities and endocrine diseases. Best Pract Res Clin Gastroenterol 27:553–563

Calabria AC, Coggins SA (2024) Disorders of neonatal mineral metabolism and metabolic bone disease. Princ Neonatol:230–240

Campbell JE, Newgard CB (2021) Mechanisms controlling pancreatic islet cell function in insulin secretion. Nat Rev Mol Cell Biol 22:142–158

Cathey AL, Watkins DJ, Rosario ZY et al (2021) Gestational hormone concentrations are associated with timing of delivery in a fetal sex-dependent manner. Front Endocrinol (Lausanne) 12. https://doi.org/10.3389/fendo.2021.742145

Chen Q, Yu W, Shi J et al (2014) Insulin alleviates the inflammatory response and oxidative stress injury in cerebral tissues in septic rats. J Inflamm (United Kingdom) 11:1–7

Chiefari E, Mirabelli M, La Vignera S et al (2021) Insulin resistance and cancer: in search for a causal link. Int J Mol Sci 22:11137

Christ-Crain M, Bichet DG, Fenske WK et al (2019) Diabetes insipidus. Nat Rev Dis Prim 5. https://doi.org/10.1038/s41572-019-0103-2
Costa MA (2016) The endocrine function of human placenta: an overview. Reprod Biomed Online 32:14–43
Da Silva XG (2018) The cells of the islets of Langerhans. J Clin Med 7:54
Daghlas SA, Mohiuddin SS (2022) Biochemistry, Glycogen. StatPearls
Daneva T, Spinedi E, Hadid R, Gaillard RC (1995) Impaired hypothalamo-pituitary-adrenal axis function in Swiss nude athymic mice. Neuroendocrinology 62:79–86
Das S, Dasari A (2021) Epidemiology, incidence, and prevalence of neuroendocrine neoplasms: are there global differences? Curr Oncol Rep 23:1–7
Daughaday WH (1979) Anterior pituitary. In: Ingbar SH (ed) Contemporary endocrinology: volume 1. Springer US, Boston, pp 21–76
Davidovici BB, Orion E, Wolf R (2008) Cutaneous manifestations of pituitary gland diseases. Clin Dermatol 26:288–295
Debela DT, Muzazu SGY, Heraro KD et al (2021) New approaches and procedures for cancer treatment: current perspectives. SAGE open Med 9:20503121211034370
DeLuca HF (1975) The kidney as an endocrine organ involved in the function of vitamin D. Am J Med 58:39–47
Dockray GJ (2004) Clinical endocrinology and metabolism. Gastrin. Best Pract Res Clin Endocrinol Metab 18:555–568
Dodé C, Rondard P (2013) PROK2/PROKR2 Signaling and Kallmann Syndrome. Front Endocrinol (Lausanne) 4:19
Dominari A, III Hathaway D, Pandav K et al (2020) Thymosin alpha 1: a comprehensive review of the literature. World J Virol 9:67–78
Drucker D, Shumak S, Angel A (1984) Schmidt's syndrome presenting with intrauterine growth retardation and postpartum addisonian crisis. Am J Obstet Gynecol 149:229–230
Dunn D, Turner C (2016) Hypothyroidism in women. Nurs Womens Health 20:93–98
Dupont J, Rideau N, Simon J (2022) Endocrine pancreas. Sturkie's Avian Physiol:915–937
Edlow AG, Norwitz ER (2019a) Endocrine diseases of pregnancy, 8th edn. Elsevier Inc
Edlow AG, Norwitz ER (2019b) Chapter 27 – endocrine diseases of pregnancy. In: Strauss JF, Barbieri RL (eds) Yen and Jaffe's reproductive endocrinology, 8th edn. Elsevier, Philadelphia, pp 662–708.e17
Emons G (2022) Hormone-dependent cancers: molecular mechanisms and Therapeutical implications. Cells 12:110
Erichsen MM, Løvås K, Skinningsrud B et al (2009) Clinical, immunological, and genetic features of autoimmune primary adrenal insufficiency: observations from a Norwegian registry. J Clin Endocrinol Metab 94:4882–4890
Favero G, Bonomini F, Rezzani R (2021) Pineal gland tumors: a review. Cancers (Basel) 13. https://doi.org/10.3390/cancers13071547
Gaitonde DY, Rowley KD, Sweeney LB (2012) Hypothyroidism: an update. Am Fam Physician 86:244–251
Gianetti E, Tusset C, Noel SD et al (2010) TAC3/TACR3 mutations reveal preferential activation of gonadotropin-releasing hormone release by neurokinin B in neonatal life followed by reversal in adulthood. J Clin Endocrinol Metab 95:2857–2867
Girling JS, Whitaker HC, Mills IG, Neal DE (2007) Pathogenesis of prostate cancer and hormone refractory prostate cancer. Indian J Urol IJU J Urol Soc India 23:35
Glazer DI, Corwin MT, Mayo-Smith WW (2021) Incidental adrenal nodules. Radiol Clin 59:591–601
Goldberg L (1972) Transmission of a vitamin-d metabolite in breast milk. Lancet 300:1258–1259
Goldstein AL, Badamchian M (2004) Thymosins: chemistry and biological properties in health and disease. Expert Opin Biol Ther 4:559–573
Gromada J, Franklin I, Wollheim CB (2007) α-Cells of the endocrine pancreas: 35 years of research but the enigma remains. Endocr Rev 28:84–116
Group ECW (2004) Hormones and breast cancer. Hum Reprod Update 10:281–293

Hååg P, Bektic J, Bartsch G et al (2005) Androgen receptor down regulation by small interference RNA induces cell growth inhibition in androgen sensitive as well as in androgen independent prostate cancer cells. J Steroid Biochem Mol Biol 96:251–258

Han L, Li B, Xu X et al (2021) Umbilical cord blood adiponectin, leptin, insulin, and ghrelin in premature infants and their association with birth outcomes. Front Endocrinol (Lausanne) 12. https://doi.org/10.3389/fendo.2021.738964

Harper M, Murnaghan GA, Kennedy L et al (1989) Phaeochromocytoma in pregnancy. Five cases and a review of the literature. BJOG An Int J Obstet Gynaecol 96:594–606

He Y, Sun MM, Zhang GG et al (2021) Targeting PI3K/Akt signal transduction for cancer therapy. Signal Transduct Target Ther 6:425

Helander HF, Fändriks L (2012) The enteroendocrine "letter cells"—time for a new nomenclature? Scand J Gastroenterol 47:3–12

Herrick EJ, Bordoni B (2023) Embryology, Placenta. [Updated 2023 May 1]. In: StatPearls [Internet]. Treasure Island (FL): StatPearls Publishing; 2024 Jan-. Available from: https://www.ncbi.nlm.nih.gov/books/NBK551634/

Hess LW, Peaceman A, O'Brien WF et al (1988) Adnexal mass occurring with intrauterine pregnancy: report of fifty-four patients requiring laparotomy for definitive management. Am J Obstet Gynecol 158:1029–1034

Huang W, Molitch ME (2019) Pituitary tumors in pregnancy. Endocrinol Metab Clin N Am 48:569–581

Huppertz B, Schleußner E (2023) The placenta: basics and clinical significance. Springer, Berlin, Heidelberg

Husami N, Jewelewicz R, Vande WRL (1977) Pregnancy in patients with pituitary tumors. Fertil Steril 28:920–925

In't Veld P, Marichal M. Microscopic anatomy of the human islet of Langerhans. Adv Exp Med Biol. 2010;654:1-19. https://doi.org/10.1007/978-90-481-3271-3_1

Karpińska M, Czauderna M (2022) Pancreas—its functions, disorders, and physiological impact on the mammals' organism. Front Physiol 13:317

Katz VL, Watson WJ, Hansen WF, Washington JL (1993) Massive ovarian tumor complicating pregnancy. A case report. J Reprod Med 38:907–910

Kempers MJE, van Trotsenburg ASP, van Rijn RR et al (2007) Loss of integrity of thyroid morphology and function in children born to mothers with inadequately treated Graves' disease. J Clin Endocrinol Metab 92:2984–2991

Key TJA (1995) Hormones and cancer in humans. Mutat Res Mol Mech Mutagen 333:59–67

Kim SY (2015) Diagnosis and treatment of hypopituitarism. Endocrinol Metab 30:443–455

Koh TJ, Chen D (2000) Gastrin as a growth factor in the gastrointestinal tract. Regul Pept 93:37–44

Kopczak A, Stieg MR, Renner U, Stalla GK (2017) Advances in understanding hypopituitarism. F1000Research 6:1–8

Krassas G, Karras SN, Pontikides N (2015) Thyroid diseases during pregnancy: a number of important issues. Hormones 14:59–69

Lacroix A, Feelders RA, Stratakis CA, Nieman LK (2015) Cushing's syndrome. Lancet 386:913–927

LaFranchi SH (2011) Approach to the diagnosis and treatment of neonatal hypothyroidism. J Clin Endocrinol Metab 96:2959–2967

Lake MG, Krook LS, Cruz SV (2013) Pituitary adenomas: an overview. Am Fam Physician 88(5):319-27. PMID: 24010395.

Laurberg P, Bournaud C, Karmisholt J, Orgiazzi J (2009) Management of Graves' hyperthyroidism in pregnancy: focus on both maternal and foetal thyroid function, and caution against surgical thyroidectomy in pregnancy. Eur J Endocrinol 160:1–8

Lazarus JH (2002) Thyroid dysfunction: reproduction and postpartum thyroiditis. In: Seminars in reproductive medicine. Copyright© 2002 by Thieme Medical Publishers, Inc., 333 Seventh Avenue, New, pp 381–388

Lazarus JH, Ammari F, Oretti R et al (1997) Clinical aspects of recurrent postpartum thyroiditis. Br J Gen Pract 47:305–308

Lechan RM, Arkun K, Toni R (2019) Pituitary anatomy and development. In: Tritos NA, Klibanski A (eds) Contemporary endocrinology. Springer International Publishing, Cham, pp 11–53
Lemaitre M, Jannin A, Chevalier B, Vantyghem M-C (2022) The heart, an endocrine gland: natriuretic peptides. Annales d'Endocrinologie Elsevier, pp. 59–62
Louis DN, Ohgaki H, Wiestler OD et al (2007) The 2007 WHO classification of tumours of the central nervous system. Acta Neuropathol 114:97–109
Luton D, Le Gac I, Vuillard E et al (2005) Management of Graves' disease during pregnancy: the key role of fetal thyroid gland monitoring. J Clin Endocrinol Metab 90:6093–6098
Mandel SJ, Brent GA, Larsen PR (1993) Levothyroxine therapy in patients with thyroid disease. Ann Intern Med 119:492–502
Mannelli M, Bemporad D (2002) Diagnosis and management of pheochromocytoma during pregnancy. J Endocrinol Investig 25:567–571
Marqusee E, Hill JA, Mandel SJ (1997) Thyroiditis after pregnancy loss. J Clin Endocrinol Metab 82:2455–2457
Massie CE, Lynch A, Ramos-Montoya A et al (2011) The androgen receptor fuels prostate cancer by regulating central metabolism and biosynthesis. EMBO J 30:2719–2733
May CL, Kaestner KH (2010) Gut endocrine cell development. Mol Cell Endocrinol 323:70–75
Mazumder T, Akter E, Rahman SM et al (2022) Prevalence and risk factors of gestational diabetes mellitus in Bangladesh: findings from demographic health survey 2017–2018. Int J Environ Res Public Health 19. https://doi.org/10.3390/ijerph19052583
McIntyre HD, Catalano P, Zhang C et al (2019) Gestational diabetes mellitus. Nat Rev Dis Prim 5:47
Moalla M, Hadj Kacem F, Al-Mutery AF et al (2019) Nonstop mutation in the Kisspeptin 1 receptor (KISS1R) gene causes normosmic congenital hypogonadotropic hypogonadism. J Assist Reprod Genet 36:1273–1280
Moleti M, Di Mauro M, Alibrandi A et al (2020) Postpartum thyroiditis in women with Euthyroid and hypothyroid Hashimoto's thyroiditis antedating pregnancy. J Clin Endocrinol Metab 105:e2421–e2428
Nagy A, Malcomson RDG (2022) The endocrine system. Keeling's Fetal Neonatal Pathol:677–708
Najafi L, Abedini A, Kadivar M et al (2019) Gestational diabetes mellitus: the correlation between umbilical coiling index, and intrapartum as well as neonatal outcomes. J Diabetes Metab Disord 18:51–57
Nelles JL, Hu W-Y, Prins GS (2011) Estrogen action and prostate cancer. Expert Rev Endocrinol Metab 6:437–451
New MI, Carlson A, Obeid J et al (2001) Extensive personal experience: prenatal diagnosis for congenital adrenal hyperplasia in 532 pregnancies. J Clin Endocrinol Metab 86:5651–5657
Norman RJ, Teede HJ (2018) A new evidence-based guideline for assessment and management of polycystic ovary syndrome. Med J Aust 209:299–300
Orrego JJ, Barkan AL (2000) Pituitary disorders: drug treatment options. Drugs 59:93–106
Osborn AG, Salzman KL, Thurnher MM et al (2012) The new World Health Organization classification of central nervous system tumors: what can the neuroradiologist really say? Am J Neuroradiol 33:795–802
Pang S, Wallace MA, Hofman L et al (1988) Worldwide experience in newborn screening for classical congenital adrenal hyperplasia due to 21-hydroxylase deficiency. Pediatrics 81:866–874
Patel P, Galoian K (2018) Molecular challenges of neuroendocrine tumors. Oncol Lett 15:2715–2725
Paul CA, Virgo KS, Wade TP et al (2000) Adrenalectomy for isolated adrenal metastases from non-adrenal cancer. Int J Oncol 17:181–188
Pentland B, Mitchell JD, Coll RE, Ford MJ (1985) Central nervous system sarcoidosis. QJM An Int J Med 56:457–465
Poloz Y, Stambolic V (2015) Obesity and cancer, a case for insulin signaling. Cell Death Dis 6:e2037–e2037
Poretsky L (2010) Principles of diabetes mellitus. Springer. https://doi.org/10.1007/978-0-387-09841-8.

Raivio T, Falardeau J, Dwyer A et al (2007) Reversal of idiopathic hypogonadotropic hypogonadism. N Engl J Med 357:863–873
Raphael MJ, Chan DL, Law C, Singh S (2017) Principles of diagnosis and management of neuroendocrine tumours. CMAJ 189:E398–E404
Rehman G, Kumari N, Bano F, Tyagi RK (2023) Thyroid hormone receptor beta: relevance in human health and diseases. Endocr Metab Sci 13:100144
Reid JR, Wheeler SF (2005) Hyperthyroidism: diagnosis and treatment. Am Fam Physician 72:623–630
Rhyu J, Yu R (2021) Newly discovered endocrine functions of the liver. World J Hepatol 13:1611–1628
Rindi G, Mete O, Uccella S et al (2022) Overview of the 2022 WHO classification of neuroendocrine neoplasms. Endocr Pathol 33:115–154
Rivera J-A (2006) Lymphocytic hypophysitis: disease spectrum and approach to diagnosis and therapy. Pituitary 9:35–45
Rodriguez AC, Blanchard Z, Maurer KA, Gertz J (2019) Estrogen signaling in endometrial cancer: a key oncogenic pathway with several open questions. Horm Cancer 10:51–63
Rodríguez-Porcel F, Hughes I, Anderson D et al (2013) Foster Kennedy syndrome due to meningioma growth during pregnancy. Front Neurol 4:183
Rongioletti F, Smoller BR (2010) Clinical and pathological aspects of skin diseases in endocrine, metabolic, nutritional and deposition disease
Saeedi M, Cao Y, Fadl H et al (2021) Increasing prevalence of gestational diabetes mellitus when implementing the IADPSG criteria: a systematic review and meta-analysis. Diabetes Res Clin Pract 172:108642
Schally AV (2000) Use of GnRH in preference to LH-RH terminology in scientific papers. Hum Reprod 15:2059–2061
Schally AV, Arimura A, Baba Y et al (1971a) Isolation and properties of the FSH and LH-releasing hormone. Biochem Biophys Res Commun 43:393–399
Schally AV, Arimura A, Kastin AJ et al (1971b) Gonadotropin-releasing hormone: one polypeptide regulates secretion of luteinizing and follicle-stimulating hormones. Science 173:1036–1038
Scharrer B (1969) Neurohumors and Neurohormones: definitions and terminology. In: Kappers JA (ed) Neurohormones and Neurohumors: structure and function of regulatory mechanisms. Springer, Berlin Heidelberg, Berlin, Heidelberg, pp 1–20
Scherer T, Sakamoto K, Buettner C (2021) Brain insulin signalling in metabolic homeostasis and disease. Nat Rev Endocrinol 17:468–483
Schlechte JA (2003) Prolactinoma. 2035–2041
Schleußner E (2023) Endocrinology of the placenta. In: Huppertz B, Schleußner E (eds) The placenta: basics and clinical significance. Springer, Berlin Heidelberg, Berlin, Heidelberg, pp 91–104
Schlumberger M-J, Filetti S, Hay ID (2011) CHAPTER 14 – nontoxic diffuse and nodular Goiter and thyroid neoplasia. In: Melmed S, Polonsky KS, Larsen PR, Kronenberg HM (eds) Williams textbook of endocrinology, 12th edn. W.B. Saunders, Philadelphia, pp 440–475
Scorsetti M, Leo F, Trama A et al (2016) Thymoma and thymic carcinomas. Crit Rev Oncol Hematol 99:332–350
Silveira LG, Latronico AC, Seminara SB (2013) Kisspeptin and clinical disorders. Adv Exp Med Biol 784:187–199
Solis-Cohen L, Steinbach M (1939) Sporadic congenital goiter causing death of an infant. Am J Dis Child 58:1067–1071
Soria B (1998) Physiology and pathophysiology of the islets of Langerhans. Springer Science & Business Media, https://doi.org/10.1007/978-1-4899-1819-2.
Stagnaro-Green A (2012) Approach to the patient with postpartum thyroiditis. J Clin Endocrinol Metab 97:334–342
Stanfield (2013) 7. Stanfield- Principles of human physiology
Sun Q, Li J, Gao F (2014) New insights into insulin: the anti-inflammatory effect and its clinical relevance. World J Diabetes 5:89

Svoboda M, Tastenoy M, Vertongen P, Robberecht P (1994) Relative quantitative analysis of glucagon receptor mRNA in rat tissues. Mol Cell Endocrinol 105:131–137

Swartz SL, Williams GH, Hollenberg NK et al (1980) Primacy of the renin-angiotensin system in mediating the aldosterone response to sodium restriction. J Clin Endocrinol Metab 50:1071–1074

Tan DX, Xu B, Zhou X, Reiter RJ (2018) Pineal calcification, melatonin production, aging, associated health consequences and rejuvenation of the pineal gland. Molecules 23. https://doi.org/10.3390/molecules23020301

Thomas CG Jr, Croom RD III (1987) Current management of the patient with autonomously functioning nodular goiter. Surg Clin North Am 67:315–328

Thomas DD, Corkey BE, Istfan NW, Apovian CM (2019) Hyperinsulinemia: an early indicator of metabolic dysfunction. J Endocr Soc 3:1727–1747

Toto KH (1994) Endocrine physiology: a comprehensive review. Crit Care Nurs Clin North Am 6:637–659

Trarbach EB, Abreu AP, Silveira LFG et al (2010) Nonsense mutations in FGF8 gene causing different degrees of human gonadotropin-releasing deficiency. J Clin Endocrinol Metab 95:3491–3496

Tsirlin A, Oo Y, Sharma R et al (2014) Pheochromocytoma: a review. Maturitas 77:229–238

Tsoli M, Chatzellis E, Koumarianou A et al (2019) Current best practice in the management of neuroendocrine tumors. Ther Adv Endocrinol Metab 10:2042018818804698

Unuane D, Tournaye H, Velkeniers B, Poppe K (2011) Endocrine disorders & female infertility. Best Pract Res Clin Endocrinol Metab 25:861–873

Vaidya RA, Desai S, Moitra P et al (2023) Hyperinsulinemia: an early biomarker of metabolic dysfunction. Front Clin Diabetes Healthc 4:1159664

Vargas E, Carrillo Sepulveda MA (2018) Biochemistry, insulin, Metabolic Effects. StatPearls

Vesely DL (2001) Atrial natriuretic peptides in pathophysiological diseases 01:1–23

Watt MJ, Miotto PM, De Nardo W, Montgomery MK (2019) The liver as an endocrine organ—linking NAFLD and insulin resistance. Endocr Rev 40:1367–1393

Wells SA Jr (2016) Progress in endocrine neoplasia. Clin Cancer Res 22:4981–4988

Williams AJ, Robson H, Kester MHA et al (2008) Iodothyronine deiodinase enzyme activities in bone. Bone 43:126–134

Wu K, Zhou Y, Ke S et al (2021) Lifestyle is associated with thyroid function in subclinical hypothyroidism: a cross-sectional study. BMC Endocr Disord 21:112

Yang X, Wang J (2019) The role of metabolic syndrome in endometrial cancer: a review. Front Oncol 9:744

Yu J (2014) Endocrine disorders and the neurologic manifestations. Ann Pediatr Endocrinol Metab 19:184

Zdrojewicz Z, Pachura E, Pachura P (2016) The thymus: a forgotten, but very important organ. Adv Clin Exp Med 25:369–375

Zhang W, Liu HT (2002) MAPK signal pathways in the regulation of cell proliferation in mammalian cells. Cell Res 12:9–18

Zhou Q, Acharya G (2022) Editorial: placental hormones and pregnancy-related endocrine disorders. Front. Endocrinol. (Lausanne). 13:905829

Zhu Q, Scherer PE (2018) Immunologic and endocrine functions of adipose tissue: implications for kidney disease. Nat Rev Nephrol 14:105–120

Zouboulis CC (2009) The skin as an endocrine organ. Dermatoendocrinol 1:250–252

Biomolecular Components of Blood and Their Role in Health and Diseases

12

Arunim Shah, Sarthak Srivastava, and C. P. Chaturvedi

Abstract

Blood is a vital, dynamic, and complex fluid that circulates through the circulatory system delivering essential substances to various tissues and organs in the body. It contains a diverse array of biomolecules that play a critical role in maintaining homeostasis, transporting oxygen and nutrients, and removing waste products. Essential components of blood include plasma, red blood cells, white blood cells, platelets, and serum. These components are supported or formed by biomolecules in the blood, namely, plasma proteins, clotting factors, hemoglobin, enzymes, lipids, glucose, electrolytes, hormones, cytokines, growth factors, nucleotides and nucleosides, and wasted products. While each component in blood has an important role and work in conjunction for proper physiological function, disruption in any of the component leads to a serious health condition such as anemia, hemoglobinopathies, bleeding disorders, leukemia, lymphoma, myeloproliferative disease, thrombocytopenia disseminated intravascular coagulation, hemolytic disease of new born, aplastic anemia, multiple myeloma, and various other hematological malignancies. In this chapter, we will try to take each biomolecular component into account and discuss its role in health and diseases.

Keywords

Biomolecules · Hemoglobinopathies · Anemia · Red blood cells · White blood cells

---

Arunim Shah and Sarthak Srivastava contributed equally with all other contributors.

A. Shah · S. Srivastava · C. P. Chaturvedi (✉)
Stem Cell Research Centre, Department of Hematology, Sanjay Gandhi Post Graduate Institute of Medical Sciences, Lucknow, Uttar Pradesh, India

R. L. Singh et al. (eds.), *Clinical Applications of Biomolecules in Disease Diagnosis*, https://doi.org/10.1007/978-981-97-4723-8_12

12.1 Introduction

Blood is a connective tissue and a vital body fluid that circulates continuously through the intricate network of blood vessels and plays essential roles in maintaining health and homeostasis in the body, serving as the lifeline of the human body (Frenette et al. 2020). Blood plays a pivotal role in sustaining life by facilitating essential physiological processes such as delivering vital substances to cells and tissues while removing metabolic waste products. This dynamic fluid performs a myriad of functions, ranging from oxygen transport to immune defense, and from clot formation to hormone distribution. Blood is composed of various cells, proteins, nutrients, and waste products (Cherian 2022). One of the most critical functions of blood is its role in oxygen transport. Red blood cells, or erythrocytes, are specialized cells that contain hemoglobin, a protein that binds to oxygen in the lungs and releases it to tissues throughout the body. This process, known as oxygenation, is fundamental to cellular respiration and energy production, ensuring the vitality of all organs and systems (West and Luks 2016). In addition to oxygen transport, blood is also integral to immune defense mechanisms. White blood cells, or leukocytes, patrol the bloodstream and tissues, identifying and neutralizing pathogens such as bacteria, viruses, and fungi. Through a complex interplay of cellular and humoral immunity, the body's immune system relies on blood to mount rapid and effective responses to infections and foreign invaders (Tamang et al. 2022).

Furthermore, blood plays a crucial role in hemostasis, the process of preventing and stopping bleeding. Platelets, or thrombocytes, are small cell fragments that aggregate at the site of vascular injury to form a temporary plug, preventing excessive blood loss. This initial hemostatic response is reinforced by the formation of a fibrin clot, which seals the wound and promotes tissue repair (Periayah et al. 2017). Beyond its role in oxygen transport, immune defense, and clotting, blood serves as a vehicle for the transportation of nutrients, hormones, and waste products throughout the body. Nutrients absorbed from the gastrointestinal tract, such as glucose, amino acids, and fatty acids, are transported in the bloodstream to cells for energy production and tissue repair. Hormones secreted by endocrine glands, such as insulin, thyroid hormone, and cortisol, are also carried by the blood to target tissues, where they regulate various physiological processes (Newsholme 2021). Conversely, metabolic waste products, such as carbon dioxide and urea, are transported by the blood to the lungs and kidneys for elimination from the body. The blood delivers carbon dioxide to the lungs for exhalation, ensuring the removal of this waste product from the bloodstream. Similarly, the kidneys filter waste products from the blood, producing urine for excretion from the body.

Understanding the intricacies of blood composition and function is fundamental to diagnosing and managing a wide range of health conditions, making it a subject of enduring interest and importance in biomedical research and clinical practice. Blood-related diseases and disorders encompass a wide range of conditions affecting the production, function, and regulation of blood cells and plasma components. From anemia and leukemia to thrombocytopenia and hemophilia, these conditions can have profound effects on health and quality of life. Timely diagnosis, appropriate treatment, and ongoing management are essential for optimizing outcomes and minimizing complications in individuals affected by blood-related diseases and disorders. Continued research and advances in medical science hold promise for improving our understanding and management of these complex conditions in the future.

12.2 Blood

Blood is a vital connective tissue in the human body, playing crucial roles in maintaining health and homeostasis. Unlike other connective tissues, blood consists of both cellular and extracellular components, with cellular elements collectively known as formed elements. These formed elements include red blood cells (RBCs), white blood cells (WBCs), and platelets. Additionally, blood stands out from other connective tissues due to its unique fluid extracellular matrix, referred to as plasma (Basu and Kulkarni 2014). Plasma, the liquid component of blood that is a straw-colored fluid, comprises approximately 55–60% of whole blood, while blood cells make up the remaining 40–45%. Plasma comprises proteins (albumin, clotting factors, antibodies, enzymes, and hormones), carbohydrates (glucose), and fat particles, in addition to being primarily water (Benjamin and McLaughlin 2012). The composition of blood varies slightly between individuals but typically accounts for about 7–8 percent of total body weight. An average-sized man typically has approximately 12 pints of blood in his body, while an average-sized woman has about nine pints. This variation in blood volume reflects differences in body size, composition, and physiological demands (Fig. 12.1).

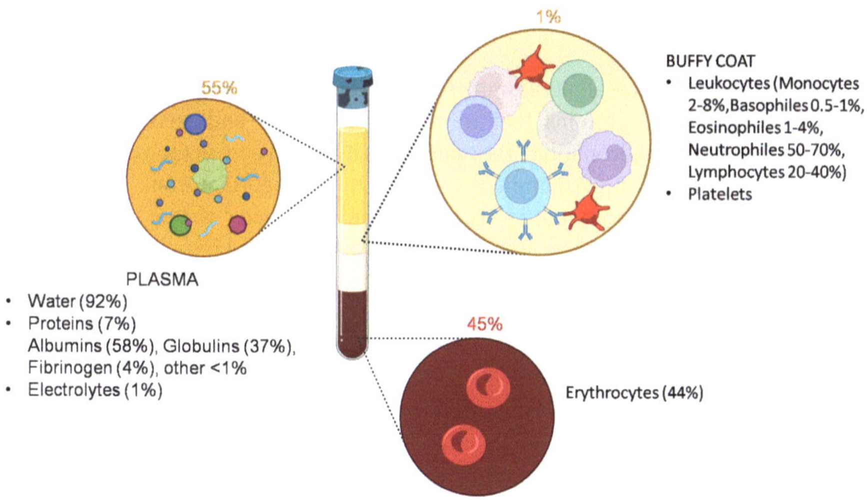

Fig. 12.1 Composition of blood

12.2.1 Characteristics of Blood

Blood is approximately five times more viscous than water and fairly sticky to the touch. Viscosity is a degree of a fluid's thickness or resistance to waft and is affected by the presence of plasma proteins and formed elements inside the blood. The viscosity of blood has a drastic effect on blood pressure and flow (Reinke et al. 1986; Alexy et al. 2022). Blood temperature is usually slightly higher than body temperature (37 °C), with daily changes of 0.5 °C considered normal. Blood arteries have relatively smooth surfaces, but as they age and lose elasticity, they undergo friction and resistance, leading to heat production. This accounts for the modest increase in temperature. The pH of blood averages around 7.4, but it can range from 7.35 to 7.45 in a healthy individual. Blood has a slightly higher pH (alkaline) than pure water (7.0).

12.2.2 Components of Blood

12.2.2.1 Blood Plasma

Plasma, like other bodily fluids, is mostly made up of water, accounting for approximately 92% of its composition. This water contains a combination of molecules, the majority of which are proteins. The plasma contains hundreds of dissolved or suspended chemicals, many of which are present in trace amounts.

12.2.2.2 Plasma Proteins

Proteins account for approximately 7% of plasma volume nearly all that is not water. The plasma contains various unique proteins, as well as a modest number of regulatory proteins such as enzymes and hormones (Williams et al. 2019).

Plasma proteins are classified into three primary classes as follows:

12.2.2.2.1 Albumin

Albumin is the most prevalent plasma protein. Albumin molecules, produced by the liver, act as binding proteins that transport fatty acids and steroid hormones. Although lipids are hydrophobic, they can be transported across the watery plasma by attaching to albumin. Albumin is also the most important contributor to the osmotic pressure of blood; that is, its presence retains water within blood vessels and attracts water from tissues, across blood vessel walls, and into the bloodstream. This helps maintain blood volume and pressure. Albumin makes up around 54% of total plasma protein concentration.

12.2.2.2.2 Globulins

The second most prevalent plasma proteins are globulins. The group consists of three subgroups: alpha, beta, and gamma globulins. Alpha and beta globulins carry iron, lipids, and the fat-soluble vitamins A, D, E, and K to cells; they, like albumin, contribute to osmotic pressure. Gamma globulins, also called antibodies or immunoglobulins, are immune-related proteins. Plasma cells are responsible for producing immunoglobulins, while the liver produces other plasma proteins. Globulins account for approximately 38% of total plasma protein volume.

12.2.2.2.3 Fibrinogen

The least prominent plasma protein is fibrinogen. Fibrinogen, like albumin, alpha, and beta globulins, is a liver-produced protein. It is necessary for blood clotting, which is discussed later in this chapter. Fibrinogen accounts for approximately 7% of total plasma protein volume. In addition to proteins, plasma contains a vast range of additional chemicals. Electrolytes (sodium, potassium, and calcium ions), dissolved gases (oxygen, carbon dioxide, and nitrogen), organic nutrients (vitamins, lipids, glucose, and amino acids), and metabolic wastes are all considered. Non-protein solutes account for roughly 1% of plasma volume.

12.2.2.3 Erythrocytes

The erythrocyte, often known as a red blood cell (or RBC), is by far the most commonly formed element. A single drop of blood contains millions of erythrocytes and just thousands of leukocytes. Males have roughly 5.4 million erythrocytes per microliter (μL) of blood, whereas females have around 4.8 million (Guyton and Hall 2015). It is estimated that erythrocytes account for around 25% of the total cells in the body. The cells are tiny, measuring approximately 7–8 μm on average. The basic activities of erythrocytes are to pick up breathed oxygen from the lungs and deliver it to the body's tissues, as well as carbon dioxide waste from tissues being transported to the lungs for expiration, accounting for approximately 24% of the total. Erythrocytes stay in the vascular network. Although leukocytes normally leave blood arteries to conduct their protective duties, the migration of erythrocytes from blood arteries is aberrant.

12.2.2.3.1 Shape and Structure of Erythrocytes

As an erythrocyte grows in the bone marrow, it extrudes its nucleus and the majority of its other organelles. An immature erythrocyte, also known as a reticulocyte, will often have organelle remnants during its first day or two in circulation (Nandakumar et al. 2016). Reticulocytes make up about 1–2% of the erythrocyte count and provide a rough indication of RBC formation. Abnormally low or high rates indicate deviations. Ribosome remnants (reticulum) are swiftly removed, leaving adult erythrocytes with minimal internal structural components. They rely on anaerobic respiration due to a lack of mitochondria. This means that they don't use any of the oxygen they're carrying, allowing them to transfer it all to tissues. They also lack endoplasmic reticula and don't produce proteins. Erythrocytes incorporate structural proteins that help blood cells keep their one-of-a-kind shape and help them skip via capillaries. This comprises spectrin, a cytoskeletal protein element (Anderson et al. 2018). Erythrocytes are biconcave disks, indicating they are plump at the periphery and very thin in the core. Because they lack most organelles, there is greater inner space for hemoglobin molecules, which, as we will see momentarily, carry gasses. The biconcave shape has a higher surface area-to-volume ratio than a sphere of the same diameter, facilitating gas exchange.

12.2.2.3.2 Hemoglobin

Hemoglobin is a big molecule composed of proteins and iron. The structure is made up of four folded chains of globin protein (alpha 1 and 2, beta 1 and 2). Each globin molecule is attached to heme, a red pigment with an iron ion (Fe^{2+}) per molecule. A single erythrocyte can hold 300 million hemoglobin molecules and over 1 billion oxygen molecules. In the lungs, hemoglobin absorbs oxygen and binds to iron ions, generating oxyhemoglobin. The bright red, oxygenated hemoglobin travels to the body's tissues, where it releases part of its oxygen molecules, resulting in darker red deoxyhemoglobin, also known as decreased hemoglobin. Oxygen release is determined by the need for oxygen in the surrounding tissues; therefore hemoglobin rarely, if ever, leaves all of its oxygen behind. In the capillaries, carbon dioxide enters the bloodstream. Approximately 76% dissolves in plasma, with some remaining as dissolved CO and the remainder producing bicarbonate ions. Approximately 23–24% of it binds to the amino acids in hemoglobin, generating a compound known as carbaminohemoglobin. Hemoglobin transports carbon dioxide from capillaries to the lungs and exchanges it for oxygen. RBC levels affect the body's ability to provide oxygen to tissues. Insufficient RBCs due to ineffective hematopoiesis can cause anemia in many ways. Polycythemia is a disorder caused by an overproduction of red blood cells. Polycythemia's main disadvantage is increased blood viscosity, which hinders the heart's ability to circulate blood to tissues.

12.2.2.3.3 Leukocytes and Platelets

The leukocyte, often known as a white blood cell (or WBC), is an important part of the body's fight against sickness. Leukocytes clear up trash and defend the body against invading microbes and altered DNA. Platelets play a crucial role in repairing damaged blood vessels and promoting healing and restoration (Prinyakupt and Pluempitiwiriyawej 2015). Leukocytes and erythrocytes are both derived from hematopoietic stem cells in bone marrow, although they differ significantly. For example, leukocytes are far less abundant than erythrocytes: Usually, there are only 5000–10,000 per μL. They are larger than erythrocytes and are the only complete cells with a nucleus and organelles. While there is only one type of erythrocyte, there are multiple types of leukocytes. These types of cells have significantly shorter life spans than erythrocytes, often as brief as a few hours or even minutes during acute infections. One of the most distinguishing features of leukocytes is their mobility. Erythrocytes circulate in blood vessels, while leukocytes perform protective tasks in tissues.

12.2.2.3.3.1 Classification of Leukocytes

Leukocytes are divided into two types based on the presence of visible granules in their cytoplasm. Granular leukocytes (granulocytes) have many granules within their cytoplasm. They consist of neutrophils, eosinophils, and basophils. Granules are present in agranular leukocytes; however, they are less prominent. Agranular leukocytes consist of monocytes, which mature into phagocytic macrophages, and lymphocytes, which originate from lymphoid stem cells (Fig. 12.2).

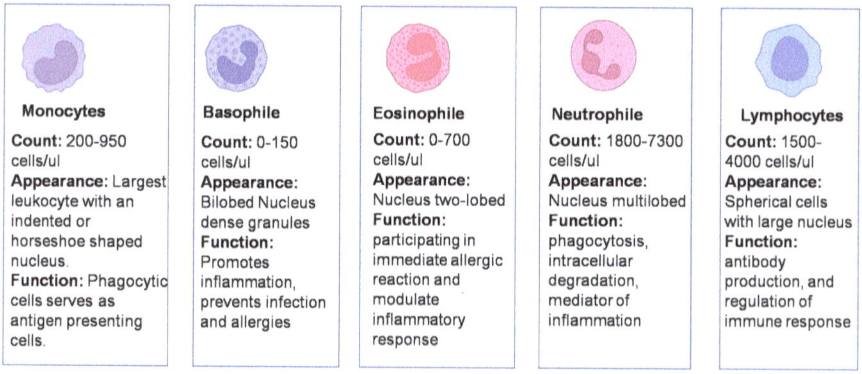

Fig. 12.2 Types of leukocytes, their count in blood per microliter, appearance, and functions

Granular Leukocytes

Granular leukocytes, formed in the red bone marrow, have a limited lifespan ranging from hours to days. They usually have a lobed nucleus and are categorized based on the type of stain that highlights their granules (Schulze and Bugert 2018).

Neutrophils

Neutrophils are the most prevalent type of leukocyte, accounting for 50–70% of the total count. They are 10–12 μm in diameter and much larger than erythrocytes. They are named neutrophils because their granules are best visible with chemically neutral stains. The granules are abundant yet very fine, and they typically seem light violet. The nucleus is lobed and can have two to five lobes, which increase with cell age. Older neutrophils contain an increasing number of lobes and are commonly known as polymorphonuclear (a nucleus with several forms) or simply "polys." Younger and immature neutrophils produce lobes, known as "bands."

Neutrophils are effective phagocytes that prefer bacteria and respond quickly to infection sites. Their granules contain lysozyme, an enzyme that can lyse, or break down, bacterial cell walls; oxidants such as hydrogen peroxide; and defensins, proteins that adhere to and pierce bacterial and fungal plasma membranes, allowing cell contents to flow out (Nauseef and Borregaard 2014). Abnormally high neutrophil counts may indicate infection or inflammation caused by pathogens but can also be seen in burn patients or those under stress. Burn damage stimulates neutrophil proliferation to fight infections caused by the breakdown of the skin's barrier. Low numbers can be caused by drug toxicity or other conditions, making them more susceptible to infections (Rosales 2018).

Eosinophils

Eosinophils typically account for 2–4% of total leukocyte count. They also measure 10–12 μm in diameter. The granules of eosinophils are best stained with the acidic stain eosin. The eosinophil nucleus normally has two to three lobes, and when correctly stained, the granules will have a characteristic red-to-orange tint. Eosinophil granules contain antihistamine molecules that inhibit the activity of histamines, which are inflammatory compounds generated by basophils and mast cells. Eosinophil granules can be poisonous to parasitic worms that enter the body through the integument or by raw or undercooked food consumption. Eosinophils can also be phagocytose and are more effective when antibodies bind to the target, forming an antigen–antibody combination. High eosinophil counts are common in people with allergies, parasitic worm infestations, and some autoimmune illnesses. Low numbers could be attributed to drug toxicity or stress (Falcone et al. 2006).

Basophils

Basophil granules respond well to basic (alkaline) stains. Basophils have enormous granules that stain dark blue and can obscure the two-lobed nucleus due to their abundance. In general, basophils amplify the inflammatory response. They share this characteristic with mast cells. Mast cells were formerly thought to be basophils that departed the body's circulation. However, it appears that the two cell types emerge from distinct lineages. Basophils emit histamines that cause inflammation and heparin, which prevents blood clotting. High basophil numbers are linked to allergies, parasite infections, and hypothyroidism. Low numbers are linked to pregnancy, stress, and hyperthyroidism (Min et al. 2012).

Agranular Leukocytes

Agranular leukocytes have smaller and less apparent granules in their cytoplasm than granular leukocytes. The nucleus has a basic form, with an indentation but no identifiable lobes. Agranulocytes are classified into two major types: lymphocytes and monocytes.

Lymphocytes

Lymphocytes are the only blood cells that develop from lymphoid stem cells. Lymphatic tissues play a significant role in the development and reproduction of these cells, which originate in the bone marrow. Lymphocytes are the second most prevalent type of leukocyte, accounting for 20–30% of all leukocytes. They are necessary for the immunological response. The size range of lymphocytes is fairly broad, with some authorities distinguishing two size classes and others three. Large cells are typically 10–14 μm in size, with a decreased nucleus-to-cytoplasm ratio and more granules. Smaller cells (6–9 μm) have a greater nucleus-to-cytoplasm ratio, giving a "halo" effect. A few cells may fall outside these ranges (14–17 μm). This finding resulted in the three-size range classification (Larosa and Orange 2008).

Lymphocytes are classified into three types: natural killer cells, B cells, and T cells. Natural killer (NK) cells can identify cells that lack "self" proteins on the plasma membrane or have alien or aberrant signals. These "non-self" cells include cancer cells, virus-infected cells, and other cells with unusual surface proteins. They offer widespread, nonspecific immunity. The bigger lymphocytes are usually NK cells. B cells and T cells, commonly known as B lymphocytes and T lymphocytes, play important roles in the body's defense against pathogens (disease-causing microbes) and particular immunity. One type of B cell (plasma cells) produces antibodies or immunoglobulins that bind to certain foreign or aberrant components of plasma membranes. This is also known as humoral (body fluid) immunity. T cells provide cellular immunity by physically fighting foreign or infected cells. A memory cell is a type of B or T cell that develops after being exposed to a disease and responds quickly to repeated infections. Memory cells, unlike other leukocytes, can live for several years. The bone marrow matures B cells, while the thymus matures T cells. This stage of the maturation process gives rise to the terms B and T cells. Lymphocytes have complicated activities, which will be discussed in depth in the chapter on the lymphatic system and immunology. Smaller lymphocytes are either B or T cells, but they cannot be distinguished on a standard blood smear.

Abnormally high lymphocyte counts are associated with both viral infections and some forms of malignancies. Abnormally low lymphocyte counts are indicative of long-term (chronic) sickness or immunosuppression, particularly those induced by HIV infection and medication regimens that frequently include steroids.

Monocytes

Monocytes derive from myeloid stem cells. They generally account for 2–8% of the overall leukocyte count. They are easily identified by their huge size (12–20 μm) and indented or horseshoe-shaped nuclei. Macrophages are monocytes that have exited the bloodstream and phagocytize detritus, foreign infections, worn-out erythrocytes, and a variety of other dead, worn, or injured cells. Macrophages also produce antimicrobial defensins and chemotactic molecules, which draw additional leukocytes to the site of infection. Some macrophages remain in certain sites, whereas others move around in the tissue fluid. Abnormally high monocyte numbers are linked to viral or fungal infections, TB, some types of leukemia, and other chronic illnesses. Bone marrow suppression is often the cause of abnormally low numbers.

Platelets

Platelets are sometimes referred to as thrombocytes; however, these persist in bone marrow tissue and eventually form platelet-precursor extensions that stretch through the walls of bone marrow capillaries, releasing hundreds of cytoplasmic fragments into the bloodstream, each surrounded by a piece of the plasma membrane. These contained particles are platelets. Throughout its existence, each megakaryocyte platelet is small (2–4 μm in diameter) but abundant, with 150,000–160,000 per μL of blood. After entering the circulation, roughly one-third move to the spleen for storage until being released in reaction to a blood artery rupture. They are then triggered to carry out their primary role, which is to prevent blood loss. Platelets last roughly 10 days before being phagocytosed by macrophages (Gremmel et al. 2016).

Platelets are necessary for hemostasis, which is the cessation of blood flow once a vessel is damaged. They also release several growth factors that are necessary for tissue growth and repair, particularly connective tissue. Some therapies now include concentrated platelet infusions to promote recovery produces between 2000 and 3000 platelets. Following platelet release, macrophages eat megakaryocyte remains, which are essentially cell nuclei. This is incorrect because it implies that they are a sort of cell. A platelet is a piece of cytoplasm from a megakaryocyte that is surrounded by a plasma membrane. Megakaryocytes, derived from myeloid stem cells, are typically 50–100 μm in diameter with an expanded, lobed nucleus. As previously stated, thrombopoietin, a glycoprotein released by the kidneys and liver, promotes the proliferation of megakaryoblasts, which develop into megakaryocytes.

12.2.3 Functions of Blood

The main function of blood is to transport nutrients and oxygen to cells and eliminate waste from them. However, it is only where the narrative starts. The distribution of heat, defense, and homeostasis maintenance are among the other unique roles of blood.

12.2.3.1 Transportation

The digestive tract absorbs nutrients from what we eat. The majorities of these molecules enters the bloodstream and go to the liver for processing before being released back into the bloodstream to reach body cells. When you breathe, oxygen enters your bloodstream and travels from your lungs to your heart, which then distributes it throughout your body. Endocrine glands throughout the body produce hormones into the bloodstream, which travel to distant target cells. Blood delivers cellular waste and byproducts to organs for disposal.

12.2.3.2 Defense

White blood cells (WBCs), also known as leukocytes, protect the body from external threats like bacteria and viruses. They also seek out and destroy internal threats like cancerous cells with mutated DNA. When blood vessels are damaged, platelets and proteins dissolve in the plasma. This keeps the body from suffering additional blood loss.

12.2.3.3 Maintenance of Homeostasis

Body temperature is controlled by a classic negative feedback loop. Exercising on a warm day activates homeostatic mechanisms, such as enhanced blood transfer from the core to the periphery. The skin's blood arteries dissipate heat to the environment, resulting in cooler blood returning to the human core. On cold days, the body's core temperature is maintained by diverting blood away from the skin. Blood also serves to maintain the body's chemical balance. Blood proteins and chemicals function as buffers, regulating pH levels. Blood additionally allows for adjustment of the water content material of frame cells with the aid of keeping osmolality (De Hert 2012).

12.2.4 Complete Blood Count (CBC)

A complete blood count (CBC) is a common blood test that provides valuable information about the cellular components of blood. It measures various parameters,

Table 12.1 Blood component and their reference range in blood

Blood component	Abbreviation used	Reference range	SI Reference range
White blood cells	WBC	4500–11,000/mm	4.5–11.0 × 10/L
Red blood cells*	RBC	Male: 4.3–5.9 million/mm Female: 3.5–5.5 million/mm	Male: 4.3–5.9 × 10/L Female: 3.5–5.5 × 10/L
Hemoglobin*	HGB	Male: 13.5–17.5 g/dL Female: 12.0–16.0 g/dL	Male: 2.09–2.71 mmol/L Female: 1.86–2.48 mmol/L
Hematocrit	HT	Male: 41%–53% Female: 36%–46%	Male: 0.41–0.53 Female: 0.36–0.46
Mean corpuscular volume	MCV	80–100 μm	80–100 fl
Mean corpuscular hemoglobin	MCH	25.4–34.6 pg/cell	0.39–0.54 fmol/cell
Mean corpuscular hemoglobin concentration	MCHC	31%–36% Hb/cell	4.81–5.58 mmol Hb/L
Platelets	Platelets	150,000–400,000/mm	150–400 10/L

including red blood cells (RBCs), white blood cells (WBCs), platelets, hemoglobin, and hematocrit, to assess overall health and detect a wide range of medical conditions (Table 12.1).

12.3 Blood and Its Related Diseases

Blood-related diseases encompass a wide range of conditions that affect the composition, function, or production of blood and its cellular components. These diseases can be broadly categorized into several groups.

12.3.1 Anemia

Anemia is a condition marked by a decline in the quantity of red blood cells or a reduction in hemoglobin levels, resulting in a diminished capacity to carry oxygen in the bloodstream. This deficiency arises from various causes:

- Insufficient Hemoglobin or Red Blood Cell Production: The body fails to generate an adequate amount of hemoglobin or red blood cells.
- Bleeding: Excessive bleeding results in the rapid loss of red blood cells and hemoglobin, surpassing the rate at which they can be replenished.
- Red Blood Cell Destruction: The body prematurely breaks down red blood cells, leading to the destruction of the hemoglobin contained within them (Altemose and Atkinson 2023).

Different types of anemia can arise from various underlying causes, as shown below.

12.3.1.1 Iron Deficiency Anemia

Iron deficiency anemia is caused by inadequate iron intake, inefficient iron absorption, or ongoing blood loss, such as through menstruation or gastrointestinal bleeding. Iron deficiency stands as the primary global cause of anemia, as iron plays a vital role in hemoglobin formation, essential for oxygen transport and carbon dioxide removal in red blood cells. The body primarily stores iron in hemoglobin, with additional reserves in ferritin and hemosiderin located in the bone marrow, spleen, and liver (Halterman and Segel 2023). Iron deficiency anemia can arise from various causes, including dietary factors, physiological changes, gastrointestinal tract abnormalities, and blood loss. Diets lacking in sufficient iron intake can contribute to this condition, as the body absorbs only a fraction of the iron ingested. Individuals with imbalanced diets may consequently experience some level of iron deficiency anemia. Additionally, certain physiological changes, such as growth spurts in children and adolescents, as well as the increased demands of pregnancy and lactation, necessitate heightened iron intake and red blood cell production, potentially leading to anemia if not met. Gastrointestinal tract abnormalities, such as malabsorption after certain surgeries or conditions affecting the upper small intestine, can impair iron absorption, exacerbating the risk of anemia. Furthermore, blood loss from various sources, including gastrointestinal bleeding, menstrual bleeding, or injury, can deplete iron levels in the body, contributing to the development of iron deficiency anemia (Halterman and Segel 2023). The most common symptoms of iron deficiency anemia may vary among individuals but often include abnormal paleness or a lack of color in the skin, irritability, fatigue or a sense of easily tiring, increased heart rate (known as tachycardia), a sore or swollen tongue, enlargement of the spleen, and unusual cravings such as a desire to eat substances like dirt or ice, a condition known as pica.

Iron is abundant in various dietary sources and plays a crucial role in numerous physiological processes within the body. Upon ingestion, iron is absorbed primarily in the duodenum and upper jejunum of the small intestine. Once absorbed, iron binds to transferrin, a plasma protein that facilitates its transport throughout the bloodstream. Transferrin-bound iron is then utilized in the synthesis of hemoglobin, the protein responsible for carrying oxygen in red blood cells. Additionally, excess iron is stored in the liver, spleen, and bone marrow in the form of ferritin and hemosiderin, serving as a reserve for future needs. Throughout the body, iron is essential for cellular functions, including energy production and DNA synthesis. Thus, ensuring an adequate intake of iron-rich foods is vital for maintaining overall health and well-being (Halterman and Segel 2023).

12.3.1.2 Vitamin Deficiency Anemias

Vitamin deficiency anemia refers to a condition where the body lacks sufficient levels of specific vitamins essential for red blood cell production. Vitamins crucial for the synthesis of healthy red blood cells include vitamin B12, also known as cobalamin, and folate, or vitamin B9. When the body lacks these vitamins, red blood cell production is impaired, leading to anemia. Vitamin B12 is primarily found in animal products such as meat, fish, eggs, and dairy, while folate is abundant in leafy greens, legumes, and fortified grains. Causes of vitamin deficiency anemia may include poor dietary intake, malabsorption disorders, certain medical conditions, or medications that interfere with vitamin absorption or metabolism. Vitamin deficiency anemia can arise from inadequate intake of foods rich in vitamins B12 or folate or from conditions that hinder the body's ability to absorb or utilize these nutrients effectively. Individuals with conditions such as biliary atresia, celiac disease, Crohn's disease, or cystic fibrosis may experience impaired absorption of these vitamins due to gastrointestinal abnormalities or malabsorption issues. Pernicious anemia, a type of vitamin deficiency anemia, occurs when the body lacks intrinsic factor, a substance necessary for the absorption of vitamin B12 from foods in the digestive tract. Without intrinsic factor, the body cannot effectively absorb vitamin B12, leading to its deficiency and subsequent anemia. Identifying and managing the underlying cause of impaired vitamin absorption or utilization is crucial in treating and preventing vitamin deficiency anemia.

12.3.1.3 Hemolytic Anemias

Hemolytic anemia is a condition characterized by the accelerated destruction of red blood cells, outpacing the body's ability to produce new ones, a process known as hemolysis. As carriers of oxygen to various tissues and organs, red blood cells play a crucial role in maintaining bodily functions. However, individuals with hemolytic anemia experience lower-than-normal levels of red blood cells, leading to anemia. Consequently, insufficient oxygen delivery to tissues and organs occurs, impairing their proper functioning. This condition can manifest into two forms: inherited and acquired. Inherited hemolytic anemia arises when the genetic mutation responsible for the disorder is passed down from parents to their offspring. Conversely, acquired hemolytic anemia develops later in life and is not present from birth (Baldwin et al. 2023).

12.3.1.4 Immune Hemolytic Anemia

Anemia, a medical condition characterized by insufficient healthy red blood cells, affects the body's ability to supply oxygen to its tissues adequately. Red blood cells, responsible for transporting oxygen throughout the body, typically have a lifespan of around 120 days before they are naturally eliminated by the body. However, in hemolytic anemia, red blood cells are destroyed prematurely, leading to a shortage of functional cells. One form of hemolytic anemia, known as immune hemolytic anemia, occurs when the body's immune system produces antibodies that mistakenly identify and attack its own red blood cells. This autoimmune response results from the immune system misrecognizing the body's cells as foreign invaders, leading to their destruction (Hill and Hill 2018).

Hemolytic anemia is caused by various factors, including exposure to certain chemicals, medications, or toxins, as well as infections. Another potential cause is the transfusion of blood from a donor whose blood type does not match that of the recipient, leading to an immune reaction against the donated red blood cells. Certain cancers can also disrupt normal red blood cell function, contributing to hemolytic anemia. In cases where antibodies form against red blood cells without a clear cause, the condition is termed idiopathic autoimmune hemolytic anemia. These antibodies may arise as a complication of another underlying disease, previous blood transfusions, or during pregnancy, particularly if there is a mismatch in blood types between the mother and the baby. The risk factors associated with hemolytic anemia are closely linked to its underlying causes, highlighting the importance of identifying and addressing these factors to manage the condition effectively (Hill and Hill 2018).

Symptoms of hemolytic anemia can vary depending on the severity of the condition. In mild cases, individuals may not experience noticeable symptoms initially. However, as the anemia progresses, common early signs may include increased feelings of weakness or fatigue, particularly during physical activity, along with headaches and difficulties concentrating or thinking clearly. As the condition worsens, additional symptoms may manifest, such as lightheadedness upon standing, a visibly pale skin color (pallor), shortness of breath, and a sore tongue. These symptoms arise due to the decreased oxygen-carrying capacity of the blood resulting from the destruction of red blood cells, highlighting the importance of timely diagnosis and management of hemolytic anemia (Baldwin et al. 2023).

To diagnose hemolytic anemia and determine its severity and underlying cause, various tests may be required. These tests include assessing the absolute reticulocyte count to evaluate red blood cell production, conducting direct or indirect Coombs tests to detect antibodies attacking red blood cells, checking for hemoglobin in the urine as a sign of hemolysis, measuring lactate dehydrogenase (LDH) levels indicative of tissue damage, and examining red blood cell count, hemoglobin, and hematocrit levels. Additionally, serum bilirubin levels and serum haptoglobin levels are assessed to evaluate red blood cell breakdown and the body's response to hemolysis. Other tests such as the Donath–Landsteiner test, cold agglutinins test, platelet count, and protein electrophoresis may also be performed to further investigate the underlying cause of hemolytic anemia (Baldwin et al. 2023).

The initial treatment for hemolytic anemia typically involves steroid medications like prednisone. If steroid therapy fails to alleviate symptoms, alternative approaches such as intravenous immunoglobulin (IVIG) therapy or splenectomy (surgical removal of the spleen) may be considered. In cases where the immune system remains overactive despite steroid treatment, immunosuppressive drugs like azathioprine (Imuran), cyclophosphamide (Cytoxan), or rituximab (Rituxan) may be prescribed to suppress immune activity. However, blood transfusions are administered cautiously due to the risk of incompatible blood types, which can exacerbate red blood cell destruction (Baldwin et al. 2023).

12.3.1.5 Hereditary Spherocytosis

Hereditary spherocytosis is a genetic blood disorder characterized by hemolytic anemia, where red blood cells are disrupted. This leads to the production of fragile red blood cells that are prone to premature destruction or destruction at a faster rate than usual. This condition arises due to a defect in the genes responsible for encoding proteins involved in maintaining the shape and flexibility of red blood cells' membranes. As a result, affected red blood cells lose their typical biconcave disk shape and instead become spherical, or spherocytes. This altered morphology impairs the cells' ability to traverse through narrow blood vessels and efficiently carry oxygen to tissues, leading to anemia and associated symptoms.

12.3.1.6 Sickle Cell Anemia

Sickle cell anemia, a form of sickle cell disease, is characterized by abnormal hemoglobin molecules, leading to formation of red blood cells that assume a sickle or crescent shape, impairing their ability to effectively transport oxygen throughout the body. Typically, red blood cells are pliable and round, allowing them to smoothly navigate blood vessels. In contrast, the rigid and adhesive nature of sickle cells can impede blood flow, leading to episodes of pain and potentially causing blockages in blood vessels (Pham et al. 2019).

Sickle cell anemia, a hereditary blood disorder, arises from a mutation in the gene responsible for producing hemoglobin, the protein within red blood cells crucial for oxygen transport throughout the body. This genetic alteration leads to the formation of abnormal hemoglobin, known as hemoglobin S (HbS), which causes red blood cells to adopt a rigid, sticky, and distorted shape. For an individual to manifest sickle cell anemia, both parents must carry and pass on a copy of the mutated gene to their child. Inheritance follows an autosomal recessive pattern, meaning that the child must inherit two copies of the mutated gene one from each parent. When only one parent transmits the mutated gene, the child inherits the sickle cell trait, characterized by one normal hemoglobin gene and one mutated hemoglobin S gene. Individuals with the sickle cell trait produce a combination of normal and sickle cell hemoglobin, leading to the presence of some sickle-shaped red blood cells in their circulation. Typically, carriers of the sickle cell trait do not exhibit symptoms of the disease but serve as carriers who can pass the mutated gene to their offspring (Belvis et al. 2018; Pham et al. 2019).

Symptoms of sickle cell anemia typically emerge around the age of 6 months and can vary in severity and presentation over time. Common symptoms include anemia, characterized by a shortage of red blood cells due to the premature breakdown of sickle cells, resulting in fatigue due to inadequate oxygen supply to tissues. Pain crises, marked by intense and episodic pain, occur when sickle-shaped red blood cells obstruct blood flow in small vessels, affecting various body parts such as the chest, abdomen, and joints. The duration and frequency of these crises vary among individuals, with some experiencing a few episodes annually while others endure numerous crises requiring hospitalization. Chronic pain may also accompany bone and joint damage, ulcers, and other complications. Swelling of the hands and feet can occur as sickle cells impede blood circulation in these areas. Individuals with sickle cell anemia are prone to frequent infections due to spleen damage caused by sickle cells, necessitating vaccinations and antibiotics, particularly in infants and children, to prevent serious infections like pneumonia. Additionally, delayed growth and puberty may result from the reduced oxygen and nutrient supply associated with insufficient red blood cells. Vision problems may arise from sickle cell-induced blockages in the tiny blood vessels supplying the eyes, potentially leading to retinal damage and visual impairment (Belvis et al. 2018).

Treatment for sickle cell anemia typically involves managing symptoms, preventing complications, and improving quality of life. This may include pain management with medications and non-pharmacological approaches, such as heat therapy and relaxation techniques. Hydroxyurea may be prescribed to reduce pain crises frequency, while blood transfusions can help manage severe anemia and prevent complications like stroke. Antibiotics and vaccinations are crucial for infection prevention, especially in children. In select cases, bone marrow transplantation may offer a potential cure. Symptom management, supportive care, and individualized treatment plans are key components of comprehensive sickle cell management (Belvis et al. 2018; Pham et al. 2019; Strouse 2016).

12.3.1.7 Thalassemia

Thalassemia, a genetic blood disorder, arises from mutations in the DNA of cells responsible for producing hemoglobin—the oxygen-carrying component within red blood cells. These mutations, inherited from parents, disrupt the normal synthesis of hemoglobin chains, leading to either alpha-thalassemia or beta-thalassemia (Cohen et al. 2004). Alpha-thalassemia's severity hinges on the number of mutated genes inherited. Individuals may have one, two, three, or rarely, four mutated genes. With one mutated gene, there are typically no symptoms, but carriers can pass the condition to their offspring. Two mutated genes result in mild symptoms, termed alpha-thalassemia trait, while three mutations lead to moderate to severe symptoms. In the exceedingly rare occurrence of inheriting four mutated genes, the outcome is often fatal, with stillbirth or infant death (Tamary and Dgany 2020).

Beta-thalassemia, on the other hand, involves mutations in two genes responsible for beta hemoglobin chains. Inheritance of one mutated gene typically causes mild symptoms, known as thalassemia minor or beta-thalassemia. However, inheriting two mutated genes results in moderate to severe symptoms, termed thalassemia major or Cooley anemia. Babies born with two defective beta hemoglobin genes may initially appear healthy but develop symptoms within the first few years of life. Thalassemia intermedia, a milder form, can also arise from two mutated genes (Langer 2024). Thalassemia presents a significant health concern globally, with varying prevalence rates across different regions and populations.

The treatment approach for thalassemia depends on the severity of the condition and its specific complications. For individuals with mild forms of thalassemia, such as thalassemia minor, treatment may not be necessary, and individuals can lead normal lives without significant medical intervention. However, for those with moderate to severe thalassemia, various treatment modalities are available to manage symptoms and improve quality of life. One of the primary treatment strategies is regular blood transfusions, which help alleviate anemia by providing healthy red blood cells. However, frequent transfusions can lead to iron overload in the body over time. To address this, individuals may undergo iron chelation therapy, a process that involves the administration of medications to remove excess iron from the body and prevent organ damage (Ali et al. 2021). Additionally, some individuals may benefit from splenectomy, the surgical removal of the spleen, particularly if it becomes enlarged and contributes to worsening anemia or reduces the lifespan of transfused red blood cells. For individuals with severe thalassemia who are unable to produce adequate amounts of hemoglobin, bone marrow transplantation may be considered as a potential curative treatment option. This procedure involves replacing the defective bone marrow with healthy stem cells from a compatible donor, allowing the production of normal red blood cells. Moreover, supportive measures such as folic acid supplementation, vaccination against infections, and regular medical monitoring are essential components of thalassemia management to optimize health outcomes and prevent complications. Overall, a multidisciplinary approach involving hematologists, transfusion specialists, and other healthcare professionals is crucial to tailor treatment plans according to individual needs and ensure comprehensive care for individuals with thalassemia (Cohen et al. 2004).

12.3.1.8 Anemia of Chronic Disease

Anemia of inflammation, also known as anemia of chronic disease (ACD), is a type of anemia characterized by reduced red blood cell count and hemoglobin levels, primarily affecting individuals with underlying conditions associated with inflammation. These conditions may include infections, autoimmune disorders, cancer, and chronic kidney disease (CKD). Anemia occurs when the blood has insufficient red blood cells or hemoglobin to adequately transport oxygen throughout the body, impairing essential cellular functions. Hemoglobin, an iron-rich protein present in red blood cells, facilitates the binding and transport of oxygen from the lungs to various tissues and organs. However, in anemia of inflammation, although individuals may have normal or elevated levels of stored iron in their body tissues, there is often a decrease in iron levels in the bloodstream. This discrepancy arises because inflammation disrupts the body's ability to utilize stored iron effectively in the production of healthy red blood cells, contributing to the development of anemia. Consequently, individuals with anemia of inflammation may experience symptoms associated with insufficient oxygen delivery to tissues, such as fatigue, weakness, and shortness of breath. Understanding the underlying inflammatory condition and addressing its management is crucial in effectively treating anemia of inflammation and restoring optimal red blood cell production and hemoglobin levels (Ganz 2019). In anemia of inflammation, alterations in the functioning of the immune system can instigate changes within the body that contribute to the development of this type of anemia. These changes can manifest in various ways such as the body may exhibit abnormalities in the storage and utilization of iron, a vital component for red blood cell production. This disruption in iron metabolism can impede the availability of iron for incorporation into hemoglobin, the protein responsible for oxygen transport in red blood cells. The kidneys, which play a crucial role in regulating erythropoiesis (the production of red blood cells), may produce reduced levels of erythropoietin (EPO). EPO is a hormone that stimulates the bone marrow to generate new red blood cells. Insufficient production of EPO can lead to decreased red blood cell formation, exacerbating the anemic state. Furthermore, even if adequate levels of

EPO are present, the bone marrow may exhibit diminished responsiveness to its signals. Consequently, the bone marrow may produce fewer red blood cells than required to maintain normal oxygen levels in the blood (Ganz 2019).

Blood tests can detect signs of anemia of inflammation, other types of anemia, or underlying health issues. Blood count tests are employed to assess various components and characteristics of the blood, including the red blood cell count, average size of red blood cells, hemoglobin levels in the blood and within the red blood cells, as well as the presence of developing red blood cells. Some of these blood count tests may be integrated into a comprehensive assessment known as a complete blood count. Additionally, a blood smear may be conducted to examine the morphology, size, and quantity of red blood cells. Furthermore, iron levels in the blood and stored in the body are evaluated through tests that measure iron concentration, transferrin levels (a protein responsible for iron transport in the blood), and ferritin levels (a protein storing iron within the body's cells). Anemia of inflammation may be diagnosed if blood test results indicate the presence of anemia, low iron levels in the blood, and normal iron levels in the body's tissues. If the cause of anemia of inflammation remains unclear based on initial blood tests, further diagnostic evaluations may be performed to identify the underlying cause (Ganz 2019).

Treatment of the underlying condition may lead to an improvement or resolution of the anemia. For instance, medications aimed at reducing inflammation in conditions like rheumatoid arthritis can alleviate anemia. Erythropoiesis-stimulating agents (ESAs) such as epoetin alpha or darbepoetin alpha may be prescribed to stimulate red blood cell production in cases related to chronic kidney disease (CKD), cancer chemotherapy, or HIV treatments. These agents are administered via injections, sometimes taught for self-administration at home. Additionally, iron supplements may be prescribed to complement the effects of ESAs. In cases of hemodialysis, intravenous ESAs and iron supplements can be administered during treatment sessions. In severe instances, blood transfusions may be necessary to rapidly boost hemoglobin levels and oxygenation in the blood (Madu and Ughasoro 2017).

12.3.1.9 Aplastic Anemia

Aplastic anemia is a disorder characterized by insufficient production of new blood cells by the body, resulting in fatigue and heightened susceptibility to infections and bleeding. This is a rare and life-threatening bone marrow failure illness defined by pancytopenia, a shortage in all three major blood cell lines: red blood cells (anemia), white blood cells (increased infection risk), and platelets (impaired coagulation). Aplastic anemia results from damage to the stem cells in the bone marrow, which are responsible for producing blood cells—red cells, white cells, and platelets. This impairment leads to a condition where the bone marrow is either depleted of blood cells (aplastic) or contains significantly fewer cells than normal (hypoplastic). The primary cause of aplastic anemia is often an autoimmune reaction, wherein the immune system targets and attacks the stem cells in the bone marrow. Additionally, various factors can contribute to bone marrow injury and affect blood cell production (Silberstein and Huyck 2008). These include exposure to radiation and chemotherapy during cancer treatments, which may inadvertently harm healthy stem cells along with cancerous cells, leading to temporary aplastic anemia. Exposure to certain toxic chemicals found in pesticides, insecticides, and benzene, a component of gasoline, has also been associated with aplastic anemia, with symptoms potentially improving upon avoiding repeated exposure. Furthermore, some medications, particularly those used to treat conditions like rheumatoid arthritis and certain antibiotics, can induce aplastic anemia as a side effect (Brodsky and Jones 2006). Autoimmune disorders, viral infections impacting bone marrow function (such as hepatitis, Epstein–Barr, cytomegalovirus, parvovirus B19, and HIV), and even pregnancy, where the immune system may inadvertently target bone marrow, are among the diverse array of factors linked to aplastic anemia. In some instances, aplastic anemia may arise without a discernible cause, labeled as idiopathic aplastic anemia. Notably, there are associations between aplastic anemia and other rare disorders, such as paroxysmal nocturnal hemoglobinuria, which accelerates the breakdown of red blood cells, potentially leading to aplastic anemia, or vice versa. Moreover, Fanconi's anemia, an inherited rare disease characterized by physical abnormalities and developmental defects, can progress to aplastic anemia (Chen 2013).

Aplastic anemia is associated with various inherited bone marrow failure syndromes, characterized by insufficient stem cell production within the bone marrow. This failure occurs due to genetic abnormalities impacting the bone marrow's ability to generate an adequate number of stem cells. Several inherited conditions are linked to this phenomenon, including Fanconi's anemia, dyskeratosis congenita, Shwachman–Diamond syndrome, Diamond–Blackfan anemia, and Pearson syndrome (Dokal 2006). These syndromes often manifest with a range of symptoms and complications related to bone marrow dysfunction, highlighting the genetic complexity underlying aplastic anemia. There are certain factors that increase the risk of its development. These include undergoing intensive radiation or chemotherapy treatments for cancer, exposure to harmful chemicals, and the use of specific prescription medications, like chloramphenicol for bacterial infections, or gold compounds for rheumatoid arthritis. Treatment approaches for aplastic anemia are tailored to individual circumstances, considering factors such as the underlying cause and severity of the condition. In cases where aplastic anemia develops as a result of ongoing cancer or autoimmune disease treatments, adjusting these therapies may effectively manage the condition. For individuals with moderate aplastic anemia, characterized by lower-than-normal blood cell levels without significant symptoms, close monitoring of overall health and blood counts may be recommended to promptly address any worsening of the condition (Dezern and Brodsky 2011).

For more severe forms of aplastic anemia, various treatment options are available. Immunosuppressant medications, such as antithymocyte globulin injection (Atgam®) and cyclosporine (Sandimmune®), may be prescribed to suppress the immune system's attack on the stem cells in the bone marrow. Blood transfusions can help replace depleted red blood cells and platelets, alleviating certain symptoms, although they do not provide a cure for aplastic anemia. Antibiotics may also be administered to manage the increased risk of bacterial infections associated with aplastic anemia (Dezern and Brodsky 2011).

12.3.2 Multiple Myeloma

Multiple myeloma is a cancer originating from a specific type of white blood cell known as a plasma cell. These cells normally aid in combating infections by producing antibodies, which target and eliminate pathogens. However, in multiple myeloma, malignant plasma cells accumulate within the bone marrow, the spongy tissue found within bones responsible for blood cell production. As a result, the cancerous cells overshadow healthy blood cell production, disrupting the balance. Rather than generating functional antibodies, the cancerous plasma cells produce abnormal proteins, leading to various complications associated with multiple myeloma. The exact cause of myeloma remains uncertain. It initiates with the transformation of a single plasma cell within the bone marrow, the spongy tissue inside bones where blood cells are produced, into a cancerous myeloma cell. This aberrant cell multiplication occurs rapidly, deviating from the usual regulated growth and lifespan of healthy cells. Unlike healthy cells, cancer cells exhibit uncontrolled proliferation and evade programmed cell death. Consequently, myeloma cells accumulate within the bone marrow, displacing healthy blood cells and leading to symptoms like fatigue and increased susceptibility to infections. Despite their malignant nature, myeloma cells persist in attempting to produce antibodies, akin to normal plasma cells. However, these antibodies, known as monoclonal proteins or M proteins, cannot be effectively utilized by the body. Instead, they accumulate, causing various complications such as kidney damage and bone deterioration, thereby heightening the risk of fractures. Notably, multiple myeloma often originates from a precursor condition termed monoclonal gammopathy of undetermined significance (MGUS), characterized by low levels of M proteins that do not inflict harm on the body.

In the early stages of multiple myeloma, individuals may not exhibit any symptoms. However, as the disease progresses, various signs and symptoms may manifest, including bone pain, particularly in areas like the spine, chest, or hips. Other common symptoms can encompass nausea, constipation, loss of appetite, mental fogginess or confusion, fatigue, susceptibility to infections, weight loss, weakness, increased thirst, and frequent urination. These symptoms often develop gradually and may vary in severity among affected individuals (Seesaghur et al. 2021). Several factors can elevate the risk of developing multiple myeloma. Primarily, advancing age plays a significant role, with the majority of diagnoses occurring in individuals in their late 60 s. Additionally, gender is a factor, as men have a higher predisposition to the disease compared to women. Furthermore, the presence of monoclonal gammopathy of undetermined significance (MGUS), which often precedes multiple myeloma, serves as an additional risk factor. Despite these identified factors, there are currently no preventive measures for multiple myeloma, as the condition is not attributed to any specific actions or behaviors. Multiple myeloma, a cancer that originates in the bone marrow's plasma cells, can lead to various complications.

12.3.3 Thrombocytopenia

Thrombocytopenia, characterized by a reduced blood platelet count, is a condition where the body has fewer colorless blood cells known as platelets, crucial for blood clotting. Platelets play a vital role in stopping bleeding by forming clumps and plugs in blood vessel injuries. This condition can arise from various causes, including bone marrow disorders like leukemia, immune system issues, or as a side effect of specific medications. Thrombocytopenia can affect individuals of all ages, encompassing both children and adults (Jinna and Khandhar 2023). Thrombocytopenia is defined by a platelet count of less than 150,000 per microliter of circulating blood. Given that each platelet has a lifespan of around 10 days, the body routinely replenishes its platelet supply by generating new platelets in the bone marrow. While thrombocytopenia is seldom inherited, it can be triggered by various medications or medical conditions. Regardless of the cause, the reduction in circulating platelets can occur due to one or more mechanisms: platelet entrapment in the spleen, diminished platelet production, or accelerated platelet destruction (Jinna and Khandhar 2023). Trapped platelets can result from spleen enlargement, a condition known as splenomegaly, which is often associated with certain disorders. An enlarged spleen can sequester an excessive number of platelets, thereby reducing the platelet count in circulation. Conversely, decreased platelet production can arise from factors such as leukemia, specific types of anemia, viral infections like hepatitis C or HIV, chemotherapy, radiation therapy, or excessive alcohol consumption.

Certain medications can also reduce platelet counts, either by directly affecting platelet production or by causing immune-mediated destruction of platelets. Examples include heparin, quinine, sulfa-containing antibiotics, and anticonvulsants. Identifying the underlying cause of thrombocytopenia is crucial for determining the appropriate treatment approach and managing potential complications (Kim 2022). Unusually heavy menstrual flows, fatigue, and an enlarged spleen may also be indicative of thrombocytopenia. Recognizing these symptoms is crucial for prompt diagnosis and appropriate management of the condition (Choudhary et al. 2023).

Treatment strategies for thrombocytopenia aim to increase platelet count, prevent bleeding complications, and address the underlying cause of the condition. In cases where thrombocytopenia is mild and does not pose significant bleeding risks, treatment may not be necessary, and close monitoring of platelet levels may suffice. Medications such as corticosteroids, immunoglobulins, or thrombopoietin receptor agonists can stimulate platelet production or prevent the immune system from destroying platelets. In emergency situations or severe cases, platelet transfusions may be administered to quickly raise platelet levels and reduce bleeding risks.

12.3.4 Disseminated Intravascular Coagulation (DIC)

Disseminated intravascular coagulation (DIC) is a seldom-encountered disorder characterized by abnormal blood clotting and potential harm to organs, coupled with unmanageable bleeding. It typically arises as a complication of various severe medical conditions, and without prompt treatment, it can pose life-threatening risks (Costello and Nehring 2023). Understanding the mechanisms behind disseminated intravascular coagulation (DIC) involves grasping the normal clotting process in the body. Typically, the liver produces clotting factors, proteins that bind to platelets in the blood, facilitating the formation of blood clots (Papageorgiou et al. 2018). These clots serve to staunch or slow bleeding, initiating the body's healing mechanisms. However, in DIC, an excessive amount of blood clots forms, surpassing the body's requirements. DIC progresses through two distinct stages: Initially, small blood clots begin to obstruct blood vessels, leading to compromised blood supply to vital organs. Subsequently, platelets and clotting factors become depleted, leaving the body incapable of managing bleeding effectively (Costello and Nehring 2023). Risk factors for disseminated intravascular coagulation (DIC) encompass severe infections, injuries, specific medical conditions, or complications arising from medical treatments. Medical conditions contributing to DIC may include sepsis, certain types of leukemia, pancreatitis, liver disease, severe tissue injuries like burns or head trauma, and significant complications during pregnancy and childbirth. Additionally, medical treatments such as blood transfusion reactions, severe immune system responses to organ transplants, recent surgeries, or anesthesia administration may also trigger DIC. If left untreated, DIC can lead to severe complications such as stroke, shock, and uncontrolled bleeding (Levi and Scully 2018).

Symptoms of disseminated intravascular coagulation (DIC) can manifest as uncontrollable bleeding from multiple sites across the body, extensive bruising, confusion, memory impairment, or alterations in behavior. Additionally, individuals may experience difficulty breathing (dyspnea) and develop a high fever exceeding 103 degrees Fahrenheit (39.4 degrees Celsius). These symptoms typically indicate a serious and potentially life-threatening medical condition requiring prompt medical attention (Costello and Nehring 2023).

12.3.5 Hemophilia

Hemophilia is an uncommon condition characterized by impaired blood clotting due to insufficient levels of clotting factors, essential proteins responsible for initiating the clotting process. Individuals with hemophilia may experience prolonged bleeding following injuries compared to those with normal clotting abilities. While minor cuts typically pose minimal concern, individuals with severe hemophilia face heightened risks of internal bleeding, particularly in joints like the knees, ankles, and elbows. Internal bleeding has the potential to cause damage to vital organs and tissues, posing life-threatening complications. These clotting factors, working in tandem with platelets, help create clots to staunch bleeding. Hemophilia is predominantly a genetic disorder, typically inherited from one's parents, and manifests in varying degrees of severity depending on the deficiency of clotting factors present (Mitchell and Walsh 2023). Hemophilia can be congenital or acquired, with congenital hemophilia being the more common form, typically inherited from parents. Congenital hemophilia is further classified based on the deficient clotting factor, with hemophilia A, characterized by low levels of factor 8, being the most prevalent, followed by hemophilia B, associated with deficient factor 9 levels (Mitchell and Walsh 2023).

On the other hand, acquired hemophilia manifests in individuals without a family history of the disorder and occurs when the immune system erroneously targets clotting factors 8 or 9 in the blood. This variation of hemophilia can be linked to various factors, including pregnancy, autoimmune conditions, cancer, multiple sclerosis, and drug reactions. In terms of inheritance, hemophilia is typically transmitted through the X chromosome, as the defective gene responsible for the condition is often located on this chromosome.

Consequently, males are predominantly affected by hemophilia, inheriting the faulty gene from their carrier mothers. While most carrier females exhibit no symptoms, some may experience bleeding issues if their clotting factor levels are moderately decreased (Mitchell and Walsh 2023). The signs and symptoms of hemophilia can vary depending on the severity of the deficiency in clotting factors. Mild reductions in clotting factor levels may result in bleeding only after surgery or trauma, while severe deficiencies can lead to spontaneous bleeding with no apparent cause. Common indications of spontaneous bleeding include excessive bleeding from cuts or injuries, large or deep bruises, and unexpected bleeding following vaccinations. Additionally, individuals with hemophilia may experience joint pain, swelling, or tightness, as well as blood in the urine or stool, unexplained nosebleeds, and irritability in infants. One of the most serious complications of severe hemophilia is bleeding into the brain, which can occur even from minor head injuries. Symptoms of this condition include prolonged headache, repeated vomiting, drowsiness, double vision, sudden weakness or clumsiness, and in severe cases, convulsions or seizures (Green 2016).

12.3.6 Myeloproliferative Neoplasms

Myeloproliferative neoplasms (MPNs) represent a group of rare blood cancers characterized by an abnormal increase in the production of blood cells within the bone marrow. These cancers involve the overproduction of various blood cell types, including red blood cells, white blood cells, and platelets, which are normally produced in the bone marrow's spongy tissue. In individuals with MPNs, there is a disruption in the normal process of blood cell production. Despite their potential life-threatening nature, MPNs typically progress slowly, often remaining asymptomatic for years before symptoms become apparent. Referred to interchangeably as chronic myeloproliferative neoplasms or myeloproliferative disorders, these conditions are chronic in nature, indicating their long-term presence. While rare, there is a possibility that a myeloproliferative neoplasm may transform into a more aggressive form of blood cancer over time (Spiers et al. 2023).

Myeloproliferative neoplasms (MPNs) are a group of blood diseases characterized by the overproduction of a specific type of blood cell by the bone marrow, leading to an imbalance in the body's blood cell composition. This excess production can have varying impacts on the body depending on the type of blood cell affected. For instance, certain MPNs increase the risk of cardiovascular events such as heart attacks or strokes, while others may result in conditions like anemia. Understanding these rare diseases involves gaining insight into the processes of bone marrow function and blood cell production. Blood cells originate from stem cells within the bone marrow, a vital tissue located in the center of bones. These stem cells can differentiate into two main types: lymphoid stem cells and myeloid stem cells (Kondo 2010).

Lymphoid stem cells give rise to white blood cells, which play a crucial role in the body's immune response against infections (Charles et al. 2001). On the other hand, myeloid stem cells have the potential to develop into various blood cell types, including red blood cells responsible for oxygen transport, white blood cells involved in immune defense, and platelets that contribute to blood clotting (Lee and Hong 2020). The production of blood cells is tightly regulated by genetic instructions encoded within the cells. Under normal circumstances, stem cells undergo controlled division and multiplication as dictated by these genetic signals. However, when certain genes mutate, aberrant instructions are sent to the stem cells, prompting uncontrolled proliferation and accumulation of mature blood cells (Edginton-White and Bonifer 2022). Over time, these abnormal blood cells accumulate in the bone marrow or enter the bloodstream in excessive numbers, disrupting normal blood flow dynamics. Consequently, blocked blood flow can lead to the development of serious medical conditions and complications associated with MPNs (Tremblay et al. 2021).

Myeloproliferative neoplasms (MPNs) are acquired genetic disorders, meaning they are not inherited from biological parents but instead develop due to mutations in genes responsible for regulating cell growth. These mutations disrupt the normal process of blood cell development, leading to the onset of MPNs (Vainchenker and Kralovics 2017). Recent advancements in medical research have shed light on the specific genetic mutations associated with different types of MPNs. For instance, mutations in the Janus kinase 2 (JAK2) gene are commonly found in individuals with polycythemia vera, primary myelofibrosis, and essential thrombocythemia. These mutations can trigger uncontrolled cell proliferation, contributing to the development of MPNs (Kralovics et al. 2005). Additionally, mutations in the MPL gene or CALR gene have been identified in individuals with essential thrombocythemia and primary myelofibrosis (Asp et al. 2016). These genetic alterations play a role in the pathogenesis of MPNs by disrupting normal cellular functions. Furthermore, chromosome errors, such as the Philadelphia chromosome abnormality, are associated with chronic myelogenous leukemia (CML). This aberration involves the exchange of genetic material between chromosomes, leading to the formation of the Philadelphia chromosome (Asp et al. 2016). While the exact triggers for these genetic changes remain unclear, these discoveries have significantly improved the understanding of MPNs and facilitated the development of targeted diagnostic approaches and treatment strategies. By identifying specific genetic mutations, healthcare providers can tailor treatment plans to address the underlying molecular abnormalities driving the disease progression (Asp et al. 2016; Vainchenker and Kralovics 2017). The three most common types of myeloproliferative neoplasms (MPNs) are polycythemia vera, essential thrombocythemia, and primary myelofibrosis, each characterized by distinct features and potential complications (Pizzi et al. 2021).

Essential thrombocythemia arises when the bone marrow generates an excess of platelets, even when they are not required. This surplus of platelets predisposes individuals to the formation of blood clots, elevating the risk of heart attacks or strokes. Symptoms of essential thrombocythemia develop gradually, often detected during routine blood tests indicating high platelet levels. While typically slow to progress, some individuals with essential thrombocythemia may develop leukemia over time. Other types of myeloproliferative neoplasms (MPNs) include chronic eosinophilic leukemia (CEL), chronic myelogenous leukemia (CML), chronic neutrophilic leukemia (CNL), and myeloproliferative neoplasm, unclassifiable (MPN-U), each presenting distinct characteristics and potential complications (Pizzi et al. 2021).

Chronic eosinophilic leukemia (CEL) is characterized by the excessive production of eosinophils, a type of white blood cell. Typically, CEL progresses slowly; however, in rare instances, it can transform into acute myeloid leukemia (AML), a more aggressive form of cancer. CEL is also referred to as hypereosinophilic syndrome (Pizzi et al. 2021). Chronic myelogenous leukemia (CML) involves the overproduction of granulocytes, a type of white blood cell. The accumulation of these cells can disrupt the bone marrow's ability to produce other essential blood cell types, leading to complications. Chronic neutrophilic leukemia (CNL) is characterized by the excessive production of neutrophils, another type of white blood cell. Similar to other MPNs, CNL can lead to various symptoms and complications associated with abnormal blood cell production (Szuber et al. 2022).

Myeloproliferative neoplasm, unclassifiable (MPN-U), encompasses cases that do not fit into the established categories of MPNs. It may involve the overproduction of different blood cell types, including white blood cells, red blood cells, or platelets. MPN-U presents unique challenges in diagnosis and management due to its heterogeneous nature and variable clinical manifestations (Pizzi et al. 2021; Wahed and Dasgupta 2015). In the initial stages of myeloproliferative neoplasms (MPNs), symptoms may not be apparent. However, as the condition progresses, signs such as an enlarged spleen (splenomegaly) may become noticeable. Splenomegaly typically presents as a feeling of fullness, pressure, or discomfort beneath the ribs on the left side, where the spleen is located. Although splenomegaly is a common symptom across most MPNs, it is less frequently observed in cases of essential thrombocythemia (Baumeister et al. 2021).

12.3.6.1 Symptoms of Specific MPNs

Chronic eosinophilic leukemia (CEL) often manifests with a rash as the primary symptom, accompanied by fatigue and fever. Additional symptoms depend on the body parts affected by elevated eosinophil levels. Chronic myelogenous leukemia (CML) and chronic neutrophilic leukemia (CNL) may present with bone pain, night sweats, fever, fatigue, easy bruising, and weight loss. In essential thrombocythemia, symptoms may include easy bruising, unexplained bleeding from the nose, mouth, or gums, gastrointestinal bleeding, and blood in the urine. Polycythemia vera (PV) can lead to headaches, dizziness, fatigue, blurred or double vision, and other visual disturbances. Primary myelofibrosis (PMF) symptoms may include those of anemia, such as fatigue, weakness, and shortness of breath. Additionally, individuals may experience pale skin, night sweats, fevers, itching, abdominal fullness, rapid satiety, weight loss, and bone pain (Geyer and Mesa 2017).

Allogeneic stem cell transplantation stands as the sole definitive cure for myeloproliferative neoplasms (MPNs). However, due to the rigorous nature of the transplant process, many individuals are ineligible candidates. In such cases, healthcare providers focus on managing the condition, alleviating symptoms, and averting complications (Choi et al. 2015). Treatment strategies aim to induce remission where possible, with approaches tailored to the specific type of MPN.

For chronic eosinophilic leukemia, therapeutic interventions may encompass chemotherapy, corticosteroids, or immunotherapy, targeting the reduction of eosinophil levels. In chronic myelogenous leukemia, targeted therapy serves as the cornerstone treatment, inhibiting aberrant cell proliferation. Additional modalities such as chemotherapy, immunotherapy, radiation therapy, and stem cell transplants may also be employed. Chronic neutrophilic leukemia management may involve chemotherapy, immunotherapy, or stem cell transplants to mitigate disease progression. Essential thrombocythemia treatment strategies vary based on symptomatology. Treatment modalities encompass targeted therapy, chemotherapy, immunotherapy, radiation therapy, and stem cell transplants (Venugopal and Mascarenhas 2020).

The prognosis for individuals diagnosed with myeloproliferative neoplasms (MPNs) varies depending on several factors, including the specific type of MPN, the timing of diagnosis, and individual response to treatment. With careful monitoring and appropriate therapy, many people live for several years following diagnosis. However, there is no one-size-fits-all prognosis for these conditions, as outcomes can differ significantly. Survival rates for specific MPNs offer some insight into expected outcomes. For instance, chronic myelogenous leukemia (CML) has seen improved survival rates due to advancements in targeted therapies, with a five-year survival rate of around 90%. Similarly, individuals with chronic neutrophilic leukemia and polycythemia vera can expect an average survival of approximately 20 years with proper management. Essential thrombocythemia often results in long-term survival, especially with medications aimed at preventing complications like blood clots. Meanwhile, most individuals with primary myelofibrosis survive for 5–10 years post-diagnosis. Understanding one's prognosis allows for informed decision-making and personalized healthcare planning with the guidance of healthcare providers (Castillo Tokumori et al. 2022; Choi et al. 2015; Venugopal and Mascarenhas 2020).

12.3.7 Leukemia

Leukemia, a hematological malignancy, is characterized by the uncontrolled proliferation of abnormal white blood cells (leukocytes) in the bone marrow and peripheral blood. This disease disrupts the normal process of blood cell production, leading to the accumulation of immature or dysfunctional cells and impairing the body's ability to fight infection (Chennamadhavuni et al. 2023; Taylor et al. 2017).

Leukemia can be broadly categorized based on two key factors: Based on maturation rate, acute leukemia is characterized by the fast and uncontrolled proliferation of immature white blood cells. These immature cells inhibit healthy blood cell synthesis, resulting in rapid progression and severe symptoms. Chronic Leukemia: A slower progression occurs, with the bone marrow producing a greater number of mature (but still aberrant) white blood cells. Symptoms can develop gradually over time. Based on cell lineage: Myeloid leukemia is a leukemia originating from myeloid stem cells and impacts the formation of numerous blood cells, including red blood cells, platelets, and certain white blood cells. Lymphoid leukemia: These leukemias originate from lymphoid stem cells and predominantly target the development of lymphocytes, a crucial component of the immune system (Taylor et al. 2017). This classification system yields four major types of leukemia:

12.3.7.1 Acute Lymphoblastic Leukemia (ALL)

It is a rapidly progressing cancer of the blood and bone marrow characterized by the uncontrolled proliferation of immature lymphoid cells, or lymphoblasts. It is the most common type of leukemia in children, although it can also affect adults. ALL arises from genetic mutations and chromosomal abnormalities that disrupt normal hematopoiesis and lead to the accumulation of malignant lymphoblasts. The disease typically presents with symptoms such as fatigue, fever, easy bruising or bleeding, bone pain, and enlarged lymph nodes. Treatment for ALL has evolved significantly over the years, with chemotherapy remaining the mainstay of therapy. Combination chemotherapy regimens, including corticosteroids, anthracyclines, vincristine, and asparaginase, are used to induce remission and eradicate leukemic cells. Central nervous system prophylaxis is also an essential component of treatment, given the high risk of CNS involvement in ALL. With advancements in risk stratification, supportive care, and targeted therapies, the prognosis for ALL has greatly improved, with high rates of cure achievable, particularly in pediatric patients.

12.3.7.2 Acute Myeloid Leukemia (AML)

It is a rapidly progressing cancer of the blood and bone marrow characterized by the abnormal proliferation of myeloid progenitor cells, leading to the accumulation of immature blasts. AML is heterogeneous, with various genetic and molecular abnormalities contributing to its pathogenesis. These include chromosomal translocations, gene mutations (such as FLT3, NPM1, and CEBPA), and aberrant expression of transcription factors critical for hematopoiesis. The disease primarily affects older adults, although it can occur at any age. Symptoms of AML may include fatigue, easy bruising or bleeding, fever, and recurrent infections. Diagnosis typically involves blood tests, bone marrow biopsy, and cytogenetic analysis to identify specific genetic abnormalities and classify the disease subtype (Percival et al. 2017). Treatment for AML typically involves induction chemotherapy to achieve remission, followed by consolidation therapy to prevent relapse. Hematopoietic stem cell transplantation may be considered for eligible patients, particularly those at high risk of relapse. Despite significant advances in treatment, the prognosis for AML remains variable, with outcomes influenced by factors such as age, cytogenetics, and response to therapy (Winer and Stone 2019).

12.3.7.3 Chronic Lymphocytic Leukemia (CLL)

Chronic lymphocytic leukemia (CLL) is a type of blood cancer characterized by the accumulation of abnormal mature B lymphocytes, or white blood cells, in the blood, bone marrow, and lymphoid tissues. It is the most common type of leukemia in adults, typically occurring in older individuals (Mukkamalla et al. 2023). CLL often progresses slowly, and some patients may not require immediate treatment upon diagnosis. However, the disease can eventually advance, leading to symptoms such as fatigue, swollen lymph nodes, and recurrent infections. CLL is heterogeneous, with varying clinical courses and outcomes among patients. Prognostic factors such as genetic abnormalities (e.g., deletion 17p, trisomy 12), immunoglobulin heavy-chain variable region (IGHV) mutation status, and expression of cell surface markers (e.g., CD38, ZAP-70) help stratify patients into different risk groups and guide treatment decisions (Editor et al. 2012). Treatment options for CLL include chemotherapy, immunotherapy (e.g., monoclonal antibodies such as rituximab), targeted therapy (e.g., BTK inhibitors like ibrutinib) (Brown et al. 2018), and stem cell transplantation for eligible patients. Recent advances in understanding the molecular biology of CLL and the development of novel targeted therapies have significantly improved outcomes for patients, offering new hope for more effective and personalized treatment approaches.

12.3.7.4 Chronic Myeloid Leukemia (CML)

Chronic myeloid leukemia (CML) is a type of blood cancer characterized by the abnormal proliferation of myeloid cells in the bone marrow and their accumulation in the blood. The hallmark of CML is the presence of the Philadelphia chromosome (Ph), resulting from a reciprocal translocation between chromosomes 9 and 22 [t(9;22)(q34;q11)]. This translocation leads to the formation of the BCR-ABL1 fusion gene, which encodes a constitutively active tyrosine kinase protein. The dysregulated activity of this protein promotes uncontrolled cell growth and inhibits apoptosis, contributing to the development of leukemia. CML typically progresses slowly through three phases: chronic phase, accelerated phase, and blast crisis. Treatment options for CML have been revolutionized by the development of tyrosine kinase inhibitors (TKIs), such as imatinib, dasatinib, and nilotinib, which specifically target the BCR-ABL1 protein and effectively induce remission in the majority of patients. With the advent of TKIs, the prognosis for patients with CML has dramatically improved, transforming it from a fatal disease to a manageable chronic condition in many cases. However, long-term monitoring and adherence to treatment are essential to ensure optimal outcomes and prevent disease progression or relapse (Jabbour and Kantarjian 2020).

Leukemia accounts for a significant proportion of cancer diagnoses worldwide, with variations in incidence rates across different geographic regions and populations. The incidence of leukemia is higher in developed countries, with an estimated 61,090 new cases and 23,660 deaths expected in the United States as per 2021 cancer statistics (Siegel et al. 2021). The etiology of leukemia is multifactorial, with genetic predisposition, exposure to carcinogens or ionizing radiation, viral infections, and immune system dysfunction playing contributing roles. Diagnosis typically involves blood tests, bone marrow biopsy, and molecular testing to characterize the specific subtype and genetic abnormalities associated with the disease. Treatment strategies for leukemia vary depending on factors such as subtype, stage, and patient age and health status and may include chemotherapy: targeted therapy, stem cell transplantation HSCT, or immunotherapy, radiation therapy. Leukemia presents a complex and multifaceted landscape. Understanding the different classifications and ongoing research efforts are crucial for navigating this disease.

12.3.8 Lymphoma

Lymphoma is a complex and heterogeneous group of malignancies originating from lymphocytes, which are a crucial component of the body's immune system. This disease encompasses two main categories: Hodgkin lymphoma (HL) and non-Hodgkin lymphoma (NHL), each with distinct subtypes and clinical characteristics. HL is characterized by the presence of Reed–Sternberg cells, whereas NHL comprises a diverse array of lymphoid neoplasms, including diffuse large B-cell lymphoma (DLBCL), follicular lymphoma (FL), mantle cell lymphoma (MCL), and peripheral T-cell lymphoma (PTCL), among others. The etiology of lymphoma is multifactorial, involving a combination of genetic predisposition, environmental factors, immune dysregulation, and infectious agents such as Epstein–Barr virus (EBV) and *Helicobacter pylori*. While the exact cause of lymphoma remains unclear in many cases, certain risk factors, such as age, gender, family history, immunodeficiency, and exposure to certain chemicals or radiation, have been implicated in its development.

Symptoms of lymphoma can vary depending on the subtype and location of the disease but often include painless swelling of lymph nodes, particularly in the neck, armpits, or groin, as well as fever, night sweats, unexplained weight loss, fatigue, and itching. Some individuals may also experience symptoms related to the involvement of other organs, such as enlarged spleen or liver, abdominal pain, chest pain, difficulty breathing, or neurological symptoms.

Diagnosis of lymphoma typically involves a combination of imaging studies, such as computed tomography (CT) scans or positron emission tomography (PET) scans, biopsy of affected tissue (e.g., lymph node or bone marrow biopsy), and molecular testing to identify specific genetic abnormalities and classify the disease subtype. Additionally, blood tests may be performed to assess blood cell counts, organ function, and markers of inflammation or immune activation (Lewis et al. 2020).

Treatment for lymphoma depends upon factors such as subtype, stage, and patient characteristics, with the goal of inducing remission, eradicating cancerous cells, and improving overall survival. The main treatment modalities include chemotherapy, immunotherapy, radiation therapy, targeted therapy, and stem cell transplantation (Min and Lee 2022). Chemotherapy, which involves the use of drugs to kill cancer cells, is often used as the initial treatment for lymphoma, either alone or in combination with other agents (Zahavi and Weiner 2020). Immunotherapy, including monoclonal antibodies such as rituximab, targets specific proteins on cancer cells to enhance the immune system's ability to recognize and destroy them (Sotomayor et al. 2007). Radiation therapy uses high-energy beams to target and destroy cancer cells in specific areas of the body, particularly for localized disease or to reduce symptoms such as pain or swelling. Targeted therapy agents, such as tyrosine kinase inhibitors (TKIs) or BCL-2 inhibitors, target specific molecular abnormalities in cancer cells, disrupting their growth and survival pathways. Stem cell transplantation, either autologous (using the patient's own stem cells) or allogeneic (using donor stem cells), may be considered for eligible patients with high-risk or relapsed disease, aiming to replace diseased bone marrow with healthy stem cells and restore normal blood cell production.

In recent years, there have been significant advancements in the treatment of lymphoma, driven by a better understanding of the molecular mechanisms underlying the disease and the development of targeted therapies and immunotherapies (Deshpande and Munoz 2022). These novel treatment approaches offer the promise of improved outcomes and reduced toxicity compared to traditional chemotherapy regimens, particularly for patients with relapsed or refractory disease (Deshpande and Munoz 2022).

12.4 Conclusions

The exploration of blood cellular components and associated diseases within this chapter sheds light on the intricate interplay between biomolecular elements and physiological functions. From plasma proteins to clotting factors, hemoglobin, and beyond, each component plays a vital role in maintaining homeostasis and ensuring the proper functioning of bodily systems. Through an in-depth examination of various hematological conditions such as anemia, leukemia, and thrombocytopenia, among others, we have gained insights into the complex mechanisms underlying these diseases and their impact on human health. By comprehensively understanding the role of blood cellular components in health and disease, healthcare professionals are better equipped to diagnose, treat, and manage hematological disorders effectively. As we continue to advance our knowledge and research in this field, we move closer to unraveling the mysteries of blood and enhancing the quality of care for patients affected by hematological conditions.

References

Alexy T, Detterich J, Connes P, Toth K, Nader E, Kenyeres P, Arriola-Montenegro J, Ulker P, Simmonds MJ (2022) Physical properties of blood and their relationship to clinical conditions. Front Physiol 13:906768

Ali S, Mumtaz S, Shakir HA, Khan M, Tahir HM, Mumtaz S, Mughal TA, Hassan A, Kazmi SAR, Sadia Irfan M, Khan MA (2021) Current status of beta-thalassemia and its treatment strategies. Mol Genet Genomic Med 9(12):1788

Altemose KE, Atkinson MA (2023) Anemia. Evidence-Based Nephrol, 2: 379–398

Anderson HL, Brodsky IE, Mangalmurti NS (2018) The evolving erythrocyte: red blood cells as modulators of innate immunity. J Immunol 201(5):1343–1351

Asp J, Andréasson B, Hansson U, Wasslavik C, Abelsson J, Johansson P, Palmqvist L (2016) Mutation status of essential thrombocythemia and primary myelofibrosis defines clinical outcome. Haematologica 101(4):e129

Baldwin C, Pandey J, Olarewaju O (2023) Hemolytic anemia. In: Pathobiology of human disease: a dynamic encyclopedia of disease mechanisms, pp 1532–1543

Basu D, Kulkarni R (2014) Overview of blood components and their preparation. Indian J Anaesth 58(5):529–537

Baumeister J, Chatain N, Sofias AM, Lammers T, Koschmieder S (2021) Progression of myeloproliferative neoplasms (mpn): diagnostic and therapeutic perspectives. Cells 10(12)

Belvis D, Henderson KJ, Benzon HA (2018) Sickle cell disease. Essen Pain Med:333–338

Benjamin RJ, McLaughlin LS (2012) Plasma components: properties, differences and uses. Transfusion 52:9S–19S

Brodsky RA, Jones R (2006) Aplastic anemia. In: Autoimmune diseases, 4th edn, pp 591–599

Brown JR, Cymbalista F, Sharman J, Jacobs I, Nava Parada P, Mato A (2018) The role of rituximab in chronic lymphocytic leukemia treatment and the potential utility of biosimilars. Oncologist 23(3):288

Castillo Tokumori F, Al Ali N, Chan O, Sallman D, Yun S, Sweet K, Padron E, Lancet J, Komrokji R, Kuykendall AT (2022) Comparison of different treatment strategies for blast-phase myeloproliferative neoplasms. Clin Lymphoma Myeloma Leuk 22(7):e521–e525

Charles A, Janeway J, Travers P, Walport M, Shlomchik MJ (2001) The components of the immune system. https://www.ncbi.nlm.nih.gov/books/NBK27092/

Chen BJ (2013) Aplastic anemia. Hematology 54

Chennamadhavuni A, Lyengar V, Mukkamalla SKR, Shimanovsky A (2023) Leukemia. https://www.ncbi.nlm.nih.gov/books/NBK560490/

Cherian VT (2022) Physiological functions of blood. In: Liu H, Kaye AD, Jahr JS (eds) Blood substitutes and oxygen biotherapeutics. Springer, Cham

Choi CW, Bang SM, Jang S, Jung CW, Kim HJ, Kim HY, Kim SJ, Kim YK, Park J, Won JH (2015) Guidelines for the management of myeloproliferative neoplasms. Korean J Intern Med 30(6):771

Choudhary MK, Mishra AK, Kumar P, Jamal I, Ahmad A, Prasad G, Prasad D (2023) Study of the aetiology and clinical manifestations of thrombocytopenia in a tertiary care centre. Cureus 15(7)

Cohen AR, Galanello R, Pennell DJ, Cunningham MJ, Vichinsky E (2004) Thalassemia Hematol 4(1):14–34

Costello RA, Nehring SM (2023) Disseminated intravascular coagulation. Hemostasis Thrombosis 4:55–59

De Hert S (2012) Physiology of hemodynamic homeostasis. Best Pract Res Clin Anaesthesiol 26(4):409–419
Deshpande A, Munoz J (2022) Targeted and cellular therapies in lymphoma: mechanisms of escape and innovative strategies. Front Oncol 12
Dezern AE, Brodsky RA (2011) Clinical management of aplastic anemia. Expert Rev Hematol 4(2):221
Dokal I (2006) Fanconi's anaemia and related bone marrow failure syndromes. Br Med Bull 78(1):37–53
Edginton-White B, Bonifer C (2022) The transcriptional regulation of normal and malignant blood cell development. FEBS J 289(5):1240–1255
Editor C, Tefferi A, Jabbour E, Kantarjian H (2012) Chronic myeloid leukemia: update on diagnosis monitoring and management. Am J Hematol 87(11):1037–1045
Falcone FH, Zillikens D, Gibbs BF (2006) The 21st century renaissance of the basophil? Current insights into its role in allergic responses and innate immunity. Exp Dermatol 15(11):855–864
Frenette PS, Strell C, Li R (2020) The endothelial niche in health and disease: a new player in cancer progression. Exp Cell Res 386(2):1117
Ganz T (2019) Anemia of inflammation. New Eng J Med 381(12):1148–1157
Geyer H, Mesa RA (2017) Approach to MPN symptom assessment. Curr Hematol Malig Rep 12(5):381
Green D (2016) Hemophilia. Linked Blood Hemophilia AIDS:55–66
Gremmel T, Frelinger AL, Michelson AD (2016) Platelet physiology. Semin Thromb Hemost 42(3):191–204
Guyton AC, Hall JE (2015) Chapter 18, the erythrocyte and its production. In: Textbook of medical physiology, 13th edn. Elsevier, Philadelphia, pp 215–226
Halterman JS, Segel GB (2023) Iron deficiency anemia. Pediatr Clin Advis:31–31
Hill A, Hill QA (2018) Autoimmune hemolytic anemia. Hematol Am Soc Hematol Educ Program 2018(1):382
Jabbour E, Kantarjian H (2020) Chronic myeloid leukemia: 2020 update on diagnosis, therapy and monitoring. Am J Hematol 95(6):691–709
Jinna S, Khandhar PB (2023) Thrombocytopenia. National Library of Medicine https://www.ncbi.nlm.nih.gov/books/NBK542208/
Kim DS (2022) Recent advances in treatments of adult immune thrombocytopenia. Blood Res 57(Suppl 1):112
Kondo M (2010) Lymphoid and myeloid lineage commitment in multipotent hematopoietic progenitors. Immunol Rev 238(1):37
Kralovics R, Passamonti F, Buser AS, Teo S-S, Tiedt R, Passweg JR, Tichelli A, Cazzola M, Skoda RC (2005) A gain-of-function mutation of JAK2 in myeloproliferative disorders. New Eng J Med 352(17):1779–1790
Langer AL (2024) Beta-Thalassemia.GeneReviews®. https://www.ncbi.nlm.nih.gov/books/NBK1426/
Larosa DF, Orange JS (2008) Lymphocytes. J Allergy Clin Immunol 121(2 Suppl):S364–9 quiz S412
Lee JY, Hong SH (2020) Hematopoietic stem cells and their roles in tissue regeneration. Int J Stem Cells 13(1):1
Levi M, Scully M (2018) How I treat disseminated intravascular coagulation. Blood 131(8):845–854
Lewis WD, Lilly S, Jones KL (2020) Lymphoma: diagnosis and treatment. Am Fam Physician 101(1):34–41
Madu AJ, Ughasoro MD (2017) Anaemia of chronic disease: an in-depth review. Med Princ Pract 26(1):1
Min HY, Lee HY (2022) Molecular targeted therapy for anticancer treatment. Exp Mol Med 54(10):1670
Min B, Brown MA, Legros G (2012) Understanding the roles of basophils: breaking dawn. Immunology 135(3):192–197
Mitchell WB, Walsh C (2023) Hemophilia. Atlas Diag Hematol:360–367

Mukkamalla SKR, Taneja A, Malipeddi D, Master SR (2023) Chronic lymphocytic leukemia Cll. 10–13. https://www.ncbi.nlm.nih.gov/books/NBK470433/
Nandakumar SK, Ulirsch JC, Sankaran VG (2016) Advances in understanding erythropoiesis: evolving perspectives. Br J Haematol 173(2):206–218
Nauseef WM, Borregaard N (2014) Neutrophils at work. Nat Immunol 7:602–611
Newsholme P (2021) Cellular and metabolic mechanisms of nutrient actions in immune function. Eur J Clin Nutr 75(9):1328–1331. https://doi.org/10.1038/s41430-021-00960-z. Epub 2021 Jun 23. PMID: 34163016; PMCID: PMC8220357
Papageorgiou C, Jourdi G, Adjambri E, Walborn A, Patel P, Fareed J, Elalamy I, Hoppensteadt D, Gerotziafas GT (2018) Disseminated intravascular coagulation: an update on pathogenesis diagnosis and therapeutic strategies. Clin Appl Thromb Hemost 24(9 Suppl):8S
Percival ME, Lai C, Estey E, Hourigan CS (2017) Bone marrow evaluation for diagnosis and monitoring of acute myeloid leukemia. Blood Rev 31(4):185
Periayah MH, Halim AS, Mat Saad AZ (2017) Mechanism action of platelets and crucial blood coagulation pathways in hemostasis. Int J Hematol Oncol Stem Cell Res 11(4):319–327
Pham PTT, Pham PCT, Lew SQ (2019) Sickle cell disease. Chronic Renal Dis:813–830
Pizzi M, Croci GA, Ruggeri M, Tabano S, Dei Tos AP, Sabattini E, Gianelli U (2021) The classification of myeloproliferative neoplasms: rationale historical background and future perspectives with focus on unclassifiable cases. Cancer 13(22):5666
Prinyakupt J, Pluempitiwiriyawej C (2015) Segmentation of white blood cells and comparison of cell morphology by linear and naïve Bayes classifiers. Biomed Eng 14:63
Reinke W, Johnson PC, Gaehtgens P (1986) Effect of shear rate variation on apparent viscosity of human blood in tubes of 29 to 94 microns diameter. Circ Res 59(2):124–132
Rosales C (2018) Neutrophil: a cell with many roles in inflammation or several cell types? Front Physiol 20(9):113
Schulze TJ, Bugert P (2018) Granulocytes: from basic research to clinical use. Transfus Med Hemother 45(5):288–289
Seesaghur A, Petruski-Ivleva N, Banks VL, Wang JR, Abbasi A, Neasham D, Ramasamy K (2021) Original research: clinical features and diagnosis of multiple myeloma: a population-based cohort study in primary care. BMJ Open 11(10):52759
Siegel RL, Miller KD, Fuchs HE, Jemal A (2021) Cancer statistics. 2021CA. Cancer J Clin 71(1):7–33
Silberstein P, Huyck T (2008) Aplastic anemia. XPharm Comprehens Pharmacol Ref:1–4
Sotomayor EM, Kwak LW, Riddell SR, Bello C (2007) Monoclonal antibodies for B-cell lymphomas: rituximab and beyond. Hematology 2007(1):233–242
Spiers J, Mahdi D, Saunders C, McLornan DP (2023) Myeloproliferative neoplasms. Mol Hematol, pp 99–111
Strouse J (2016) Sickle cell disease. Handb Clin Neurol 138:311–324
Szuber N, Elliott M, Tefferi A (2022) Chronic neutrophilic leukemia: 2022 update on diagnosis genomic landscape prognosis and management. Am J Hematol 97(4):491–505
Tamang T, Baral S, Paing MP (2022) Classification of White blood cells: a comprehensive study using transfer learning based on convolutional neural networks. Diagnostics 12:2903
Tamary H, Dgany O (2020) Alpha-thalassemia. GeneReviews®. https://www.ncbi.nlm.nih.gov/books/NBK1435/
Taylor J, Xiao W, Abdel-Wahab O (2017) Diagnosis and classification of hematologic malignancies on the basis of genetics. Blood 30(4):410
Tremblay D, Yacoub A, Hoffman R (2021) Overview of MPNs: history pathogenesis diagnostic criteria and complications. Hematol/Oncol Clin North Am 35(2):159
Vainchenker W, Kralovics R (2017) Genetic basis and molecular pathophysiology of classical myeloproliferative neoplasms. Blood 129(6):667–679
Venugopal S, Mascarenhas J (2020) Novel therapeutics in myeloproliferative neoplasms. J Hematol Oncol 13(1):1–13
Wahed A, Dasgupta A (2015) Myeloid neoplasms. Hematol Coag:93–115

West JB, Luks AM (2016) Gas transport by the blood: how gases are moved to and from the peripheral tissues. West's Respirat Physiol Essent:94–115

Williams SA, Kivimaki M, Langenberg C, Hingorani AD, Casas JP, Bouchard C, Jonasson C, Sarzynski MA, Shipley MJ, Alexander L, Ash J, Bauer T, Chadwick J, Datta G, DeLisle RK, Hagar Y, Hinterberg M, Ostroff R, Weiss S, Ganz P, Wareham NJ (2019) Plasma protein patterns as comprehensive indicators of health. Nat Med 12:1851–1857

Winer ES, Stone RM (2019) Novel therapy in acute myeloid leukemia (AML): Moving toward targeted approaches. Ther Adv Hematol 10:204062071986064

Zahavi D, Weiner L (2020) Monoclonal antibodies in cancer therapy. Antibodies 9(3):1–20

Free Radicals and Their Relation to Diseases and Protection Against Them

13

Saheem Ahmad, Hamda Khan, Zeeshan Rafi, Uzma Shahab, Jalaluddin Mohammad Ashraf, Mohammad Kaleem Ahmad, Kirtanjot Kaur, Ramendra Pati Pandey, Safia Habib, and Moinuddin

Abstract

Free radicals are highly reactive molecules that possess unpaired electrons, making them unstable and causing damage to macromolecules such as proteins, lipids, and DNA. The accumulation of free radicals and oxidative stress has been linked to the

The original version of the chapter has been revised: Affiliation of the authors Hamda Khan, Safia Habib, Moinuddin and Zeeshan Rafi have been updated. A correction to this chapter can be found at https://doi.org/10.1007/978-981-97-4723-8_17

S. Ahmad (✉)
Department of Medical Laboratory Sciences, College of Applied Medical Sciences, University of Hail, Hail, Saudi Arabia
e-mail: s.ansari@uoh.edu.sa

H. Khan · S. Habib · Moinuddin
Department of Biochemistry, J.N. Medical College, Aligarh Muslim University, Aligarh, India

Z. Rafi
Department of Bioengineering, Integral University, Lucknow, Uttar Pradesh, India

U. Shahab · M. K. Ahmad
Department of Biochemistry, King George Medical University, Lucknow, Uttar Pradesh, India

J. M. Ashraf
Department of Biochemistry, Faculty of Medicine, Samarkand State Medical University, Samarkand, Uzbekistan

K. Kaur
University Centre for Research and Development, Chandigarh University, Mohali, Punjab, India

R. P. Pandey
School of Health Sciences and Technology (SOHST), UPES, Dehradun, Uttarakhand, India

R. L. Singh et al. (eds.), *Clinical Applications of Biomolecules in Disease Diagnosis*, https://doi.org/10.1007/978-981-97-4723-8_13

development of various metabolic disorders, including cardiovascular diseases, neurodegenerative disorders, diabetes, and cancer. Understanding the mechanisms by which free radicals damage macromolecules is crucial in elucidating the pathogenesis of these diseases. This chapter focuses on the mechanisms through which free radicals induce damage to macromolecules and the subsequent implications for multiple metabolic disorders. Free radicals initiate chain reactions, leading to lipid peroxidation, protein oxidation, and DNA damage, which disrupt cellular functions. Reactive oxygen species (ROS) and reactive nitrogen species (RNS) are the major types of free radicals involved in these processes. Moreover, endogenous and exogenous antioxidants play a crucial role in protecting against oxidative damage. Endogenous antioxidants, such as superoxide dismutase, catalase, and glutathione peroxidase, function as defense mechanisms within the body to neutralize free radicals and maintain redox balance. Exogenous antioxidants, obtained through dietary sources or supplements, also contribute to reducing oxidative stress and preventing the onset or progression of various diseases. This chapter further discusses the interplay between free radicals, macromolecular damage, and multiple metabolic disorders, highlighting the role of antioxidants in disease prevention and management. Understanding the mechanisms underlying oxidative stress-induced damage and the protective effects of antioxidants is essential for the development of therapeutic strategies to mitigate the detrimental effects of free radicals.

Keywords
Free radicals · Oxidative stress · Macromolecular damage · Metabolic disorders · Antioxidants

13.1 Introduction

In chemistry, radicals are molecules that contain one or more unpaired electrons in the shell around the atomic nucleus. Some exist in a stable form and perform essential functions, while some exist in an accessible form ready to interact with tissues and termed "free radicals." The biochemistry of cellular metabolism includes the generation of free radicals via ATP formation by the mitochondria (Tao et al. 2019). Once a free radical interacts with a tissue, then the changes occurs in terms of injury, and is called as oxidative stress. It is involved in the toxicities and pathology of several human diseases, such as cardiovascular diseases, cancer, and neurological disorder (Singh et al. 2008; Singh and Singh 2018; Kehrer and Klotz 2015). The potentially harmful effects of free radicals are the formation of a series of compounds known as reactive oxygen species (ROS), which tend to donate oxygen to other substances. These are highly reactive substances and can damage the integrity of various biomolecules such as DNA, proteins, carbohydrates, and lipids leading to various diseases, including diabetes, neurodegenerative diseases, atherosclerosis, respiratory diseases, rheumatoid arthritis, hypertension, and aging process (Singh et al. 2013, 2014; Phaniendra et al. 2015). The free radical reaction involves initiation, propagation, and termination step. We have illustrated this chemical reaction in Fig. 13.1.

ROS generation is supposed to involve in the oxidative stress which causes the base lesions in DNA such as guanine modification or

7,o-dihydro-8-oxo-2-deoxyguanosine (8-OH-dG); it alters the DNA repair pathway (Huang and Zhou 2021). Similarly, ROS generation leads to the formation of lipid peroxidation, and these are closely related to the development of many metabolic disorders. Therefore, it acts as a potential biomarker in diseases (Peña-Bautista et al. 2019). ROS deregulates the redox homeostasis and its associating signaling pathways, such as mitochondrial pathway, the Keap1-Nrf2-ARE pathway, the ERS pathway, or endoplasmic reticulum stress pathway (Wu et al. 2020). The hyperglycemic condition is also known to induce ROS in to the endoplasmic reticulum (ER), mitochondrial electron transport chain (ETC), phagocytic cells, and peroxisomes. Therefore, hyperglycemic condition is another pathway of generation of ROS, further leading to diabetes mellitus and cardiovascular diseases (Bhatti et al. 2022). Besides, ROS can be produced naturally or from external factors such as smoking, air pollutants, X-rays, and industrial chemicals. Oxidative stress is a normal process of our body, but when it is in an excess way or in an imbalance way, it becomes harmful. The factors involved in oxidative stress and excessive free radical production are pollution, bad habitual diet, especially junk food, radiation, and sedentary lifestyle. The body's natural immune response can also temporarily cause oxidative stress. Our body has two defense systems for free radicals that are enzymes and antioxidants (Shukla et al. 2011; Gupta et al. 2023). The prior one provides protection from the oxidative stress while the later one blocks the free radicals to react with other molecule and make it unreactive or stable. Thousands of chemicals act as antioxidants. Vitamins C and E, glutathione, beta-carotene, and phytoestrogens, called phytoestrogens, are among the many antioxidants that can neutralize the effects of free radicals. Due to the high prevalence of chronic diseases, it makes sense to use plants, especially herbal plants rich in phenols and flavonoids, to provide the antioxidants to the body (Akbari et al. 2022). The free radical oxidants such as ROS, reactive nitrogen species (RNS), and hydroxyl radical (OH) affects our body through lipid peroxidation, DNA damage, and protein denaturation. As a result, diseases developed like stroke, heart disease, atherosclerosis, and cancer (Fig. 13.2). Recent studies on oxy-radicals associated with ischemia and hypoxia

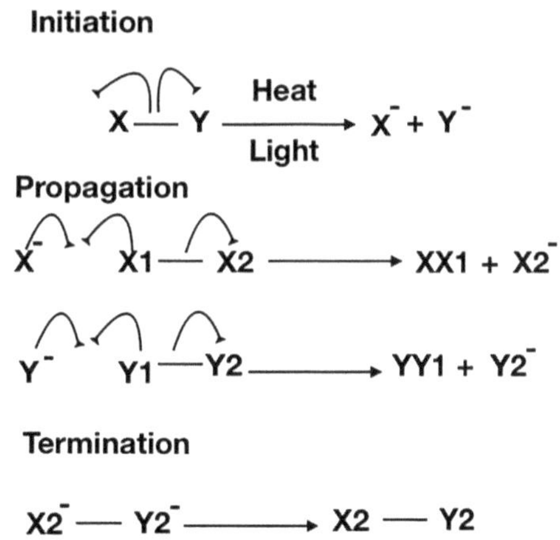

Fig. 13.1 *Initiation step*, generation of free radical; *propagation step*, free radical reacts with stable molecules to form a new free radical; *termination step*, two free radicals react with each other to form a stable compound

highlight the formation of oxy-radicals during reperfusion of ischemic tissues (Rifkind et al. 2022). ROS are found to be pervasive in the environment and play a critical role in the carbon cycle globally. In general, the oxidation of reducing substances contributes to the formation of free radicals through Haber–Weiss, Fenton, and Fenton-like reactions in alternating aerobic and anaerobic environments such as water, soil, and sediment (Xing et al. 2022). Oxygen-radicals contribute significantly to the development of osteoporosis. The presence of oxy-radicals also increases lipid damage. In fact, unsaturated lipids undergo a process called lipid peroxidation (Fig. 13.2). This is a chain reaction initiated by reactive free radicals such as ·OH that leads to the formation of lipid peroxides and other derivatives in the presence of oxygen (Di Meo and Venditti 2020).·OH is known to produce protein aggregations via creating fragmentation in amino acids through covalent bonding. After exposure to oxy-radical, amino acids present in proteins get oxidized, causing the formation of protein-protein crosslinks, denaturation, loss of function of proteins, loss of enzymatic activity, and loss of function of receptor and transport proteins (Ahmad et al. 2017). Despite being packaged with proteins as defensive as nuclear chromatin, mitochondrial DNA (mtDNA) is particularly vulnerable to ROS generated by the respiratory chain because of its proximity. mtDNA mutations can cause various diseases. A significant advance in carcinogenesis research over the past two decades has been the discovery of profound DNA damage originating from

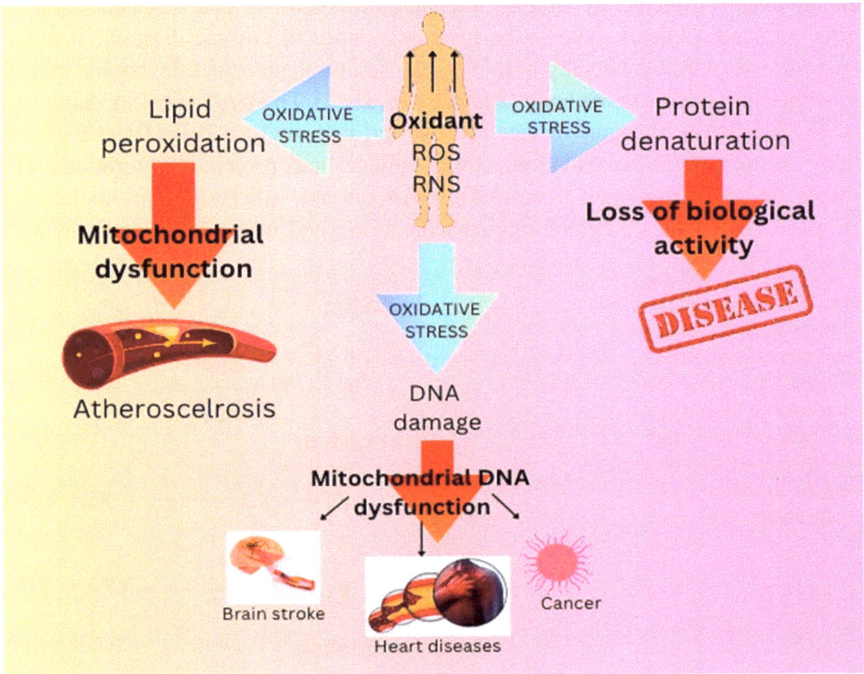

Fig. 13.2 Impact of free radicals in our daily life

endogenous cellular sources. The same is true for the aging process and medical conditions in the elderly.

There is evidence of a link between aging and mitochondrial genome dysfunction. 8-Oxo-dG has been detected in DNA treated with various oxidizing agents. Therefore, the availability of a simple method for its detection has made it the adduct of choice for measuring DNA oxidation (Marnett 2000). The body has many mechanisms to minimize the damage caused by free radicals and to repair the damage caused. It protects cells from toxicity and contributes to disease prevention. A diet rich in antioxidant, avoiding smoking, and limiting UV exposure can reduce the number of free radicals in the body.

13.1.1 Free Radicals

Free radicals, those elusive troublemakers, are molecules brimming with reactivity and armed with unpaired electrons. These rebels emerge through biological processes or external influences like radiation, pollution, and smoking. Their assault knows no bounds, infiltrating vital cellular elements like DNA, proteins, and lipids, leaving chaos in their wake. This unleashes a cascade of events known as lipid peroxidation, a destructive chain reaction (Oberley 1988; Djordjevic 2004). These radicals, born from the very essence of life itself, arise during physiological activities like oxidative phosphorylation, immune responses, and enzymatic reactions. Reactive oxygen species (ROS), the natural by-products of cellular metabolism, are prime examples. External factors such as ionizing and ultraviolet radiation, environmental pollutants, and tobacco smoke also fuel the creation of free radicals (Halliwell and Gutteridge 2015). Their treacherous reactivity arises from the unpaired electrons, rendering them as unstable entities that eagerly engage with other molecules. When these radicals encounter DNA, proteins, and lipids, they wreak havoc through oxidative damage. This malevolence, known as oxidative stress, alters the normal structure and function of these vital biomolecules. Understanding the mechanisms of free radical damage and the consequences on cellular processes is crucial for elucidating disease pathogenesis. By enhancing antioxidant defenses, supporting protein repair mechanisms, and implementing strategies to mitigate the damages, it might possible to minimize the harmful effects of free radicals on macromolecules structure and function.

13.1.2 Types of Free Radicals

The type of free radical includes both reactive nitrogen and oxygen species which might act as radical or non-radical compound. Table 13.1 explains the types of free radical (Kohen and Nyska 2002; Halliwell 2006) (Table 13.1). Reactive oxygen species include the three chemical species of the Fenton/Haber–Weiss pathway ($O_2^{\bullet-}$ superoxide radical, H_2O_2 hydrogen peroxide and $HO^{\bullet}$ hydroxyl radical), products of the partial reduction of oxygen. $O_2^{\bullet-}$ and $HO^{\bullet}$ have unpaired electrons in external

Table 13.1 Types of free radicals.

Reactive Oxygen Species(ROS) Reactive Nitrogen Species (RNS)					
Name symbol half-life (s) name symbol half-life (s)					
Radicals					
Superoxide	$O^{\bullet-}$	10^{-6} s	Nitricoxide	$No^{\bullet}$	s
Hydroxyl	$^{\bullet}Oh$	10^{-10} s	Nitrogen dioxide	$NO_2^{\bullet}$	s
Hydroperoxyl	$HO_2^{\bullet}$	s	Nitrate radical	$NO_3^{\bullet}$	s
Peroxyl	$ROO^{\bullet}$	17 s			
Alkoxyl organic	$Ro^{\bullet}$	10^{-6} s			
Hydroperoxide	ROOH	Stable			
Non-radicals					
Hydrogen peroxide	H_2O_2	Stable	Nitrous acid	HNO_2	s
Ozone	O3	s	Nitrosonium cation	No^+	s
Singlet oxygen	(^1O_2Dg)	10^{-6} s	Nitroxyl anion	NO^-	s
Hypochlorous acid	HOCl	Stable(min)	Peroxynitrite	$ONOO^-$	10^{-3} s
			Dinitrogen trioxide	N_2O_3	s
			Dinitrogen tetraoxide	N_2O_4	s
			Peroxynitrous acid	ONOOH	Fair stable

**s* seconds, *min* minutes

orbitals and are therefore defined as free radicals, whereas H_2O_2 does not contain unpaired electrons and is therefore not a radical (Moncada et al. 1991; Halliwell and Gutteridge 2007). Collectively, the radicals include superoxide ($O_2^{\bullet-}$), oxygen radical ($O_2^{\bullet\bullet}$), hydroxyl ($^{\bullet}OH$), alkoxy radical ($RO^{\bullet}$), peroxyl radical ($ROO^{\bullet}$), nitric oxide (or nitrogen monoxide) ($NO^{\bullet}$), and nitrogen dioxide ($NO_2^{\bullet}$) (Kohen and Nyska 2002; Halliwell 2006). The non-radical species include hydrogen peroxide (H_2O_2), hypochlorous acid (HOCl), hypobromous acid (HOBr), ozone (O_3), singlet oxygen ($^1O_2$), nitrous acid (HNO_2), nitrosyl cation (NO^+), nitroxyl anion (NO^-), dinitrogen trioxide (N_2O_3), dinitrogen tetroxide (N_2O_4), nitronium (nitryl) cation (NO^{2+}), organic peroxides (ROOH), aldehydes (HCOR), and peroxynitrite ($ONOO^-$) (Kohen and Nyska 2002; Halliwell 2006). However, radicals generated from oxygen mainly damage biological systems.

13.1.3 Formation of Free Radicals

The formation of free radicals can occur within the body as part of regular metabolic processes and can also be influenced by external factors (Pham-Huy et al. 2008) (Fig. 13.3). These free radicals contribute to oxidative stress and play a role in initiating DNA damage, protein damage, and lipid peroxidation leading to potential harm.

13.1.3.1 Endogenous Formation of Free Radicals

Free radicals are formed as natural by-products of cellular metabolism, including mitochondrial respiration and enzymatic reactions (Comporti 1993). One of the

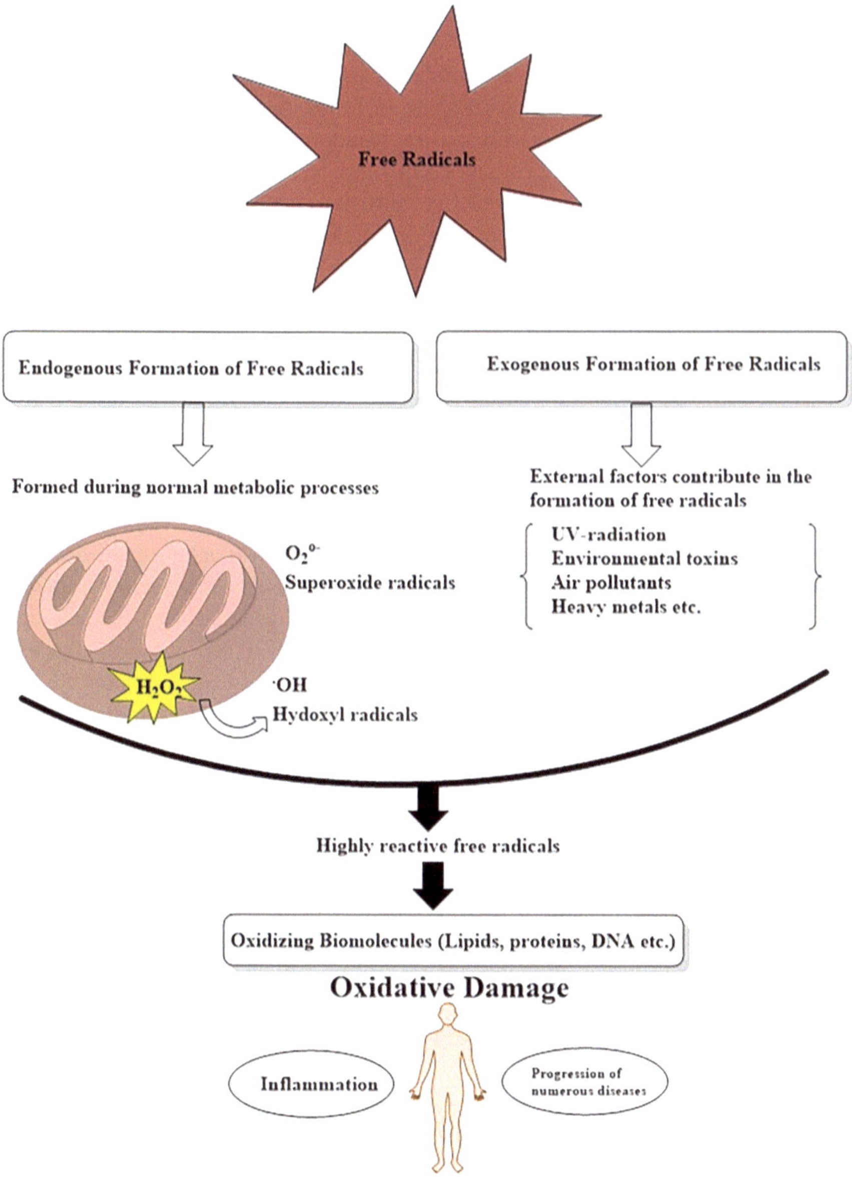

Fig. 13.3 Mechanism of the formation of free radicals

major sources of endogenous free radicals is the electron transport chain, which occurs within the mitochondria during aerobic respiration (Kovacic et al. 2005). During this process, electrons are transferred from electron donors to acceptors, generating energy in the form of adenosine triphosphate (ATP). However, a small percentage of electrons leak from the electron transport chain, leading to the formation of reactive oxygen species (ROS), such as superoxide anion (O2•-) and hydrogen peroxide (H_2O_2) (Pham-Huy et al. 2008). Enzymatic reactions involving transition metals, such as iron and copper, can also generate free radicals (Bergendi et al. 1999). These metals can participate in redox reactions and generate ROS through Fenton and Haber–Weiss reactions, resulting in the formation of hydroxyl radicals (•OH). Nitric oxide synthases, enzymes responsible for the synthesis of nitric oxide (NO), can generate reactive nitrogen species (RNS) as by-products (Sies 2015). Nitric oxide can react with superoxide anion to form peroxynitrite (ONOO-), a highly reactive RNS that can cause protein nitration and damage (Juan et al. 2022).

13.1.3.2 Exogenous Generation of Free Radicals

In addition to endogenous sources, external factors can contribute to the formation of free radicals (Lobo et al. 2010). Environmental toxins, radiation, and pollutants can increase free radical production, leading to protein damage (Pacifici and Davies 1991; Ahmad et al. 2017). For example, exposure to ultraviolet (UV) radiation from the sun can generate ROS in the skin cells (Valko et al. 2006). UV radiation induces the formation of singlet oxygen and other ROS, which can attack proteins and other biomolecules (Das and Roychoudhury 2014). Cigarette smoke contains numerous chemicals, including reactive components, which can generate free radicals upon metabolism (Church and Pryor 1985; Pham-Huy et al. 2008). These free radicals can initiate oxidative damage to proteins in the lungs and other organs. Certain drugs and chemicals can also contribute to the generation of free radicals. Some medications, such as certain antibiotics and chemotherapeutic agents, can undergo redox reactions and generate ROS as by products (Dharmaraja 2017). Environmental pollutants, such as heavy metals, air pollutants, and industrial chemicals, can also promote the formation of free radicals (Pham-Huy et al. 2008). These exogenous sources of free radicals can contribute to increased protein damage and cellular oxidative stress. The formation of free radicals, both endogenously and exogenously, poses a challenge to cellular health and protein integrity (Lushchak 2007). Understanding the sources and mechanisms of free radical generation is crucial for developing strategies to mitigate their harmful effects and protect proteins from oxidative damage (Choudhari et al. 2014).

13.2 Free Radicals Associated with DNA Damage

Various free radicals involved in DNA damage are superoxide radical, hydroxyl radical, alkoxyl radical, peroxyl radical, nitric oxide, nitrogen dioxide, and peroxynitrite. These can induce oxidative stress in DNA which might be involved in the progression of diabetes mellitus, neurodegenerative diseases, rheumatoid arthritis, cancer, cardiovascular diseases, respiratory diseases, and cataract development (Chatgilialoglu and O'Neill 2001).

13.2.1 Superoxide Radical

Since 1994, the effect of superoxide radical on DNA has been investigated thoroughly. It is believed to be less reactive and cell damage is attributed to HO produced by the Haber–Weiss reaction. In biological tissues, superoxide radicals are produced under inflammatory conditions and are known to cause oxidative DNA damage. Mitochondrial DNA is more prone to attack by superoxide radicals than nuclear DNA due to its proximity to ROS generation sites (Chung 2017).

13.2.2 Hydroxyl Radical

Among the reactive oxygen species (ROS) produced by cellular metabolism in aerobic organisms, hydroxyl radicals (OH) are the most reactive to DNA. Contrary to frequent reports in the literature, hydroxyl radicals are not the key species involved in endogenous oxidative DNA damage. It damages purines and produces 8-oxo-purine (8-oxo-pu) (Fleming and Burrows 2020). Hydroxyl radical generation by the adriamycin–Fe complex in the presence of DNA correlates with its DNA cleavage ability. Thus, hydroxyl radicals are involved as ROS in DNA damage caused by the adriamycin-Fe3+ complex (Muindi et al. 1984).

13.2.3 Alkoxyl Radical and Peroxyl Radical

The reaction of lipid alkoxyl (LO') with peroxyl radical (LOO') is slower than that of HO', and the reaction of lipid radicals is more specific. Peroxidation of polyunsaturated lipids produces a number of substances that can damage DNA. These include lipid hydroperoxides and various species containing unpaired electrons such as alkoxyl and peroxyl radicals (Burcham 1998).

13.2.4 Nitric Oxide and Nitrogen Dioxide

Nitric oxide can also react with oxygen species to produce a variety of reactive nitric oxides that can damage DNA and cause DNA sequence mutations (Routledge 2000). Previous study suggested that gaseous nitric oxide (NO) and nitrogen dioxide (NO2) were verified for their impending activity to persuade DNA single-strand breaks in Chinese Hamster (V79) cells (Görsdorf et al. 1990). Although there is no evidence that nitric oxide (NO) reacts directly with DNA, NO can alter biopolymers in various indirect ways.

13.2.5 Peroxynitrite

The destruction instigated by peroxynitrite harmfully affects biological effects of DNA. The DNA damage profile induced by peroxynitrite differs significantly from that induced by hydroxyl radicals and singlet molecular oxygen. Previous studies have suggested that peroxynitrite induces DNA strand breaks in a dose-dependent manner (Salgo et al. 1995). Peroxynitrite is a relatively long-lived oxidant, formed in vivo by the reaction of nitric oxide and superoxide anion radicals. The mechanisms underlying the oxidation of biomolecules by HOONO are not yet fully understood. The effect of free radicals linked with DNA impairment has been listed in the Table 13.2.

Table 13.2 Free radicals induced DNA damage and its biological effect

Free radical	Damage induced	Biological effect	Disease associated	Reference
Hydroxyl radical	Purine DNA lesions like 8-oxo-purine (8-oxo-Pu) are generated	Cellular damage	Cockayne syndrome (Krokidis et al. 2020)	Chatgilialoglu et al. (2021)
	Generates DNA phenotypes with various metastatic potentials	Metastatic tumor DNA	Breast cancer	Malins et al. (1996)
	DNA hyper- and hypomethylation	Chromosomal aberrations and sister chromatid exchange	Skin cancer	Shi et al. (2004)
Superoxide radical	DNA fragmentation	Aerobic metabolism	Reproductive dysfunction	Gosalvez et al. (2017)
	Mitochondrial DNA dysfunction	Retinal mitochondrial dysfunction	Diabetic nephropathy	Madsen-Bouterse et al. (2010)
	Oxidation	Mutation	Cancer	Wu and Cederbaum (2003)
Peroxynitrite	Oxidation	Oxidative modifications in both nucleobases and sugar-phosphate backbone	Carcinogenesis	Pacher et al. (2007)
	Oxidation	DNA strand breakage	Systemic lupus erythematosus (SLE)	Ahmad et al. (2009)
Nitric oxide	Oxidation	Apoptotic death	Necrosis	Król and Kepinska (2020)

13.2.6 Mechanism of Free Radical-Induced DNA Damage

DNA damage is a constant threat because nucleic acids are chemically unstable under physiological conditions and susceptible to attack by endogenous and environmental factors. Persistent DNA damage triggers signaling cascades that lead cells to apoptosis or senescence, preventing replication of damaged genomes (Yousefzadeh et al. 2021). Targeting DNA damage and its associated mechanisms with signaling pathways will provide a rationale for the development of unified interventions to combat age-related disorders and diseases (Schumacher et al. 2021). DNA damage in neuron increases risk factors for neurodegeneration. The oxidative damage and cytogenetic damage of DNA has been extensively stated in the clinical and preclinical stages of Alzheimer's disease and is assumed to contribute to the neurodegenerative process (Coppedè and Migliore 2015). It can impair the mitochondrial energy metabolism and ion homeostasis in neurons (Konopka and Atkin 2022). This causes apoptotic cell death in mitotic cells and terminally differentiated cells (Yang et al. 2008).

The neuronal DNA instability occurs in the presence of ROS. This instability activates the MAPK pathway which further leads the regulation of the cytokines and p53 pathway (Fig. 13.4). These pathways are involved in the neurodegeneration diseases. Similarly, the cardiovascular disease is the most common clinical disease characterized by angina pectoris, arrhythmias, myocardial injury, atrial fibrillation (AF), ischemic heart disease (IHD), and cardiac dysfunction. The leading cause, morbidity and mortality of the cardiovascular disease impose a large socioeconomic burden on society globally (Hu et al. 2021). An imbalance in ROS levels causes morphological and functional changes that further affected human cardiomyocytes, leading to a vicious cycle of oxidation. Genomic instability caused by incorrect or defective DNA repair can adversely affect the genetic code, leading to accumulation of mutations and ultimately increase the incidence of cardiovascular diseases (Fig. 13.5). Accumulation of DNA damage in human cardiomyocytes can trigger

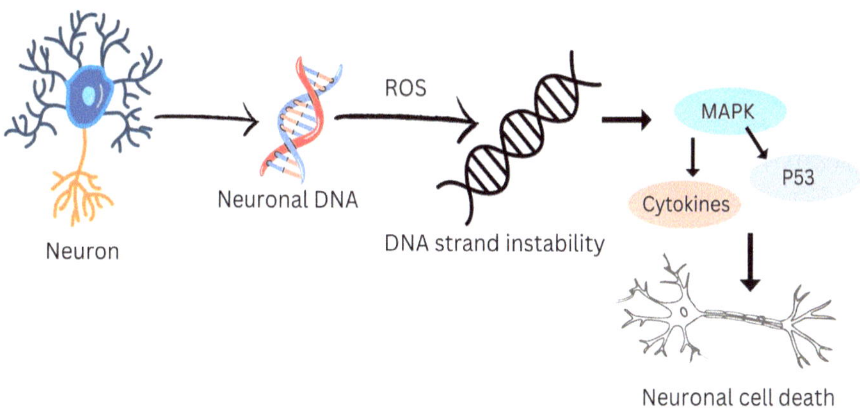

Fig. 13.4 Neuronal cell death with ROS

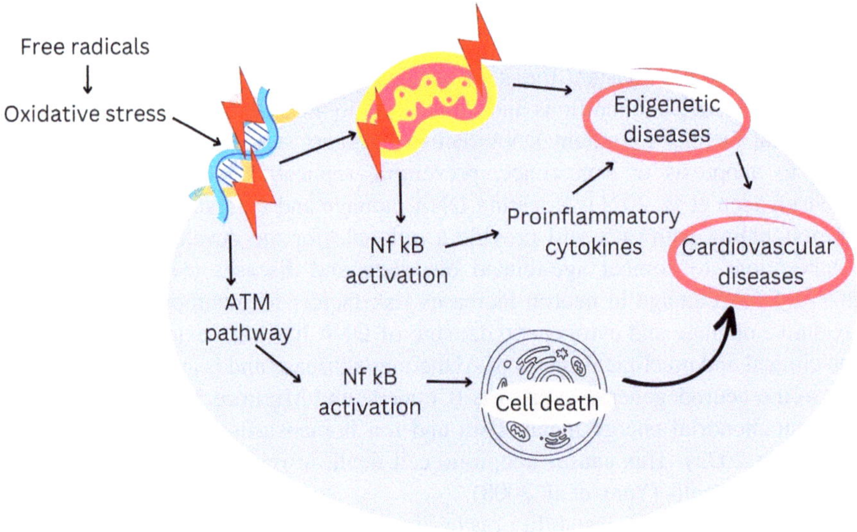

Fig. 13.5 ROS involved in cardiovascular diseases

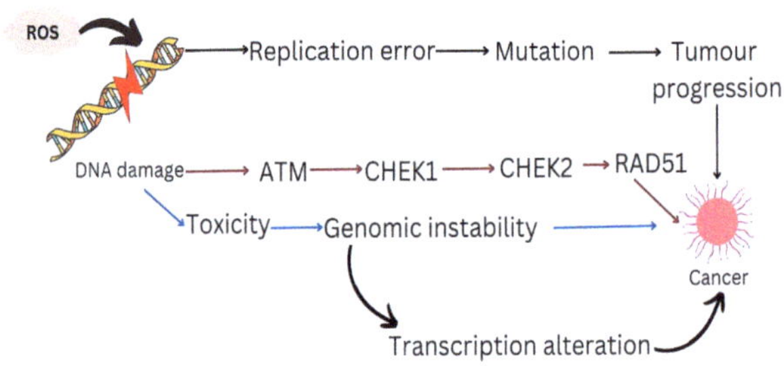

Fig. 13.6 DNA damage causes cancer

apoptosis and lead to heart failure and other cardiovascular diseases. In particular, mtDNA damage is deeply involved in cardiac pathophysiology (Nikfarjam and Singh 2023). Studies showed that the mtDNA damage associated with mitochondrial dysfunction appears to be involved in heart failure in both humans and animal models (Lowran et al. 2021).

Genomic unpredictability is one of the most common characteristics of tumor cells and is caused by DNA damage, tumor-specific DNA repair defects, and the inability to arrest the cell cycle due to damaged DNA (Alhmoud et al. 2020). This is deliberated to be the combined effect of damage to DNA replication and can lead to gene mutations, further resulting in protein changes. Mutations in oncogenes, tumor suppressor genes, or genes controlling the cell cycle can generate clonal cell populations with distinct advantages in proliferation. Oxidative stress is known to cause DNA damage and mutations in tumor suppressor genes, which are key events during the early stages of cancer development (Fig. 13.6) (Kang 2002). Interestingly, the previous results of the study suggested that environmental factors such as high psychological stress and a poor dietary profile (such as lack of antioxidants and high fat intake) increased ROS production (Bernstein et al. 2013).

13.3 Free Radicals and Lipid Peroxidation

Lipid peroxidation, a consequential outcome of this mayhem, commences when free radicals target polyunsaturated fatty acids within cell membranes. This ignites a chain reaction, with the free radical's unpaired electron reacting with a fatty acid and generates a lipid radical. This lipid radical in turn encounters molecular oxygen, birthing a lipid peroxyl radical, thereby perpetuating the chain reaction. This destructive process unfolds through initiation, propagation, and termination phases. The initiation phase commences when lipid radicals form as a result of a free radical's union with a fatty acid. The propagation phase ensues as lipid peroxyl radicals interact with neighboring lipids, fueling an unrelenting chain reaction (Halliwell 1989). The termination phase intervenes, employing the body's antioxidant systems to neutralize lipid peroxyl radicals, halting their malevolence. Understanding the genesis and impact of these free radicals assumes paramount importance, unlocking the key to crafting strategies that thwart their pernicious influence and uphold our precious well-being (Phaniendra et al. 2015).

13.3.1 Mechanism of Lipid Peroxidation

Oxidative damage to cell membranes occurs through a chain reaction known as lipid peroxidation, which is instigated when free radicals assault and impair polyunsaturated fatty acids (PUFAs). PUFAs, especially those comprising multiple double bonds, exhibit high vulnerability to oxidation due to their molecular composition (Comporti 1993). The assault of free radicals on PUFAs triggers a sequence of reactions that culminate in the creation of lipid peroxides (Gahalain et al. 2011). These lipid peroxides, including hydroperoxides, perpetuate the cycle of destruction by targeting adjacent molecules, thereby perpetuating the chain reaction (Halliwell 1989).

13.3.1.1 Initiation, Propagation, and Termination Phases

The process of lipid peroxidation comprises three distinct phases: initiation, propagation, and termination, each with its unique characteristics and mechanisms.

In the initiation phase, the formation of free radicals occurs, including hydroxyl radicals (•OH) or lipid peroxyl radicals (LOO•) (Joshi and Praticò 2014). These radicals can be generated through various processes, such as the interaction of reactive oxygen species (ROS) with PUFAs or the reaction between PUFAs and transition metal ions (Halliwell 1989). This initial phase sets the foundation for the subsequent propagation phase.

During the propagation phase, lipid peroxyl radicals react with adjacent lipids, initiating a chain reaction. The lipid peroxyl radicals abstract hydrogen atoms from neighboring PUFAs, leading to the creation of new lipid radicals and the regeneration of lipid peroxyl radicals. This process continues, allowing lipid peroxidation to spread throughout the cell membrane (Milkovic et al. 2015).

The termination phase involves the elimination of lipid peroxyl radicals through diverse antioxidant systems present in the body (Joshi and Praticò 2014). Antioxidants, both enzymatic and non-enzymatic, play a crucial role in neutralizing and scavenging the lipid peroxyl radicals, thereby interrupting the chain reaction. Enzymatic antioxidants, such as superoxide dismutase (SOD), catalase, and glutathione peroxidase, work synergistically to detoxify the lipid peroxyl radicals (Gahalain et al. 2011). Non-enzymatic antioxidants, including vitamins C and E, glutathione, and various phytochemicals, also contribute significantly to the termination of the lipid peroxidation process (Joshi and Praticò 2014).

13.3.2 Consequences of Lipid Peroxidation

Lipid peroxidation can have severe consequences on cellular structures and functions. The oxidative damage caused by lipid peroxidation affects the integrity and fluidity of cell membranes (Sies 2015). It can disrupt the arrangement of phospholipids and membrane proteins, leading to alterations in membrane permeability and fluidity. This, in turn, can affect vital cellular processes, including cellular signaling, transport of nutrients and ions, and receptor functions.

Furthermore, lipid peroxidation products, such as malondialdehyde (MDA) and 4-hydroxynonenal (4-HNE), can react with cellular macromolecules, including DNA and proteins, causing additional damage (Pham-Huy et al. 2008). These reactive lipid peroxidation products can form adducts with DNA bases, leading to DNA strand breaks, mutations, and chromosomal aberrations. They can also modify proteins, impairing their structure and function and triggering inflammatory responses (Sies 2015).

The consequences of lipid peroxidation extend beyond cellular damage and can contribute to the pathogenesis of various diseases (Hubel et al. 1989). It has been implicated in the development of cardiovascular diseases, neurodegenerative disorders, cancer, and inflammatory conditions (Pham-Huy et al. 2008). The accumulation of lipid peroxidation products and the resulting cellular dysfunction contribute to disease progression and the aging process.

Lipid peroxidation is a complex chain reaction initiated by free radicals that can lead to severe consequences on cellular structures and functions (Hubel et al. 1989). Understanding the mechanisms and consequences of lipid peroxidation is crucial in comprehending its role in disease development and aging. Strategies aimed at preventing or mitigating lipid peroxidation, such as maintaining optimal antioxidant defense systems, are essential for preserving cellular health and overall well-being.

13.4 Free Radicals and Protein Damage

Free radicals cause protein damage through oxidative modifications, nitrosative modifications, and carbonylation, disrupting protein structure and function, leading to cellular dysfunction and disease development (Fig. 13.7).

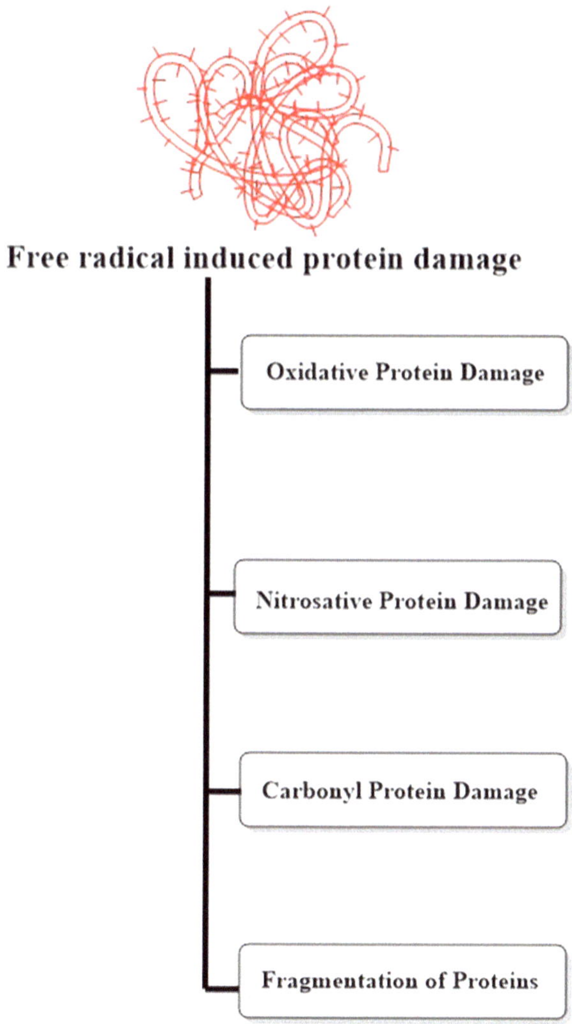

Fig. 13.7 Free radical induced protein damage

13.4.1 Oxidative Protein Damage

Free radicals can oxidize amino acid residues within proteins, leading to oxidative modifications (Stadtman 2003). This includes the oxidation of sulfur-containing amino acids such as cysteine and methionine, as well as aromatic amino acids like tryptophan, tyrosine, and phenylalanine. The oxidation of these amino acids can result in the formation of reactive intermediates and reactive oxygen species (ROS), such as hydroxyl radicals (•OH), which can react with proteins and induce oxidative modifications (Girotti 1985; Chetyrkin et al. 2008).

Oxidative modifications can have significant effects on protein structure and function. They can cause conformational changes, disrupt protein folding, impair enzymatic activity, and alter protein–protein interactions (Pham-Huy et al. 2008; Ahmad et al. 2017). These modifications can lead to loss of protein function and contribute to cellular dysfunction. Additionally, oxidative modifications can initiate further oxidative reactions and propagate protein damage through a chain reaction known as oxidative cascade.

13.4.2 Nitrosative Protein Damage

Reactive nitrogen species (RNS), particularly nitric oxide (NO) and peroxynitrite (ONOO-), can nitrosate protein tyrosine residues, leading to the formation of nitrotyrosine (Patel et al. 1999). Nitrosative modifications can modulate protein function and signaling pathways. Nitrotyrosine formation can disrupt protein–protein interactions, alter enzymatic activity, and induce protein misfolding (Pham-Huy et al. 2008; Nakamura et al. 2021). Furthermore, nitrosative modifications can contribute to the formation of protein aggregates and the development of neurodegenerative diseases (Sies 2015).

13.4.3 Carbonyl Protein Damage

Oxidative stress can result in the formation of carbonyl groups on proteins, a process known as protein carbonylation (Akagawa 2021). Carbonylated proteins are often dysfunctional and susceptible to aggregation (Tola et al. 2021). Carbonylation can occur through the direct reaction of proteins with reactive oxygen species (ROS) or through the reaction of proteins with reactive aldehydes derived from lipid peroxidation (Pham-Huy et al. 2008; Tola et al. 2021). Carbonylated proteins can impair cellular processes and contribute to the development of various diseases, including neurodegenerative disorders and cardiovascular diseases.

13.4.4 Fragmentation of Proteins Damage

Free radicals can directly break peptide bonds within proteins, resulting in protein fragmentation. This fragmentation can disrupt protein structure, impair function, and generate protein fragments with altered activities and toxic properties (Zhang et al. 2013). Protein fragmentation can lead to the loss of protein function and the generation of proteolytic fragments that can accumulate and contribute to cellular dysfunction.

The mechanisms of protein damage caused by free radicals are diverse and can have profound effects on protein structure and function. Oxidative modifications, nitrosative modifications, carbonylation, and protein fragmentation contribute to protein damage and can lead to cellular dysfunction and disease development.

13.4.5 Consequences of Protein Damage

13.4.5.1 Altered Protein Function

Protein damage induced by free radicals can have significant consequences on protein function. Oxidative modifications, nitrosative modifications, and carbonylation can alter the structure and conformation of proteins, leading to changes in enzymatic activity, protein–protein interactions, and signaling pathways (Blois 1958; Milkovic et al. 2015). These alterations can disrupt cellular homeostasis and contribute to disease progression. For example, impaired enzymatic activity due to protein damage can hinder critical metabolic processes, while disrupted protein–protein interactions can interfere with cellular signaling and communication.

13.4.5.2 Protein Aggregation and Misfolding

Protein damage caused by free radicals can promote protein aggregation and misfolding. When proteins are subjected to oxidative modifications or other forms of damage, they can lose their native conformation and adopt abnormal structures (Nagendrappa 2005; Pham-Huy et al. 2008). These damaged proteins are prone to self-association and can form insoluble aggregates, amyloid fibrils, or inclusion bodies. Protein aggregates can interfere with cellular functions and disrupt the normal activities of proteins, leading to cellular toxicity and dysfunction. The accumulation of protein aggregates is a characteristic feature of various neurodegenerative diseases, including Alzheimer's and Parkinson's disease.

13.5 Impact of Free Radicals and Its Relation to Various Metabolic Disorders

Free radical-induced protein damage has been implicated in the pathogenesis of numerous diseases. Neurodegenerative disorders, such as Alzheimer's disease, Parkinson's disease, and Huntington's disease, are associated with the accumulation of protein aggregates and the impairment of protein quality control systems

(Yoshikawa et al. 1997; Pham-Huy et al. 2008; Sies 2015). Protein damage can initiate pathological cascades, leading to the formation of toxic species and the disruption of cellular processes (Pham-Huy et al. 2008). Cardiovascular diseases, including atherosclerosis and heart failure, are also linked to protein damage, oxidative stress, and inflammation. Additionally, protein damage caused by free radicals can contribute to the development and progression of cancer by altering cellular signaling pathways and promoting genomic instability (Valko et al. 2006). Understanding the consequences of protein damage is essential for unravelling the mechanisms of disease pathogenesis. Free radical-induced protein damage can initiate detrimental processes that disrupt cellular functions and contribute to the development of various diseases (Reiter et al. 2000). By elucidating the relationship between protein damage, disease progression, and cellular dysfunction, new therapeutic strategies can be developed to mitigate the harmful effects of free radicals and preserve protein integrity. Excessive production of free radicals can overwhelm the body's antioxidant defense systems, resulting in a state known as oxidative stress. Oxidative stress arises when there is an imbalance between the generation of free radicals and the body's capacity to counteract them with antioxidants (Pham-Huy et al. 2008). Free radicals, particularly reactive oxygen species (ROS), possess the ability to harm crucial cellular components, including lipids, proteins, and DNA, thereby contributing to the emergence and advancement of various diseases (Lagouge and Larsson 2013).

13.5.1 Cardiovascular Diseases

Excessive production of free radicals can overwhelm the body's antioxidant defense systems, resulting in a state known as oxidative stress. Oxidative stress arises when there is an imbalance between the generation of free radicals and the body's capacity to counteract them with antioxidants (Pham-Huy et al. 2008). Free radicals, particularly reactive oxygen species (ROS), possess the ability to harm crucial cellular components, including lipids, proteins, and DNA, thereby contributing to the emergence and advancement of various diseases (Lagouge and Larsson 2013). Cardiovascular disease is associated with the pathogenesis of oxidative stress. Free radicals can oxidize low-density lipoproteins (LDL) present in the bloodstream, leading to the formation of oxidized LDL (oxLDL) (Yoshikawa et al. 1997). This oxLDL promotes inflammation, impairs endothelial function, and fosters the development of atherosclerotic plaques, thereby contributing to cardiovascular conditions like atherosclerosis and coronary artery disease.

Chronic diseases, such as diabetes, hypertension, and inflammatory conditions, have been linked to elevated levels of oxidative stress and the resulting damage caused by free radicals (Gutteridge and Halliwell 1993). Oxidative stress can worsen inflammation, impair insulin signaling, and contribute to tissue damage (Verdile et al. 2015; Ahmad et al. 2018). In the case of diabetes, chronic exposure to high levels of glucose and free fatty acids can increase oxidative stress, leading to dysfunction of pancreatic beta cells and insulin resistance (Rains and Jain 2011). Hypertension, characterized by high blood pressure, is also associated with oxidative stress-induced vascular dysfunction and inflammation (Rains and Jain 2011; Verdile et al. 2015). Inflammatory conditions like rheumatoid arthritis and inflammatory bowel disease are connected to heightened production of free radicals and oxidative stress. Free radicals can activate inflammatory signaling pathways, perpetuating the inflammatory response and resulting in tissue damage and chronic inflammation (Piechota-Polanczyk and Fichna 2014).

13.5.2 Neurodegenerative Diseases

Oxidative stress also plays a significant role in neurodegenerative disorders. Free radicals can cause damage to neurons and disrupt crucial neuronal signaling pathways (Phaniendra et al. 2015). This disruption can result in the accumulation of protein aggregates, neuronal cell death, and the progression of diseases such as Alzheimer's and Parkinson's disease. The development and progression of cancer are likewise linked to oxidative stress (Yoshikawa et al. 1997). Free radicals have the potential to induce DNA damage and mutations, which can initiate the formation of cancer (Chatgilialoglu and O'Neill 2001). Additionally, oxidative stress can facilitate tumor growth, angiogenesis, and metastasis by modulating signaling pathways involved in cell proliferation and survival (Von-Sonntag 2006).

13.5.3 Aging

Free radicals have been implicated in the aging process (Harman 1988). The gradual accumulation of oxidative damage can disrupt cellular functions and hasten the aging process. Free radicals have the ability to directly impair DNA, proteins, and lipids, leading to cellular dysfunction and compromised physiological processes associated with aging (Stohs 1995). The DNA damage caused by free radicals can result in mutations and genomic instability. These changes can impact vital biological processes involved in aging, including DNA repair mechanisms and telomere maintenance. Telomeres, which are protective caps found at the ends of chromosomes, naturally shorten with each round of cell division. Oxidative stress and the detrimental effects of free radicals can expedite telomere shortening, ultimately leading to cellular senescence and accelerated aging (Getoff 2007). Additionally, free radicals can hinder mitochondrial function, which plays a pivotal role in energy production and cellular metabolism (Chen et al. 2003). Mitochondrial DNA is particularly vulnerable to oxidative damage, resulting in mitochondrial dysfunction and increased production of free radicals, establishing a detrimental cycle of oxidative stress and cellular harm (Beal 1996).

13.5.4 Respiratory Diseases

There is increasing evidence that air pollution increases oxidative stress and contributes to several diseases, including respiratory diseases. COPD is a chronic inflammatory disease characterized by progressive and partially reversible airway obstruction. Oxidative stress plays a central role in the development of COPD. Oxidative stress activates lung epithelial cells and produces inflammatory mediators involved in macrophage activation and regulation of gene expression and transcription factors. Asthma is an inflammatory disease, and the inflammatory process in asthma is said to be associated with the control of airway hyperresponsiveness and airway remodeling. Furthermore, oxidative stress is present in the airways and bloodstream of asthma patients. The lungs are often targeted by free oxygen radicals because they come into contact with oxygen in two different ways (perfusion and ventilation), and a variety of lung diseases appear to be affected by these molecules.

13.6 Scavengers of Free Radicals

The implementation of antioxidant therapies has shown promise in mitigating the progression of chronic diseases (Pereira et al. 2015). Antioxidants possess the ability to neutralize free radicals and reduce oxidative stress, thereby helping restore the equilibrium between free radicals and antioxidants (Lobo et al. 2010). Antioxidant-rich diets, antioxidant supplementation, and lifestyle modifications that reduce oxidative stress all play significant roles in managing chronic diseases and enhancing overall health (Valko et al. 2006; Lobo et al. 2010).

13.6.1 Antioxidant Enzymes

The body possesses a robust defense system against free radicals, which includes various antioxidant enzymes (Pham-Huy et al. 2008). These enzymatic antioxidants neutralize free radicals and prevent their damaging effects (Singh et al. 2022). Three key enzymatic antioxidants are superoxide dismutase (SOD), catalase, and glutathione peroxidase. Superoxide dismutase (SOD) is an essential antioxidant enzyme that catalyzes the conversion of superoxide radicals (O_2•-) into hydrogen peroxide (H_2O_2) and molecular oxygen (O2). This enzymatic reaction is crucial as superoxide radicals are highly reactive and can lead to the formation of more damaging species. Catalase, another important antioxidant enzyme, facilitates the decomposition of hydrogen peroxide into water and oxygen, thus preventing the formation of hydroxyl radicals (•OH) (Pham-Huy et al. 2008). Glutathione peroxidase is responsible for the reduction of hydrogen peroxide and lipid peroxides, utilizing reduced glutathione (GSH) as a cofactor to maintain cellular redox balance (Sies 2015).

13.6.2 Non-enzymatic Antioxidants

Non-enzymatic antioxidants play a critical role in the scavenging of free radicals and the prevention of oxidative damage (Pham-Huy et al. 2008). Working in conjunction with enzymatic antioxidants, they help maintain a delicate equilibrium between free radicals and antioxidants. Vitamin C (ascorbic acid), a water-soluble antioxidant, directly neutralizes free radicals by donating electrons, thereby safeguarding biomolecules from oxidative damage (Mehta and Gowder 2015). Vitamin E (α-tocopherol), a lipid-soluble antioxidant, protects cell membranes from lipid peroxidation by scavenging lipid peroxyl radicals (Girotti 1985). Glutathione, a tripeptide consisting of cysteine, glycine, and glutamic acid, acts as a potent antioxidant by participating in redox reactions and maintaining the reduced state of other antioxidants (Engwa et al. 2022). Carotenoids, such as β-carotene and lycopene, are plant pigments that possess strong antioxidant properties, particularly in scavenging singlet oxygen (Mehta and Gowder 2015). These non-enzymatic antioxidants work synergistically with enzymatic antioxidants to scavenge free radicals and shield cells from oxidative damage. They can be obtained through dietary sources, including fruits, vegetables, nuts, and seeds, or through supplementation when necessary (Lobo et al. 2010). The combined action of enzymatic and non-enzymatic antioxidants ensures a robust defense system against free radicals, contributing to the maintenance of cellular homeostasis. The body employs a multifaceted defense system against free radicals. Antioxidant enzymes such as SOD, catalase, and glutathione peroxidase neutralize free radicals and prevent their detrimental effects. Non-enzymatic antioxidants, such as vitamins C and E, glutathione, and carotenoids, work in tandem with enzymatic antioxidants to scavenge free radicals and protect against oxidative damage. Maintaining an appropriate balance between free radicals and antioxidants is vital for cellular health and overall well-being.

13.7 Strategies to Mitigate Free Radicals Induced Damage

13.7.1 Antioxidant Therapies

One strategy to mitigate protein damage caused by free radicals is to enhance the antioxidant defense system. This can be achieved through dietary antioxidants or antioxidant supplementation. Consuming a diet rich in fruits, vegetables, whole grains, and nuts provides the body with essential nutrients, including vitamins C and E, carotenoids, and polyphenols, which possess antioxidant properties (Valko et al. 2006; Pham-Huy et al. 2008; Phaniendra et al. 2015). These dietary antioxidants scavenge free radicals and help to minimize oxidative damage to proteins. In certain cases, antioxidant supplementation may be beneficial, particularly for individuals with increased oxidative stress due to certain medical conditions or inadequate dietary intake of antioxidants (Meulmeester et al. 2022). Antioxidant supplements, such as vitamin C, vitamin E, and coenzyme Q10, can help bolster the antioxidant defense system and provide additional protection against free radicals (Yoshikawa et al. 1997). However, it is important to note that antioxidant supplementation should be approached with caution and under the guidance of a healthcare professional, as excessive antioxidant levels can have adverse effects (Pham-Huy et al. 2008).

13.7.2 Diet and Nutrition

Incorporating an antioxidant-rich diet is a highly effective approach to counteracting the detrimental effects of free radical damage (Poljsak 2011). Antioxidant-rich foods, such as fruits, vegetables, whole grains, and nuts, provide the body with essential nutrients that can help maintain optimal antioxidant levels and minimize oxidative stress. These foods are abundant in vitamins (e.g., vitamins C and E), minerals (e.g., selenium and zinc), and phytochemicals (e.g., flavonoids and carotenoids), which possess potent antioxidant properties (Pham-Huy et al. 2008). Including a variety of antioxidant-rich foods in the diet ensures a diverse range of antioxidants that can work synergistically to combat free radicals.

13.7.3 Lifestyle Modifications

Adopting certain lifestyle modifications can significantly reduce free radical production and oxidative damage. Regular physical exercise promotes the production of endogenous antioxidants, enhances mitochondrial function, and improves overall cellular health. Managing stress levels through techniques such as meditation, mindfulness, and relaxation exercises can also help reduce oxidative stress, as chronic stress is known to contribute to free radical production (Sies 2015). Avoiding smoking and excessive alcohol consumption is essential, as both tobacco smoke and alcohol metabolism generate free radicals and deplete antioxidant levels (Pham-Huy et al. 2008). Additionally, limiting exposure to environmental toxins, such as air pollutants and harmful chemicals, can reduce oxidative stress and free radical damage (Valko et al. 2006).

13.7.4 Antioxidant Supplementation

In certain cases, antioxidant supplementation may be beneficial, particularly for individuals with increased oxidative stress due to certain medical conditions or inadequate dietary intake. Antioxidant supplements, such as vitamins C and E, coenzyme Q10, and alpha-lipoic acid, can help augment the body's antioxidant defense system. Implementing strategies to mitigate free radical damage is crucial for maintaining optimal health. In certain cases, antioxidant supplementation may be beneficial, but it should be approached with caution and under professional guidance. By adopting these strategies, individuals can enhance their antioxidant defense mechanisms and minimize the harmful effects of free radicals.

13.7.5 Modulation of Protein Quality Control

Another approach to mitigating protein damage is through the modulation of protein quality control mechanisms. Protein quality control ensures the proper folding, maintenance, and degradation of proteins within cells (Vazquez-Calvo et al. 2020). Strategies that promote protein folding, prevent protein aggregation, and enhance protein degradation pathways can help minimize protein damage and maintain cellular proteostasis (Cuanalo-Contreras et al. 2013). Chaperones, such as heat shock proteins (HSPs), play a crucial role in facilitating proper protein folding and preventing the accumulation of misfolded or damaged proteins (Pham-Huy et al. 2008). Inducing the expression or activity of chaperones can enhance protein folding and reduce the formation of protein aggregates. This can be achieved through various means, including heat stress, pharmacological agents, or genetic manipulation.

13.8 Conclusions and Future Perspective

Free radicals can induce significant damage to proteins, leading to altered protein function, aggregation, and disease development. Understanding the mechanisms of protein damage and the consequences on cellular processes is crucial for elucidating disease pathogenesis. By enhancing antioxidant defenses, supporting protein repair mechanisms, and implementing strategies to mitigate protein damage, it may be possible to minimize the harmful effects of free radicals on protein structure and function. In conclusion, free radicals and lipid peroxidation pose significant threats to our physiological processes and disease development. Their damaging effects on lipids, proteins, and DNA underscore the urgent need to comprehend their mechanisms and consequences. By gaining insight into these processes, we can develop strategies to combat oxidative damage and safeguard our health. A balanced diet, comprising antioxidant-rich foods like fruits, vegetables, whole grains, and nuts, equips our bodies with vital nutrients to counteract free radicals. Embracing a healthy lifestyle with regular exercise, stress management, no smoking, limited alcohol consumption, and reduced exposure to toxins dramatically reduces free radical production and oxidative damage. In specific cases, antioxidant supplementation may prove beneficial, but it's crucial to seek professional guidance beforehand. By implementing these measures, we empower ourselves to fortify our body's defense mechanisms against free radicals and promote overall well-being. This chapter provides an in-depth exploration of free radical-associated damages to macromolecules, their role in various physiological processes, and their impact on human health. It also discusses the sources of free radicals and the body's defense mechanisms against their harmful effects. Finally, it concludes with potential strategies to mitigate free radical damage and maintain optimal health. Let us take proactive steps in mitigating the harmful effects of oxidative damage and optimizing our health.

References

Ahmad R, Rasheed Z, Ahsan H (2009) Biochemical and cellular toxicology of peroxynitrite: implications in cell death and autoimmune phenomenon. Immunopharmacol Immunotoxicol 31(3):388–396

Ahmad S, Khan H, Shahab U, Rehman S, Rafi Z, Khan MY, Ansari A, Siddiqui Z, Ashraf JM, Abdullah SM (2017) Protein oxidation: an overview of metabolism of Sulphur containing amino acid, cysteine. Front Biosci (Schol Ed) 9:71–87

Ahmad S, Khan MY, Rafi Z, Khan H, Siddiqui Z, Rehman S, Shahab U, Khan MS, Saeed M, Alouffi S, Khan MS (2018) Oxidation, glycation and glycoxidation—the vicious cycle and lung cancer. Semin Cancer Biol 49:29–36

Akagawa M (2021) Protein carbonylation: molecular mechanisms, biological implications, and analytical approaches. Free Radic Res 55(4):307–320

Akbari B, Baghaei-Yazdi N, Bahmaie M, MahdaviAbhari F (2022) The role of plant-derived natural antioxidants in reduction of oxidative stress. Bio Factors 48(3):611–633

Alhmoud JF, Woolley JF, Al Moustafa AE, Malki MI (2020) DNA damage/repair management in cancers. Cancers. 12(4), 1050

Beal MF (1996) Mitochondria, free radicals, and neurodegeneration. Curr Opin Neurobiol 6(5):661–666

Bergendi L, Beneš L, Ďuracková Z, Ferenčik M (1999) Chemistry, physiology and pathology of free radicals. Life Sci 65(18–19):1865–1874

Bernstein C, Prasad AR, Nfonsam V, Bernstein H (2013) DNA damage, DNA repair and cancer. InTech, Rijeka Croatia, pp 413–465

Bhatti JS, Sehrawat A, Mishra J, Sidhu IS, Navik U, Khullar N, Reddy PH (2022) Oxidative stress in the pathophysiology of type 2 diabetes and related complications: current therapeutics strategies and future perspectives. Free Radic Biol Med 184:114–134

Blois MS (1958) Antioxidant determinations by the use of a stable free radical. Nature 181(4617):1199–1200

Burcham PC (1998) Genotoxic lipid peroxidation products: their DNA damaging properties and role in formation of endogenous DNA adducts. Mutagenesis 13(3):287–305

Chatgilialoglu C, O'Neill P (2001) Free radicals associated with DNA damage. Exp Gerontol 36(9):1459–1471

Chatgilialoglu C, Ferreri C, Krokidis MG, Masi A, Terzidis MA (2021) On the relevance of hydroxyl radical to purine DNA damage. Free Radic Res 55(4):384–404

Chen Q, Chai Y, Mazumder S, Jiang C, Macklis RM, Chisolm GM, Almasan A (2003) The late increase in intracellular free radical oxygen species during apoptosis is associated with cytochrome c release, caspase activation, and mitochondrial dysfunction. Cell Death Differ 10(3):323–334

Chetyrkin SV, Mathis ME, Ham AJL, Hachey DL, Hudson BG, Voziyan PA (2008) Propagation of protein glycation damage involves modification of tryptophan residues via reactive oxygen species: inhibition by pyridoxamine. Free Radic Biol Med 44(7):1276–1285

Choudhari SK, Chaudhary M, Gadbail AR, Sharma A, Tekade S (2014) Oxidative and antioxidative mechanisms in oral cancer and precancer: a review. Oral Oncol 50(1):10–18

Chung WH (2017) Unraveling new functions of superoxide dismutase using yeast model system: beyond its conventional role in superoxide radical scavenging. J Microbiol 55:409–416

Church DF, Pryor WA (1985) Free-radical chemistry of cigarette smoke and its toxicological implications. Environ Health Perspect 64:111–126

Comporti M (1993) Lipid peroxidation. An overview. In: Poli, G., Albano, E., Dianzani, M.U. (eds) Free Radicals: from Basic Science to Medicine. 65–79. https://doi.org/10.1007/978-3-0348-9116-5_6

Coppedè F, Migliore L (2015) DNA damage in neurodegenerative diseases. Mutat Res Fundam Mol 776:84–97

Cuanalo-Contreras K, Mukherjee A, Soto C (2013) Role of protein misfolding and proteostasis deficiency in protein misfolding diseases and aging. Int J Cell Biol 2013:1
Das K, Roychoudhury A (2014) Reactive oxygen species (ROS) and response of antioxidants as ROS-scavengers during environmental stress in plants. Front. Environ Sci 2(DEC):121942
Dharmaraja AT (2017) Role of reactive oxygen species (ROS) in therapeutics and drug resistance in cancer and bacteria. J Med Chem 60(8):3221–3240
Di Meo S, Venditti P (2020) Evolution of the knowledge of free radicals and other oxidants. Oxidative Med Cell Longev 2020:2020
Djordjevic VB (2004) Free radicals in cell biology. Int Rev Cytol 237(237):57–89
Engwa GA, Nweke FN, Nkeh-Chungag BN (2022) Free radicals, oxidative stress-related diseases and antioxidant supplementation. Altern Ther Health Med 28(1):114
Fleming AM, Burrows CJ (2020) On the irrelevancy of hydroxyl radical to DNA damage from oxidative stress and implications for epigenetics. Chem Soc Rev 49(18):6524–6528
Gahalain N, Chaudhary J, Kumar A, Sharma S, Jain A (2011) Lipid peroxidation: an overview. Int J Pharm Sci Res 2(11):2757
Getoff N (2007) Anti-aging and aging factors in life. The role of free radicals. Radiat Phys Chem 76(10):1577–1586
Girotti AW (1985) Mechanisms of lipid peroxidation. J Free Radic Biol Med 1(2):87–95
Görsdorf S, Appel KE, Engeholm C, Obe G (1990) Niltrogen dioxide induces DNA single-strand breaks in cultured Chinese hamster cells. Carcinogenesis 11(1):37–41
Gosalvez J, Tvrda E, Agarwal A (2017) Free radical and superoxide reactivity detection in semen quality assessment: past, present, and future. J Assist Reprod Genet 34(6):697–707
Gupta R, Shukla S, Gupta A, Singh RL, Singh P (2023) Estimation of phytochemical constituents and analysis of antioxidant activity in different parts of *Cassia fistula* plant. Int J Pharma Chem Ana 10(3):194–201
Gutteridge JMC, Halliwell B (1993) Invited review free radicals in disease processes: a compilation of cause and consequence. Free Radic Res Commun 19(3):141–158
Halliwell B (1989) Protection against oxidants in biological systems. The superoxide theory of oxygen toxicity. J Free Radic Biol Med. 86–123
Halliwell B (2006) Reactive species and antioxidants. Redox biology is a fundamental theme of aerobic life. Plant Physiol 141:312–322
Halliwell B, Gutteridge JMC (2007) Free radicals in biology and medicine, 4th edn. Oxford University Press, New York
Halliwell B, Gutteridge JM (2015) Free radicals in biology and medicine. Oxford University Press
Harman D (1988) Free radicals in aging. Mol Cell Biochem 84:155–161
Hu L, Wang Z, Carmone C, Keijer J, Zhang D (2021) Role of oxidative DNA damage and repair in atrial fibrillation and ischemic heart disease. Int J Mol Sci 22(8):3838
Huang R, Zhou PK (2021) DNA damage repair: historical perspectives, mechanistic pathways and clinical translation for targeted cancer therapy. Signal Transduct Target Ther 6(1):254
Hubel CA, Roberts JM, Taylor RN, Musci TJ, Rogers GM, McLaughlin MK (1989) Lipid peroxidation in pregnancy: new perspectives on preeclampsia. Am J Obstet Gynecol 161(4):1025–1034
Joshi YB, Praticò D (2014) Lipid peroxidation in psychiatric illness: overview of clinical evidence. Oxidative Med Cell Longev 2014
Juan JM, Plou CA, Pérez-Lebeña FJ, De La Lastra JMP, Juan CA, Plou FJ, Pérez-Lebeña E (2022) The nitration of proteins, lipids and DNA by peroxynitrite derivatives-chemistry involved and biological relevance. Stresses 2(1):53–64
Kang DH (2002) Oxidative stress, DNA damage, and breast cancer. AACN Adv Crit Care 13(4):540–549
Kehrer JP, Klotz LO (2015) Free radicals and related reactive species as mediators of tissue injury and disease: implications for health. Crit Rev Toxicol 45(9):765–798
Kohen R, Nyska A (2002) Oxidation of biological systems: oxidative stress and antioxidants. Toxicol Pathol 30:620–630
Konopka A, Atkin JD (2022) The role of DNA damage in neural plasticity in physiology and neurodegeneration. Front Cell Neurosci 16:836885

Kovacic P, Pozos RS, Somanathan R, Shangari N, O'Brien PJ (2005) Mechanism of mitochondrial uncouplers, inhibitors, and toxins: focus on electron transfer, free radicals, and structure -activity relationships. Curr Med Chem 12(22):2601–2623
Krokidis MGD, Errico M, Pascucci B, Parlanti E, Masi A, Ferreri C, Chatgilialoglu C (2020) Oxygen-dependent accumulation of purine DNA lesions in Cockayne syndrome cells. Cells 9(7):1671
Król M, Kepinska M (2020) Human nitric oxide synthase—its functions, polymorphisms, and inhibitors in the context of inflammation, diabetes and cardiovascular diseases. Int J Mol Sci 22(1):56
Lagouge M, Larsson N (2013) The role of mitochondrial DNA mutations and free radicals in disease and ageing. J Intern Med 273(6):529–543
Lobo V, Patil A, Phatak A, Chandra N (2010) Free radicals, antioxidants and functional foods: impact on human health. Phcog Rev 4(8):118
Lowran K, Wu C, Petersen I (2021) Abstract P487: impact of oxidative Dna damage on heart health. In: Circulation research, vol 129 No. Suppl_1. Lippincott Williams & Wilkins, Hagerstown, p AP487-AP487
Lushchak VI (2007) Free radical oxidation of proteins and its relationship with functional state of organisms. Biochem Mosc 72(8):809–827
Madsen-Bouterse SA, Zhong Q, Mohammad G, Ho YS, Kowluru RA (2010) Oxidative damage of mitochondrial DNA in diabetes and its protection by manganese superoxide dismutase. Free Radic Res 44(3):313–321
Malins DC, Polissar NL, Gunselman SJ (1996) Progression of human breast cancers to the metastatic state is linked to hydroxyl radical-induced DNA damage. Proc Natl Acad Sci USA 93(6):2557–2563
Marnett LJ (2000) Oxyradicals and DNA damage. Carcinogenesis 21(3):361–370
Mehta SK, Gowder SJT (2015) Members of antioxidant machinery and their functions. Basic Principles and Clinical Significance of Oxidative Stress 11:59–85
Meulmeester FL, Luo J, Martens LG, Mills K, van Heemst D, Noordam R (2022) Antioxidant supplementation in oxidative stress-related diseases: what have we learned from studies on alpha-tocopherol? Antioxidants 11(12):2322
Milkovic L, CipakGasparovic A, Zarkovic N (2015) Overview on major lipid peroxidation bioactive factor 4-hydroxynonenal as pluripotent growth-regulating factor. Free Radic Res 49(7):850–860
Moncada S, Palmer R, Higgs E (1991) Nitric oxide: physiology, pathophysiology and pharmacology. Pharmacol Rev 43:109–141
Muindi JR, Sinha BK, Gianni L, Myers CE (1984) Hydroxyl radical production and DNA damage induced by anthracycline-iron complex. FEBS Lett 172(2):226–230
Nagendrappa G (2005) An appreciation of free radical chemistry 3. Free radicals in diseases and health. Resonance 10:65–74
Nakamura T, Oh C, Zhang X, Lipton SA (2021) Protein S-nitrosylation and oxidation contribute to protein misfolding in neurodegeneration. Free Radic Biol Med 172:562–577
Nikfarjam S, Singh KK (2023) DNA damage response signaling: a common link between cancer and cardiovascular diseases. Cancer Med 12(4):4380–4404
Oberley LW (1988) Free radicals and diabetes. Free Radic Biol Med 5(2):113–124
Pacher P, Beckman JS, Liaudet L (2007) Nitric oxide and peroxynitrite in health and disease. Physiol Rev 87(1):315–424
Pacifici RE, Davies KJA (1991) Protein, lipid and DNA repair systems in oxidative stress: the free-radical theory of aging revisited. Gerontology 37(1–3):166–180
Patel RP, McAndrew J, Sellak H, White CR, Jo H, Freeman BA, Darley-Usmar VM (1999) Biological aspects of reactive nitrogen species. Biochim Biophys Acta Bioenerg 1411(2–3):385–400
Peña-Bautista C, Baquero M, Vento M, Cháfer-Pericás C (2019) Free radicals in Alzheimer's disease: lipid peroxidation biomarkers. Clin Chim Acta 491:85–90
Pereira C, Grácio D, Teixeira JP, Magro F (2015) Oxidative stress and DNA damage: implications in inflammatory bowel disease. Inflamm Bowel Dis 21(10):2403–2417

Pham-Huy LA, He H, Pham-Huy C (2008) Free radicals, antioxidants in disease and health. Int J Biomed Sci: IJBS 4(2):89
Phaniendra A, Jestadi DB, Periyasamy L (2015) Free radicals: properties, sources, targets, and their implication in various diseases. Indian J Clin Biochem 30:11–26
Piechota-Polanczyk A, Fichna J (2014) The role of oxidative stress in pathogenesis and treatment of inflammatory bowel diseases. Naunyn Schmiedeberg's Arch Pharmacol 387:605–620
Poljsak B (2011) Strategies for reducing or preventing the generation of oxidative stress. Oxidative Med Cell Longev 2011:1
Rains JL, Jain SK (2011) Oxidative stress, insulin signaling, and diabetes. Free Radic Biol Med 50(5):567–575
Reiter RJ, Tan D, Osuna C, Gitto E (2000) Actions of melatonin in the reduction of oxidative StressA review. J Biomed Sci 7(6):444–458
Rifkind JM, Abugo O, Levy A, Monticone R, Heim J (2022) Formation of free radicals under hypoxia. Surviving Hypoxia:509–525 CRC Press
Routledge MN (2000) Mutations induced by reactive nitrogen oxide species in the supF forward mutation assay. Mutat Res Fundam Mol 450(1–2):95–105
Salgo MG, Bermudez E, Squadrito GL, Pryor W (1995) Peroxynitrite causes DNA damage and oxidation of thiols in rat thymocytes (corrected). Arch Biochem Biophys 322(2):500–505
Schumacher B, Pothof J, Vijg J, Hoeijmakers JH (2021) The central role of DNA damage in the ageing process. Nature 592(7856):695–703
Shi H, Hudson LG, Ding W, Wang S, Cooper KL, Liu S, Liu KJ (2004) Arsenite causes DNA damage in keratinocytes via generation of hydroxyl radicals. Chem Res Toxicol 17(7):871–878
Shukla M, Singh SV, Singh P, Singh U, Vishwakerma SP, Khanna AA, Sexana JK, Singh RL (2011) Antidyslipidimic and antioxidant activity of hydroethanolic fruit extract of *Ficus glomerota*. Asian J Pharm Clin Res 4(2):145–148
Sies H (2015) Oxidative stress: a concept in redox biology and medicine. Redox Biol 4:180–183
Singh P, Singh RL (2018) Quantification of phytochemicals imparting antioxidant activities in commonly used vegetables. Int J Appl Sci Biotech 6(2):107–112
Singh P, Singh U, Shukla M, Singh RL (2008) Antioxidant activity imparting biomolecules in *Cassia fistula*. Adv Life Sci 2(1&2):23–28
Singh P, Vishwakarma SP, Singh RL (2013) Evaluation of antioxidant, oxidative DNA damage protective and antimicrobial activities of *Foeniculum vulgare* plant. J Med Plant Res 7(35):2551–2563
Singh RL, Vishwakarma SP, Singh P (2014) Vitamins and minerals: roles and plant sources. In: Prakash D (ed) Phytochemicals of nutraceutical importance and their role in human health. CABI-UK, pp 248–265
Singh P, Singh RL, Pathak N, Singh PK, Tripathi M, Mondal S (2022) Phytochemistry and nutraceutical properties of *Carica papaya* (Linn.): a review. Diet Suppl Nutr 1(9):1–15
Stadtman ER (2003) Oxidation of free amino acids and amino acid residues in proteins by radiolysis and by metal-catalyzed reactions. Annu Rev Biochem 62:797–821
Stohs SJ (1995) The role of free radicals in toxicity and disease. J Basic Clin Physiol Pharmacol 6(3–4):205–228
Tao G, Song G, Qin S (2019) Molecular hydrogen: current knowledge on mechanism in alleviating free radical damage and diseases. ACTA Bioch Bioph Sin 51(12):1189–1197
Tola AJ, Jaballi A, Missihoun TD (2021) Protein carbonylation: emerging roles in plant redox biology and future prospects. Plan Theory 10(7):1451
Valko M, Rhodes CJB, Moncol J, Izakovic MM, Mazur M (2006) Free radicals, metals and antioxidants in oxidative stress-induced cancer. Chem Biol Interact 160(1):1–40
Vazquez-Calvo C, Suhm T, Büttner S, Ott M (2020) The basic machineries for mitochondrial protein quality control. Mitochondrion 50:121–131
Verdile G, Keane KN, Cruzat VF, Medic S, Sabale M, Rowles J, Wijesekara N, Martins RN, Fraser PE, Newsholme P (2015) Inflammation and oxidative stress: the molecular connectivity between insulin resistance, obesity, and Alzheimer's disease. Mediat Inflamm 2015:1
Von-Sonntag C (2006) Free-radical-induced DNA damage and its repair. Springer

Wu D, Cederbaum AI (2003) Alcohol, oxidative stress, and free radical damage. Alcohol Res Health 27(4):277
Wu L, Xiong X, Wu X, Ye Y, Jian Z, Zhi Z, Gu L (2020) Targeting oxidative stress and inflammation to prevent ischemia-reperfusion injury. Front Mol Neurosci 13:28
Xing R, Chen Z, Sun H, Liao H, Qin S, Liu W, Zhou S (2022) Free radicals accelerate in situ ageing of microplastics during sludge composting. J Hazard Mater 429:128405
Yang JL, Weissman L, Bohr VA, Mattson MP (2008) Mitochondrial DNA damage and repair in neurodegenerative disorders. DNA Repair 7(7):1110–1120
Yoshikawa T, Naito Y, Kondo M (1997) Free radicals and diseases. Food Free Radic, pp 11–19
Yousefzadeh M, Henpita C, Vyas R, Soto-Palma C, Robbins P, Niedernhofer L (2021) DNA damage—how and why we age? elife 10:e62852
Zhang W, Xiao S, Ahn DU (2013) Protein oxidation: basic principles and implications for meat quality. Crit Rev Food Sci Nutr 53(11):1191–1201

Clinical Biochemistry of Cancer

14

Shalini and Vinod Pravin Sharma

Abstract

Clinical chemistry is a multidisciplinary field in medical sciences that concentrates on key qualitative and quantitative tests of critical analytes or markers in bodily fluids and tissues. It employs advanced analytical tools and techniques, along with specialized instruments, to carry out these tests. This interdisciplinary domain incorporates knowledge from toxicology, medicine, biology, chemistry, biomedical engineering, informatics, and other applied sciences, holding direct relevance to both plant and animal life. The hepatic system, responsible for metabolism and excretion, plays a key role in managing toxicants, substances, or chemicals that may cause harm through adverse effects or toxicity. Hepatotoxicants are chemicals or metabolites that can lead to liver injury, often resulting from biotransformation processes. Carcinogenicity, the ability to cause cancer, is a critical concern, with scientists assessing its risk through studies and regulatory bodies relying on data to guide cancer prevention and treatment. Toxicity denotes a substance's harmful potential, affecting various organisms and cell types, leading to hepatotoxicity, nephrotoxicity, or cytotoxicity. Toxicity is dose dependent, varying with age, sex, and species. The Drug Toxicity Index (DTI) predicts clinical outcomes with insights beyond traditional dose–response. Tackling toxicity is crucial for ecosystem health and addressing environmental challenges.

Keywords

Carcinogenic · Environmental · Implications · Risk · Preventive · Toxicity

Shalini
National Forensic Science University, Gandhinagar, Gujarat, India

V. P. Sharma (✉)
CSIR-Indian Institute of Toxicology Research, CSIR-IITR, Lucknow, UP, India

R. L. Singh et al. (eds.), *Clinical Applications of Biomolecules in Disease Diagnosis*, https://doi.org/10.1007/978-981-97-4723-8_14

14.1 Introduction

Worldwide, cancer remains one of the greatest health hazards. According to World Health Organization (WHO) estimates, there would be 77% more new instances of cancer globally by 2050 than there were in 2022 (20 million cases). According to projections, the countries with the highest Human Development Index (HDI) will observe the greatest increase in cases of cancer in absolute terms. That is to say, by 2050, an additional 4.8 million instances are projected (relative to 2022 forecasts). The illness involves the rapid growth of abnormal cells, which can breach their usual boundaries, infect adjacent tissues, spread to distant organs, and eventually take over the entire body. We must focus on providing fair access to top-notch medical treatments for patients, regardless of where they live or their financial resources. Cancer is a term that covers various illnesses resulting from abnormal cell growth (Cooper 2000). Rather than a single disease, it includes over 100 distinct conditions. Typically, each type of cancer is named based on the specific type of cell where it begins. Lung cancer starts in the lungs, while melanoma begins in *melanocytes*, which are the skin's pigment-producing cells. Tumour refers to an abnormal growth of cells, which can be classified as either benign or malignant (Kalaiyarasi et al. 2020). The characteristics differences between benign and malignant tumours are summarized in Table 14.1 (Fig. 14.1).

Cancers, which are malignant tumours, possess the ability to grow and metastasize to other body regions (Martin et al. 2013). Their designation as "cancers" stems from the Latin term for *crab*, reflecting their tissue-invasive characteristics reminiscent of a crab's movement. Benign tumours can grow but do not spread beyond their original location. Certain types of cancer, such as leukaemia, mostly *lymphomas* and *myeloma*, do not lead to the formation of tumours. The patient's age, general health, the kind of sarcoma, the tumour's location, and the extent of its dissemination all influence the course of treatment. Even though receiving a cancer diagnosis

Table 14.1 Differences between benign and malignant tumours

Characteristics	Benign tumour	Malignant tumour
Growth rate	Typically grows slowly and are usually well-defined	Malignant grow rapidly and may invade surrounding tissues
Cell characteristic	Tumour cells resemble normal cells and are well differentiated	Malignant tumour cells may appear abnormal and can vary in differentiation
Metastasis	Tumours remain localized and do not metastasize to other areas of the body	Have the potential to metastasize, spreading to distant organs and tissues
Mitotic activity	Tumour has low mitotic activity, with fewer dividing cells	Exhibit high mitotic activity, with numerous dividing cells
Encapsulation	Tumours are often encapsulated, meaning they are surrounded by a fibrous capsule	Lacks a well-defined capsule and may infiltrate nearby tissues

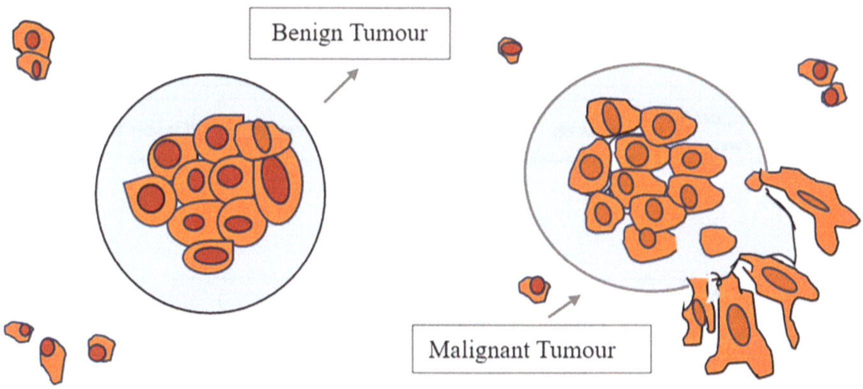

Fig. 14.1 Tumour: benign and malignant

can change your life, it's crucial to maintain hope. Most often, a strong positive attitude will create more miracles than any wonder drug. Cancer survivors have demonstrated immense courage and resilience in their journey to overcome cancer. Few of the cancer survivors may serve as an inspiration to many and their stories of triumph provide hope and strength to those still fighting the battle. Moreover, cancer patients need the support and love of a community to help them through their treatment.

14.2 Types of Cancer

Cancer comprises various diseases marked by uncontrolled cell growth. The World Health Organization (WHO) classifies cancer based on the site of origin and its characteristics. Bladder cancer affects the bladder, causing urinary issues. Brain and central nervous system cancers impact the brain or spine, leading to neurological symptoms. Breast cancer is a common in women, while cervical cancer, often HPV-related, affects the cervix (Burd 2003; Khodabandehlou et al. 2019). Childhood cancers include leukaemia and neuroblastoma. Colorectal cancer targets the colon or rectum, while endometrial cancer arises from the uterus lining. Gallbladder cancer, rare but aggressive, is less common. Other cancers include lung, liver, kidney, oesophageal, pancreatic, skin, prostate, ovarian, testicular, and thyroid cancers, along with Hodgkin and non-Hodgkin lymphoma (Fig. 14.2). Variation in cancers like origins, risk factors, and symptoms are summarized in Table 14.2. Each type presents distinct diagnosis, treatment, and prognosis challenges. The WHO emphasizes early detection, prevention, and varied treatment approaches to reduce cancer's global impact.

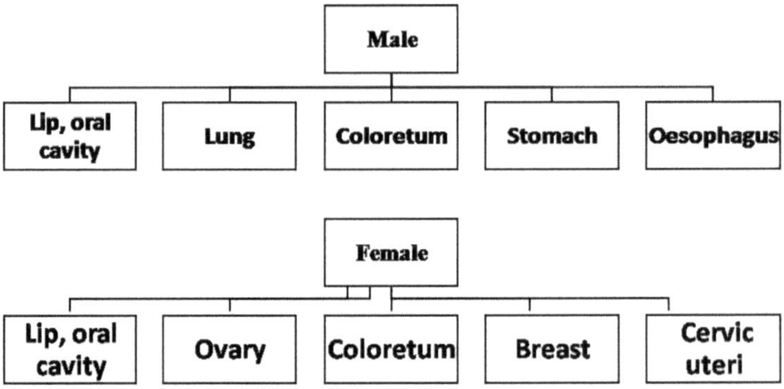

Fig. 14.2 Common cancer types in male and female in India

Table 14.2 Variation in cancers: origins, risk factors, and symptoms

Cancer	Origin	Risk factors	Symptoms
Lip, oral cavity	Cancer that affects the lips and other parts of the mouth, such as the tongue, gums, cheeks, and the roof or floor of the mouth	Risk factors include using tobacco products, binge drinking, and having HPV	Symptoms can involve mouth sores, pain, and trouble swallowing or chewing
Breast cancer	This cancer starts in breast tissue, often in the ducts or lobules. Although it can strike men as well, it more frequently affects women	Symptoms may encompass mouth sores, pain, and difficulty swallowing or chewing	Symptoms might be a breast lump, changes in breast shape, or unusual discharge
Cervix uteri	The cervix, or bottom part of the uterus, is where cervical cancer begins	It's mainly caused by human papillomavirus (HPV) infection, and early detection is possible with pap smears and HPV testing	Symptoms can include unusual bleeding, pelvic pain, or discomfort during intercourse
Lung cancer	Lung cancer is a leading cause of cancer-related death because it starts in the lung tissue	It is mainly linked to smoking, though pollutants and radon gas are also risk factors	Symptoms are chronic cough, shortness of breath, chest pain, and blood in the cough
Oesophagus	This cancer impacts the oesophagus, the tube that connects the stomach to the throat	It has a high mortality rate due to late detection. Utilizing tobacco products, drinking too much alcohol, and chronic acid reflux (GERD)	Symptoms can involve difficulty swallowing, chest pain, and unintended weight loss
Ovary	Ovarian cancer develops in the ovaries and is often diagnosed at a late stage because symptoms are subtle	Risk factors encompass a hormone replacement therapy, family history, and hereditary abnormalities including BRCA1 and BRCA2	Symptoms include bloating, pelvic pain, and a frequent urge to urinate

(continued)

Table 14.2 (continued)

Cancer	Origin	Risk factors	Symptoms
Stomach	Also known as gastric cancer, it begins in the stomach lining	Common risk factors include *helicobacter pylori* infection, dietary habits, and smoking	Symptoms might be stomach pain, indigestion, loss of appetite, and unexplained weight loss
Colorectum	Colorectal cancer affects the colon or rectum and often starts from benign polyps that become cancerous	Risk factors include age, family history, a diet high in processed or red meat, and obesity	Symptoms might include alterations in bowel habits, rectal bleeding, and abdominal discomfort

14.3 Other Types of Cancer

14.3.1 Adenocarcinoma

It is characterized by the production of fluids and mucus. Tissues with these specific epithelial cells are sometimes called glandular tissues, as they produce fluids to maintain tissue moisture. Adenocarcinomas are commonly found in lung, breast, colon, and pancreatic cancers. They originate from epithelial cells; epithelial cells cover both internal and external body surfaces and during observation under a microscope, and epithelial cells generally display a columnar shape.

14.3.1.1 Symptoms

14.3.1.1.1 Breast Cancer

Signs may include the presence of a lump in the breast and discharge from the nipple.

14.3.1.1.2 Lung Cancer

Common indications consist of coughing up bloody mucus, chest discomfort, and difficulties with breathing.

14.3.1.1.3 Pancreatic Cancer

Symptoms might involve jaundice, dark-coloured urine, weight loss, and pain in the back or abdomen.

14.3.1.1.4 Colon Cancer

Signs may comprise alterations in bowel movements, bleeding from the rectum, abdominal discomfort, and unexplained weight loss.

14.3.1.2 Treatment

1. Surgery is a typical treatment for breast cancer that involves removing malignant cells or tumours from the breast. There are other surgical procedures that can be

used, like reconstructive surgery, mastectomy which removes the entire breast, and lumpectomy, which removes the tumour. Usually, after surgery, radiation therapy is utilized to target any cancer cells that remain. To destroy cancer cells, high-energy radiation is used. Chemotherapy is a multidisciplinary approach to treating breast cancer that employs medications to either eradicate or halt the spread of cancer cells.

2. It can decrease tumours before the surgery, remove any remaining cancer cells following surgery, or control the spread of the cancer in more advanced cases.
3. Hormonal therapy may be used to destroy hormone-sensitive cancer cells which are hormone sensitive and might have spread or metastasized to other parts of the body. It may be used to ease a cancer patient's symptoms from pain due to presence of a large tumour. It helps to inhibit cancer cell growth.
4. Immunotherapy strengthens the immune system of the body to fight cancer cells by improving its capacity to recognize and target them.
5. The results of treatment for several types of breast cancer have been encouraging, particularly for HER2-positive and hormone receptor-negative varieties.

Research and treatment for cancer are developing quickly. Different cancer kinds have distinct biomarkers that can be used to target therapies more precisely and extend remission times. Even though there is mounting evidence that a targeted, methodical approach to cancer treatment may be the way of the future, chemotherapy is still a frequently employed therapeutic option despite its acknowledged detrimental effects on a patient's physical and mental health. For advanced-stage tumours, chemotherapy remains the first-line treatment when radiation therapy or surgery is not practical for a particular reason.

Chemotherapeutic medications have been a mainstay of cancer treatment for decades.Treatment for cancer uses a variety of tactics to fight the illness. The primary goal of surgery is to remove malignant tissue through incisions in radiation therapy, and high-energy radiation is utilized to kill cancer cells. It can be used as a stand-alone treatment or in combination with surgery. The use of medications to prevent or reduce the growth of cancer cells is known as chemotherapy. In order to stop tumours that are hormone-sensitive from developing, hormone treatment modifies the amounts of hormones. Immunotherapy increases immunity, which facilitates the identification and removal of cancer cells.

Several combinations of these treatments may be employed, contingent upon the nature and stage of the malignancy. To increase the likelihood of a local recurrence-free survival, patients with localized well-differentiated and grade 1–2 dedifferentiated liposarcoma may receive pre-operative radiation therapy. It is not advisable to propose radiation therapy for other histologic subtypes. Additionally, patients with retroperitoneal sarcoma who are receiving radiation therapy prior to surgery should be informed of the higher risk of significant side effects during the peri-operative phase.

14.3.1.3 Risk Factors

1. Alcohol consumption: Excessive drinking can elevate the risk of specific cancers.
2. Diet: The risk of cancer may be increased by poor dietary habits that include consuming low amounts of fruits and vegetables and excessive amounts of processed foods and red meat.
3. Obesity is associated with an increased risk of several malignancies, independent of weight.
4. Smoking: Tobacco use poses a significant risk for various cancers, particularly lung cancer.
5. Hormones: Hormonal factors, like hormone replacement therapy or reproductive history, can affect cancer risk.
6. Family history: Having family members with a history of these cancers can raise one's own risk.

14.3.2 Skin Cancer

For skin cancer, basal cell carcinoma (BCC) is the most common type, and it typically begins in the basal layer or bottom portion of the epidermis. It commonly impacts regions that are frequently exposed to sunlight, such as the face.

14.3.2.1 Symptoms

1. Persistent open sores that fail to heal.
2. Darkened patches of skin.
3. Shiny raised bumps or nodules.
4. Elevated areas resembling scar tissue.

14.3.2.2 Treatment

1. Surgery: Removing the tumour surgically is a common approach for treating BCC. This process involves excising to guarantee total eradication and remove the malignant tissue together with a margin of good skin.
2. Chemical Peel: Chemical peels use a chemical solution to induce blistering and subsequent peeling of the skin. This method is effective for removing superficial BCC lesions.
3. Targeted Cancer Cells Therapy: Targeted therapy utilizes drugs crafted to precisely target cancer cells by disrupting specific molecules essential for tumour growth. This approach is frequently applied in cases of advanced or metastatic basal cell carcinoma (BCC).
4. Radiation Therapy: High-energy X-rays are utilized in radiation therapy to eradicate cancer cells, often applied when removing any residual malignant cells after surgery is not a possibility.
5. Chemotherapy: In this treatment, drugs are used to either kill or inhibit the growth of cancer cells.
6. It is commonly used for advanced or metastatic cases of basal cell carcinoma (BCC) when surgery or other local therapies are not feasible.

7. Photodynamic Therapy: Photodynamic therapy (PDT) utilizes a photosensitizing agent activated by light to selectively destroy cancer cells while sparing healthy tissue.
8. Immune Therapy System: The mechanism of immunotherapy involves invigorating the immune system to identify and combat cancerous cells. When used to treat advanced or metastatic basal cell carcinoma, it strengthens the immune system's capacity to combat the cancer (BCC).

14.3.2.3 Risk Factors

1. Fair skin: People with lighter skin tones have lower levels of melanin, which offers less defence against UV radiation, thereby heightening the vulnerability to skin cancer.
2. Sun exposure: Prolonged exposure to UV radiation, notably from sunlight, significantly raises the likelihood of developing basal cell carcinoma.
3. Family history: Skin cancer runs in families, most likely due to inherited genetic predispositions or comparable environmental variables.
4. Age: As people age, their chance of developing basal cell carcinoma increases due to the accumulation of additional risk factors and years of continuous sun exposure.
5. Skin cancer risk is increased by immune-system-weakening conditions and treatments, such as certain diseases or drugs, which lower the body's ability to fight against malignant cells.

14.3.3 Squamous Cell Carcinoma

Squamous cells, which are epithelial cells found immediately below the skin's surface, are the source of squamous cell carcinomas. Moreover, the lining of the lungs, kidneys, bladder, intestines, and stomach are composed of these cells. Squamous cells seem flat and look like fish scales when viewed under a microscope. Another term for squamous cell carcinomas is epidermoid carcinomas. A few examples include breast, prostate, intestinal, and lung cancers.

14.3.3.1 Symptoms

1. Scaly and dark skin patches: These patches may have a rough texture and appear darker than the surrounding skin.
2. Open sores with raised borders: Sores on the skin with raised edges that may take longer to heal.
3. Firm growths: Raised areas on the skin that feel firm and rough.
4. Spots resembling age spots: Spots on the skin that resemble age spots but may be larger or have irregular borders.
5. Wart-like growths: Growths on the skin with a rough surface texture resembling warts.
6. Horn-like growths: Growths on the skin that protrude like horns.
7. Sores growing in scars: Sores that develop within existing scars on the skin.

14.3.3.2 Treatment

1. Excision: Surgically removing the cancerous tissue.
2. Mohs surgery: Surgical procedure involving the removal of thin layers of cancerous tissue and examination under a microscope until all cancer cells are removed.
3. Curettage and electrodesiccation: The cancerous tissue is scraped off, and then an electric current is applied to eradicate any remaining cancer cells.
4. High Radiation:
5. High-dose radiation treatment: Employing high-energy radiation to eradicate cancer cells. Cryosurgery: Freezing the cancerous tissue to destroy it.

14.3.3.3 Risk Factors

1. Fair skin: Lighter skin tones are more vulnerable to UV radiation damage, increasing the likelihood of developing skin cancer.
2. Sun exposure: Skin cancer risk can be increased by prolonged exposure to sunlight, especially in the absence of protection.
3. Family history of skin cancer: Having relatives with a history of skin cancer can elevate one's own risk.
4. Being over 50 years old: The likelihood of developing skin cancer increases with age.
5. Immune system deterioration: Skin cancer risk may be elevated by specific diseases or drugs that weaken the immune system.
6. Being infected with human papillomavirus (HPV): Particular strains of HPV are correlated with an elevated risk of developing skin cancer.

14.3.4 Sarcoma

"Sarcoma," stemming from the Greek term "*sarkōma*" which means "fleshy", describes a cancerous tumour that arises from connective tissues. These tumours can form in different regions of the body, as connective tissues are distributed throughout. Approximately 60% of sarcomas initiate in the arms or legs, 30% in the torso or abdomen, and the remaining 10% in the head or neck region. Sarcomas can affect both children and adults. Soft tissue sarcomas are cancers originating from the fat, muscles, blood vessels, or nerves in the body. These cancers occur when normal cells within these tissues undergo changes and become malignant. Unlike some other cancer types, the exact cause of most sarcomas remains unidentified. Factors such as cigarette smoke, alcohol consumption, and diet have not been linked to sarcoma development. Patients often notice sarcomas as lumps or bumps, referred to as tumours. While not all tumours are cancerous, it's essential to consult a doctor promptly if a tumour or suspicious area is large, growing rapidly, or causing pain, as it may indicate the presence of a sarcoma.

14.3.4.1 Types of Sarcomas

14.3.4.1.1 Bone Sarcomas

Bone sarcoma encompasses various types of cancer, such as osteosarcoma, chondrosarcoma, Ewing sarcoma, haemangioendothelioma, angiosarcoma, fibrosarcoma/myofibrosarcoma, and chordoma.

14.3.4.1.2 Soft Tissue Sarcomas

Represent a group of cancers originating in the supportive and connecting connective tissues throughout the body, comprising blood vessels, adipose cells, joint linings, lymph vessels, muscle, nerves, and tendons.

14.3.4.2 Causes of Sarcoma

1. Uncertain Origin: Despite extensive research, the precise cause of sarcoma remains unknown in some cases. While genetic mutations or environmental factors may contribute, the specific triggers remain unclear.
2. Exposure to Ionizing Radiation: Prolonged exposure to ionized radiation, whether through cancer treatment or occupational contact, can heighten the risk of developing sarcomas.
3. Chemotherapeutic Medications with Alkylation Agents: Certain chemotherapy drugs containing alkylating agents, utilized in cancer treatment, may elevate the risk of sarcoma development as a long-term side effect.
4. Inherited Genetic Syndromes: Inherited genetic mutations, such as those associated with Li–Fraumeni syndrome, hereditary retinoblastoma, or neurofibromatosis type 1, can predispose individuals to sarcoma development.
5. Paget's Disease of Bone: Sarcomas can sometimes occur inside the diseased bones. The condition is characterized by aberrant remodelling of the bone.
6. Exposure to Vinyl Chloride and Arsenic: Occupational exposure to chemicals like vinyl chloride (used in plastic production) or arsenic (found in certain industrial processes and contaminated water) is linked to an increase sarcoma risk.
7. Kaposi's Sarcoma: This cancer, triggered by human herpesvirus 8 (HHV-8) infection, is more frequently observed in individuals with compromised immune systems, as those suffering from HIV/AIDS.
8. Lymphedema: Chronic limb swelling due to lymphedema, often post-cancer treatment, can predispose individuals to a rare sarcoma type known as lymphangiosarcoma, affecting the lymphatic vessels.

14.4 Characteristics of Cancer Development and Biomarkers

Characteristics of cancer development are presented in the Fig. 14.3.

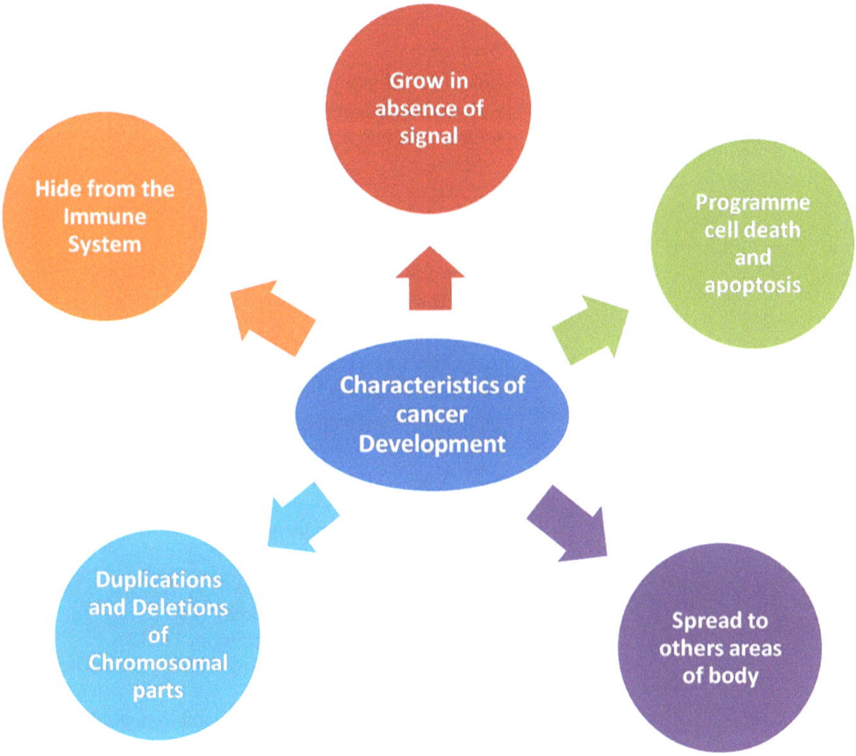

Fig. 14.3 Characteristics of cancer development

14.4.1 Cancer Biomarkers

There are a number of biomarkers available that diagnose cancers (Table 14.3).

14.4.2 Clinical Classification of Cancer Biomarkers

14.4.2.1 Screening and Diagnostic Biomarkers

- PSA, or prostate-specific antigen, is used to detect prostate cancer.
- The colorectal cancer faecal occult blood test (FOBT).
- For colorectal, lung, and other cancers, carcinoembryonic antigen (CEA) is used.
- Human papillomavirus (HPV) testing for cervical cancer.
- Breast imaging and mammography for breast cancer.
- CA-125 for ovarian cancer.

Table 14.3 Types of cancer biomarkers

Types of biomarkers	Biomarkers
1. Genetic biomarkers	Mutation and gene alterations
	• Gene expression profiles • Oncotype DX in breast cancer • Mamma print in breast cancer • Prosigna in breast cancer • Decipher in prostate cancer • VeriStrat in lung cancer • Gene expression classifier in colon cancer
	DNA as a cancer biomarker
	RNA as a cancer biomarker
	Epigenetics as a cancer biomarkers
2. Protein biomarkers	Proteins as a cancer biomarkers
3. Metabolic biomarkers	Metabolites and metabolic pathways: • Glycolysis • TCA cycle (citric acid cycle) • Lipid metabolism • Amino acid metabolism • Nucleotide metabolism
	Metabolic imaging techniques • Positron emission tomography (PET) • Magnetic resonance spectroscopy (MRS) • Magnetic resonance imaging with hyperpolarized substrates (HP-MRI) • Single-photon emission computed tomography (SPECT) • Dynamic contrast-enhanced magnetic resonance imaging (DCE-MRI) • Optical imaging
	Molecular probes and contrast agents Fluorescent probes Magnetic resonance imaging (MRI) contrast agents Positron emission tomography (PET) tracers Ultrasound contrast agents Near-infrared (NIR) imaging probes
4. Cells as cancer biomarker	Circulating tumour cells as cancer biomarkers
	Immune cells as cancer biomarkers
	Cancer stem cells as cancer biomarkers
5. Lamins as cancer biomarkers	
6. Galectins as cancer biomarkers	
7. Carbohydrate antigens as cancer biomarkers	
8. Viruses as cancer biomarkers	
9. Exosomes as a cancer biomarker	
10. Lipids as cancer biomarkers	

14.4.2.2 Prognostic Biomarkers

Assessing the course of a disease and the prognosis of patients:

- Hormone receptor status, Ki-67, and Oncotype DX in breast cancer.
- Microsatellite instability (MSI), carcinoembryonic antigen (CEA), and BRAF V600E mutation in colorectal cancer.
- EGFR mutation, ALK rearrangement, and PD-L1 expression in lung cancer.
- Prostate-specific antigen (PSA), Gleason score, and expression of androgen receptor (AR) in prostate cancer.

14.4.2.3 Tumour Staging and Grading Systems

- Breast Cancer: Human Epidermal Growth Factor Receptor 2 (HER2) expressions, Ki-67, and the status of oestrogen and progesterone receptors.
- Colorectal Cancer: Microsatellite instability (MSI), BRAF V600E mutation, KRAS mutation.
- Lung Cancer: NM staging rearrangement and mutation of the epidermal growth factor receptor (EGFR).
- Protein Specific Antigen Test- The test measures the amount of antigen in blood. PSA is a protein produced by prostate gland and score greater than 2.5 ng/ml is considered abnormal (for men in their 40 and 50s while median PSA is 0.6–0.7 ng/ml). For men in the 60s, normal range is beteeen 1.0–1.5ng/ml; PSA score greater than 4.0 is considered abnormal. These values are indicative and help to assess risks while biopsy helps to confirm the cancer. The gleason gradiing system (GGS) helps to evaluate the progonosis of men with cancer based on microscopic appearance,biopsy results corelation and therapeutic pathways accordingly.

14.4.2.4 Predictive Biomarkers

- Breast cancer HER2 status.
- ALK rearrangements in lung cancer; EGFR mutations in non-small cell lung cancer (NSCLC).
- Expression of ERCC1 in lung cancer; BRAF V600E mutation in melanoma; and MSI-H/dMMR status in colorectal cancer.

Adjuvant and advanced illness settings have seen improved survival rates thanks to novel treatment medicines, albeit at a higher cost and toxicity risk. Unfortunately, the long-term prognosis for metastatic disease is still poor, and there are still major obstacles because of treatment failure and late-stage diagnosis. Biomarkers have a crucial role in early identification, prognosis, survival, and therapy response prediction. The last 30 years have witnessed significant progress in the fields of genomics and molecular pathology of cancer biomarkers, which has improved survival rates by enabling more personalized treatment plans (Ogunwobi et al. 2020).

14.5 General Aspects of Cancer

Carcinogenesis, the process of cancer development, particularly in the realm of chemical carcinogenesis, delves into understanding how chemicals induce cancer. It also involves devising experimental models to ascertain the carcinogenic potential of substances. A pivotal aspect of toxicology lies in identifying potential human carcinogens. However, cancer is not a single illness, rather a broad spectrum of illnesses distinguished by the unrestricted growth of abnormal cells. These cells form a population with the capacity to proliferate unchecked and invade nearby tissues, rendering cancer lethal. Epidemiological investigations have shown that cancer incidence rises exponentially with age. This phenomenon is interpreted to signify those multiple critical mutations often ranging from three to seven that are requisite for cancer initiation within a single cell. Analyses of human tumours at the molecular level have corroborated the accumulation of mutations in vital genes during cancer progression.

Most cancers originate from a solitary cell and arise from the accumulation of myriad critical mutations within it over time. Initially, a somatic mutation in a crucial gene confers a growth advantage to the cell, prompting the expansion of the mutant clone. Subsequent mutations in other pivotal genes within this clone further bolster its growth advantage. This cycle of mutation and selection persists, culminating in a clone of cells harbouring mutations in numerous critical genes.

14.6 Oncogene and Tumour Suppressor Genes

Genes that usually encourage cell growth, division, or survival are known as proto-oncogenes. A proto-oncogene may become an oncogene through improper activation resulting from mutation or from having too many copies of the gene (Table 14.4). This mutation may cause unchecked cell proliferation, which may eventually result in cancer. When a proto-oncogene is in its natural condition, it controls cell division and growth much like the accelerator in an automobile. Conversely, an oncogene causes unchecked cell division by functioning as a gas pedal that is locked in the "on" position.There are various methods in which oncogenes can be activated or switched on in cells.

Table 14.4 Representation of oncogenes (Gillies et al. 2008)

Incidence	Class	Example
Simian sarcoma	Growth factors	Sis/PDGF
Lung cancer, breast cancer	Receptor tyrosine kinases	EGFR, HER2
Colon cancer, head and neck cancer	Cytoplasmic tyrosine kinases	Src, Syk, Abl
Melanoma, colon	Cytoplasmic serine/threonine kinases	BRaf
Pancreatic cancer	21-kDa GTPase	H-Ras, N-Ras, K-Ras
Burkitt's lymphoma	Transcription factors	Myc

14.6.1 Gene Variants/Mutations

Some individuals have mutations in their genes that result in the chronic activation of an oncogene. These changes in DNA can be inherited from one's parents or develop during one's lifetime as a result of mistakes made during the replication of genes during cell division.

14.6.2 Epigenetic Modifications

Gene activity can be regulated by cells without affecting the actual gene sequences. This is accomplished by adding different chemical groups to genetic material, such as DNA or RNA, which can change a gene's activity state. Oncogenes can occasionally become activated due to epigenetic modifications, which may result in aberrant cell proliferation. To learn more about these alterations, see "Gene Changes and Cancer". Oncogenes can be turned on or triggered in cells in a number of ways. Examples are shown below.

14.6.2.1 Gene Mutations and Variants

Certain people have genetic mutations that cause an oncogene to become chronically activated. These genetic alterations can arise during a person's lifespan due to errors occurring during gene replication during cell division, or they can be inherited from one's parents.

14.6.2.2 Changes in Epigenetics

Cells have the ability to control gene activity without changing the underlying gene sequences. This is achieved by introducing distinct chemical groups into genetic material, including DNA or RNA, which have the ability to alter the active state of a gene. Oncogenes can occasionally be activated by epigenetic modifications, which results in aberrant cell development. The cell may expand out of control as a result of this oncogene transition. Gene duplication occurs when a gene is duplicated in some cells. This results in the overproduction of a particular protein and may cause unchecked cell division.

Genes that naturally slow down cell division or cause programmed cell death, or apoptosis, when necessary are known as tumour suppressor genes (Table 14.5).

Table 14.5 Representation of tumour suppressor genes (Gillies et al. 2008)

Incidence	Class	Example
Breast, colon	Phosphatase	PTEN
Colon cancer, neurofibromatosis type 2	Cell-cell and extracellular matrix	APC, GP43/Merlin
Breast and ovarian cancer, retinoblastoma	DNA repair and cell cycle checkpoints	BRCA1/2, pRb, p53
Neurofibromatosis type 1	G- protein (ras) inhibitor	Neurofibromin 1
Renal cell cancer	Ubiquitin ligase	VHL
Pheochromocytoma	Dehydrogenases	Succinate dehydrogenases B and D

Tumour suppressor gene failure can cause uncontrolled cell growth, which may result in cancer.

Like the brake pedal on an automobile, a tumour suppressor gene helps stop cells from proliferating too quickly, just as brakes stop an automobile from speeding. Cell division may become uncontrollably large if a tumour suppressor gene has a malfunction, such as a pathogenic mutation that makes it nonfunctional.

Tumour suppressor gene mutations that are inherited have been related to certain familial cancer syndromes, increasing the risk of acquiring particular cancers in families.

However, during the course of a person's lifetime, the majority of mutations in genes that suppress tumours are acquired rather than inherited. Consider the crucial tumour suppressor gene TP53, which codes for the p53 protein that controls cell division. Li–Fraumeni syndrome, a disorder where family members are at elevated risk of multiple cancers due to the hereditary TP53 mutation, can result from TP53 mutations that are inherited. Mutations in the TP53 gene are also frequently detected in cancer cells from people without a family history of the illness. These mutations occur during a person's life and contribute to cancer cell growth, but they're only present in cancer cells and not in the rest of the body, so they can't be inherited by children.

14.7 Factors Responsible for Cancer Development

Cancer, marked by irregular cell growth and dissemination, can be impacted by various elements or conditions. These encompass chemical carcinogens, substances in the environment or workplaces that increase cancer risk, like tobacco smoke, asbestos, benzene, and specific pesticides. Being exposed to radiation, like X-rays and UV rays from the sun, can damage DNA, which raises the risk of developing cancer. Moreover, specific infectious agents, like viruses, bacteria, and parasites, are associated with particular cancers, such as HPV with cervical cancer and hepatitis viruses with liver cancer. Dietary factors also contribute, with certain components like red meat, saturated fats, and limited fruit and vegetable intake linked to higher cancer risk, but diets high in antioxidants and plant-based foods may lower risk. Lifestyle choices like tobacco use, excessive alcohol consumption, sedentary habits, and obesity can also elevate cancer risk. Additionally, genetic predisposition, encompassing inherited mutations and family history, can increase susceptibility to certain cancers. Understanding these factors is essential for implementing prevention strategies, assessing risk, and designing targeted interventions to lower cancer incidence and enhance public health outcomes.

14.7.1 Coherent Factors/Environmental Contaminants

Air pollution exposure to various contaminants—like particulate matter (PM 1.0, PM 2.5, PM 10.0), nitrogen oxides, sulphur dioxide, and emissions from vehicles and industries—is linked to elevated cancer rates, particularly respiratory and lung cancers.

14.7.1.1 Radiation Exposure

Especially from ionizing radiation sources such as medical procedures, nuclear accidents, and radioactive materials, radiation exposure can cause genetic damage to DNA/RNA, raising the risk of developing cancer.

14.7.1.2 Chemical Pollutants

Harmful substances found in our environment, including air, water, soil, and consumer products, can elevate the risk of cancer. These may consist of industrial chemicals, pesticides, heavy metals, and volatile organic compounds (VOCs).

14.7.1.3 Occupational Hazards

Workers in certain industries face risks from exposure to carcinogens like asbestos, benzene, formaldehyde, and silica dust, leading to cancers related to their occupation. Tobacco smoke contains cancer-causing substances and is correlated with elevated rates of lung cancer and other malignancies.

14.7.1.4 Endocrine-Disrupting Chemicals (EDC)

Some environmental contaminants can disrupt hormone function, potentially contributing to hormone-related cancers such as breast and prostate cancer.

14.8 Stages of Cancer

Staging is a method for describing the size of a cancer and its extent of growth. Tests are performed when a cancer is first diagnosed in order to assess the tumour's size, detect any tissue invasion, and ascertain whether the disease has spread to various parts of the body. The reason staging is important is that it aids the treatment team in selecting the best course of action.

If the cancer is confined to a single location, doctors might suggest local treatment such as surgery or radiotherapy, which can sometimes be sufficient to eliminate the cancer entirely. Local treatment focuses solely on a specific area of the body. However, if the cancer has spread, systemic treatment is usually required. Systemic treatments are designed to work throughout the entire body and encompass chemotherapy, hormone therapy, targeted cancer drugs, and immunotherapy. These broader approaches are necessary when the cancer has metastasized beyond the initial site. Cancer staging systems help doctors categorize cancers based on their size and spread (Fig. 14.4). The two main systems are the tumour, node, metastasis (TNM) system and the numeric system, offering common language for physicians to describe cancer characteristics, compare treatment outcomes in research, and maintain consistent treatment guidelines across different hospitals and clinics. Some blood and lymphatic cancers use their own distinct staging systems.

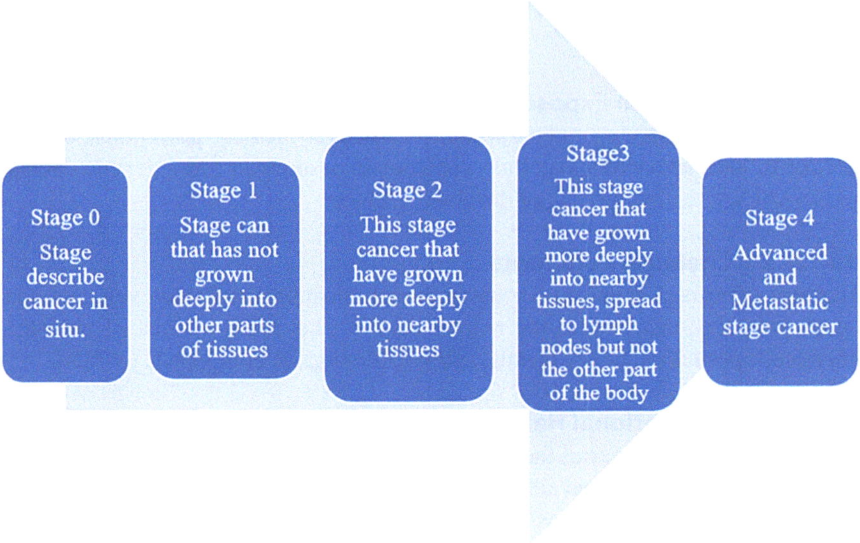

Fig. 14.4 Different stages of cancer

The TNM staging system stands for tumour, node, metastasis:

1. "T" indicates the size of the tumour and its depth of invasion into surrounding tissue. It's graded on a scale of 1 to 4, with 1 being the smallest and 4 the largest.
2. "N" indicates whether cancer has spread to lymph nodes. This is rated from 0 to 3, where 0 means no cancer in the lymph nodes and 3 signifies a significant number of cancerous lymph nodes.
3. "M" represents if the cancer has metastasized to other parts of the body. It is marked as 0 or 1, with M0 meaning no metastasis and M1 indicating that the cancer has spread.

For instance, a small tumour that has spread to lymph nodes but not to other parts of the body might be staged as T2 N1 M0, while an advanced cancer with metastasis could be T4 N3 M1. Occasionally, additional letters like "a", "b", or "c" are used to further categorize the stages. In lung cancer, for example, M1a indicates the cancer has spread to the other lung, M1b signifies spread to one other area, and M1c represents metastasis to multiple locations.

The prefix "p" is used for pathological staging, which is determined after surgery by examining the cancer cells in a lab. An example is pT4. The prefix "c" stands for clinical staging, based on what the doctor observes and learns from tests prior to surgery. An example of this would be cT2. This system provides a structured way to describe cancer progression and guide treatment plans. Cancer staging systems are essential for describing the size and spread of cancer. Two common systems are the TNM system and the numerical system, allowing doctors to use a shared language to classify cancer, compare treatment results, and establish standard treatment protocols across hospitals and clinics. Certain blood and lymphatic cancers have distinct staging systems.

14.9 Mechanisms of Action

Human epithelial tissue and connective tissue work together to maintain the stability of biological structures in the body. In human cells, biological factors like organic cations and anions play a role in keeping this organization intact through a balance of forces. Cancer cells, however, disrupt this balance when heavy metals or heavy metal compounds replace normal tissue cells, affecting the organic cation factors. This can cause the stable intermolecular structure to break down, leading to molecular imbalances and bifurcation. Genetic variations resulting from this process can direct the bifurcation toward malignant cancer cell division.

This generally occurs near arterial or venous access channels, suggesting a few key points:

- Directional Cell Division: Cancer cells tend to divide in specific directions, leading to tumour growth.
- Heavy Metal Influence: Heavy metal cations contribute to cancer activity by damaging nearby cells, potentially due to larger sizes and disruptive properties.
- Nutrient Uptake by Cancer Cells: Cancer tissues often grow rapidly because the surrounding blood vessels' diameter is too large, allowing increased nutrient intake, which disrupts the normal basal metabolic rate.
- Damage to Nearby Normal Cells: This disruption can cause a potential difference and increased ion movement, leading to further damage to healthy cells.
- Metabolic Issues in Cancer: As cancer cells replace normal cells with organic compounds, their larger diameter can prevent normal metabolic processes, such as waste excretion through capillaries.

Moreover, nuclear radiation can replace blood heme iron ions with oxygen-containing acids, but these compounds may have diameters too large to pass through normal capillaries, potentially causing issues with energy release. Because cancer cells are typically fat-soluble at room temperature, they do not easily degrade, and their smaller amino acids may not easily substitute back into the normal structure.

14.10 Statistics: International and National

The estimates provided by the IARC highlight the growing worldwide burden of cancer, its disproportionate effect on marginalized populations, and the urgent need to address cancer disparities. They are predicated on the top data sources that each country will have in 2022. In 2022, it was expected that there will be 20 million new cases of cancer and 9.7 million cancer-related deaths. It was estimated that 53.5

million persons with cancer were still alive 5 years following their diagnosis. One in five people will have cancer at some point in their lives; one in nine men and one in 12 women will die from the disease. Only 39% of participating nations in the worldwide WHO study on UHC and cancer included the fundamentals of cancer management in their "health benefit packages", or publicly funded core health services available to all people (HBP). Merely 28% of the participating nations also provided coverage for palliative care, which includes pain management in general and not simply that related to cancer. As of 2022, over two-thirds of new cases and deaths worldwide were attributable to ten different forms of cancer, according to updated estimates accessible on the IARC's Global Cancer Observatory. Data pertains to 36 cancers and 185 nations.

Lung cancer accounted for 2.5 million new cases, or 12.4% of all new cases, making it the most frequent cancer worldwide. Lung cancer accounted for 1.8 million of all cancer deaths, making it the leading cause of death from the disease. The next most common causes of cancer-related mortality were colorectal cancer (900,000 deaths, 9.3%), liver cancer (760,000 deaths, 7.8%), breast cancer (670,000 deaths, 6.9%), and stomach cancer (660,000 deaths, 6.8%). Asia's long history of tobacco use is likely contributing to the rise in lung cancer cases as the most common cancer. There were differences in the global total for both sexes in terms of incidence and death based on gender. Breast cancer was the most common cancer in women, but lung cancer was the most common in males. Cervical cancer was the ninth leading cause of cancer-related deaths and the eighth most common disease globally in terms of new cases and fatalities (348186). In 25 nations, mostly in sub-Saharan Africa, it is the most common cancer among women. Even though there are differences in incidence, cervical cancer can be eliminated as a public health concern by expanding the WHO Cervical Cancer Elimination Initiative.

14.11 Multiple Omics Approaches in Cancer Research

Living organisms constantly exchange millions of signals between cells, tissues, organs, and external stimuli. The body's finely tuned responses, occurring at various levels and scales, play a crucial role in maintaining balance (homeostasis) and dealing with potential threats like pathogens, toxins, and genetic mutations. To comprehend the fundamental aspects of cancer biology and the development of tumours, it's essential to adopt a comprehensive approach that examines genomic and epigenetic abnormalities, alongside environmental factors. Multi-omics studies, which analyse diverse high-dimensional datasets at multiple levels, are instrumental in uncovering the intricate interplay between cells and their surroundings, shedding light on the complexities of cancer onset and progression (Fig. 14.5). Multiple omics methodologies have been employed in various clinical investigations to enhance the identification of clinical subtypes or resistance to drugs, forecast effective combination therapies, and pinpoint predictive biomarkers aimed at improving the response rate to targeted treatments.

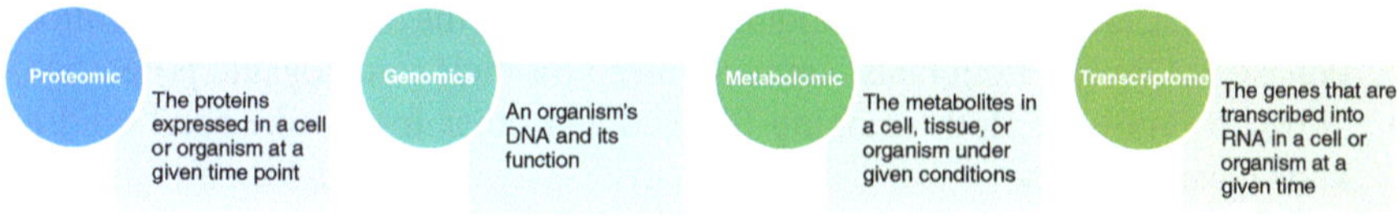

Fig. 14.5 Multiple omics approaches in cancer research

14.12 Health and Toxicity

14.12.1 Carcinogenic Chemicals

Carcinogens are substances or forms of exposure that can cause cancer. They may be natural, such as toxins from fungi, or synthetic, like asbestos and various chemicals used in industry. Common carcinogens include benzene, formaldehyde, and asbestos. You can encounter carcinogens in workplaces, environmental pollution, consumer products, and even some foods and drinks. Depending on the chemical and how you are exposed to it, prolonged exposure to these compounds can raise your chance of developing certain cancers. Chemicals known as "endocrine disruptors" can cause disruptions to an animal's or human's hormonal system. Examples include phthalates, some herbicides like DDT, and bisphenol A (BPA). They are commonly found in plastics, personal care products, industrial chemicals, and pesticides. These disruptors can imitate or obstruct hormones, potentially leading to reproductive problems, developmental issues, and other health complications.

14.12.2 PFOS and PFAS

PFOS (perfluorooctanesulfonic acid) belongs to the category of synthetic chemicals known as PFAS (per- and polyfluoroalkyl substances) is resistant to heat, grease, and water. PFAS can be found in industrial products such as firefighting foams, water-resistant clothes, and non-stick cookware. They have a lengthy half-life in the environment due to their chemical stability. The impact of PFOS and other PFAS on the immune system, thyroid issues, development issues poses an increased risk of specific carcinogenecity, as toxicants. Furthermore, these chemicals can cause significant environmental pollution, contaminating water sources and soil.

The following chemical agents have been identified as both endocrine-disrupting chemicals and known carcinogens that pose threats to human health:

- Metals: These include arsenic, cadmium, nickel, and chromium.
- Industrial Chemicals: Substances like vinyl chloride, benzidine (used in dye production), and solvents such as benzene and trichloroethylene.
- Hydrocarbons: This category covers polycyclic aromatic hydrocarbons (PAHs), often found in combustion byproducts, as well as dioxins and polychlorinated biphenyls (PCBs).
- Fibres and Dust: Includes silica, asbestos, and wood dust.

- Pesticides: Certain pesticides, notably those listed in the Stockholm Convention's Persistent Organic Pollutants, are recognized for their carcinogenic properties.
- Tobacco Products: Both smoking tobacco and other forms like betel quid are known carcinogens.
- Pharmaceuticals: There are also numerous pharmaceuticals known to be carcinogenic to humans.

These chemicals are commonly found in various industrial, environmental, and consumer products, contributing to human health risks.

The Stockholm Convention is an international accord designed to safeguard human health and the environment against the harmful effects of persistent organic pollutants (POPs). POPs are chemicals that remain stable in the environment over long periods, disperse widely across the globe, build up in the fatty tissues of living organisms, and pose hazards posed by its toxicity to both people and wildlife.

The goals of the Stockholm Convention are to:

- Encourage the use of safer alternatives.
- Identify and address additional POPs.
- Clear up old stockpiles and equipment containing POPs.
- Collaborate for a future without POPs.

The Global Environmental Facility (GEF) serves as the designated interim financial support system for the Stockholm Convention. EDCs include other confirmed or suspected carcinogens. Hormone-active substances including certain insecticides, phthalates, and BPA are recognized or believed to increase the risk of cancer, which is not surprising given the high number of malignancies linked to hormones (Fucic et al. 2010; Rebuzzini et al. 2022). The best EDCs to use, as well as the stages of life (pregnancy, childhood, and adulthood) at which EDC exposure most strongly increases the risk of cancer, are still open concerns. Early-life exposure to chemicals such as BPA, phthalates, perfluorinated compounds, PCBs, and pesticides like malathion and DDT has been shown to raise the risk of cancer in later life by research using cellular and animal models (Fucic et al. 2010; Rebuzzini et al. 2022). For several of these substances in humans, emerging epidemiological research is starting to reveal correlative correlations Cohn et al. 2012; O'Shaughnessy et al. 2021; Soto and Sonnenschein 2010).

14.13 Surgical Intervention for Cancer Treatment

Surgery is a common approach to treating cancer, aiming to remove cancerous tissue from the body. Risks of cancer surgery are mentioned in Fig. 14.6. Its role and depending on the type and stage of cancer and objectives may change.

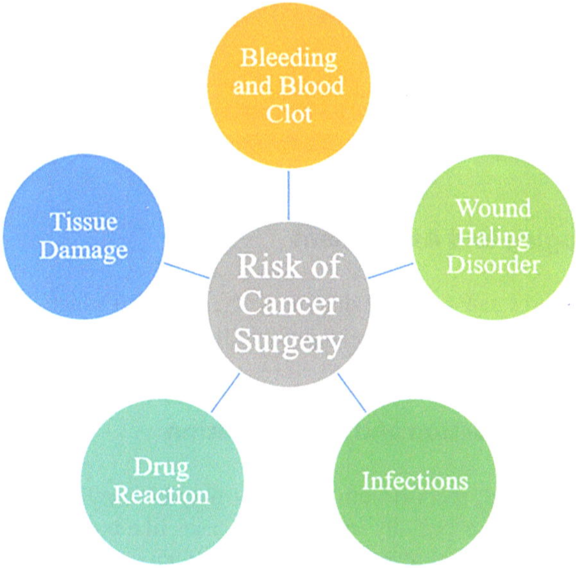

Fig. 14.6 Risks of cancer surgery

14.13.1 Removal of Entire Tumours

Surgery is used to remove cancer that is localized in one area. Surgeons aim to excise the entire tumour along with a margin of healthy tissue to ensure no cancer cells are left behind. The method of removal for solid tumours which have not metastasized to other areas of the body is commonly undertaken by the oncologists.

14.13.2 Debulking Tumours

Surgery can be done to reduce the size of a tumour when removing it entirely could harm vital organs or other parts of the body. In order to improve the effectiveness of other therapies, such as chemotherapy or radiation therapy, on the remaining cancer, a portion of a large tumour must be removed. Debulking process of tumours may reduce symptoms caused by the tumour's pressure or bulk.

14.13.3 Reduction of Implications of Cancer Impact

Surgery can relieve symptoms caused by cancerous tumours, such as pain or pressure. This is often the case when tumours compress nerves, block organs, or cause other physical issues. By removing or reducing the size of these tumours, surgery can improve a patient's comfort and quality of life.

14.13.4 Diagnostic Procedures

In addition to removing tumours, surgery is sometimes used to obtain tissue samples for diagnostic purposes, a process known as a biopsy. This helps to identify the type and stage of cancer, which informs treatment decisions.

14.13.5 Staging and Assessment

Surgical procedures can also help determine the extent to which cancer has spread within the body, aiding in staging and guiding further treatment planning.

14.13.6 Reconstruction and Rehabilitation

After tumour removal, reconstructive surgery may be used to restore appearance and function, which is especially important in cases like breast cancer. This can be crucial for a patient's physical recovery and emotional well-being.

14.13.7 Combination with Other Treatments

Surgery is often part of a broader treatment plan, used alongside chemotherapy, radiation therapy, or immunotherapy. This combined approach can increase the effectiveness of cancer treatment by targeting any cancer cells that may still be present and lowering the likelihood of recurrence. Overall, surgery is a versatile tool in the treatment of cancer, with specific applications depending on the nature and stage of the cancer, the patient's general condition, and the intended course of treatment.

14.14 Surgical Operations and Complications

During surgery, surgeons make incisions in the body using sharp equipment and tiny, thin blades known as scalpels. Incisions through skin, muscle, and occasionally bone are frequently made during this procedure. These incisions might hurt after surgery, and healing takes time. During surgery, anaesthesia is used to minimize pain. It involves drugs or other chemicals that impair awareness or induce loss of feeling. Three categories of anaesthesia exist:

- A small, targeted area of the body is rendered numb by local anaesthesia.
- An arm or leg, for example, can be rendered completely numb by regional anaesthesia.
- Complete loss of awareness and sensation is the outcome of general anaesthesia, which is similar to a very deep slumber.

There are other means of performing surgery that do not entail cuts with scalpels. Among these, a few are mentioned below.

14.14.1 Cryosurgery

Cryosurgery or cryotherapy is a medical procedure that uses extremely cold temperatures produced by argon gas or liquid nitrogen to remove aberrant tissue. It is frequently used to treat precancerous lesions on the skin or cervix, retinoblastoma, and early-stage skin cancer.

14.14.2 Lasers

A type of treatment that uses powerful light beams to sever tissue is called a laser. Because they can focus precisely on small areas, lasers are useful for precise surgical treatments. Tumours or growths with the potential to develop into cancer can also be reduced in size or eliminated by lasers. They are usually used to treat malignancies that are either internal or external to the body, including basal cell carcinoma; precancerous cervical alterations; oesophageal, cervical, and vaginal cancers; as well as non-small cell lung cancer.

14.14.3 Hyperthermia

High heat is applied to particular body parts to cause hyperthermia. In addition to making cancer cells more vulnerable to radiation and some chemotherapy medications, this heat can harm and even kill cancer cells. Hyperthermia may be effective in hypoxic and nutrient deprived areas where radiotherapy and chemotherapy are less effective. It is a minimal, invasive technique to shrink the size of tumours, nodules or other growth in the body. Local heating improves tumour blood supply, resulting in a decrease in radiation resistance and helps to enhance the effectiveness of radiation/chemotherapy to destroy tumors. The degree of pain relief through radiofrequency ablation (RFA) may vary based on the cause and location as it is used to heat up a small area of nerve tissue. It may be applied for both benign and malignant cancers.

14.15 Salient Approaches for Effective Management of Critical Care Patients

14.15.1 Early Detection

14.15.1.1 Screening Programme

Implementing screening methods to detect common cancers at an early stage, PSA tests for prostate cancer, Pap smears for cervical cancer, colonoscopies for colorectal cancer, and mammograms for breast cancer are a few examples.

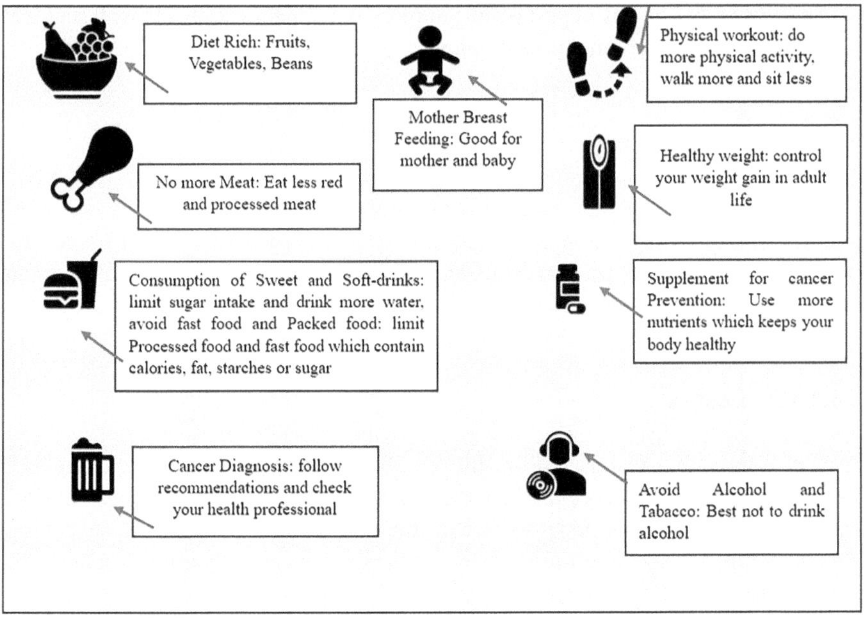

Fig. 14.7 Recommendation of preventive approaches

14.15.1.2 Genetic Testing

Genetically altered individuals may benefit from specialist screening and preventive actions for their families, as they are associated with an increased risk of cancer.

14.15.2 Prevention

1. Lifestyle modifications: Promoting healthy lifestyle habits like balanced nutrition, consistent physical activity, abstaining from tobacco, and moderating alcohol intake and practicing sun safety to reduce the risk of certain cancers (Fig. 14.7).
2. Vaccination: Administering vaccines to prevent infections known to increase the likelihood of developing specific cancers, such as cervical cancer, which is prevented with the HPV vaccine and other cancers associated with the human papillomavirus (HPV).
3. Chemoprevention: Chemoprevention involves the utilization of medications or natural compounds, such as selective oestrogen receptor modulators (SERMs) to lessen the risk of cancer development in high-risk individuals, including breast cancer.

14.15.3 Treatment

1. Surgery: Removing localized tumours and surrounding tissues, often as the initial treatment for solid tumours.
2. Drugs are used in chemotherapy to either kill or limit the growth of cancer cells, usually having an impact on the entire body.
3. High-energy beams are used in radiation therapy to destroy cancer cells or reduce tumour size. CAR-T cell therapy, checkpoint inhibitors, and cancer vaccinations are examples of immunotherapy that boost immunity to fight and destroy cancer cells.
4. Medicines are used in targeted therapy designed to focus on specific molecular changes that drive cancer growth, like tyrosine kinase inhibitors or monoclonal antibodies.

14.15.4 Supportive Care

1. Pain management involves delivering efficient pain alleviation to enhance the standard of living for those with cancer, frequently employing a blend of medications, procedures, and complementary treatments.
2. Psychosocial support: Addressing the emotional, social, and psychological requirements of cancer patients and their families via means of therapy, support groups, and more forms of assistance.
3. Palliative care: Incorporating specialized medical care that prioritizes symptom management and patients with advanced or terminal cancer in terms of their quality of life and supportive care.

14.16 Computational Toxicology and Its Benefit for Cancer Therapy

Computational methods play crucial roles in various aspects of cancer research, including predicting chemical toxicity and assessing their potential to cause cancer. These methods utilize molecular characteristics, structural attributes, and biological information to evaluate carcinogenicity (Fig. 14.8). Additionally, computational screening techniques like virtual screening and molecular docking are employed to pinpoint potential anti-cancer agents by predicting their interactions with molecular targets relevant to cancer pathways. Furthermore, computational systems biology integrates diverse omics data to construct network models elucidating the intricate molecular mechanisms driving carcinogenesis. Machine learning algorithms, such as random forests and neural networks, analyse extensive omics datasets to uncover biomarkers linked to cancer initiation and progression. Moreover, computational strategies aid in identifying existing drugs with potential anti-cancer properties, paving the way for drug repurposing in cancer treatment. In oncology, randomized clinical trials are the premier method for evaluating the benefits and risks of new

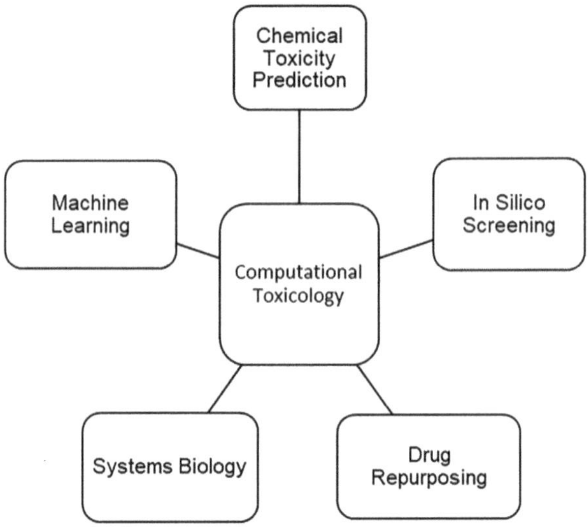

Fig. 14.8 Application of computational toxicology in cancer research

treatments compared to conventional care. With advancements in our understanding of cell biology, immunotherapeutic agents have been developed that can replicate normal cell growth processes. However, since these treatments can harm both healthy and cancer cells, there's a preference for targeted therapies. Targeted therapies work by impacting pathways that are crucial for cancer cell growth, division, and spread.

The Frank R. Lautenberg Chemical Safety for the twenty-first Century Act, which was passed in 2016, was the first piece of legislation in the United States to use innovative testing methods to decrease the number of vertebrate animals used in chemical safety studies. The advancement of artificial intelligence and computational toxicology approaches to applying novel testing methodologies is central to this objective. The phrases volume (quantity of data), velocity (growing of data), and variety (diversity of sources) have been used in the present big data era to characterize the chemical, in vitro, and in vivo data that are currently available for toxicity modelling (Ciallella and Zhu 2019).

Computational approaches in toxicology have been strengthened by the development of vast toxicological databases and advancements in machine learning algorithms. Working with these enormous databases of regulatory data has made it possible to examine animal models' repeatability, highlighting flaws in earlier in vivo techniques. This should make it easier to introduce new strategies and provide a benchmark that can be satisfied by any alternate way that has been verified against these techniques. Based on these large databases, Quantitative Structure Activity Relationships (QSAR) models for skin sensitization, ocular irritation, and other human health hazards have also highlighted some of the difficulties associated with computational modelling, such as validation problems, model interpretation problems, and model selection problems (Luechtefeld et al. 2018).

With ongoing scientific and technological progress, new treatment strategies have come into focus, including gene therapy; stem cell therapy, immunotherapy, and nanotechnology-based therapies. A cutting-edge method in cancer treatment involves combining chemotherapy with nanoparticles. Enhanced imaging technologies and diagnostic methods now allow for earlier detection of many cancers. As academicians and oncologists unfold cancer signalling pathways with better understanding, diagnostics and therapies are poised to become even more effective.

14.17 Artificial Intelligence (AI) and Cancer

An effective course of treatment for cancer requires early and accurate detection. When it comes to identifying tiny cancer indications that human observers could miss, AI algorithms show potential in this field. They do this by analysing medical imaging such as mammograms, MRIs, and CT scans. These algorithms aid radiologists by highlighting suspicious areas, enabling further assessment, and ultimately resulting in earlier diagnoses and better patient outcomes. Precision medicine seeks to customize medical care based on each patient's unique attributes, encompassing genetics, lifestyle, and environmental influences. AI is integral to this approach in oncology, as it sifts through extensive genomic and clinical data to pinpoint personalized treatment plans. By merging molecular profiling with AI-generated predictive models, healthcare providers can enhance treatment decisions, dosage, and timing to optimize treatment effectiveness while minimizing potential side effects. AI-driven methods have transformed the drug discovery and development process by speeding up the discovery of new therapeutic compounds and repurposing existing drugs for treating cancer. Machine learning algorithms scrutinize extensive biological and chemical data to forecast the pharmacological characteristics of potential compounds, prioritize promising candidates for further study, and refine drug design. This streamlined approach has the potential to cut costs and expedite the timeline for introducing new cancer treatments to market.

AI Applications for Common Cancers in Medical Imaging: Three broad categories can be used to classify artificial intelligence technologies for image-based clinical tasks in oncology:

- Identifying abnormalities.
- Determining the shape or volume of a suspected lesion, as well as its histopathologic diagnosis, disease stage, and molecular profile.
- Evaluation of prognosis or response to treatment over the course of observation. The abbreviations 2D, 3D, and CNS represent two, three, and central neural systems, respectively.

Precision medicine and artificial intelligence (AI) together have the potential to completely transform the medical field. Precision medicine techniques pinpoint the characteristics of patients with uncommon treatment outcomes or distinct medical requirements. Artificial intelligence (AI) uses complex calculation and inference to produce insights, gives the system the ability to reason and learn, and enhances human decision-making in clinicians. The most challenging problems facing precision medicine, particularly those where nongenomic and genomic determinants, along with data from patient symptoms, clinical histories, and lifestyles, will facilitate personalized diagnosis and prognostication, appear to be best addressed by translational research examining this convergence, according to recent literature (Bi et al. 2019; Johnson et al. 2021).

14.17.1 Monitoring and Management of Cancer Patients

Continuous monitoring of cancer patients' well-being is critical for evaluating treatment effectiveness, identifying disease advancement, and addressing treatment-related complications. AI-driven monitoring platforms combine information from electronic health records, wearable gadgets, and biomarker assessments to offer immediate insights into patients' medical journeys. Through the analysis of longitudinal data, machine learning algorithms forecast disease progression, pinpoint individuals at heightened risk, and recommend tailored intervention plans, ultimately enhancing patient care and prognosis.

14.18 Forensic Issues with Cancer Patients

Forensic toxicological and medico-legal assessments are vital in cases where drugs are administered inconsistently to terminal cancer patients. These patients, often on complex drug regimens, face risks related to dosing errors and potential drug interactions that can lead to fatal outcomes. Best practices emphasize the need to carefully evaluate drug-to-drug interactions in palliative care for terminal cancer patients. A case study illustrated the importance of analysing toxicological data and assessing professional liability when drugs are inappropriately administered to terminal cancer patients. Forensic evaluations play a key role in determining the cause of death and assessing liability in such cases. Toxicological analyses shed light on the pharmacological interactions between medications like midazolam and morphine in terminal cancer patients. Guidelines recommend conducting forensic toxicological and medico-legal evaluations to ensure the safe management of terminal cancer patients and prevent fatal incidents. Forensic assessments can provide insights into whether the administered therapy was appropriate and help understand the patient's clinical condition at the time (Basilicata et al. 2021).

14.19 Spices as Potential Agents for Cancer Prevention and Treatment

There is a long history of using spices in both cooking and medicine (Singh et al. 2013). Several spices have been connected in recent research to the prevention and treatment of cancer because of their immunomodulatory, antioxidant, and anti-inflammatory properties (Gupta et al. 2015). Spices like turmeric, black cumin, ginger, garlic, saffron, black pepper, and chili pepper include bioactive compounds called curcumin, thymoquinone, piperine, and capsaicin, which have been related to malignancies of the prostate, lung, liver, breast, stomach, colorectal, and cervical regions (Fig. 14.9). These substances work by triggering the death of cells in tumours; impeding their growth, migration, and invasion; and increasing the susceptibility of tumours to chemotherapy and radiation. To learn more about how effective spices are in treating and preventing cancer, more research is needed.

In the realm of Indian traditional medicine, turmeric has been esteemed for its healing properties, including its application in addressing various health issues,

Fig. 14.9 Effective traditional spices for prevention/treatment and management of cancer

including cancer. The bioactive compounds having antioxidant properties have potential to reduce the free radicals which can cause cancer (Singh et al. 2016; Singh and Singh 2018). Ayurvedic texts suggest that curcumin, the primary active component in turmeric, harbours potent anti-cancer attributes. Black cumin, alternatively known as kalonji or black seed, holds a significant place in traditional Indian medicine. Preclinical investigations have indicated encouraging anti-cancer effects of thymoquinone, the principal bioactive compound found in black cumin. Ginger, a staple spice in Indian culinary traditions and traditional medicine, is believed to possess anti-inflammatory properties that could potentially mitigate the likelihood of developing certain cancer types. Garlic, a commonly used ingredient in Indian cuisine and traditional medicine, is valued for its diverse health benefits. It contains organosulfur compounds thought to exhibit anti-cancer properties. Saffron, referred to as "kesar" in Hindi, holds a revered position in Indian culture and cuisine, primarily owing to its medicinal attributes, which include its perceived potential in cancer prevention and treatment. Black pepper, a ubiquitous spice in Indian cooking and traditional medicine, contains piperine, its active compound that has been scrutinized for its anti-cancer effects, including its capacity to heighten the bioavailability of other anti-cancer compounds (Turrini et al. 2020). Chili peppers, frequently incorporated into Indian dishes and traditional remedies, boast capsaicin as their active component responsible for their fiery heat. Preclinical studies suggest capsaicin's potential anti-cancer properties.

Complementary and alternative medicine (CAM) is increasingly being used by those suffering from cancer. This is a reflection of the various challenges and requirements that traditional medicine is now failing to address. Alternative methods of treatment are used by a sizable percentage of cancer patients in developed nations to enhance their quality of life in addition to traditional symptom management (Ayele 2018). However, in countries such as India, the situation is drastically different. Over 80% of individuals with cancer have an incurable late-stage disease when they are first diagnosed. As a result, therapy options become more complicated and healing becomes more difficult. Strict laws and regulations make it more difficult to obtain oral morphine for the purpose of relieving pain in individuals with terminal cancer. Patients in rural parts of the country often lack access to traditional medical care and are forced to experiment with alternative therapies such as homeopathic medications, naturopathy, hydrotherapy, homeopathic remedies, acupuncture, auto urine therapy, osteopathy, and vipassana.

Individuals with diverse racial and ethnic backgrounds, age ranges, and sexual orientations are susceptible to different diseases due to unique physical, genetic, and environmental factors. Oncologists use CAR-T cell therapy in combination with immunotherapy to target genetic or protein abnormalities that have been identified in a patient's tumour, allowing the individual's treatment to be tailored to their unique needs. It functions as monoclonal antibodies, immunological modulators, therapeutic vaccines, and check point inhibitors. Many cancer types are responding quite well to targeted immunotherapy.

14.20 Futuristic Approaches for Drug Delivery System

The global markets for cancer therapy and innovative drug delivery systems include a variety of techniques such as liquid emboli, polymers, micelles, drug-eluting beads, embolization particles, and nanoparticles. There is an increasing trend towards the use of smart nanoparticles, which can react to or be steered by biological signals, offering a more precise method for targeting cancer treatment. Smart nanoparticles and biomolecules created to improve the efficacy and safety of cancer treatments are the consequence of significant advancements in oncology, nanotechnology, and biomedicine. The modern cancer treatment approaches may include precision medicine and hybrid, translational models for surgical interventions. Computer-aided robotic surgery may be another option for complicated carcinogenic situations of specific target organs.

Nowadays, many cancer treatments are with high level of precision and minimally invasive techniques, supported by molecular imaging, and even robot- or AI-assisted procedures. Chemotherapy, a common treatment method, has seen improvements through refined dosing strategies, as well as its use in neoadjuvant or adjuvant settings, and as part of combination therapies. Radiation oncology has also progressed, with a focus on precision radiotherapy. First-line treatment recommendations rely on the cancer's type and stage at diagnosis and are continually updated with new therapeutic options. The emergence of targeted therapy and immunotherapy has transformed cancer treatment, aided by advanced diagnostic and molecular characterization tools.

Developing effective and safe cancer nanomedicine is the main goal of current oncology techniques (Pathak et al. 2022). Stem cell therapy has shown promise in healing and restoring diseased or damaged tissues, addressing both primary and metastatic cancer sites. New medicinal and diagnostic possibilities have been made possible by nanoparticles. Due to its ability to minimize harm to healthy cells and stop the growth and spread of particular cancer cells, targeted treatment has the potential to be a game-changer. A minimally invasive technique called ablation therapy has made it possible to burn or freeze tumours without requiring open surgery. Natural antioxidants have the capacity to identify free radicals and neutralize their harmful effects, which may help treat or prevent cancer, according to research (Vishwakarma et al. 2013; Singh et al. 2023). Numerous of these novel technologies are currently undergoing clinical trial research, with several of them having previously received approval. High pressure, liquid oxygen, liquid air, and liquid CO2 are used in cryoablation techniques, which are based on the Joule–Thomson phenomenon. This cooling action causes ice crystals to form, which are then used to treat lesions, warts, and keratosis.

Radiofrequency ablation (RFA) is a minimally invasive therapy guided by imaging that kills cancer cells by subjecting them to hyperthermic (high-frequency electrical currents) conditions. Needle electrodes are inserted into a tumour cell using imaging methods including magnetic resonance imaging (MRI), CT, or ultrasound. Small-size tumours of this type are generally best treated with RFA. Compared to alternative treatment modalities with few limitations, endoscopic ultrasonography-guided radiofrequency ablation is preferable because it typically allows for selective tissue ablation with the least amount of damage to the surrounding tissues.

Gene therapy is the process of repairing a broken gene by introducing a healthy copy into the genome to treat that particular condition. The production of wild-type tumour suppressor genes, the manufacture of proapoptotic and chemo-sensitive genes, genes that elicit certain anti-tumour immune responses, and the targeted silencing of oncogenes are among the cancer gene therapy strategies.

Radiation-induced oxidative damage to deoxyribonucleic acid (DNA) and other bimolecular species can produce radical oxygen species (ROS), which can profoundly alter transcription factor control. Vitamins, polyphones, and plant-derived bioactive substances are examples of natural antioxidants that are employed as preventative and restorative strategies (Gupta et al. 2023; Singh et al. 2014).

An immune system-blocking protein termed PD-L1 has been engineered to be inhibited by a medicine such as pembrolizumab, ipilimumab, nivolumab, atezolizumab, etc. an immune checkpoint inhibitor, allowing immune cells to eradicate cancer. After severe autoimmune toxicity, immune checkpoint inhibitor therapy may be continued with the targeted control of B cell-mediated autoimmunity with certain medications. This medicine is being used successfully for the management of melanomas and other skin cancers, as well as breast, pulmonary, endometrial, renal, oesophageal, and many other malignancies. It also has the ability to induce dormancy in many extremely advanced, possibly deadly disorders. Immune checkpoint inhibitor therapy may continue to the point of severe autoimmune damage, utilizing specific medications to target B cell-mediated inflammation.

As we are aware, oestrogen receptor alpha is primarily the target for treatment of breast cancer. Anti-oestrogen drugs, e.g., tamoxifen, toremifene, and fulvestrant, are being used for the treatment, but subsequently the reports indicate resistance to such drugs and recently according to an article published in current science (2024). Aspirin may help as a potential anti-cancer drug due to its correlation findings for being sensitive to oestrogen receptor alpha. Aspirin may bind to a specific region of oestrogen receptor alpha and inhibit the normal function under variable dose conditions it may also combine with existing therapy to enhance their efficacy or even help to reduce dosage subject to standardization and validation by the clinicians.

14.21 Conclusions

The market for cancer drugs worldwide was estimated to be worth \$135.49 billion in 2020. From 2021 to 2030, it is projected to increase at a 7.5% CAGR to reach \$274.40 billion. Monoclonal antibody (mAbs)-based biological medications are becoming more popular for the treatment of different kinds of cancer. Growing cancer rates, a boom in cutting-edge treatments, and an aging population are the main drivers of this trend. The market is expanding due to other factors such as increased cancer awareness, earlier screenings, and improved access to cancer medications. The traditional spices, viz., turmeric, cinnamon, pepper fenugreek, cloves, etc., are rich in antioxidants and help to neutralize the free radicals generated within the body, aid in digestion, produce digestive enzymes with reduction of oxidative stress and inflammation, help in regulation of blood pressure, increase metabolism, and prevent salient diseases. One innovative approach to treating cancer is the creation of therapeutically valuable nanoparticles by artificial intelligence (AI). The term "computer-aided nanoparticles design" refers to a range of theoretical and computational techniques used in the development of medications that are now being used in clinical settings or are undergoing human testing. This subject combines a number of scholarly fields, including theoretical chemistry, cheminformatics, and molecular modelling. Typical computer approaches used in this process include molecular docking, molecular dynamics, similarity searches, and quantitative structure-activity relationships (QSAR). These advanced computational tools are being continuously researched and developed to help in the creation of novel and effective cancer treatments. One of the most important components of veterinary medicine related to animal feed and food manufacturing for livestock is antibiotics. It is essentially required to show the significance of antibiotics since they are used to treat bacterial illnesses.

While they were outlawed for a short while as an evolutionary factor, their usage in animals may result in residues in food, which poses dangers to the well-being of consumers, the environment, technology, and animal health. Water contamination is currently a major environmental problem that affects not only biodiversity but also human health (Morris 1995; Iredell 2023). The two most significant types of pollutants among them are inorganic and organic ones, which are difficult to treat with currently available methods due to their high toxicity and persistence. Concerns about the potential direct effects on public health are highlighted by the need to address the challenges associated with treating industrial wastewater and eliminating dangerous organic pollutants, such as azo dyes and carcinogenic methyl orange (MO), and colorants from various industries. Scientists, policy experts, and government organizations in India are collaborating to support cancer research studies aimed at achieving large-scale, low-tech, economically feasible health results. The complex environment encompassing cancer research is made more so by the need to balance public policy across a number of competing agendas, including research competency, capacity, and training needs from clinical researchers. Furthermore, there is a need for parallel enhanced collaborative funding from the government to develop strong infrastructure across a range of domains, such as clinical trials, tissue banking, and registries, with the goal of streamlining the regulatory environment for public health.

Acknowledgements The authors are thankful to CSIR-IITR, Lucknow, and RML Avadh University for mutually sharing the infrastructure and key resources of knowledge resource centre in the development of the manuscript and encouragement.

Conflict of Interest There is no conflict of interest and this book chapter is especially for usage by academicians to serve as baseline information in recent perspectives.

References

Ayele TT (2018) A review on traditionally used medicinal plants/herbs for cancer therapy in Ethiopia: current status, challenge and future perspectives. Org Chem Curr Res 07(02)

Basilicata P, Giugliano P, Vacchiano G, Simonelli A, Guadagni R, Silvestre A, Pieri M (2021) Forensic toxicological and medico-legal evaluation in a case of incongruous drug administration in terminal cancer patients. Toxics 9(12):356

Bi WL, Hosny A, Schabath MB, Giger ML, Birkbak NJ, Mehrtash A et al (2019) Artificial intelligence in cancer imaging: clinical challenges and applications. Cancer J Clin 69(2):127–157

Burd EM (2003) Human papillomavirus and cervical cancer. Clin Microbiol Rev 16(1):1–17

Ciallella HL, Zhu H (2019) Advancing computational toxicology in the big data era by artificial intelligence: data-driven and mechanism-driven modeling for chemical toxicity. Chem Res Toxicol 32(4):536–547

Cohn BA, Terry MB, Plumb M, Cirillo PM (2012) Exposure to polychlorinated biphenyl (PCB) congeners measured shortly after giving birth and subsequent risk of maternal breast cancer before age 50. Breast Cancer Res Treat 136(1):267–275

Cooper GM (2000) The cell: a molecular approach, 2nd edn. Sinauer Associates; The Development and Causes of Cancer, Sunderland. https://www.ncbi.nlm.nih.gov/books/NBK9963/

Fucic A, Gamulin M, Ferencic Z, Katic J, Krayer Von Krauss M, Bartonova A, Merlo DF (2010) Environmental exposure to xenoestrogens and oestrogen related cancers: reproductive system, breast, lung, kidney, pancreas, and brain. Environ Health 11(Suppl 1):S8

Gillies RJ, Robey I, Gatenby RA (2008) Causes and consequences of increased glucose metabolism of cancers. J Nucl Med 49(Suppl 2):24S–42S

Gupta R, Singh RL, Singh P (2015) Quantification of phytochemicals and evaluation of antioxidant potential of ethanolic leaf extract of *Terminalia bellirica, Terminalia chebula and Emblica officinalis* Vis-a-Vis Triphala. Int J Pharm Sci Rev Res 32(2):14–22

Gupta R, Shukla S, Gupta A, Singh RL, Singh P (2023) Estimation of phytochemical constituents and analysis of antioxidant activity in different parts of *Cassia fistula* plant. Int J Pharma Chem Ana 10(3):194–201

Iredell S (2023) Water pollution and cancer: an updated review. Sci Insights 43(4):1079–1086

Johnson KB, Wei WQ, Weeraratne D, Frisse ME, Misulis K, Rhee K, Zhao J, Snowdon JL (2021) Precision medicine, AI, and the future of personalized health care. Clin Trans Sci 14(1):86–93

Kalaiyarasi M, Dhanasekar R, Sakthiya Ram S, Vaishnavi P (2020) Classification of benign or malignant tumor using machine learning. Mater Sci Eng 995:012028

Khodabandehlou N, Mostafaei S, Etemadi A et al (2019) Human papilloma virus and breast cancer: the role of inflammation and viral expressed proteins. BMC Cancer 19:61

Luechtefeld T, Rowland C, Hartung T (2018) Big-data and machine learning to revamp computational toxicology and its use in risk assessment. Toxicol Res 7(5):732–744

Martin TA, Ye L, Sanders AJ et al (2013) Cancer invasion and metastasis: molecular and cellular perspective. In: Madame Curie bioscience database [Internet]. Landes Bioscience, Austin, pp 2000–2013. https://www.ncbi.nlm.nih.gov/books/NBK164700/

Morris RD (1995) Drinking water and cancer. Environ Health Perspect 103(Suppl 8):225/231

O'Shaughnessy KL, Fischer F, Zenclussen AC (2021) Perinatal exposure to endocrine disrupting chemicals and neurodevelopment: how articles of daily use influence the development of our children. Best Pract Res Clin Endocrinol Metab 35(5):101568

Ogunwobi OO, Mahmood F, Akingboye A (2020) Biomarkers in colorectal cancer: current research and future prospects. Int J Mol Sci 21(15):1–20

Pathak N, Singh P, Singh PK, Sharma S, Singh RP, Gupta A, Mishra R, Mishra VK, Tripathi M (2022) Biopolymeric nanoparticles based effective delivery of bioactive compounds toward the sustainable development of anticancerous therapeutics. Front Nutr 9:963413

Rebuzzini P, Fabozzi G, Cimadomo D, Ubaldi FM, Rienzi L, Zuccotti M, Garagna S (2022) Multi- and transgenerational effects of environmental toxicants on mammalian reproduction. Cells 11(19):3163

Singh P, Singh RL (2018) Quantification of phytochemicals imparting antioxidant activities in commonly used vegetables. Int J Appl Sci Biotech 6(2):107–112

Singh P, Vishwakarma SP, Singh RL (2013) Evaluation of antioxidant, oxidative DNA damage protective and antimicrobial activities of *Foeniculum vulgare* plant. J Med Plant Res 7(35):2551–2563

Singh RL, Vishwakarma SP, Singh P (2014) Vitamins and minerals: roles and plant sources. In: Prakash D (ed) Phytochemicals of nutraceutical importance and their role in human health. CABI-UK, pp 248–265

Singh U, Shukla M, Singh P, Kakkar P, Singh RL (2016) Role of *Vicia faba* fruit extract against cytotoxicity induced by acetaminophen in primary cultured rat hepatocytes. Int J Pharm Pharm Sci 8(8):71–78

Singh U, Singh P, Singh PK, Shukla S, Singh RL (2023) Evaluation of antioxidant, oxidative DNA damage and hepatoprotective activities of ethanolic leaf extract of *Lagenaria siceraria* plant against acetaminophen induced hepatotoxicity. Func Foods Health Dis 13(3):117–134

Soto A, Sonnenschein C (2010) Environmental causes of cancer: endocrine disruptors as carcinogens. Nat Rev Endocrinol 6(7):363–370

Turrini E, Sestili P, Fimognari C (2020) Overview of the anticancer potential of the "king of spices" Piper nigrum and its main constituent piperine. Toxins 12(12):747

Vishwakarma SP, Singh P, Singh U, Singh RL (2013) Antioxidant activities of some tuberous plants leaves. Int J Pharm Sci Rev Res 20(1):28–33

Outbreak of Deadly Coronavirus (COVID-19) Pandemic: Diagnosis and Management

15

Ankit Tiwari, Divyanshu Singh, Dharmendra Kumar, Vikas Chandra, Naveen Kumar Vishvakarma, Dhananjay Shukla, Harit Jha, and Rajat Pratap Singh

Abstract

An outbreak of COVID-19 disease was caused by a novel severe acute respiratory syndrome coronavirus 2 (SARS-CoV-2) in 2019. The health crisis during the COVID-19 pandemic has put the world at risk and posed significant challenges to healthcare systems around the world. The SARS-CoV-2 virus is spread by inhalation or contact with infected droplets resulting in a wide spectrum of symptoms, such as fever, cough, breathlessness, fatigue, loss of taste and smell, etc., or may be asymptomatic. During pandemic, diagnostic techniques including molecular and serological testing have been of great help in identifying infected persons and guiding preventive measures. Antigen-antibody and polymerase chain reaction (PCR) assays have been widely employed for early identification of infected persons. Management of the SARS-CoV-2 spread has largely depended on early case detection, contact tracing, and isolation of the infected person. Healthcare professionals, public health organizations, and governments have worked together to implement a variety of measures that have proven successful in reducing transmission rates. Several therapeutic strategies have been investigated for the treatment of COVID-19, such as immunomodulatory therapy, plasma therapy, antiviral drugs, and cell-based therapy. Remdesivir (an antiviral drug) and hydroxychloroquine (an antimalarial drug) has shown potential in reducing hospitalization and improving recovery rates in complicated patients. Different types of vaccines including inactivated virus vaccines, protein-based vaccines, viral vector vaccines, and RNA and DNA vaccines against COVID-19 have been developed. Several types of herbal remedies play a major role in boosting immunity and minimizing the symptoms of COVID-19. Additionally, social

A. Tiwari · D. Singh · D. Kumar · V. Chandra · N. K. Vishvakarma · D. Shukla · H. Jha · R. P. Singh (✉)
Department of Biotechnology, Guru Ghasidas Vishwavidyalaya, Bilaspur, Chhattisgarh, India

R. L. Singh et al. (eds.), *Clinical Applications of Biomolecules in Disease Diagnosis*, https://doi.org/10.1007/978-981-97-4723-8_15

distancing, mask-wearing, and vaccination campaigns have played pivotal roles in preventing the virus's spread and lessening its effects on communities. We might be more capable of managing pandemics by implementing future strategies into practice such as improving diagnostic facilities, hospital infrastructure, and public health measures.

Keywords

COVID-19 · RT-PCR · Antiviral drugs · Vaccines · Herbal remedies

15.1 Introduction

The COVID-19 pandemic caused by the novel coronavirus SARS-CoV-2 (severe acute respiratory syndrome coronavirus 2) had a significant impact on the world's economy, society, and healthcare system. The disease has affected millions of people and caused a significant amount of fatalities due to its rapid spread worldwide. The pandemic was introduced in 2020 as pneumonia of unknown causes in China. Later on, the causing agent was identified as a coronavirus and renamed as SARS-CoV-2, and the disease was named as coronavirus disease 19 (COVID-19). Since then, 770,437,327 people all over the world have been affected by this virus and 6,956,900 deaths have been reported (WHO 2020). The fatality rates of Middle East respiratory syndrome coronavirus (MERS-CoV), SARS-CoV, and SARS-CoV-2 are 2.9%, 9.6%, and 36%, respectively. The other human coronaviruses OC43, NL63, HKU1, and 229E normally causes self-limited symptoms with mild illness (Corman et al. 2018). The symptoms of SARS-CoV infection can be divided into mild, moderate, and severe. The mild and moderate symptoms such as fever, cold, dry cough, and loss of smell and taste may be treated at home and by isolating themselves (da Rosa et al. 2021), but severe infections of COVID-19 can lead to hypoxia because of high viral titres. Severe symptoms had been reported in people having comorbid conditions such as diabetes, obesity, cardiovascular disease, and neurological disorders (Vardhan et al. 2022). Children were less clinically affected by the virus than adults. Different methods have been employed for diagnosis of COVID-19 virus (Fig. 15.1). Molecular diagnostics such as reverse transcription polymerase chain reaction (RT-PCR) is still more reliable with high sensitivity and specificity for the detection of viral genes from respiratory samples of infected patients (Corman et al. 2020). Antibodies produced in response to SARS-CoV-2 infection were detected by serological assays. Serological test was essential in determining the disease's prevalence and locating people who have already been infected with it (Guo et al. 2020a). Antiviral drugs such as remdesivir have shown potential in lowering hospitalization and enhancing clinical outcomes (Polack et al. 2020). The worldwide vaccination played a crucial role in reducing the effects of the COVID-19 pandemic. The effectiveness of vaccines in avoiding serious illness and lowering transmission rates is essential (Polack et al. 2020). Moreover, many clinical therapeutic strategies such as plasma therapy, circulatory and blood purification support, immunotherapy, invasive mechanical ventilation, oxygen therapy, antiviral therapy,

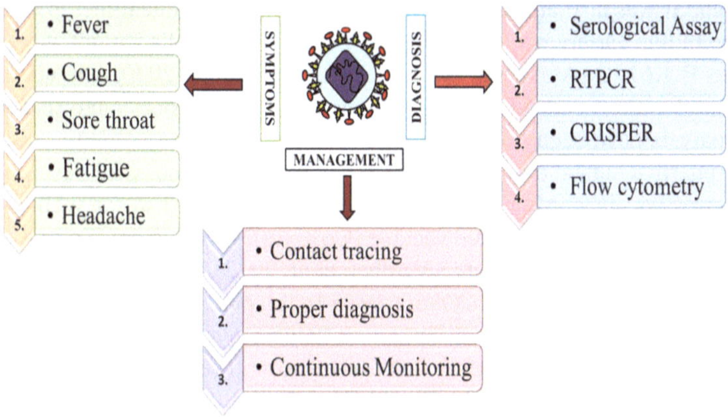

Fig. 15.1 Symptoms, management, and diagnosis of COVID-19

and additional treatment including use of conventional medication have been used for the treatment of COVID-19 (Sheahan et al. 2020).

15.2 Outbreak of COVID-19

The city of Wuhan, which is located in the Chinese province of Hubei, reported a few cases of pneumonia of unidentified origin on December 31, 2019, to the WHO (Gabutti et al. 2020). The agent responsible for the cases was identified as a novel coronavirus (SARS-CoV-2) on January 7, 2020, which was isolated from a throat swab sample of a patient in Wuhan (Wang et al. 2020a). Many initial cases were linked to Huanan Seafood Wholesale Market suggesting that SARS-CoV-2 was transmitted from animals to humans. However, a genomic study has provided evidence that the virus was introduced from another, yet unknown, location, into the market where it spread more rapidly, although human-to-human transmission may have occurred earlier. It was also suggested that coronavirus from bats (*Rhinolophus* bats) and SARS-CoV-2 are closely related (Benvenuto et al. 2020). The number of COVID-19 cases was increasing exponentially due to human-to-human transmission. On January 30, 2020, the COVID-19 outbreak was declared as Public Health Emergency of International Concern (PHEIC) by the World Health Organization (WHO) (Burki 2020). The COVID-19 pandemic has affected people's lives all over the world with large number of fatalities (Wang et al. 2020a). The emergence of new SARS-CoV-2 variants including alpha, beta, gamma, and delta presented significant difficulties in controlling the outbreak. These variants showed different levels of disease severity, greater transmissibility, and a potential immunological escape (Singh et al. 2022a). The COVID-19 delta variant (B.1.617.2) was initially identified in India. This variant was identified as most dangerous and contagious variants among the so far known variants by the WHO. Delta variant was spread more readily among adults and children, and symptoms of this variant were varied and severe in patients. It was also appearing to be more resistant to clinical therapeutic strategies. The Indian variety was originally designated for the delta variant, which is also known as B.1.617.2 (Alexandar et al. 2021). Since January 2020, a number of studies and publications on SARS-CoV-2 whole genome sequences analysis and mutational analysis have been initiated in collaborations between academic researchers and bio-pharmaceutical companies which has led to the successful development of a number of vaccines. Many vaccines have been developed so far such as mRNA vaccine (BNT16b2b), inactivated vaccine (Covaxin), viral vectors (Sputnik V), peptide vaccine (EpiVacCorona), etc. (Vardhan et al. 2022).

During the second wave of COVID-19 in India, patients showing comorbidities were reported a histopathological feature of "mucormycosis" or black fungus. The fungal invasion of blood vessels results in mycotic thrombosis, ischaemic infarction, and ultimately necrosis of affected host tissues leading to mucormycosis, a potentially lethal infection (Afroze et al. 2017). The group of moulds known as mucormycetes including *Rhizopus* and *Mucor* was responsible for mucormycosis (Bhatia 2022). The overall worldwide incidence of mucormycosis has been estimated to range from 0.005 to 1.7 per million people (Jeong et al. 2019). However, the prevalence of mucormycosis in India has been reported to 0.14 per 1000 people (Prakash and Chakrabarti 2019).

The current global situation of COVID-19 requires continued vigilance, research, and international cooperation. By understanding the multifaceted impacts of the pandemic on healthcare, society, and the economy, we can better navigate the path forward. As of the latest available data, the world continues to grapple with the challenges posed by this pandemic. The threat that SARS-CoV-2 poses to the health and well-being of millions of people worldwide must be urgently addressed if it persists. At the same time, it's critical to understand that the situation with the virus may be better than it has ever been. This reality forces us to refrain from using alarmist rhetoric and to present reasonable and workable answers in order to guide people toward a new, nonemergency phase of the pandemic. According to the World Health Organization (WHO), there have been 770,437,327 confirmed cases of COVID-19, including 6,956,900 deaths that have been reported, and a total of 13,500,122,024 doses of the COVID-19 vaccine have been given out globally (WHO 2020).

15.3 Transmission Modes of SARS-CoV2 (COVID-19)

The possible ways of transmission of SARS-CoV2 are as follows:

15.3.1 Droplets/Aerosol

The SARS-CoV-2 virus spread through the respiratory droplets and aerosols produced when an infected person coughs or sneezes. These respiratory droplets and aerosols may reach another person's nose, mouth, or eyes and infect them (Li et al. 2020). The droplet's size is crucial in determining whether it can carry the virus or not. Since they can remain in the air for a long time and travel distances greater than one meter, droplet nuclei particles of <5 μm in diameter are regarded as the carriers (Mukhra et al. 2020). The infectious virus persisted in the aerosols for around 3 hours under controlled settings, gradually declining from $10^{3.5}$ to $10^{2.7}$$TCID_{50}$ per millilitre of air (Van Doremalen et al. 2020). According to the WHO, droplets smaller than 5 microns produced after medical procedures might spread and "cannot be ruled out" which is the airborne mode of transmission of coronavirus (Mukhra et al. 2020).

15.3.2 Fomite Transmission

Inanimate objects such as a dish, doorknob, or article of clothing or surfaces that have been exposed to infectious agents are known as fomites. These fomites can act as a source of transmission of SARS-CoV-2 virus. The persistence of SARS-CoV-2 on various materials has revealed that the virus can stay up to 72 hours on plastic and stainless steel surfaces (Chin et al. 2020). In a study, samples were collected from toilets and sinks of a clinical facility treating COVID-19 patients. These samples were found positive for viral RNA indicating that these surfaces may act as source for virus transmission (Ong et al. 2020). Similarly, SARS-CoV-2 RNA was detected in the samples taken from the floors of the intensive care unit (70%), general wards (15.4%), and the hospital pharmacy (100%) at the Huoshenshan Hospital in Wuhan (Guo et al. 2020b). Nevertheless, these findings imply that fomites may help in the nosocomial and community transmission of SARS-CoV-2.

15.3.3 Saliva

Saliva plays a greater role in the transmission of SARS-CoV-2 virus. The SARS-CoV-2-laden saliva may spread the virus into the throat, lungs, or other systems. The cellular entry of SARS-CoV-2 is mediated by ACE-2 (angiotensin converting enzyme-2) receptor. ACE-2 is expressed in primary cells of all types of salivary glands and in tissues lining the oral cavity. ACE-2 is prevalent in the oral mucosa epithelial cells and spread through saliva. According to reports, the tongue (95.86%) has higher expression of these receptor than the buccal and gingival tissues (Xu et al. 2020). Culture test has been suggested in order to determine the presence of viral RNA in saliva for potential transmission.

15.3.4 Vertical Transmission

The vertical transmission of the virus from mother to foetus or newborn is a contentious point regarding the spread of SARS-CoV-2. According to Cao et al., a vaginal delivery of an infant from a mother who has been diagnosed with COVID-19 increases the risk of perinatal infection (Cao et al. 2020).

15.3.5 Conjunctival Transmission

The nasolacrimal duct connects the respiratory system to the conjunctival mucosa, and both tissues share an ACE-II receptor (Sun et al. 2020). The COVID-19 transmission might be possible through the ocular surface. There has only ever been one report of conjunctivitis and viral RNA in tear fluid. However, lack of the virus in the patients' conjunctival sac suggested a peculiar route of SARS-CoV-2 transmission (Mukhra et al. 2020; Xia et al. 2020).

15.3.6 Sexual Transmission

Sexual activity may be a further likely way of transmission of SARS-CoV-2. Since the gastrointestinal tract may continue to shed the viral RNA days after recovery, it is hypothesized that some sexual behaviour may even result in additional ways of transmission by both direct and indirect exposure of oral-anal contact (Patrì et al. 2020). However, more research should concentrate on determining whether SARS-CoV-2 infection can affect the reproductive system. Testicles and orchitis have both been known to get infected by SARS coronavirus. The possibility of the novel virus infecting the human testes increases because SARS-CoV-2 and SARS-CoV-1 share the ACE-II receptor. As a result, it's suggested that COVID-19 male patients who have recovered should be assessed for their reproductive behaviour because the human testis is a vulnerable organ for the virus infection (Wang and Xu 2020).

15.4 Structure of SARS-CoV-2

The family *Coronaviridae* has four genera: alpha-coronavirus, beta-coronavirus, gamma-coronavirus, and delta-coronavirus. Coronaviruses are members of the *Coronavirinae* subfamily. They are enveloped single-stranded RNA viruses. The genomic organization and expression pattern are similar in all coronaviruses. Unlike other RNA viruses, CoVs have a larger single-stranded positive-sense RNA (+ssRNA) with genome size 27–32 kb. Outside of the genome, the capsid is formed by the nucleocapsid protein (N). The genome is enclosed by an envelope that attaches to three structural proteins: membrane protein (M), spike protein (S), and envelope protein (E) (Wang et al. 2020c).

15.4.1 Spike(S) Protein

The host receptor angiotensin-converting enzyme 2 (ACE2) is recognized by the transmembrane S protein that protrudes from the viral surface, mediating coronavirus entrance into host cells (Walls et al. 2020). The spike protein is a type I-TM protein with a clove-like structure having ectodomain (ED) region, transmembrane (tm) region, and intracellular domain (Prajapat et al. 2020). The receptor-binding S1

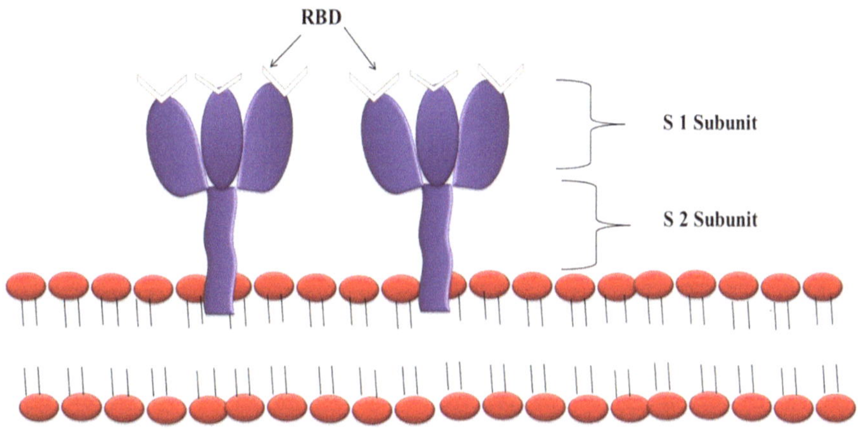

Fig. 15.2 Spike glycoprotein

domain (three S1 heads) and the membrane fusion subunit S2 (trimeric stalk) on the C-terminal make up the ED (Fig. 15.2). The virion is known as CoV because the spike proteins that make up its outer surface congregate in the trimeric form, giving the appearance of virions a tiara (Belouzard et al. 2012). The S2 subunit is made up of the fusion peptide (FP), the fusion-peptide proximal region (FPPR), the heptad repeat 1 (HR1), the central helix (CH), the connector domain (CD), the heptad repeat 2 (HR2), the transmembrane segment (TM), and the cytoplasmic tail (CT) (Zhang et al. 2021). RBD is a key region for coronavirus invasion of host cells, which is located at the apex of the S1 subunit. The RBD encounters a hinge-like conformational change both before and after S protein attaches to ACE2, which causes it to momentarily rotate between open (or up) and closed (or down) states (Walls et al. 2020).

Small molecule inhibitors, antibodies, and vaccinations are only a few of the antiviral treatments that have been developed which target the S protein. Inhibitors of S glycoproteins typically prevent the fusion of virus with a membrane by competitively reducing the interaction between RBD and ACE2 (Yan et al. 2022). Ivermectin was shown to have broad-spectrum antiviral activity in vitro (de Oliveira et al. 2021). Arbidol causes the viral glycoprotein to become structurally rigid, which restricts the conformational rearrangements required to facilitate membrane fusion and virus penetration (Padhi et al. 2021).

15.4.2 Nucleocapsid (N) Protein

Nucleocapsid (N) protein participates in SARS-CoV viral replication and assembly. It disrupts the host's interferon system. N protein received greater attention for therapeutic targets because it has a number of advantages including higher sequence and reduced mutation risk and induces stronger host's immune response. N protein of SARS-CoV2 has shown similarity with SARS-CoV N protein (Kannan et al. 2020). The main role of N protein is to wrap the viral genome into a long helical ribonucleocapsid (RNP) complex (Bai et al. 2021). N protein consists of N terminal domain (NTD), C terminal domain (CTD), and three intrinsically disordered regions, an N-arm, a C-tail, and a flexible linker that connects the NTD and CTD. The NTD and CTD are primarily associated with RNA binding and dimerization/oligomerization, respectively.

A protective immune response can be elicited by using the N protein as an immunodominant antigen (Yan et al. 2022). A monoclonal antibody nCoV396 from COVID-19 patients showed interaction with NTD of the N protein (Kang et al. 2021). Based on specific structural details, this antibody could be changed in a site-directed manner to enable the production of monoclonal antibodies with higher efficacy and a wider spectrum. Furthermore, a small compound PJ34, which target the ribonucleotide-binding site in NTD, exhibits strong suppression of the RNA-binding activity of the HCoV-OC43 N protein and can stop HCoV-OC43 from replicating (Peng et al. 2020).

15.4.3 Envelope Protein

The transmembrane E protein of SARS-CoV-2 has N-terminal ectodomain, transmembrane domain (TMD), and C-terminal domain (Yan et al. 2022). A pentameric ion channel (PDB ID 7K3G) makes up the structure of TMD which is made up of five helices. The stacking (Phe23 and Phe26), van der Waals packing (among the Val29-Leu31-Ile33 trio), and significant hydrophobic contacts (among the numerous hydrophobic residues at the pore of the channel) all contribute to the stability of the nearby helix and helical bundle (Mandala et al. 2020). Two ion channel medicines, hexamethyleneamiloride (HMA) and amantadine (AMT), restrict the ion channel activity of the E protein by interacting with the polar amino acids near the ion channel's entrance and occupying the amino-terminal lumen of the channel (Pervushin et al. 2009). Apical cell polarity complex made of PALS1, Crumbs, and Pals1-associated tight junction protein (PATJ) is disrupted by the postsynaptic density-95 (PSD-95), zonaoccludens 1 (ZO-1) (PDZ)-binding motif at the C-terminus of the E protein. This motif recognizes the PDZ domain of the human cell junction protein PALS1. As a result, the lung epithelial connections become loose and porous, which promotes the proliferation and spread of viruses (Chai et al. 2021). It has been speculated that the BCG vaccine may be an effective barrier against SARS-CoV-2 infection and/or could decrease the severity of COVID-19 sickness. Ecological research suggests that regions and countries that require BCG immunization of individuals have less COVID-19 infections and death (O'Neill and Netea 2020). The SARS-CoV-2 E protein might be the target of an alternative COVID-19 therapeutic approach that induces specific host immunity. A portion of the E protein consisting of residues 17 to 29 showed a significant amount of sequence identity with proteins that mimic tuberculins. These proteins may induce the host immune response (Nuovo et al. 2020).

15.4.4 Membrane (M) Glycoprotein

The structure of the viral envelope is determined by the membrane (M) protein. M protein is a type III glycoprotein which is the most prevalent structural protein of SARS-CoV-2 (de Haan et al. 2000). SARS-CoV-2 M protein has 90.5% similarity with SARS-CoV M protein (Mahtarin et al. 2022). SARS-CoV-2 M protein consists of short N-terminus, larger C-terminal cytoplasmic domain, and triple-transmembrane domain (Thomas 2020). M protein plays a crucial role in coronavirus assembly by interacting with other structural viral proteins. The M protein may be a potential target for vaccine development because humoral and cellular immune responses recognize the M protein as a significant immunogen (Liu et al. 2010). Remdesivir has shown strong binding affinity with M protein (Khan et al. 2021).

15.5 Clinical Manifestation

Eighty per cent of COVID-19 patients have moderate illness, while the severity of their infections varies (Rahman et al. 2022). Severe illness had been reported in 15% of cases that causes dyspnoea, hypoxia, and lung abnormalities, whereas in 5% of cases, the patients were critically ill with respiratory failure by acute respiratory distress syndrome(ARDS), shock, and/or multi-organ dysfunction (Yang et al. 2020). As ACE-2 appears in the lungs, heart, endothelium, renal tubular epithelium, intestinal epithelium, and pancreas, SARS-CoV-2 might have the ability to pass through these tissues, multiply, and affect these organs, resulting in multiple organ dysfunction syndrome (MODS) (Zhao et al. 2020). Increased pro-inflammatory mediators and excessive lymphocyte activation in COVID-19 patients induce immune-mediated damages. Patients with comorbid conditions were more vulnerable to serious infections. Acute ARDS, septic shock, metabolic acidosis, coagulation problems, and MODS had been reported in severe illness (Mehta et al. 2021). Children were less vulnerable to SARS-CoV-2 infection and usually don't have any symptoms. Gastrointestinal problems are another potential symptom of COVID-19. In a study, it was found that 17.6% of patients with COVID-19 had gastrointestinal problems. Furthermore, even after respiratory specimens were negative for SARS-CoV-2 infection, 48.1% of patients had viral shedding in their faeces, which might last for up to or more than 33 days after disease onset (Cheung et al. 2020). Numerous researches had documented COVID-19's neurological symptoms. A brain MRI demonstrated signal intensities in the right mesial temporal lobe and hippocampus along the wall of the right lateral ventricle raising the probability of SARS-CoV-2 meningitis. There are possibilities of SARSCoV-2 infection in unconscious patients, which could result in a horizontal infection (Moriguchi et al. 2020). Additionally, COVID-19 was associated to cutaneous abnormalities. Cutaneous abnormalities can be categorized into five major categories—morbilli form rash (containing macules and papules, resembling measles), urticaria (itchy red welts), vesicles (small blisters), pseudo-chilblains (also known as "COVID toes," painful inflammation of

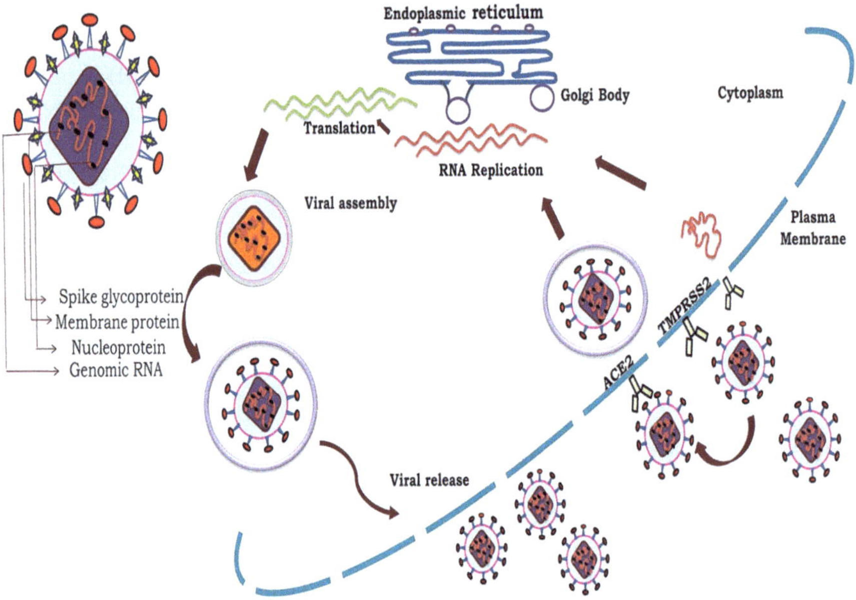

Fig. 15.3 SARS-CoV-2 entries into the host cell

the digits in response to cold), vaso-occlusive lesions (due to thrombosis and occlusion of small arteries, with subsequent ischaemia) (Polly and Fernandez 2022).

15.5.1 Host Immune Response

The host serine protease TMPRSS2 plays an essential role in proteolytic stimulation of the spike protein during the binding of SARS-CoV-2 with receptor and cellular entry (Hoffmann et al. 2020). Angiotensin-converting enzyme 2 (ACE2) is the main cellular receptor for the SARS-CoV-2 spike protein (Zhou et al. 2020; Soni et al. 2020). After binding of SARS-CoV-2 to the cell surface receptor, the viral genomic RNA is released into the host cytoplasm by fusion of viral and host membrane (Fig. 15.3) (Emrani et al. 2021). SARS-CoV-2 internalized into endosome in some cells following cathepsin-mediated cleavage triggered by low pH which facilitate the nucelocapsid entry into the cytoplasm (Bayati et al. 2021). After entry into the cells, host cells generate innate and adaptive immune response against the virus.

15.5.1.1 Innate Immune Response

The first line of defence consists of neutophiles, dedritic cells, monocytes, natural killer (NK) cells, and macrophages. These cells are equipped with different pattern recognition receptors (PRRs) such as toll like receptors (TLRs), nucleotide binding domain, nucleotide binding leucine-rich repeat receptors, absent in melanoma 2

(AIM2), C-type lectin receptors (CLRs), Z-DNA binding protein 1 (ZBP1), retinoic acid inducible gene 1 (RIG-1), and melanoma differentiation-associated protein 5 (MDA5) (Kanneganti 2020). These PRRs recognize pathogen-associated molecular patterns (PAMPs) to induce immune response. SARS-CoV-2 genome is recognized by TLR3, TLR7, and TLR8 or by RIG-1 and MDA5 which activates NF-kappaB and IRF 3 response through mitochondrial antiviral signalling (MAVS) (Wang et al. 2022b). SARS-CoV-2 infections also trigger the induction of NLRP3 inflammososme assembly which causes caspase-1 activation, releasing IL-1β and cleave gasdermin (GSDM) D. This leads to pore formation in plasma membrane and pyroptotic cell death (Ferreira et al. 2021).

15.5.1.2 Adaptive Immune Response

Adaptive immunity generates long-term memory and might recall immunological responses. B cell produces IgM, IgA, IgG, and IgE antibodies which are essential against antiviral response. IgD and IgM is the first antibody to be generated when exposed to SARS-CoV-2 infection (Li et al. 2022). IgG antibodies are produced 7–14 days of post-infection as an immune response due to their high avidity, affinity maturation, and efficacy for neutralizing infections. IgG activate complement system and neutralize infected cells by the mechanism of antibody-dependent cellular cytotoxicity (ADCC) (Galipeau et al. 2020). IgA antibodies, which have higher serum levels than IgM, are produced later. Specific IgA predominates the early neutralizing reaction, and it can be detected before the IgM (Sterlin et al. 2021). IgA play a vital role in mucosal immunity to SARS-CoV-2 infection by aggregation and limiting initial host cell infection (Li et al. 2022). T cell response can also be detected in SARS-CoV-2 infection with CD_4^+ being more noticeable than CD_8^+ T cell response (Grifoni et al. 2020). According to Braun et al. (2020), 83% of infected people had CD_4^+ T cell responses to the spike glycoprotein in their peripheral blood when compared to healthy controls and patients with established SARS-CoV-2 infections. Moreover, 35% of healthy donors possessed CD_4^+ T cells that were spike-reactive and focused on the C-terminal epitopes of the spike protein. According to these findings, the SARS-CoV-2-reactive T cells in normal donors most likely developed as a result of prior exposure to seasonal coronaviruses. This brings up important concerns like the possible impact of cross-reactive immune responses on COVID-19 clinical phenotypes and the resistance of children and young adults to symptomatic SARS-CoV-2 infection (Braun et al. 2020).

15.6 Diagnosis

15.6.1 Lateral Flow Immunoassay

Lateral flow immunoassays (LFIA) have been widely used in the diagnosis of COVID-19, particularly for the detection of viral antigens or antibodies in patient samples. LFIA is a rapid, less expensive, and relatively simple diagnostic method for direct detection of SARS-CoV-2 viral antigens that provides quick results, making it valuable for point-of-care testing and mass screening (Andryukov 2020). It is also known as rapid diagnostic tests (RDTs) (Andryukov 2020).

15.6.1.1 Antigen Lateral Flow Tests

Antigen LFIAs are designed to detect the specific viral antigens, typically the nucleocapsid protein or the spike protein of SARS-CoV-2in nasopharyngeal swab, nasal swab specimens, and other respiratory secretions. The LFIA test strip contains antibodies specific to the viral antigens (Somborac Bačura et al. 2021). When the sample is applied, any viral antigens present in the sample bind to these antibodies, forming a visible line on the test strip, indicating a positive result. The appearance of a control line ensures the validity of the test (Mahmoudinobar et al. 2021).

15.6.1.2 Antibody Lateral Flow Tests

Antibody LFIA is used to detect the specific antibodies produced by an individual's immune system in response to a SARS-CoV-2 infection. These tests are designed to detect either IgM antibodies or IgG antibodies. The LFIA device contains viral antigens that are immobilized on the test strip. When a blood sample is applied to the device, if any specific antibodies present in the blood sample, it will bind to the antigens, resulting in the appearance of coloured lines in the corresponding regions on the test strip. The presence of IgM antibodies indicates a recent or active infection, while the presence of IgG antibodies suggests a past infection (Ragnesola et al. 2020).

Lateral flow immunoassays for SARS-CoV-2 are generally less sensitive as compared to nucleic acid amplification tests (NAATs) such as RT-PCR.

15.6.2 Chest X-Ray

Chest X-ray (CXR) is an imaging technique which was used for diagnosis of COVID-19 patients during the pandemic (Wong et al. 2020). It is less sensitive than computed tomography (CT) but widely used in diagnosis of COVID-19 due to its low cost and widespread availability. SARS-CoV-2 predominantly affects the respiratory system resulting in lung damage and inflammation. With the help of chest X-ray, abnormalities in the lungs of COVID-19 patients like ground glass opacities, consolidation, and interstitial patterns can be detected (Hakami et al. 2023). Such variations have been seen as increased or decreased density on X-ray imaging. The findings of chest X-ray in individual with probable COVID-19 infection can be (1) normal, which is not rare in the earlier stage of the infection, thus cannot be rule out the infection; (2) ordinary, in which patchy distribution and reticular pattern can be seen with higher percentage in lower areas, the spread usually appears bilateral and peripheral; (3) moderate, includes consolidation and ground glass opacities; and (4) anomalous, includes lobar consolidation, masses and lung nodules, pleural effusion, and cavitation (Chamorro et al. 2021).

15.6.3 Chest CT Scan

A computed tomography (CT) scan employs X-rays to generate detailed cross-sectional images of the body. CT scans are also used for the diagnosis COVID-19. It provides valuable insights into lung abnormalities associated with the infection. These insights include the visualization of characteristic features like ground glass opacities (GGOs) and consolidations, allowing for early detection even before symptoms manifest, assessing disease severity, and monitoring disease progression (Wen et al. 2020). Patients with COVID-19 are divided into groups as mid (3–6 days) and late (6–12 days) on the basis of symptoms and the first CT. In certain groups, there is an increased sensitivity to CT scans due to a higher prevalence of COVID-19 pneumonia. The "late" groups were found to have a positive projected value and a significantly higher sensitivity than the earlier (Axiaq et al. 2021). However, it is imperative to utilize CT scans carefully and interpret them in conjunction with other clinical information by experienced radiologists.

15.6.4 RT-PCR

RT-PCR (reverse transcription polymerase chain reaction) is one of the most widely used diagnostic methods for detecting the presence of the SARS-CoV-2 virus infection. It is considered as gold standard for detecting the presence of the SARS-CoV-2 virus (Mohammadi et al. 2022). The first step in RT-PCR-based diagnosis is the collection of samples from the patients. The most common sample for COVID-19 testing is nasopharyngeal swabs. Other samples like saliva, sputum, or respiratory tract aspirates may also be used for detection of SARS-CoV-2 (Xiao et al. 2020). The collected samples are subjected for isolation of RNA. The extracted RNA is converted into complementary DNA (cDNA) with the help of enzyme reverse transcriptase by the process known as reverse transcription (Mohammadi et al. 2022). The cDNA is subjected for amplification with the help of polymerase chain reaction (PCR) by using specific primers. These primers are designed to bind unique regions of the viral genetic material, allowing selective amplification of the target viral RNA. The results are interpreted on the basis of cycle threshold (Ct) value. The cycle threshold (Ct) value represents the number of PCR cycles required for the detection of the target gene. The Ct value provides the information about the viral load in the sample. RT-PCR is widely used for diagnosing COVID-19 due to its high sensitivity and specificity. This process allows the detection of even a small amount of viral genetic material present in the sample (Xiao et al. 2020). However, the accuracy of the test also depends on the quality of the collected sample and the timing of the test in relation to symptom onset and the stage of infection.

15.6.5 Serological Assay

Serological assays also known as antibody tests have been used as diagnostic tools for COVID-19. These tests detect the presence of specific antibodies (IgM or IgG) produced by the immune system in response to the SARS-CoV-2 virus (Whitman et al. 2020). Serological assays can be performed by using different techniques, including enzyme-linked immunosorbent assay (ELISA) and chemiluminescent immunoassay (CLIA) (Ernst et al. 2021). The assay involves the coating of specific SARS-CoV-2 antigens (spike protein or nucleocapsid protein) in microplate well (ELISA) or attached to magnetic beads or microparticles (CLIA). The collected blood sample (serum) is added to the coated surface or mixed with the antigen-coated beads. The sample is incubated to allow the binding of specific antibody in the blood with antigen. A secondary antibody labelled with an enzyme (ELISA) or chemiluminescent or fluorescent marker (CLIA) is added. The secondary antibody is specific and binds with antigen-bound primary antibody. After binding of secondary antibody, a chromogenic substrate is added. The enzyme labelled with secondary antibody generates colour (signal) by acting on chromogenic substrate in ELISA. The signal is proportional to the amount of bound antibodies, indicating the presence of specific antibodies in the sample (Meyer et al. 2020).

15.6.6 CRISPER-Based Diagnosis

CRISPR (clustered regularly interspaced short palindromic repeats) is a powerful gene-editing technology that has revolutionized the field of molecular biology. While CRISPR is primarily known for its gene-editing capabilities, it has also been explored for diagnostic purposes, including the detection of the SARS-CoV-2 virus (Carter et al. 2020). One of the most commonly used CRISPR-based diagnostic methods for COVID-19 is known as SHERLOCK (Specific High-sensitivity Enzymatic Reporter unLOCKing), which combines CRISPR and isothermal amplification methods (Rahimi et al. 2021). SHERLOCK relies on the Cas13 enzyme, which is guided by a CRISPR RNA molecule to specifically bind and cleave viral RNA sequences resulting in a detectable signal confirming the presence of the virus (Rahimi et al. 2021). Another CRISPR-based diagnostic technique is known as DETECTR (DNA Endonuclease-Targeted CRISPR Trans Reporter), which utilizes the CRISPR-Cas12a protein (Broughton et al. 2020). Similar to SHERLOCK, DETECTR detects viral RNA sequences through CRISPR RNA-guided binding and cleavage, leading to a signal generation that indicates the presence of the virus. Both SHERLOCK and DETECTR have shown promising results in terms of sensitivity and specificity for detecting SARS-CoV-2 in patient samples (Safari et al. 2021).

It is important to note that while CRISPR-based diagnostics for COVID-19 hold great potential, their development and validation require rigorous testing and regulatory approval before they can be widely used in clinical settings. Researchers and scientists are working diligently to refine and optimize these CRISPR-based diagnostic approaches to ensure their accuracy, sensitivity, and specificity.

15.7 Treatments and Management of COVID-19

Although there is no specific treatment for SARS-CoV-2 virus, prevention, management, and supportive healthcare to control disease may provide relief in the COVID-19 outbreak.

15.7.1 Antiviral Agents

The antiviral agents effective against the other infections were used for reducing the severity and treatment of COVID-19 during the COVID-19 outbreak.

15.7.1.1 Lopinavir/Ritonavir(LPV/r)

LPV/r (a co-formulation of lopinavir and ritonavir) is the first ritonavir booster. The combination of lopinavir-ritonavir is one of the repurposed drugs currently used in the treatment of COVID-19. Lopinavir (LPV) is a peptidomimetic HIV type 1 aspartate protease inhibitor used in combination with ritonavir (booster) in HIV infection therapy. LPV binds to the catalytic site of aspartate protease inhibitors and prevent the maturation of viral polyprotein precursors into mature, functional proteins, which are required for viral replication. LPV is typically administered along with low booster doses of ritonavir, which slows the metabolism of LPV in the liver by inhibiting the cytochrome P450 3A4 enzyme (Meini et al. 2020).

15.7.1.2 Remdesivir

Remdesivir (GS-5734) was developed by Gilead Sciences along with the US Centres for Disease Control and Prevention (CDC) and the US Army Medical Research Institute of Infectious Diseases (USAMRIID) for treating RNA-based viruses with global pandemic potentials such as Ebola virus (Eastman et al. 2020). Remdesivir belongs to the class of adenosine nucleotides analogue. It inhibits the viral replication by binding with the RNA-dependent RNA polymerase (Warren et al. 2016). Remdesivir has broad-spectrum activity against several virus families, including filoviruses (Ebola) and coronaviruses (SARS/MERS-CoV), and exhibited prophylactic and therapeutic efficacy in nonclinical models of coronaviruses. In vitro, remdesivir exhibits anti-SARS-CoV-2 action with an EC_{50} value of 1.76 μM in Vero E6 cells (Wang et al. 2020b). Due to its broad-spectrum activity, remdesivir is also used for the treatment of COVID-19.

15.7.1.3 Favipiravir

Favipiravir (T-705; 6-fluoro-3-hydroxy-2-pyrazinecarboxamide) is an antiviral agent that inhibits RNA-dependent RNA polymerase (RdRp) of RNA viruses. Intracellular phosphoribosylation of favipiravir leads to the formation of favipiravir

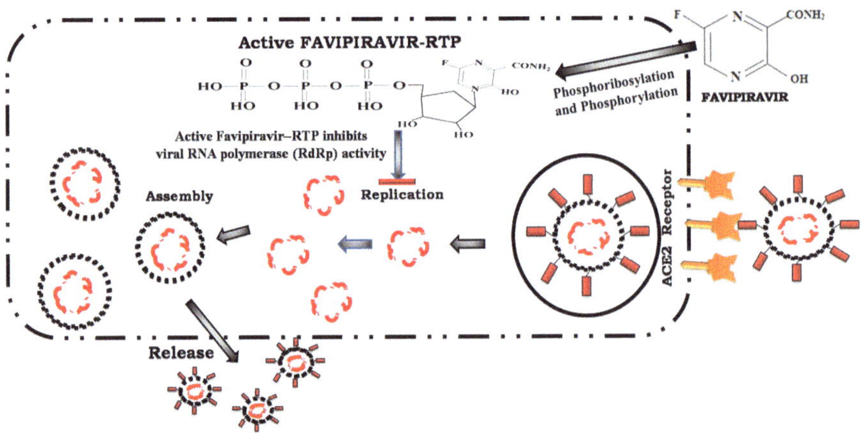

Fig. 15.4 Mode of action of favipiravir

ribofuranosyl-5B-triphosphate (favipiravir-RTP). Favipiravir-RTP inhibits the RNA polymerase activity by participating in viral replication (Fig. 15.4). RdRp of the influenza virus is affected by favipiravir-RTP (with an IC50 of0.022 μg/mL); however, it does not affect the human DNA polymerases α, β, and γ subunits up to a concentration of 100 μg/mL. Not only influenza virus, favipiravir also exhibits inhibitory effects on a wide range of RNA viruses, such as arena-, bunya-, flavi-, and filoviruses causing haemorrhagic fevers (Du and Chen 2020). The genome sequence analysis revealed that beta-coronavirus (2019-nCoV) contains the RdRp gene similar to those of SARS-CoV and MERS-CoV. Therefore, favipiravir is one the possible options for the treatment of COVID-19.

15.7.2 Antimalarial Drugs

The synthetic antimalarial drugs hydroxychloroquine (HCQ) and chloroquine (CQ) were used as therapeutic agents against COVID-19. Several experimental studies have shown positive outcomes due to immunomodulatory properties and in vitro antiviral effect of HCQ/CQ against SARS-CoV-2 (Gautret et al. 2020). Both CQ and HCQ are lysosomotropic and get deposited in acidic vesicles. It was believed that antiviral efficacy of CQ and HCQ against the coronaviruses was due to the well-known lysosomotropic property of these drugs, causing high endosomal pH and interfering with virus cell fusion (Khuroo 2020). Clinical trial has also been performed in which HCQ is used in a combination with azithromycin, zinc supplement, nitazoxanide, lopinavir, ritonavir, or ivermectin. The Government of India, Ministry of Health and Family Welfare Guidelines on clinical management of COVID-19 recommend the administration of hydroxychloroquine in combination with azithromycin (Bajpai et al. 2022). However, recent studies suggest that HCQS and AZ combination have no effect on mortality of COVID-19 patients (Goldman et al. 2021).

15.7.3 COVID-19 Vaccines

The first SARS-CoV-2 genome sequence was published on January 11, 2020, and this data considerably helped in the rapid and major development of COVID-19 vaccine (Kyriakidis et al. 2021). This was the first time in the history of vaccine development that initiate clinical trials for COVID-19 vaccines were initiated in less than 6 months and conditional approval was obtained in less than 10 months. It was made possible in short timeframe due to the quick genome sequencing of SARS-CoV-2, international collaboration among researchers, market demand, and ample financing (Li et al. 2021).

15.7.3.1 Whole Virus Vaccine

The whole virus vaccines can be developed using attenuated or inactivated forms of virus to induce protective immunity. Whole virus vaccines have been developed for immunization against the SARS-CoV-2. There are two forms of whole virus vaccines, live attenuated vaccines and inactivated vaccines. The virus used in live attenuated vaccines is weakened but still capable of growth and replication without being infectious (Nagy and Alhatlani 2021). In inactivated vaccines, the genetic material of the virus is inactivated by heat, chemicals, or radiation. As a result, they cannot infect cells but are able to elicit an immune response (Gao et al. 2020). The inactivated vaccines can only induce humoral responses against SARS-CoV-2 virus whereas live attenuated vaccines may induce both cellular and humoral response (Rahman et al. 2022). In response to inactivated vaccines, host immune system neutralizes and removes pathogens and their toxins by increasing the production of antibodies so that they cannot bind with the cell surface receptor (Zhang et al. 2022). The live attenuated vaccines trigger immune response and facilitate long-term immunity (Fan et al. 2020). Examples of inactivated and live attenuated vaccines against SARS-CoV-2 are COVI VAC (Serum Institute of India), Corona Vac (SinoVac Biotech), VLA 2001 (French biotechnology company Valneva SE), Covaxin (Bharat Biotech Limited) (Table 15.1).

15.7.3.2 Nucleic Acid Vaccines

Nucleic acid vaccines are produced using genetic material (DNA or RNA) of a specific virus in order to induce protective immunity. The DNA vaccines can be developed by inserting an antigen-coding gene into a bacteria-derived plasmid under the control of powerful promoter. The antigen-coding gene can be expressed in the host cell and induce humoral and cellular immunity (Frederiksen et al. 2020). In RNA vaccines, antigen of interest is encoded in mRNA. The RNA vaccines deliver mRNA to the cells where cellular factories utilize this mRNA to make viral proteins (antigens) which elicit an immune response (Ndwandwe and Wiysonge 2021). The first authorized DNA and RNA vaccine for humans against SARS-CoV-2 is the ZyCoV-D

Table 15.1 Ayurvedic, Unani, and Siddha remedies suggested by the Ministry of Ayush

Remedies	Formulation	Herbal plants	References
Ayurvedic	Samshamani Vati	Giloy/Guduchi (*Tinospora cordifolia*) stem	Nesari et al. (2022)
Ayurvedic	Anu Taila	Bela (*Aegle marmelos*), devdaru (*Cedrus deodara*), daruharidra (*Berberis aristata*), teipatra (*Cinnamomum tamala*), Nagar motha (*Cyperus scariosus*), jivanti (*Leptadenia reticulata*), sweta candana (*Santalum album*), jala (*Pavonia odorata*), prishnaparni (*Uraria picta*), kamala kesara (*Nelumbo nucifera*), dalchini (*Cinnamomum verum*), sevya (*Chrysopogon zizanioides*), vidanga (*Embelia ribes*), utpala (*nymphaea nouchali*), *agaru* (*Aquilaria agallocha*), satavar (*Asparagus racemosus*), palwa (*Cyperus platyphyllus*), tilatailya (*Sesamum indicum*), mulethi (*Glycyrrhiza glabra*), brihati (*Solanum indicum*), Kantakari (*Solanum surattense*), surbhi (*Pluchea lanceolata*), salaparni (*Desmodium gangeticum*), renuka (*Vitex agnus-castus*), truti (*Elettaria cardamomum*)	Duraipandi and Selvakumar (2020)
Ayurvedic	Ayurvedic Kadha	Tulsi (*Ocimum tenuiflorum*) Dalchini (*Cinnamomum verum*) Black pepper (*Piper nigrum*) Ginger (*Zingiber officinale*)	Nesari et al. (2022)
Ayurvedic	Ayush-64	*P. kurroa* Royle ex. Benth (*Katuki*) root, *S. chirata* Pexbex. Karst (*Kiratatikta*) whole-plant, *C. crista* L. (*Kuberaksha*) seed, *A. scholaris* R. Br (Saptaparna) bark	Panda et al. (2022)
Ayurvedic	Ayush Kwath	*Tulsi* (*Ocimum sanctum*), *Dalchini* (*Cinnamomum zeylanicum*), *Sunthi*/ginger (*Zingiber officinale*) and *Marich*/black pepper (*Piper nigrum*)	Gautam et al. (2022)
Unani	Arq-e-ajeeb	Camphor Menthol Thymol (*Thymus vulgaris*)	Chen et al. (2013) Taylor et al. (2020) Al-Khalaf (2013)
Unani	Habb-e-Mubarak	*Myrica esculenta* Buch.-ham. ex D. Don *Caesalpinia bonducella* (L.) Roxb. (cotyledon)	Kabra et al. (2019) Arunadevi et al. (2015)
Unani	Khamira-e-Banafsa	*Viola odorata* flower	Gerlach et al. (2019)
Unani	Laooq-e-Katan	*Linum usitatissimum* L. seed	Liang et al. (2019)
Siddha	Ahatodai Manapagu	*Adhatoda vasica* Nees leaves	Singh et al. (2010)

Table 15.1 (continued)

Remedies	Formulation	Herbal plants	References
Siddha	Kabasura kudineer	*Tragia involucrata* L. leaves	Kumar et al. (2020)
Siddha	Thonthasura kudineer	*Trichosanthes cucumerina* L. whole plant	Kumar et al. (2020)

vaccine developed by Cadila in Ahmedabad, Gujarat, India, and BioNTech vaccine developed by Pfizer, respectively (Mallapaty 2021). Other nucleic acid vaccines developed against COVID-19 are ZyCov-D (Zydus Cadila), INO-4800 (Inovio), AG0302-COVID-19 (AnGes), etc. (Table 15.1).

15.7.3.3 Viral Vector Vaccines

Viral vector vaccines use modified version of an unrelated virus as a vector to stimulate the instruction for making antigens into the host cells in order to induce an immune response. These vaccines induce the antibody production and T cell responses because their mode of action mimics a real infection. One of the viral vector vaccine developed by University of Oxford in collaboration with Astra Zeneca for COVID-19 is Vaxzevria (also known as AZD1222, ChAdOx1 nCoV-19, or ChAdOx1). Covishield is also a viral vector vaccine produced by the Serum Institute of India. It is made up of a recombinant, replication-impaired chimpanzee adenovirus (ChAdOx1) vector containing the full-length wild-type SARS-CoV-2 spike protein with an N-tPA (terminal tissue plasminogen activator) signal sequence that transports the protein to the cell surface (Folegatti et al. 2020; Chi et al. 2022). The viral vector vaccines developed for COVID-19 are ChAdOX1-S-/AZD1222/Oxford-Astra Zeneca vaccine (Astra Zeneca and the University of Oxford), Ad5-nCoV/Convidecia (Can Sino Biological Inc.), Brilife (IIBR-100) (NeuroRx, Inc. in collaboration with Cromos, Brilife Georgia, Israel Institute of Biological Research), DelNS1-2019-nCoV-RBD-OPT1 (Beijing Wantai Biological Pharmacy, Xiamen University), etc. (Table 15.1).

15.7.3.4 Protein-Based Vaccines

Protein-based subunit vaccines contain the viral antigenic fragments that are combined with an adjuvant to elicit an immune response (Callaway 2020). These vaccines are more tolerable than mRNA and viral vector vaccines and considerably safer than inactivated virus vaccines (Chi et al. 2022). The first WHO approved protein-based COVID-19 vaccine is Novavax vaccine (NVX-CoV2373) (recombinant spike protein nanoparticle vaccine) sold as Nuvaxovid by Novavax or Covovax by Serum Institute of India (Table 15.2). It consists of recombinant 2P and fully mutated furin cleavage site spike protein. These proteins were combined with a saponin-based adjuvant (Matrix-M) to produce nanoparticles (Dunkle et al. 2022).

Table 15.2 Different types of SARS-CoV-2 vaccine candidates

Type of vaccine	Vaccine name	Company/organization	Target	Efficacy	References
Inactivated virus	CoronaVac	Sinovac biotech China	Inactivated SARS-CoV-2 with aluminium hydroxide	Showed almost 60% efficacy	Tanriover et al. (2021)
Inactivated virus	Covaxin	Bharat biotech limited (India)	Inactivated SARS-CoV-2 with Algel-IMDG	Exhibited 77.8% efficacy against symptomatic individuals and also effectively protect against VOCs	Sapkal et al. (2021)
Inactivated virus	VLA2001	French biotech company Valneva	Inactivated SARS-CoV-2 with alum and CpG 1018	Exhibited humoral and cellular immune response	Lazarus et al. (2021)
Attenuated	COVI-VAC	Serum Institute of India	Attenuated through deoptimization	Stimulates serum and mucosal immune responses	Wang et al. (2021a)
DNA vaccine	ZyCoV-D	Zydus Cadila, India	Plasmid DNA	Exhibited 66.6% efficacy	Khobragade et al. (2022)
DNA vaccine	INO-4800	Inovio pharma, international vaccine institute	PGX0001 expression, antigen: S protein	100% efficacy in phase I US	Sanicas et al. (2020)
DNA vaccine	AG0302 COVID-19	An Ges, Osaka University	Plasmid DNA with alum		Chauhan et al. (2021)
RNA vaccine	BNT162b2	Pfizer BioNTech	LNP	Exhibited 95% based on phase 2/3	Sahin et al. (2021)
RNA vaccine	mRNA-1273	Moderna Tx Inc.	LNP	100% against B.1.1.7, 95.7% against B.1.3.5 in Qatar and 94.1% in the USA	Baden et al. (2021)
Non-replicating vector	ChAdOx1-S-/AZD1222/Covishield	Astra Zeneca and university of Oxford/serum institute of India	Chimpanzee adenovirus vector encoding S protein	66.7% based on phase 4 clinical trial	Voysey et al. (2021)
Non-replicating vector	Ad5nCoV/Convidecia	Cansino biological Inc. and Beijing institute of biotech	Adenovirus type 5 vector expressing S protein	63.7% to 57.5% after phase 3 clinical trial	Halperin et al. (2022)

Non-replicating vector	Sputnik V	Gamaleya research institute of epidemiology and microbiology	Non-replicating vector of adeno (Ad26)	Exhibited 91.6% efficacy after second dose and showed almost 90% efficacy against B.1.617.2	Chavda et al. (2022)
Replicating vector	Brilife (IIBR-100)	CanSino biological Inc. and Beijing Institute of Biotech	RVsV with S protein	NR	Levin et al. (2021)
Replicating vector	DeINS1-2019 nCoV-RBD OPT1	Beijing Wantai biological pharmacy	H1N1 influenza with nonspecified RBD of Sprotein	NR	Chavda et al. (2022)
Recombinant protein vaccine	NVX-CoV2373	America	S protein with matrix-M	Exhibited 89.7% efficacy against SARS-CoV-2 in phase 3	Heath et al. (2023)
Recombinant protein vaccine	NanoCovax	Nanogen pharmaceutical biotechnology JNC. (Vietnam)	Recombinant S protein with alum	Exhibited approx 89.1% efficacy	Nguyen et al. (2022)

15.7.4 Plasma Therapy

Plasma constitutes of water, inorganic salts, and organic molecules including thousands of proteins, such as albumin, immunoglobulins, coagulation and antithrombotic factors, complement components, etc. (Benjamin and McLaughlin 2012). It has been demonstrated that plasma from healthy donors has immunomodulatory effects by inhibiting complement activation, inflammatory cytokines, and autoantibodies (Lünemann et al. 2015). Convalescent plasma (CP) therapy is a type of adaptive immunotherapy that has been used to treat many infectious diseases such as SARS, MERS, and H1N1 with excellent efficacy and safety (Duan et al. 2020). Preliminary studies have reported that the administration of human convalescent plasma is effective in the management of COVID-19 cases (Shen et al. 2020).

15.7.5 Cell-Based Therapy

Several cell-based therapy approaches are being tested to develop effective therapeutics for COVID-19 patients. So far, stem cells, natural killer (NK), dendritic cells (DN), and exosomes have been identified as the most promising candidates for COVID-19 treatment. Natural killer (NK) cells are important because of their ability to lyse virus-infected cells and regulate the immune response (Cancio et al. 2020).The mesenchymal stem cells (MSCs) are a promising therapeutic agent due to their pro/anti-inflammatory and immune-modulating properties (Golchin et al. 2020).

15.8 Herbal Remedies

A phytopharmaceutical preparation derived entirely from a whole plant or portions of a plant (flower, leaves, bark, stem, rhizome, and root) or its exudates (resins, latex, and gums) is referred to herbal intervention (Mehta et al. 2015). Alkaloids, flavonoids, terpenoids, phenols, polyphenols, tannins, saponins, polysaccharides, proteins, lipids, and peptides are some of the active phytochemical constituents which are responsible for the pharmacological effects of herbal formulations (Singh et al. 2014; Singh et al. 2023). These phytochemicals have antipyretic, anti-inflammatory, expectorant, anti-asthmatic, antitussive, and antiviral properties that hold numerous functions against invasion, penetration, reproduction, and expression of the virus (Singh et al. 2013, 2022b). Examples of possible SARS-CoV-2 protease inhibitors include *Azadirachta indica*, *Nigella sativa*, *Allium cepa*, *Allium sativum*, *Aloe vera*, *Mentha piperita*, *Cannabis sativa*, *Curcuma longa*, *Glycyrrhiza glabra*, *Nyctanthes arbor-tristis*, *Ocimum sanctum*, *Withania somnifera*, and *Zingiber officinale* (Abiri et al. 2021).

The main clinical symptom of COVID-19 is fever. Neem (*Azadirachta indica*) leaf tree is traditionally used to treat fevers. The animal studies have shown that it also has anti-inflammatory properties (Eze et al. 2022). Ethanolic extracts of

N. sativa seeds showed antiviral activity by lowering the viral load and alpha-fetoprotein levels and improving liver function parameters in hepatitis C patients (Abdel-Moneim et al. 2013). Garlic (*Allium sativum*) and onion (*Allium cepa*) are common home remedies in Ethiopia for a variety of diseases. The avian influenza virus (H9N2) was treated by using onion, which has long been used in traditional medicine to treat a variety of conditions and infections (Ahmadi et al. 2018). Onion has also been suggested as a possible treatment for COVID-19 patients due to its anti-inflammatory, immunomodulatory, antithrombotic, and antiviral properties (Dorsch and Ring 2021). Some of the herbal remedies suggested by Ayush mantralaya for the prevention of COVID-19 are listed in Table15.1.

15.9 Lesson from COVID-19 Pandemic

The emergence of the COVID-19 pandemic was highlighting the need for rapid response and advancements in the field of healthcare system. In the wake of the COVID-19 pandemic, the world has witnessed unprecedented challenges that have pushed healthcare systems, governments, and societies to their limits. The SARS-CoV-2 virus has proven to be highly contagious and adaptable, requiring innovative and flexible strategies to combat its impact (Ciotti et al. 2020).

15.9.1 Vaccine Development

One of the most significant achievements during the COVID-19 pandemic was the development of vaccine at remarkable speed. This is possible through the utilization of innovative biotechnological techniques. The m-RNA vaccine technology used in the Pfizer-BioNTech and Moderna vaccines represents an innovative approach. The successful deployment of these vaccines underscores the potential of biotechnology in responding to emerging infectious diseases (Wang et al. 2021b).

15.9.2 Diagnostic Tools

Another crucial lesson from the pandemic is the importance of rapid and accurate diagnostic tools. Molecular biology and biotechnology played pivotal roles in the development of various diagnostic tests, such as PCR-based assays and rapid antigen tests. These tools not only enabled efficient screening but also helped in tracking the spread of the virus (Afzal 2020).

15.9.3 Drug Discovery

The search for effective treatments of COVID-19 prompted intensive research towards drug discovery. Biotechnology platforms like CRISPR and high-throughput screening methods facilitated the identification of potential therapeutic candidates. This area also highlighted the need for more robust antiviral drug development pipelines (Safari et al. 2021).

15.9.4 Data Science and Bioinformatics

The pandemic underscored the importance of data science and bioinformatics in understanding the virus's genetic makeup and its evolution. Researchers worldwide collaborated to sequence the virus's genome and share this data openly. This transparency accelerated our understanding of SARS-CoV-2 and enabled the development of diagnostic tools and vaccines (Ma et al. 2021).

15.9.5 Vaccine Distribution

Effective distribution of vaccines presented significant logistical challenges globally. The companies worked on innovative solutions for vaccine storage, transportation, and distribution, ensuring that vaccines reached even remote areas (Asundi et al. 2021).

15.10 Strategies for Future Pandemics

The COVID-19 pandemic has forced the world to rethink its approach to managing global health crises. The COVID-19 pandemic has highlighted the need for a comprehensive and adaptable approach to pandemic management. As we move forward, the strategies employed to combat SARS-CoV-2 will continue to evolve. Vaccination campaigns, advanced testing and contact tracing, public health awareness, international collaboration, and adaptive policy-making are essential components of an effective pandemic response (Sachs et al. 2022). By learning from our experiences with COVID-19, we can better prepare for future challenges and safeguard global health.

15.11 Conclusion

Over the course of the pandemic, considerable improvements have been achieved with COVID-19 diagnosis. The polymerase chain reaction (PCR) test, which detects the virus' genetic material, was initially the main diagnostic method. Further, fast antigen tests were developed for quicker identification and mass screening. Serological testing has also proved crucial in identifying previous infections and monitoring the frequency of antibodies in communities. A multimodal approach involving medical professionals, decision-makers, and the general public was necessary for the management of COVID-19. Vaccination has played a critical role in curbing the virus's spread. Public health initiatives including social isolation, wearing masks, and proper hygiene have been crucial in minimizing transmission rates during the pandemic. The appearance of new variations and the potential for upcoming pandemics show the necessity of continued research, surveillance, and prevention. We can improve our global healthcare systems and response capabilities in order to build a more resilient future by taking lessons from the difficulties and overcomes of the COVID-19 pandemic.

References

Abdel-Moneim A, Morsy BM, Mahmoud AM et al (2013) Beneficial therapeutic effects of *Nigella sativa* and/or *Zingiber officinale* in HCV patients in Egypt. EXCLI J 12:943

Abiri R, Abdul-Hamid H, Sytar O et al (2021) A brief overview of potential treatments for viral diseases using natural plant compounds: the case of SARS-CoV. Molecules 26(13):3868

Afroze SN, Korlepara R, Rao GV, Madala J (2017) Mucormycosis in a diabetic patient: a case report with an insight into its pathophysiology. Contemporary clinical dentistry 8(4):662. https://doi.org/10.4103/ccd.ccd_558_17

Afzal A (2020) Molecular diagnostic technologies for COVID-19: limitations and challenges. J Adv Res 26:149–159

Ahmadi S, Rajabi Z, Vasfi-Marandi M (2018) Evaluation of the antiviral effects of aqueous extracts of red and yellow onions (*Allium Cepa*) against avian influenza virus subtype H9N2. Iranian Journal of Veterinary Science and Technology 10(2):23–27

Alexandar S, Ravisankar M, Kumar RS, Jakkan K (2021) A comprehensive review on Covid-19 delta variant. Int J Clin Pharmacol Res 5:83–85. https://doi.org/10.1101/2021.06.20.21259195v1

Al-Khalaf MI (2013) Thyme and thymol effects on induced bronchial asthma in mice. Life Sci J 10(2):693–699

Andryukov BG (2020) Six decades of lateral flow immunoassay: from determining metabolic markers to diagnosing COVID-19. AIMS Microbiol 6(3):280

Arunadevi R, Murugammal S, Kumar D, Tandan SK (2015) Evaluation of Caesalpinia bonducella flower extract for anti-inflammatory action in rats and its high performance thin layer chromatography chemical fingerprinting. Indian Journal of pharmacology 47(6):638

Asundi A, O'Leary C, Bhadelia N (2021) Global COVID-19 vaccine inequity: the scope, the impact, and the challenges. Cell Host Microbe 29(7):1036–1039

Axiaq A, Almohtadi A, Massias SA, Ngemoh D, Harky A (2021) The role of computed tomography scan in the diagnosis of COVID-19 pneumonia. Curr Opin Pulm Med 27(3):163–168

Baden LR, El Sahly HM, Essink B, Kotloff K et al (2021) Efficacy and safety of the mRNA-1273 SARS-CoV-2 vaccine. N Engl J Med 384(5):403–416

Bai Z, Cao Y, Liu W, Li J (2021) The SARS-CoV-2 Nucleocapsid Protein and Its Role in Viral Structure Biological Functions and a Potential Target for Drug or Vaccine Mitigation. Viruses 13(6):1115. https://doi.org/10.3390/v13061115

Bajpai J, Pradhan A, Verma AK, Kant S (2022) Use of hydroxychloroquine and azithromycin combination to treat the COVID-19 infection. World journal of experimental medicine 12(3):44

Bayati A, Kumar R, Francis V, McPherson PS (2021) SARS-CoV-2 infects cells after viral entry via clathrin-mediated endocytosis. J Biol Chem 296:100306

Belouzard S, Millet JK, Licitra BN, Whittaker GR (2012) Mechanisms of coronavirus cell entry mediated by the viral spike protein. Viruses 4(6):1011–1033. https://doi.org/10.3390/v4061011

Benjamin RJ, McLaughlin LS (2012) Plasma components: properties, differences, and uses. Transfusion 52(1):9S–19S. https://doi.org/10.1111/j.1537-2995.2012.03622.x

Benvenuto D, Giovanetti M, Ciccozzi A, Spoto S et al (2020) The 2019-new coronavirus epidemic: evidence for virus evolution. J Med Virol 92(4):455–459. https://doi.org/10.1002/jmv.25688

Bhatia M (2022) The rise of mucormycosis in Covid-19 patients in India. Expert Rev Anti-Infect Ther 20(2):137–138. https://doi.org/10.1080/14787210.2021.1960822

Braun J, Loyal L, Frentsch M, Wendisch D et al (2020) SARS-CoV-2-reactive T cells in healthy donors and patients with COVID-19. Nature 587(7833):270–274

Broughton JP, Deng X, Yu G, Fasching CL et al (2020) CRISPR–Cas12-based detection of SARS-CoV-2. Nat Biotechnol 38(7):870–874

Burki TK (2020) Coronavirus in China. Lancet Respir Med 8(3):238. https://doi.org/10.1016/S2213-2600(20)30056-4

Callaway E (2020) The race for coronavirus vaccines: a graphical guide. Nature 580(7805):576–577. https://doi.org/10.1038/d41586-020-01221-y

Cancio M, Ciccocioppo R, Rocco PR, Levine BL et al (2020) Emerging trends in COVID-19 treatment: learning from inflammatory conditions associated with cellular therapies. Cytotherapy 22(9):474–481

Cao Q, Chen YC, Chen CL, Chiu CH (2020) SARS-CoV-2 infection in children: transmission dynamics and clinical characteristics. J Formos Med Assoc 119(3):670–673. https://doi.org/10.1016/j.jfma.2020.02.009

Carter LJ, Garner LV, Smoot JW, Li Y et al (2020) Assay techniques and test development for COVID-19 diagnosis. ACS Cent Sci 6(5):591–605

Chai J, Cai Y, Pang C, Wang L et al (2021) Structural basis for SARS-CoV-2 envelope protein recognition of human cell junction protein PALS1. Nat Commun 12(1):1–6. https://doi.org/10.1038/s41467-021-23533-x

Chamorro EM, Tascón AD, Sanz LI, Vélez SO, Nacenta SB (2021) Radiologic diagnosis of patients with COVID-19. Radiología (English Edition) 63(1):56–73

Chauhan N, Soni S, Gupta A, Aslam M, Jain U (2021) Interpretative immune targets and contemporary position for vaccine development against SARS-CoV-2: a systematic review. J Med Virol 93(4):1967–1982

Chavda VP, Bezbaruah R, Athalye M, Parikh PK et al (2022) Replicating viral vector-based vaccines for COVID-19: potential avenue in vaccination arena. Viruses 14(4):759

Chen W, Vermaak I, Viljoen A (2013) Camphor—a fumigant during the black death and a coveted fragrant wood in ancient Egypt and Babylon—a review. Molecules 18(5):5434–5454

Cheung KS, Hung IFN, Chan PPY, Lung KC et al (2020) Gastrointestinal manifestations of SARS-CoV-2 infection and virus load in fecal samples from a Hong Kong cohort: systematic review and meta-analysis. Gastroenterology 159(1):81–95. https://doi.org/10.1053/j.gastro.2020.03.065

Chi WY, Li YD, Huang HC, Chan TE et al (2022) COVID-19 vaccine update: vaccine effectiveness, SARS-CoV-2 variants, boosters, adverse effects, and immune correlates of protection. J Biomed Sci 29(1):1–27

Chin AWH, Chu JTS, Perera MRA, Hui KPY et al (2020) Stability of SARS-CoV-2 in different environmental conditions. The Lancet Microbe 1(1):e10. https://doi.org/10.1016/S2666-5247(20)30003-3

Ciotti M, Ciccozzi M, Terrinoni A, Jiang WC et al (2020) The COVID-19 pandemic. Crit Rev Clin Lab Sci 57(6):365–388

Corman VM, Muth D, Niemeyer D, Drosten C (2018) Hosts and sources of endemic human coronaviruses. Adv Virus Res 100:163–188. https://doi.org/10.1016/bs.aivir.2018.01.001

Corman V, Landt O, Kaiser M, Molenkamp R et al (2020) Detection of 2019 -nCoV by RT-PCR. Euro Surveill 25(3):1–8

da Rosa MR, Francelino Silva Junior LC, Santos Santana FM, Farias de Oliveira T et al (2021) Clinical manifestations of COVID-19 in the general population: systematic review. Wien Klin Wochenschr 133(7–8):377–382. https://doi.org/10.1007/s00508-020-01760-4

de Haan CAM, Vennema H, Rottier PJM (2000) Assembly of the coronavirus envelope: homotypic interactions between the M proteins. J Virol 74(11):4967–4978. https://doi.org/10.1128/jvi.74.11.4967-4978.2000

de Oliveira OV, Rocha GB, Paluch AS, Costa LT (2021) Repurposing approved drugs as inhibitors of SARS-CoV-2 S-protein from molecular modeling and virtual screening. J Biomol Struct Dyn 39(11):3924–3933. https://doi.org/10.1080/07391102.2020.1772885

Dorsch W, Ring J (2021) Anti-inflammatory, anti-thrombotic and antiviral substances from onions could be an option for the treatment of COVID-19: a hypothesis. J Bacteriol Parasitol 12:1000387

Du YX, Chen XP (2020) Favipiravir: pharmacokinetics and concerns about clinical trials for 2019-nCoV infection. Clinical Pharmacology & Therapeutics 108(2):242–247

Duan K, Liu B, Li C, Zhang H et al (2020) Effectiveness of convalescent plasma therapy in severe COVID-19 patients. Proc Natl Acad Sci 117(17):9490–9496

Dunkle LM, Kotloff KL, Gay CL, Áñez G et al (2022) Efficacy and safety of NVX-CoV2373 in adults in the United States and Mexico. N Engl J Med 386(6):531–543

Duraipandi S, Selvakumar V (2020) Reinventing nano drug delivery systems for hydrophilic active ingredients in Ayurvedic lipid based formulations containing poly herbal decoction. Journal of Ayurveda and integrative medicine 11(3):224–227

Eastman RT, Roth JS, Brimacombe KR, Simeonov A et al (2020) Remdesivir: a review of its discovery and development leading to emergency use authorization for treatment of COVID-19. ACS central science 6(5):672–683

Emrani J, Ahmed M, Jeffers-Francis L, Teleha JC et al (2021) SARS-COV-2, infection, transmission, transcription, translation, proteins, and treatment: a review. Int J Biol Macromol 193:1249–1273

Ernst E, Wolfe P, Stahura C, Edwards KA (2021) Technical considerations to development of serological tests for SARS-CoV-2. Talanta 224:121883. https://doi.org/10.1016/j.talanta.2020.121883

Eze MO, Ejike CE, Ifeonu P, Udeinya IJ et al (2022) Anti-COVID-19 potential of *Azadirachta indica* (neem) leaf extract. Scientific African 16:e01184

Fan X, Cao D, Kong L, Zhang X (2020) Cryo-EM analysis of the post-fusion structure of the SARS-CoV spike glycoprotein. Nat Commun 11(1):3618

Ferreira AC, Soares VC, de Azevedo-Quintanilha IG, Dias SD et al (2021) SARS-CoV-2 engages inflammasome and pyroptosis in human primary monocytes. Cell death discovery 7(1):43

Folegatti PM, Ewer KJ, Aley PK, Angus B et al (2020) Safety and immunogenicity of the ChAdOx1 nCoV-19 vaccine against SARS-CoV-2: a preliminary report of a phase 1/2, single-blind, randomised controlled trial. The Lancet 396(10249):467–478

Frederiksen LS, Zhang Y, Foged C, Thakur A (2020) The long road toward COVID-19 herd immunity: vaccine platform technologies and mass immunization strategies. Front Immunol 11:1817

Gabutti G, d'Anchera E, Sandri F, Savio M, Stefanati A (2020) Coronavirus: update related to the current outbreak of COVID-19. Infect Dis Ther 9(2):241–253. https://doi.org/10.1007/s40121-020-00295-5

Galipeau Y, Greig M, Liu G, Driedger M, Langlois MA (2020) Humoral responses and serological assays in SARS-CoV-2 infections. Front Immunol 11:610688

Gao Q, Bao L, Mao H, Wang L et al (2020) Development of an inactivated vaccine candidate for SARS-CoV-2. Science 369(6499):77–81

Gautam S, Gautam A, Chhetri S, Bhattarai U (2022) Immunity against COVID-19: potential role of Ayush Kwath. Journal of Ayurveda and integrative medicine 13(1):100350

Gautret P, Lagier JC, Parola P, Meddeb L et al (2020) Hydroxychloroquine and azithromycin as a treatment of COVID-19: results of an open-label non-randomized clinical trial. Int J Antimicrob Agents 56(1):105949. https://doi.org/10.1016/j.ijantimicag.2020.105949

Gerlach SL, Chandra PK, Roy U, Gunasekera S et al (2019) The membrane-active phytopeptide cycloviolacin O2 simultaneously targets HIV-1-infected cells and infectious viral particles to potentiate the efficacy of antiretroviral drugs. Medicines 6(1):33

Golchin A, Seyedjafari E, Ardeshirylajimi A (2020) Mesenchymal stem cell therapy for COVID-19: present or future. Stem Cell Rev Rep 16:427–433

Goldman JD, Diaz G, Urba WJ (2021) Use of hydroxychloroquine in combination with azithromycin for patients with COVID-19 is not supported by recent literature. Int J Antimicrob Agents 57(1):106174. https://doi.org/10.1016/j.ijantimicag.2020.106174
Grifoni A, Weiskopf D, Ramirez SI, Mateus J et al (2020) Targets of T cell responses to SARS-CoV-2 coronavirus in humans with COVID-19 disease and unexposed individuals. Cell 181(7):1489–1501
Guo L, Ren L, Yang S, Sharma L et al (2020a) Profiling early humoral response to diagnose novel coronavirus disease (COVID-19). Clin Infect Dis:1–28
Guo ZD, Wang ZY, Zhang SF, Li X et al (2020b) Aerosol and surface distribution of severe acute respiratory syndrome coronavirus 2 in hospital wards, Wuhan, China. Emerg Infect Dis 26(7):1586–1591. https://doi.org/10.3201/eid2607.200885
Hakami AM, Altubayqi MH, Altubayqi TH, Refaei EA et al (2023) Radiological coronavirus disease 2019 findings in chest imaging techniques: an update of current evidence. Hail Journal of Health Sciences 5(2):37–45
Halperin SA, Ye L, MacKinnon-Cameron D, Smith B, Cahn PE et al (2022) Final efficacy analysis, interim safety analysis, and immunogenicity of a single dose of recombinant novel coronavirus vaccine (adenovirus type 5 vector) in adults 18 years and older: an international, multicentre, randomised, double-blinded, placebo-controlled phase 3 trial. Lancet 399(10321):237–248
Heath PT, Galiza EP, Baxter DN, Boffito M et al (2023) Safety and efficacy of the NVX-CoV2373 coronavirus disease 2019 vaccine at completion of the placebo-controlled phase of a randomized controlled trial. Clin Infect Dis 76(3):398–407
Hoffmann M, Kleine-Weber H, Schroeder S, Krüger N et al (2020) SARS-CoV-2 cell entry depends on ACE2 and TMPRSS2 and is blocked by a clinically proven protease inhibitor. Cell 181(2):271–280
Jeong W, Keighley C, Wolfe R, Lee WL et al (2019) The epidemiology and clinical manifestations of mucormycosis: a systematic review and meta-analysis of case reports. Clin Microbiol Infect 25(1):26–34. https://doi.org/10.1016/j.cmi.2018.07.011
Kabra A, Martins N, Sharma R, Kabra R, Baghel US (2019) Myrica esculenta Buch.-ham. Ex D. Don: a natural source for health promotion and disease prevention. PLANTS-BASEL 8(6)
Kang S, Yang M, He S, Wang Y et al (2021) A SARS-CoV-2 antibody curbs viral nucleocapsid protein-induced complement hyperactivation. Nat Commun 12(1):1–11. https://doi.org/10.1038/s41467-021-23036-9
Kannan S, Subbaram K, Ali S, Kannan H (2020) Molecular characterization and amino acid homology of nucleocapsid (N) protein in SARS-CoV-1, SARS-CoV-2, MERS-CoV, and bat coronavirus. J Pure Appl Microbiol 14(1):757–763. https://doi.org/10.22207/JPAM.14.SPL1.13
Kanneganti TD (2020) Intracellular innate immune receptors: life inside the cell. Immunol Rev 297(1):5
Khan FI, Kang T, Ali H, Lai D (2021) Remdesivir strongly binds to RNA-dependent RNA polymerase, membrane protein, and Main protease of SARS-CoV-2: indication from molecular modeling and simulations. Front Pharmacol 12(July):1–14. https://doi.org/10.3389/fphar.2021.710778
Khobragade A, Bhate S, Ramaiah V, Deshpande S et al (2022) Efficacy, safety, and immunogenicity of the DNA SARS-CoV-2 vaccine (ZyCoV-D): the interim efficacy results of a phase 3, randomised, double-blind, placebo-controlled study in India. Lancet 399(10332):1313–1321
Khuroo MS (2020) Chloroquine and hydroxychloroquine in coronavirus disease 2019 (COVID-19). Facts, fiction and the hype: a critical appraisal. Int J Antimicrob Agents 56(3):106101. https://doi.org/10.1016/j.ijantimicag.2020.106101
Kumar MP, Sundaram KM, Ramasamy MS (2020) Coronavirus spike (S) glycoprotein (2019-Ncov) targeted siddha medicines Kabasura Kudineer and Thonthasura Kudineer in silico evidence for Corona viral drug. Asian Journal of Pharmaceutical Research and Health Care 19:20–27
Kyriakidis NC, López-Cortés A, González EV, Grimaldos AB, Prado EO (2021) SARS-CoV-2 vaccines strategies: a comprehensive review of phase 3 candidates. npj Vaccines 6(1):8. https://doi.org/10.1038/s41541-021-00292w

Lazarus R, Taucher C, Brown C, Corbic I et al (2021) Immunogenicity and safety of inactivated whole virion Coronavirus vaccine with CpG (VLA2001) in healthy adults aged 18 to 55: a randomised phase 1/2 clinical trial. MedRxiv 17:2021-08

Levin Y, Balakirski NM, Caraco Y, Ben-Ami E et al (2021) Ethics and execution of developing a 2nd wave COVID vaccine–our interim phase I/II VSV-SARS-CoV2 vaccine experience. Vaccine 39(21):2821–2823

Li Q, Guan X, Wu P, Wang X et al (2020) Early transmission dynamics in Wuhan, China, of novel coronavirus–infected pneumonia. N Engl J Med 382(13):1199–1207. https://doi.org/10.1056/nejmoa2001316

Li Y, Tenchov R, Smoot J, Liu C et al (2021) A comprehensive review of the global efforts on COVID-19 vaccine development. ACS Central Science 7(4):512–533. https://doi.org/10.1021/acscentsci.1c00120

Li Q, Wang Y, Sun Q, Knopf J et al (2022) Immune response in COVID-19: what is next? Cell Death & Differentiation 29(6):1107–1122

Liang S, Li X, Ma X, Li A et al (2019) A flaxseed heteropolysaccharide stimulates immune responses and inhibits hepatitis B virus. Int J Biol Macromol 136:230–240

Liu J, Sun Y, Qi J, Chu F et al (2010) The membrane protein of severe acute respiratory syndrome coronavirus acts as a dominant immunogen revealed by a clustering region of novel functionally and structurally defined cytotoxic T-lymphocyte epitopes. J Infect Dis 202(8):1171–1180. https://doi.org/10.1086/656315

Lünemann JD, Nimmerjahn F, Dalakas MC (2015) Intravenous immunoglobulin in neurology—mode of action and clinical efficacy. Nat Rev Neurol 11(2):80–89

Ma L, Li H, Lan J, Hao X et al (2021) Comprehensive analyses of bioinformatics applications in the fight against COVID-19 pandemic. Comput Biol Chem 95:107599

Mahmoudinobar F, Britton D, Montclare JK (2021) Protein-based lateral flow assays for COVID-19 detection. Protein Eng Des Sel 34

Mahtarin R, Islam S, Islam MJ, Ullah MO et al (2022) Structure and dynamics of membrane protein in SARS-CoV-2. J Biomol Struct Dyn 40(10):4725–4738. https://doi.org/10.1080/07391102.2020.1861983

Mallapaty S (2021) India'S DNA COVID vaccine is a FIRST—more are coming. Nature 597:161

Mandala VS, McKay MJ, Shcherbakov AA, Dregni AJ et al (2020) Structure and drug binding of the SARS-CoV-2 envelope protein transmembrane domain in lipid bilayers. Nat Struct Mol Biol 27(12):1202–1208. https://doi.org/10.1038/s41594-020-00536-8

Mehta P, Shah R, Lohidasan S, Mahadik KR (2015) Pharmacokinetic profile of phytoconstituent (s) isolated from medicinal plants—a comprehensive review. J Tradit Complement Med 5(4):207–227

Mehta A, Ratre YK, Sharma K et al (2021) Interplay of nutrition and Psychoneuroendocrineimmune modulation: relevance for COVID-19 in BRICS nations. Front Microbiol 12:769884. https://doi.org/10.3389/fmicb.2021.769884

Meini S, Pagotto A, Longo B, Vendramin I et al (2020) Role of Lopinavir/ritonavir in the treatment of Covid-19: a review of current evidence, guideline recommendations, and perspectives. J Clin Med 9(7):2050

Meyer B, Torriani G, Yerly S, Mazza L et al (2020) Validation of a commercially available SARS-CoV-2 serological immunoassay. Clin Microbiol Infect 26(10):1386–1394. https://doi.org/10.1016/j.cmi.2020.06.024

Mohammadi MR, Omidi AH, Sabati H (2022) Current trends and new methods of detection of SARS-CoV-2 infection. Cell Mol Biomed Reports 2(3):138–150

Moriguchi T, Harii N, Goto J, Harada D et al (2020) A first case of meningitis/encephalitis associated with SARS-Coronavirus-2. Int J Infect Dis 94:55–58. https://doi.org/10.1016/j.ijid.2020.03.062

Mukhra R, Krishan K, Kanchan T (2020) Possible modes of transmission of novel coronavirus SARS-CoV-2: a review. Acta Biomed 91(3):1–11. https://doi.org/10.23750/abm.v91i3.10039

Nagy A, Alhatlani B (2021) An overview of current COVID-19 vaccine platforms. Comput Struct Biotechnol J 19:2508–2517

Ndwandwe D, Wiysonge CS (2021) COVID-19 vaccines. Curr Opin Immunol 71:111–116

Nesari T, Kadam S, Vyas M, Huddar VG et al (2022) AYURAKSHA, a prophylactic Ayurvedic immunity boosting kit reducing positivity percentage of IgG COVID-19 among frontline Indian Delhi police personnel: a non-randomized controlled intervention trial. Front Public Health 10:920126

Nguyen TP, Do Q, Phan LT, Dinh DV et al (2022) Safety and immunogenicity of Nanocovax, a SARS-CoV-2 recombinant spike protein vaccine: interim results of a double-blind, randomised controlled phase 1 and 2 trial. The lancet regional health–Western Pacific 24

Nuovo G, Tili E, Suster D, Matys E, Hupp L, Magro C (2020) Strong homology between SARS-CoV-2 envelope protein and a mycobacterium sp. antigen allows rapid diagnosis of mycobacterial infections and may provide specific anti-SARS-CoV-2 immunity via the BCG vaccine. Ann Diagn Pathol 48:151600. https://doi.org/10.1016/j.anndiagpath.2020.151600

O'Neill LAJ, Netea MG (2020) BCG-induced trained immunity: can it offer protection against COVID-19? Nat Rev Immunol 20(6):335–337. https://doi.org/10.1038/s41577-020-0337-y

Ong SWX, Tan YK, Chia PY, Lee TH et al (2020) Air, surface environmental, and personal protective equipment contamination by severe acute respiratory syndrome coronavirus 2 (SARS-CoV-2) from a symptomatic patient. JAMA - J Am Med Assoc 323(16):1610–1612. https://doi.org/10.1001/jama.2020.3227

Padhi AK, Seal A, Masood J, Ahamed M, Tripathi T (2021) Unraveling the mechanism of arbidol binding and inhibition of SARS-CoV-2 : Insights from atomistic simulations. Eur J Pharmacol 894(December 2020):173836. https://doi.org/10.1016/j.ejphar.2020.173836

Panda AK, Kar S, Rai AK, Rao BC, Srikanth N (2022) AYUSH-64: a potential therapeutic agent in COVID-19. Journal of Ayurveda and Integrative Medicine 13(2):100538

Patrì A, Gallo L, Guarino M, Fabbrocini G (2020) Sexual transmission of severe acute respiratory syndrome coronavirus 2 (SARS-CoV-2): a new possible route of infection? J Am Acad Dermatol 82(6):e227. https://doi.org/10.1016/j.jaad.2020.03.098

Peng Y, Du N, Lei Y, Dorje S et al (2020) Structures of the SARS -CoV-2 nucleocapsid and their perspectives for drug design. EMBO J 39(20):1–12. https://doi.org/10.15252/embj.2020105938

Pervushin K, Tan E, Parthasarathy K, Lin X et al (2009) Structure and inhibition of the SARS coronavirus envelope protein ion channel. PLoS Pathog 5(7):e1000511. https://doi.org/10.1371/journal.ppat.1000511

Polack FP, Thomas SJ, Kitchin N, Absalon J et al (2020) Safety and efficacy of the BNT162b2 mRNA Covid-19 vaccine. N Engl J Med 383(27):2603–2615. https://doi.org/10.1056/nejmoa2034577

Polly S, Fernandez AP (2022) Common skin signs of COVID-19 in adults: an update. Cleve Clin J Med 89(3):161–167. https://doi.org/10.3949/ccjm.89a.21126

Prajapat M, Sarma P, Shekhar N, Avti P et al (2020) Drug targets for corona virus: a systematic review. Indian journal of pharmacology 52(1):56

Prakash H, Chakrabarti A (2019) Global epidemiology of mucormycosis. J Fungi 5(1). https://doi.org/10.3390/jof5010026

Ragnesola B, Jin D, Lamb CC, Shaz BH et al (2020) COVID19 antibody detection using lateral flow assay tests in a cohort of convalescent plasma donors. BMC Res Notes 13(1):1–7

Rahimi H, Salehiabar M, Barsbay M, Ghaffarlou M et al (2021) CRISPR systems for COVID-19 diagnosis. ACS sensors 6(4):1430–1445

Rahman MM, Masum MHU, Wajed S, Talukder A (2022) A comprehensive review on COVID-19 vaccines: development, effectiveness, adverse effects, distribution and challenges. Virus 33(1):1–22. https://doi.org/10.1007/s13337-022-00755-1

Sachs JD, Karim SS, Aknin L, Allen J et al (2022) The lancet commission on lessons for the future from the COVID-19 pandemic. Lancet 400(10359):1224–1280

Safari F, Afarid M, Rastegari B, Borhani-Haghighi A et al (2021) CRISPR systems: novel approaches for detection and combating COVID-19. Virus Res 294:198282

Sahin U, Muik A, Vogler I, Derhovanessian E et al (2021) BNT162b2 vaccine induces neutralizing antibodies and poly-specific T cells in humans. Nature 595(7868):572–577

Sanicas M, Sanicas M, Diop D, Montomoli E (2020) A review of COVID-19 vaccines in development: 6 months into the pandemic. Pan Afr Med J 37(1)

Sapkal GN, Yadav PD, Ella R, Deshpande GR et al (2021) Inactivated COVID-19 vaccine BBV152/ COVAXIN effectively neutralizes recently emerged B. 1.1. 7 variant of SARS-CoV-2. J Travel Med 28(4):taab051

Sheahan TP, Sims AC, Leist SR, Schäfer A et al (2020) Comparative therapeutic efficacy of remdesivir and combination lopinavir, ritonavir, and interferon beta against MERS-CoV. Nat Commun 11(1):222. https://doi.org/10.1038/s41467-019-13940-6

Shen C, Wang Z, Zhao F, Yang Y et al (2020) Treatment of 5 critically ill patients with COVID-19 with convalescent plasma. JAMA 323(16):1582–1589

Singh KP, Upadhyay B, Pra R, Kumar A (2010) Screening of Adhatoda vasica Nees as a putative HIV-protease inhibitor. J Phytology 2(4)

Singh P, Vishwakarma SP, Singh RL (2013) Evaluation of antioxidant, oxidative DNA damage protective and antimicrobial activities of *Foeniculum vulgare* plant. J Med Plant Res 7(35):2551–2563

Singh RL, Vishwakarma SP, Singh P (2014) Vitamins and minerals: roles and plant sources. In: Prakash D (ed) Phytochemicals of nutraceutical importance and their role in human health. CABI-UK, pp 248–265

Singh H, Dahiya N, Yadav M, Sehrawat N (2022a) Emergence of SARS-CoV-2 new variants and their clinical significance. Can J Infect Dis Med Microbiol 2022:1. https://doi.org/10.1155/2022/7336309

Singh P, Singh RL, Pathak N, Singh PK, Tripathi M, Mondal S (2022b) Phytochemistry and nutraceutical properties of *Carica papaya* (Linn.): a review. Diet Suppl Nutr 1(9):1–15

Singh U, Singh P, Singh PK, Shukla S, Singh RL (2023) Evaluation of antioxidant, oxidative DNA damage and hepatoprotective activities of ethanolic leaf extract of *Lagenaria siceraria* plant against acetaminophen induced hepatotoxiocity. Funct Foods Health Dis 13(3):117–134

Somborac Bačura A, Dorotić M, Grošić L, Džimbeg M, Dodig S (2021) Current status of the lateral flow immunoassay for the detection of SARS-CoV-2 in nasopharyngeal swabs. Biochem Med 31(2):230–239

Soni VK, Mehta A, Ratre YK et al (2020) Curcumin, a traditional spice component, can hold the promise against COVID-19? Eur J Pharmacol 886:173551. https://doi.org/10.1016/j.ejphar.2020.173551

Sterlin D, Malaussena A, Gorochov G (2021) IgA dominates the early neutralizing antibody response to SARS-CoV-2 virus. Medicine sciences: M/S 37(11):968–970

Sun CB, Wang YY, Liu GH, Liu Z (2020) Role of the eye in transmitting human coronavirus: what we know and what we do not know. Front Public Heal 8. https://doi.org/10.3389/fpubh.2020.00155

Tanriover MD, Doğanay HL, Akova M, Güner HR et al (2021) Efficacy and safety of an inactivated whole-virion SARS-CoV-2 vaccine (CoronaVac): interim results of a double-blind, randomised, placebo-controlled, phase 3 trial in Turkey. Lancet 398(10296):213–222

Taylor DJ, Hamid SM, Andres AM, Saadaeijahromi H et al (2020) Antiviral effects of menthol on Coxsackievirus B. Viruses 12(4):373

Thomas S (2020) The structure of the membrane protein of sars-cov-2 resembles the sugar transporter semisweet. Pathog Immun 5(1):342–363. https://doi.org/10.20411/pai.v5i1.377

Van Doremalen N, Bushmaker T, Morris DH, Holbrook MG et al (2020) Aerosol and surface stability of SARS-CoV-2 as compared with SARS-CoV-1. N Engl J Med 382(16):1564–1567

Vardhan A, Swetha H, Padmavathi P, Sreelatha P (2022) COVID - 19 : comprehensive review on mutations and current vaccines. Arch Microbiol 204(1):1–17. https://doi.org/10.1007/s00203-021-02606-x

Voysey M, Clemens SA, Madhi SA, Weckx LY et al (2021) Single-dose administration and the influence of the timing of the booster dose on immunogenicity and efficacy of ChAdOx1 nCoV-19 (AZD1222) vaccine: a pooled analysis of four randomised trials. Lancet 397(10277):881–891

Walls AC, Park YJ, Tortorici MA, Wall A et al (2020) Structure, function, and antigenicity of the SARS-CoV-2 spike glycoprotein. Cell 181(2):281–292.e6. https://doi.org/10.1016/j.cell.2020.02.058

Wang Z, Xu X (2020) scRNA-seq profiling of human testes reveals the presence of the ACE2 receptor, a target for SARS-CoV-2 infection in spermatogonia, Leydig and Sertoli cells. Cells 9(4):920
Wang D, Hu B, Hu C, Zhu F et al (2020a) Clinical characteristics of 138 hospitalized patients with 2019 novel coronavirus-infected pneumonia in Wuhan, China. JAMA - J Am Med Assoc 323(11):1061–1069. https://doi.org/10.1001/jama.2020.1585
Wang D, Huang J, Yeung AWK, Tzvetkov NT et al (2020b) The significance of natural product derivatives and traditional medicine for COVID-19. PRO 8(8):937
Wang MY, Zhao R, Gao LJ, Gao XF et al (2020c) SARS-CoV-2: structure, biology, and structure-based therapeutics development. Front Cell Infect Microbiol 10(November):1–17. https://doi.org/10.3389/fcimb.2020.587269
Wang Y, Yang C, Song Y, Coleman JR et al (2021a) Scalable live-attenuated SARS-CoV-2 vaccine candidate demonstrates preclinical safety and efficacy. Proc Natl Acad Sci 118(29):e2102775118
Wang Y, Zhang Z, Luo J, Han X et al (2021b) mRNA vaccine: a potential therapeutic strategy. Mol Cancer 20(1):33
Wang H, Paulson KR, Pease SA, Watson S et al (2022a) Estimating excess mortality due to the COVID-19 pandemic: a systematic analysis of COVID-19-related mortality, 2020–21. Lancet 399(10334):1513–1536. https://doi.org/10.1016/S0140-6736(21)02796-3
Wang M, Zhao Y, Liu J, Li T (2022b) SARS-CoV-2 modulation of RIG-I-MAVS signaling: potential mechanisms of impairment on host antiviral immunity and therapeutic approaches. MedComm–future Medicine 1(2):e29
Warren TK, Jordan R, Lo MK, Ray AS, Mackman RL et al (2016) Therapeutic efficacy of the small molecule GS-5734 against Ebola virus in rhesus monkeys. Nature 531(7594):381–385
Wen Z, Chi Y, Zhang L, Liu H et al (2020) Coronavirus disease 2019: initial detection on chest CT in a retrospective multicenter study of 103 Chinese patients. Radiology: cardiothoracic. Imaging 2(2):200092
Whitman JD, Hiatt J, Mowery CT, Shy BR et al (2020) Test performance evaluation of SARS-CoV-2 serological assays. Nat Biotechnol 38(10):1174
Wong HYF, Lam HYS, Fong AHT, Leung ST et al (2020) Frequency and distribution of chest radiographic findings in patients positive for COVID-19. Radiology 296(2):E72–E78
World Health Organization (2020) A report about health. https://covid19.who.int/. Accessed 20 June 2023
Xia J, Tong J, Liu M, Shen Y, Guo D (2020) Evaluation of coronavirus in tears and conjunctival secretions of patients with SARS-CoV-2 infection. J Med Virol 92(6):589–594. https://doi.org/10.1002/jmv.25725
Xiao AT, Tong YX, Zhang S (2020) Profile of RT-PCR for SARS-CoV-2: a preliminary study from 56 COVID-19 patients. Clin Infect Dis 71:2249
Xu H, Zhong L, Deng J, Peng J et al (2020) High expression of ACE2 receptor of 2019-nCoV on the epithelial cells of oral mucosa. Int J Oral Sci 12(1):1–5. https://doi.org/10.1038/s41368-020-0074-x
Yan W, Zheng Y, Zeng X, He B, Cheng W (2022) Structural biology of SARS-CoV-2: open the door for novel therapies. Signal Transduct Target Ther 7(1):26. https://doi.org/10.1038/s41392-022-00884-5
Yang X, Yu Y, Xu J, Shu H et al (2020) Clinical course and outcomes of critically ill patients with SARS-CoV-2 pneumonia in Wuhan, China: a single-centered, retrospective, observational study. Lancet Respir Med 8(5):475–481. https://doi.org/10.1016/S2213-2600(20)30079-5
Zhang J, Xiao T, Cai Y, Chen B (2021) Structure of SARS-CoV-2 spike protein. Curr Opin Virol 50:173–182. https://doi.org/10.1016/j.coviro.2021.08.010
Zhang Z, Shen Q, Chang H (2022) Vaccines for COVID-19: a systematic review of immunogenicity, current development, and future prospects. Front Immunol 13:843928
Zhao Y, Zhao Z, Wang Y, Zhou Y et al (2020) Single-cell RNA expression profiling of ACE2, the receptor of SARS-CoV-2. Am J Respir Crit Care Med 202(5):756–759. https://doi.org/10.1164/rccm.202001-0179LE
Zhou P, Yang XL, Wang XG, Hu B et al (2020) A pneumonia outbreak associated with a new coronavirus of probable bat origin. Nature 579(7798):270–273

Techniques for the Identification and Characterization of Biomolecules

16

Niharika and Minal Garg

Abstract

Biomolecules or organic molecules including carbohydrates, lipids, proteins and nucleic acids are the important constituents of cellular organelles. These molecules are typically involved in fundamental biological processes such as cell division, development, morphogenesis, maintenance and metabolic activities of different tissues in the body. The primary structures of the above mentioned four principal organic biomolecules comprise the chains of building blocks or monomeric units. These building blocks are arranged and joined together in a peculiar fashion and stabilized by a number of factors including hydrogen bonding, covalent bonding, ionic, stacking and van der Waals interactions. Identification and characterization of biomolecules require simple, cost-effective, less labour-intensive, reliable, reproducible extraction/isolation and purification methods with high throughputs. Advent of automated high-throughput cost-effective technologies such as chromatographic, mass spectrometric, nucleic acid sequencing methods, etc. offers better selectivity and sensitivity with lower limit of quantification and thus enables the surge in growth and scope of analytical techniques in biological research. The present chapter discusses the use of various high-throughput bioanalytical techniques in assessing the sample composition and determination of complex macromolecules.

Keywords

Analytical techniques · Biomolecules · Identification and characterization of biomolecules · Low- and high-throughput techniques · Selectivity and sensitivity

Niharika · M. Garg (✉)
Department of Biochemistry, University of Lucknow, Lucknow, Uttar Pradesh, India
e-mail: garg_minal@lkouniv.ac.in

R. L. Singh et al. (eds.), *Clinical Applications of Biomolecules in Disease Diagnosis*, https://doi.org/10.1007/978-981-97-4723-8_16

16.1 Introduction

Complex biomolecules or organic macromolecules including carbohydrates, lipids, proteins and nucleic acids (RNA, DNA) are the important constituents of cellular organelles. Owing to their vital role in fundamental biological processes such as cell division, development, morphogenesis, maintenance and metabolic activities of different tissues in the body, deducing their structure and properties would help in better understanding of functional attributes in various cellular and metabolic processes. In the early nineteenth century, simple methods of fractionation, separation and addition of precipitation agents were widely used to extract and purify the biological macromolecules in order to determine their structure and characteristics. These methods were complex, time-consuming and labour-intensive and had limited throughputs. Significant progress in the development of advanced analytical technologies results in faster monitoring and detection of macromolecules with more accuracy. These high-throughput analytical techniques are not only automated but less time-consuming, reliable, reproducible, cost-effective and sensitive.

Development of advanced preparative and analytical approaches for separation, identification and quantification of biomolecules is of huge importance in many research fields. It can help the scientists to address many challenging biological issues related to the identification of potential therapeutic and diagnostic biomarkers for personalized treatment, drug discovery and development and food processing operations/industries to improve nutritional composition and thus can bring about a revolution in human health sector. The present chapter discusses the use of various bioanalytical techniques in assessing the sample composition and determination of complex macromolecules.

16.2 Carbohydrates

Carbohydrates are the most abundant class of organic molecules. They are chemically composed of carbon, hydrogen and oxygen in the ratio Cn:H2n:On. Carbohydrates are identified as an important biochemical substance with their central role in the metabolism of all the living organisms. Further as glycoproteins, they are essential in immune response in humans while they possess antioxidant properties as the part of secondary metabolites like flavonoids (Rudd et al. 2001; Davies et al. 2020). They are the important component of human nutrition and contribute to the sweetness, appearance and textural characteristics of many foods. Being digestible, it is an important source of energy whereas indigestible carbohydrate is identified as dietary fibre. The balanced intake of both the types of carbohydrates is beneficial for human nutrition. However, its excess intake results in diseases including diabetes, dental problems and attention deficit hyperactivity disorder (ADHD) (Pieper et al. 2019; Farsad-Naeimi et al. 2020).

Major classes of carbohydrates relevant for human nutrition include sugars (lactose, glucose, maltose, sucrose, fructose, etc.), sugar polyols (mannitol, sorbitol), oligosaccharides (galacto-oligosaccharides, fructo-oligosaccharides), polysaccharides (glycogen, starch) and conjugated carbohydrates (glycoproteins, glycolipids, glycophenols) (Jane et al. 2009). Basic crystalline water-soluble monomeric unit (with five or six carbon atoms) of carbohydrate also called as monosaccharide has carbonyl and hydroxyl groups as reaction centres and exists in aldehyde and or keto forms. Glycosidic linkages among the monosaccharide units form low molecular weight oligosaccharides (monomeric units less than 20) and high molecular weight polysaccharides (monomeric units more than 20). Owing to their major importance and applications in food science, nutrition, human health, fermentation industries, bioanalytics and clinical research, carbohydrates deserve structural and functional characterization. A number of biochemical and analytical methods are developed over the recent period of few decades to achieve exhaustive characterization, concentration and type of simple and complex mixture of carbohydrates.

Extreme polarity and variability observed in carbohydrates make their analysis difficult. Determination of the native sequence of the saccharides requires the cleavage of complex saccharides/oligo- or polysaccahrides into monosacharide units. Besides due to the presence of isomers of monosaccharides, their identification by various analytical methods is more challenging. The subsequent section discusses the various analytical approaches differing in terms of separation performance, sensitivity and reproducibility in the analysis of sugar composition and its profile.

16.2.1 Methods of Carbohydrate Extraction and Purification

Method of carbohydrate extraction or isolation depends upon its type and food matrix type. Traditional methods of extraction are based on solid-liquid extraction (SLE) and liquid-liquid extraction (LLE) and these procedures are not for separating specific bioactive carbohydrate (Fig. 16.1).

The sample is dried under vacuum, ground to fine powder and finally defatted by solvent extraction. Low molecular weight monosaccharides and oligosaccharides are generally extracted by boiling the sample with 80% alcohol solution. The use of microwave heat energy to heat the solvents which are in contact with samples is the principle behind microwave-assisted extraction of carbohydrates. It improves the partitioning of target compounds from the matrix into the solvent and helps in the fractionation of complex saccharides. Fractionation is required in order to produce bioactive simple oligosaccharides and monosaccharides from the complex mixture of carbohydrates. Advancements in carbohydrate fractionation are made over the few years with the aim to reduce the treatment time and enhance the selectivity between oligosaccharides even with the same degree of polymerization and obtain bioactive carbohydrates. Soluble fractions are separated from insoluble fractions by filtration. Both the filtrate and retentate are dried and weighed to determine the concentrations. Treatment of solution with clarifying agents (heavy metal salts like lead acetate) followed by filtration or centrifugation can help in the efficient removal of other components including organic acids, pigments, amino acids, minerals, vitamins, etc. Membrane processes including ultrafiltration, nanofiltration,

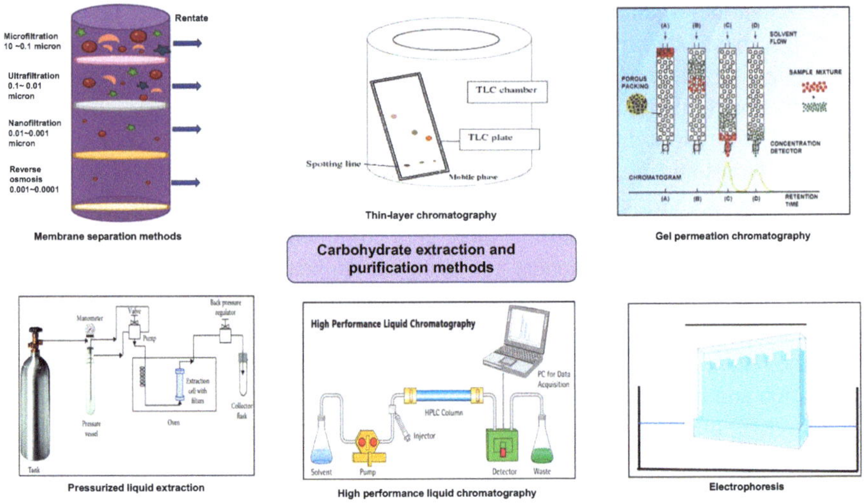

Fig. 16.1 Schematic representation of methods to extract and purify carbohydrates

microfiltration and reverse osmosis are largely used for the purification and or fractionation of bioactive carbohydrates (Akin et al. 2012). These membrane processes employ membranes of varied pore size and work under different operating pressure. Membrane separation process is driven by a pressure gradient, depending upon the solvated size and structure, membrane fractionates and components of the liquid mixture. Certain operating conditions like sample pretreatment, concentration, ionic environment, temperature, pressure and membrane type influence the membrane filtration/separation process. Among the available membrane techniques, ultrafiltration and nanofiltration are the most widely used membrane processes to purify, concentrate and fractionate carbohydrates. The important operating differences between ultrafiltration and nanofiltration are in terms of membrane pore size (1–50 nm versus 2 nm) and operating pressure (1–15 bar versus 10–50 bar).

Solvent-based extraction methods are developed over the last few years. These methods are not only environment friendly but at the same time selectively purify specific bioactive carbohydrate in higher yields. Electrophoresis method can be used to separate the carbohydrates. Prior to gel run, saccharides are first derivatized by reacting them with borates to make them electrically charged. A solution of the derivatized carbohydrates when electrophoresed are separated on the basis of their size where the smaller its size, the faster it runs in an electrical field.

16.2.2 Carbohydrate Purification and Detection Methods

Chromatographic methods like ion exchange resin columns are efficiently used in the separation of polar non-charged carbohydrates from charged molecules. Other chromatographic methods like gas chromatography (GC), high-performance liquid chromatography (HPLC) and thin layer chromatography (TLC) are commonly used to separate and identify different types of carbohydrates (Fig. 16.2a).

Separation of carbohydrates by chromatographic methods depends upon their differential adsorption characteristics, polarities or sizes, partition coefficients and type of column. For the separation of highly polar monosaccharides by reverse phase liquid chromatography (RPLC), monosaccharides require prior derivatization either with 1-naphthylamine or 1-phenyl-3-methyl-5-pyrazolone followed by their subsequent analysis using C18 stationary phases (Rakete and Glomb 2013; Xu et al. 2017). Due to low thermal instability and low volatility of saccharides, various

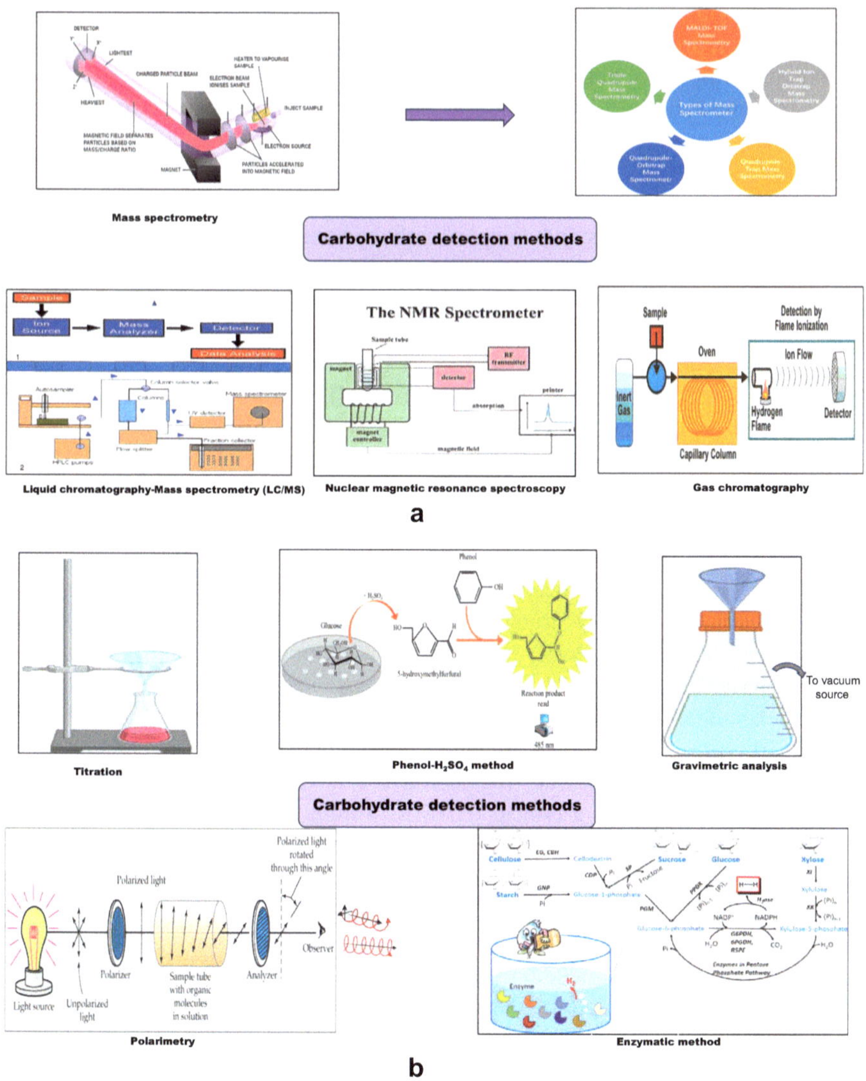

Fig. 16.2 (**a**) Methods to detect carbohydrates. (**b**) Methods to detect carbohydrates

derivatization approaches, including silylation or methylation, are applied prior to their GC analysis (Ruiz-Matute et al. 2011). Supercritical fluid chromatography (SFC), a hybrid of liquid chromatography (LC) and gas chromatography, is commonly employed in the analysis of polar metabolites including carbohydrates. This SFC method is similar to normal phase techniques and requires no derivatization. In addition to it, the use of carbon dioxide as a mobile phase reduces environmental impact (Desfontaine et al. 2018; van de Velde et al. 2020; Losacco et al. 2020). Detectors to detect the separated carbohydrates are specific for the respective chromatographic method. Mass spectrometers are the universal detectors for LC and GC whereas flame ionization detector (FID) is used for GC and SFC. Lack of chromophore in carbohydrates limits their detection by ultraviolet/visible (UV/Vis) detectors. However, prior addition of chromophore by pre- or post-column derivatization or oxidation improves their detection efficacy by UV/Vis detectors (Rühmann et al. 2014; Domingues et al. 2014). Similarly labelling of carbohydrates with fluorescent tags allows them to be detected by fluorescent-based detector systems. Detection by refractive index (RI) is limited to isocratic separations.

Detailed and complete analysis of glycosidically linked oligosaccharides, polysaccharides, glycoproteins and complex carbohydrates requires acid hydrolysis for their release into monosaccharides. Non-target approaches are required to determine the size, sequence and linkages. Matrix-assisted laser desorption/ionization coupled to mass spectrometry and size exclusion chromatography (SEC)/gel permeation chromatography (GPC) are frequently used to determine the size of saccharides (Ras et al. 2011; Ciric et al. 2012). Ion mobility mass spectrometry (IM-MS) and tandem mass spectrometry (MS/MS) are available methods for studying the sequence of the fragments of complex carbohydrates (Mariño et al. 2010; Hofmann et al. 2015). Identification of glycosidically bound complex saccharides in samples can be done by LC-IM-MS coupling method followed by database matching (Gray et al. 2019) (Fig. 16.2a).

Nuclear magnetic resonance (NMR) is currently a popular method for accurate characterization of primary structure, conformation, stoichiometry of substituents, ratios of saccharides and purity of carbohydrates (monosaccharides, oligosaccharides, polysaccharides and glycoconjugates) extracted from natural or synthetic sources. A suite of 1D, 2D and 3D experiments are generally run in a modern high field NMR typically 500 MHz or higher to determine the structure of carbohydrates. Chemical shifts for carbohydrate ring photons are observed as 3–6 ppm (4.5–5.5 ppm for anomeric carbons) in a typical ^{1}HNMR. Simple mono- and oligosaccharides show separate proton signals while bigger and complex molecules exhibit significant overlap in the signals especially in the non-anomeric region (3–4 ppm). Nevertheless, ^{13}CNMR overcomes this limitation by providing wider range of chemical shifts, blocking carbon-proton spin coupling and allowing all carbon signals high and narrow singlets to distinguish from each other. Stereo-orientation of protons relative to other protons in a sugar ring can be examined by studying the vicinal proton-proton coupling constants ($^{3}J_{HH}$) whereas vicinal hetero-nuclear H-C-O-C coupling constants tell about the molecular conformation by providing information about the torsional angles along glycosidic bond between sugars or along exocyclic fragments. Further, nuclear Overhauser effects (NOEs) which are sensitive to interatomic distances help in the optimization of molecular geometry (Fig. 16.2a).

Titrations, gravimetric and colorimetric are the simple traditional chemical methods employed to quantify the saccharides (Fig. 16.2b). Lane-Eynon method of titration is based upon the titration of carbohydrate with copper sulfate solution in the presence of methylene blue as an indicator. It helps in the determination of concentration of reducing sugars in the sample. This method does not distinguish different types of reducing sugars and is also subjected to interference from other types of reducing agents. Munson and Walker gravimetric method is based upon the formation of copper oxide precipitate when carbohydrate is oxidized in the presence of heat and treated with an excess of copper sulfate and alkaline tartrate. The precipitate is filtered, dried and weighed, and its amount is directly related to the concentration of reducing sugar in the sample. Treatment of sugars with anthrone reagent under acidic conditions yields a blue green colour product. Absorbance at 620 nm tells about the concentration of total sugars in the sample. Phenol-sulfuric acid method is another common colorimetric method where absorbance at 420 nm provides information about total sugar concentration in the sample.

Enzymatic methods are rapid, highly specific and sensitive to low concentrations and are used to determine carbohydrates based upon the type of enzyme and substrate. This method depends upon measuring the concentration of the product upon reaction completion where product concentration is directly related to initial substrate in the sample. Enzymatic method also depends on measuring the initial rate of the enzyme catalysis as the rate is proportional to the concentration of substrate in the sample (Fig. 16.2b). Commonly used physical methods to determine the concentration of carbohydrates in food are polarimetry, refractive index, IR and density. These methods are based upon reading the changes in some physicochemical properties with change in carbohydrate concentrations. Polarimeter determines the extent of polarization of an optically active carbohydrate in terms of angle of rotation which varies with the concentration of optically active sugar residues. Refractive index of the given sugar sample can be identified by measuring the angle of refraction and angle of incidence at a boundary between the analyte and another material of known refractive index. This method is quick and simple to perform and is subjected to vary with changes in temperature and wavelength. Density of aqueous carbohydrate increases with its concentration, and this is the routine method to determine carbohydrate concentrations in juices and beverages. Carbohydrates have the property to absorb infrared radiation where none of the other major food constituents absorb it; hence, the concentration of carbohydrate is determined by measuring the intensity of an infrared wave reflected from the surface of a sample. It is a simple, non-destructive technique popularly used for on-line analysis and in a quality control laboratory to routinely analyse many samples. NMR and mass spectrometry as mentioned above are more sophisticated methods to determine the structure and concentration of saccharides in a sample.

16.3 Proteins, Antigens, Antibodies and Enzymes

Proteins are the large complex biomolecules made up of long chains of amino acids and are important for the structure, function and regulation of cells, tissues and organs. Protein folding giving rise to specific secondary, tertiary and quaternary structure determines its activity and function. Many proteins as enzymes catalyse vital metabolic reactions, while few proteins perform structural and mechanical functions like actin and myosin in muscles and in cytoskeleton that maintains the cell shape by forming a system of scaffolding. Other vital functions of proteins include cell adhesion, cell signalling, cell cycle and immune responses.

16.3.1 Methods to Isolate and Purify Proteins

Antoine Fourcroy and coworkers in the eighteenth century identified proteins as distinct class of biomolecules which had the ability to coagulate upon treatment with acid and heat. Continual expansion in the protein-based applications and commercial exploitations require the need for the development of effective and efficient methods of separation and purification (Marichal-Gallardo and Alvarez 2012). Separation and purification of proteins while maintaining biological activity and chemical integrity of the polypeptide are very challenging. The choice of separation and purification methods depends upon the differences in structural, functional and chemical properties of the target proteins and other proteinaceous/non-proteinaceous components. Based on the features like amino acid sequence, primary, secondary, tertiary and quaternary structures, density, solubility, size, shape, charge distribution, isoelectric point, hydrophobicity, metal binding, ligand-binding affinity, reversible association and posttranslational modifications, several fractionation and chromatographic procedures are exploited to purify the proteins in a crude mixture (Kallberg et al. 2012) (Fig. 16.3).

Initial stages of purification procedures such as fractional precipitation and two-phase partition systems are high-capacity, low-resolution methods that yield large amounts of proteins. Low-capacity (low protein yield) and high-resolution chromatographic methods are generally employed at later stages of purification scheme (Chon and Zarbis-Papastoitsis 2011; Rosa et al. 2011). Protein extraction requires the disintegration of the sample source (tissues or cells) which can be achieved by various non-mechanical method sand mechanical methods. Non-mechanical methods include the use of chemical (acid/alkali), detergents (Triton X-100, sodium dodecyl sulfate (SDS) and ethyl dimethyl ammonium bromide (EDAB)), enzymatic treatment (like cellulases, pectinases, xylanases, chitinases and lysozyme) and osmotic shock (hypertonic/hypotonic). Mechanical methods include the use of solid shear methods (bead mill method and French press) or liquid shear methods (repeated freezing and thawing, sonication, homogenization by high pressure). Following disintegration, it is extracted and precipitated for further purification. Phenol and chloroform/methanol dissolve hydrophobic proteins and lipids and leave water-soluble matter (carbohydrates, nucleic acids, etc.) in the aqueous layer.

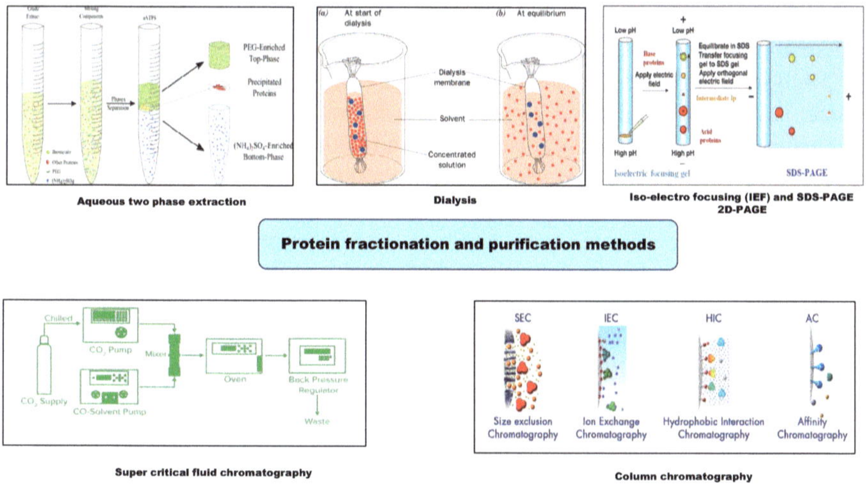

Fig. 16.3 Protein fractionation and purification methods

Soluble liquid extracts are separated from the solid cellular debris or subcellular organelles by centrifugation. Size, shape, density of the particles, viscosity of the medium and the speed of rotation influence the separation of lighter particles from the heavier ones by centrifugation method.

Early purification of proteins is carried out by precipitating the solubilized proteins followed by re-solubilization and desalting. Precipitation is based upon the principle of altering the solvation potential of the solvent, thus lowering the solubility of solute by adding the reagent. Knowledge of amino acid composition of a protein (high hydrophobic or hydrophilic amino acids) helps in selecting the ideal precipitation solvent and method. Depending upon the solubility characteristics, many organic solvents/salts including acetone, ethanol, polyethylene glycol, polyethyleneimine and ammonium sulfate are used to fractionate the proteins by precipitation (Bell et al. 1982). Degree of precipitation depends upon salt concentration, pH and concentration of the material to be precipitated. Owing to low cost, high solubility (up to 4 M) and stabilizing effect on the activity of many enzymes, ammonium sulfate is the preferred choice. The charged polymer polyethyleneimine (PEI) at slightly alkaline pH precipitates macromolecules containing acidic domains by forming ionic complexes and removes nucleic acids and nucleoproteins from crude extracts. This is followed by ammonium sulfate precipitation and chromatography. Early extraction/purification of enzymes/proteins is done on a large scale using aqueous two-phase systems. Depending upon the behaviour of hydrophilic polymers in solution, cell debris and the desired proteins are partitioned into opposite phases. Partition selectivity is increased by covalent attachment of an affinity ligand to the target protein. Affinity extraction using PEG-dextran systems at low ionic strength is the popular method for the large-scale extraction of proteins (Cordes and Kula 1986). Finally, addition of salt dissociates the complex leaving ligand to the top phase and cleaner target protein to the bottom phase. Separation and purification of proteins using two aqueous two-phase extraction systems seems to be promising alternative to expensive chromatography (Fig. 16.3).

Dialysis, a purification method based on the concentration gradient as the driving force, is generally combined with other separation and purification methods like centrifugal separation and chromatography and salting out to improve protein purity (Balashova et al. 2018; Vidya et al. 2020). Protein solution suspended in a larger volume of buffered solution is placed inside a semi-permeable membrane (dialysis bag). Membrane permeates water and ions but retains protein of interest. Another method for protein desalination, dealcoholization and enrichment is ultrafiltration which utilizes pressure or centrifugal force to purify the proteins through selective filtration of solute by a membrane (Resende et al. 2013; Zhou et al. 2015; Thammasena et al. 2020). This method is limited to isolate proteins with relatively close molecular weight (Doyen et al. 2011). Hydrophobic membranes like polysulfone and polyethersulfone are commonly used for ultrafiltration and are commercially available in the market. Due to the adsorption and deposition of protein on the membrane surface and subsurface, the permeation flux of the membrane is reduced which affects the efficacy of protein separation (Rajesh et al. 2011; Xing et al. 2012). Nevertheless, surface coating, blending inorganic fillers, graft copolymerization or hydrophilic polymers can be used to modify the hydrophobic membrane and improve the antifouling performance of the membrane (Aktas and Acar 2019) (Fig. 16.3). The most powerful method of protein fractionation and its purification from the mixture components is column chromatography. Size exclusion or gel permeation, ion exchange (anion exchangers or cation exchangers) and affinity chromatography are the three most common methods which work on the principle of size, charge and affinity, respectively. Different proteins exhibit differential migration patterns depending on their properties and those of the resin (solid phase) when passes through the column along with mobile phase. Separation and purification of the eluted protein fractions take place according to the retention time of different proteins. This can be followed by their detection at 280 nm absorbance.

Gel permeation or size exclusion chromatography separates the proteins differing in sizes and shapes. Proteins with smaller average pore size than the gel beads take long movement paths and suffer great resistance from inside the gel beads, hence eluting in greater time period compared to large proteins (Caltabiano et al. 2018; Burgess 2018). High binding capacity of hydroxylapatite, a crystallized derivative of calcium phosphate, makes hydroxylapatite chromatography a popular method to isolate proteins including enzymes, glycoproteins, lipoproteins, chromatin proteins, phosphoproteins and lectins. Adsorption of proteins on hydroxylapatite matrix via ionic interactions between carbonyl and phosphate groups on the surface of native proteins followed by its elution by increasing gradients of phosphate concentrations (e.g., 0.01 to 0.50 M) is the basis of protein separation and purification by hydroxylapatite chromatography (Fig. 16.3). The matrix of ion exchanger (commonly used resins or cellulose) grouped as anion (diethylaminoethyl (DEAE), quaternary aminoethyl (QAE)) or cation (carboxymethyl (CM), sulfoxypropyl (SP)) ionically interacts with the protein fractions with opposite charges present in the buffer solution. Gradient of increasing salt concentration (disrupts ionic and hydrogen binding) in the buffer or changing the pH of the buffer when applied to the column allows the most weakly interacting proteins to release and elute first and subsequently followed by the more strongly interacting proteins.

Affinity chromatography is based on the principle of purifying the specific protein of interest among the mixture of protein fractions. It allows the binding of specific protein with a tag (tag can be engineered into the sequence) with the specific ligand already coupled to the column resin/matrix. Binding of the most popular engineered tag, 6xHis tag, 6 His residues at either the N- or C-termini of the protein with the divalent metal as a ligand on the matrix is the basis of metal affinity chromatography. Matrix can be coupled to other commonly used ligands such as antibody or NAD+. Protein sample when loaded onto the matrices (bound with ligands) are specifically retained due to strong interactions while other non-interacting proteins are removed from the column. Altering the pH, adding salt, adding a metal chelator or a high concentration of the ligand further helps in the elution of protein of interest. Supercritical fluid chromatography (SFC) is considered as an indispensable purification technology in food and pharmaceutical industries (Manjare and Dhingra 2019). This technique of separation of chiral and achiral small molecules of drugs and proteins is similar to that of gas or liquid chromatography. SFC facilitates the use of supercritical fluid like CO_2 as a mobile phase as it behaves like a compressible fluid with low viscosity and allows the use of high linear velocities. As a result, mass transfer operates rapidly which shortens the time of analysis without loss of efficiency (Fig. 16.3).

Sodium dodecyl sulfate-polyacrylamide gel electrophoresis (SDS-PAGE) separates the protein components of the mixture based on their size and/or charge followed by their visualization either using silver stain or Coomassie Brilliant Blue dye. SDS not only imparts negative charges to the amino acids in the proteins in a uniform charge/mass ratio but also denatures the protein by removing 2°, 3° and 4° structures. Protein sample is most often also treated with ß-mercaptoethanol (ß-ME) to break any existing disulfide bonds and retain only the primary structure of the proteins. Sometimes Western blotting, the more sensitive method of detection of proteins, is performed where the gels are transferred to a solid support (nitrocellulose) and then probed with specific antibodies. Paper electrophoresis is generally used to separate amino acids or peptides of differing charge where the highly charged species move the furthest (Fig. 16.3).

Other than Western blotting, ELISA (enzyme linked immunosorbent assay) and immunohistochemical staining detect the presence and quantity of target antigens based on their binding with specific antibodies. Direct ELISA/indirect ELISA/sandwich ELISA/competitive ELISA rely on the principle of immobilization of antigen on the solid surface either directly or via the use of capture antibody followed by the binding of antigen to detection of antibody conjugated with molecule amenable for detection such as an enzyme or fluorophore. It is widely used to detect and quantify the presence of antigens in serum, plasma, cellular and tissue extracts and other body fluids including urine and saliva. Radioimmunoassay works on the principle of competitive binding of radioactive antigen (tracer) with non-radioactive antigen for the fixed number of antibody or receptor binding sites. It detects and quantitates the amount of analytes/antigen in the sample. Like ELISA and RIA, immunohistochemistry (IHC) is a powerful technique that exploits the specific binding of antigen with detection of antibody. It helps in the detection of localization of specific/target antigens in cells and tissues by examining under the bright-field microscope (Fig. 16.4a).

Isoelectric focusing (IEF) electrophoresis is usually performed on proteins in a gel and separates the protein components on the basis of their isoelectric point (pI) which is a function of the pH of its surroundings. Protein lying in a pH region above

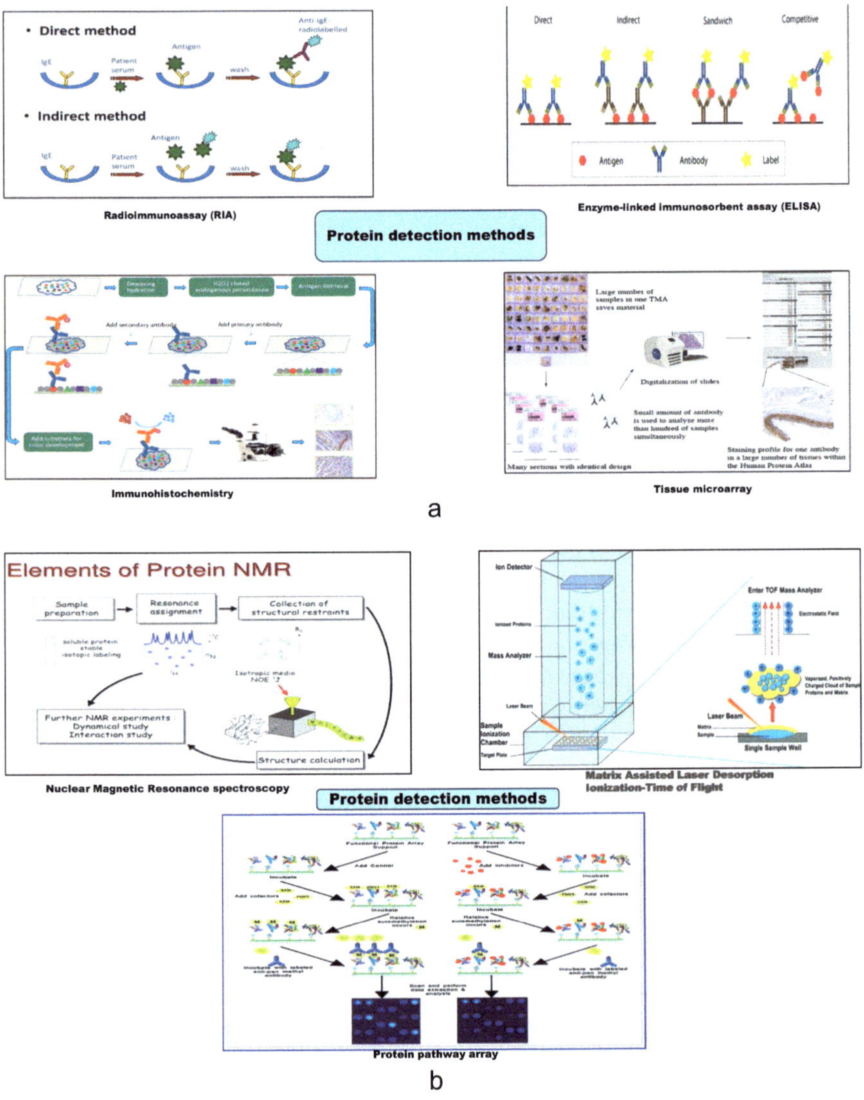

Fig. 16.4 (**a**) Methods to detect proteins, (**b**) Methods to detect proteins

its pI is negatively charged and will migrate towards the positive anode until it reaches the pH region that corresponds to its pI. At this point migration ceases and proteins with differing pIs form sharp stationary bands. 2D gel electrophoresis is a widely used method for protein analysis where it separates and fractionates thousands of proteins in one time. This method utilizes two properties (pIs and molecular weight) of proteins and separates them in two dimensions on gels. It begins with an IEF gel (in a tube/column) to separate proteins based on their pIs followed by separation on the basis of mol. wt. by SDS-PAGE gel (placing IEF gel 90 degrees to SDS-PAGE gel). Proteins labelled with different cyanine dyes such as Cy2b, Cy3 and Cy5 as reporter fluorophores when separated on 2D-PAGE are detected by fluorescence scanner and the image analysis (Morak et al. 2009). Significant spots in the gel are cut out and enzymatically digested to peptides and are taken for further analysis by MS (Fig. 16.3).

16.3.2 High-Throughput Methods to Detect Proteins/Amino Acids

High-performance liquid chromatography or reverse phase liquid chromatography with strong versatility and strong analytical capabilities is widely used to quantify small molecules and ions and to separate and purify large molecules. The basis of chromatographic separation and purification of proteins is the differential affinity between the protein mixture in the solution (mobile phase: acetonitrile, methanol, isopropanol or tetrahydrofuran) and the packing characteristics of the chromatographic column (stationary phase: butyl, C-4; octyl, C-8; octadecyl, C-18). The time taken by the different proteins to flow out of the chromatographic column is different, and thus they are separated from the rest of the mixture components. HPLC is one of preferred methods of peptide separation. Minor differences in their amino acid composition greatly affect their elution profile; nevertheless, prediction of elution order is often limited by secondary and tertiary structures as it may prevent some hydrophobic amino acid residues to interact with the support (Hermodson and Mahoney 1983). The other main drawbacks include poor reproducibility of separations and loss of native structure and biological activity of protein due to use of standard organic solvents and acidic conditions of the mobile phase and hydrocarbon bonding in the stationary phase (Benedek et al. 1984; Cohen et al. 1985).

Rapid development of high-throughput techniques over the recent years including mass spectrometry, protein pathway array and next-generation tissue microarrays are widely used in proteomic and metabolomics. Mass spectrometry is one of the most essential and popular tool to detect an intact protein or a subset of composite or surrogate peptides. It identifies proteins, their isoforms, fragments and its associated specific proteolytic activity and quantifies posttranslational modifications (Nilsson et al. 2010; Li et al. 2017; Macklin et al. 2020). Many multiple separations and prefractionation techniques like 2D-PAGE and LC are combined with MS analysis for improved detection accuracy and yields. Matrix-assisted laser desorption/ionization-time of flight (MALDI-TOF) MS is used to identify the protein when treated with either trypsin or any other proteases by means of peptide fingerprinting (PMF) (Fig. 16.4b).

LC can be combined with tandem MS equipped with ion trap, quadrupole chamber for fragmentation or hybrid quadrupole-time of flight (Q-TOF) apparatus (LC-MS/MS) for high throughputs. Based on different strategies of processing, MS methods can be of top-down, bottom-up and shotgun type. Top-down proteomics method utilizes the fragmentation of full-length protein inside the MS followed by its analysis. Peptides obtained after chemical or enzymatic digestion of proteins serve as input to the MS equipment in bottom-up method. In shotgun proteomics technique, whole proteins in a complex mixture, such as serum, urine and cell lysates are cut into peptides followed by its HPLC-MS analysis to generate a global profile of protein mixtures as genome "shotgun" sequencing (Sechi and Oda 2003) (Fig. 16.4b).

Protein pathway array (PPA), a gel-based high-throughput technique has been applied to many diseases including essential thrombocythaemia and papillary thyroid carcinoma (Hui et al. 2013; Huang et al. 2016). It detects multiple antigens in a protein sample extracted from biopsy or tissue using antibody mixtures (Zhang et al. 2009). The immunofluorescence signals of antibody-antigen reactions are converted into numeric data to quantify the protein expression (Fig. 16.4b). Due to non-specific stain or background noise, IHC offers qualitative or semiquantitative assessment of protein expression. Besides this, it is a subjective process and time-consuming. With technology advancements, tissue microarray (TMA) is gradually replacing IHC and gaining popularity in research and clinical fields (Jawhar 2009; Kim et al. 2016). TMA is based on the large-scale antibody-based molecular analysis simultaneously on small representative tissue cores of formalin-fixed paraffin-embedded (FFPE) or frozen blocks from hundreds of cases/samples assembled in an array fashion on a single histologic slide (Jawhar 2009) (Fig. 16.4b).

16.4 Lipids

Lipids are the major cellular constituents that play important roles in cell recognition, membrane formation, cell signalling and energy storage. A variety of lipid molecules present in the cell differ in the nature of the backbone, head groups, number of fatty acids, length of the chain, number of double bonds and the chemical moieties they are modified with. They display high structural diversity and complexity, and in combination with proteins they can be used to identify and differentiate microbial species.

Gram-positive and Gram-negative bacterial species display major differences in the cell wall structure and the membrane composition. Nevertheless, species belonging to the same Gram type exhibit differences in lipid composition also (Sohlenkamp and Geiger 2016). Glycerophospholipids such as phosphatidylethanolamine and phosphatidylglycerol are the commonly found lipids in many organisms. In addition to it, bacteria possess species-specific lipids with high level of structural variability. Lipopolysaccharide (LPS), a glycolipid, is a major structural component of the outer wall in Gram-negative species. The three parts of LPS include (i) hydrophobic lipid A (contributes toxicity to some extent), (ii) hydrophilic core oligosaccharide and (iii) O antigen as the outermost domain of LPS. The O antigen, a polysaccharide, is considered as the most diverse part of LPS in terms of structure and composition among the strains of species and confers antigenicity/pathogenicity (Liu et al. 2014). Other than glycolipids and glucolipids, lipoteichoic acid (LTA) is a typical constituent of the cell membrane in Gram-positive bacteria. LTA is an alditol phosphate-containing polymer linked by a lipid anchor to the membrane. Among the five types of LTA (types I–V), type I LTA is the most frequently encountered polymer. The inner leaflet of the external mycomembrane in mycobacteria contains mycolic acids which are unique long-chain fatty acids. Owing to high structural diversity displayed by mycolic acids like variations observed in the chain length (60 to 90 carbon atoms), level of unsaturation and the presence of chemical groups, such as ketones and methoxys, they are characterized as important taxonomic markers (Song et al. 2009; Marrakchi et al. 2014). Other lipids like sulfolipid 1 and polyacyltrehalose are specific to *Mycobacterium tuberculosis*, while trehalose polyphleate is specific to non-tuberculosis mycobacteria.

Glycosylinositol-phosphorylceramides (GIPCs) are the class of glycosphingolipids that play important roles in adhesion, signal transduction, modulation of the host immune response and apoptosis and are uniquely found in plant and fungal species. The lipids with conserved core structure made up of ceramide moiety is attached to a glucuronic acid-inositol phosphate group and help in establishing infection in the vertebrate host by filamentous fungi such as *Aspergillus fumigatus*. Membrane lipids in archaebacteria are composed of isoprenoid and hydroisoprenoid hydrocarbons and glycerol isopropyl ether. These lipids are used in the identification and classification of different types of archaebacteria (Tornabene et al. 1980).

16.4.1 Pretreatments and Lipid Extraction

Energy-rich fatty acids/lipids are the preferred raw materials for the biodiesel production, food, nutraceutical, pharmaceutical and cosmetic industries. Robust and cost-effective microbial lipid production system by genetic/metabolic engineering of the strains and establishing the favourable cultivation conditions is a challenge. Following its production, lipid extraction from biological tissues is the most crucial step. It depends upon the appropriate selection of solvents based on the chemical and structural characteristics of lipids. Pretreatments to disrupt the rigid cell wall of the biomass for efficient lipid extraction are done by various physical, chemical, mechanical and enzymatic methods. Physical and mechanical methods include the use of expeller press, bead milling, osmotic shock, grinding with liquid nitrogen, ultrasonication, microwave, high-speed and high-pressure homogenizer, laser, hydrodynamic cavitation, autoclaving and pulsed electric field. Chloroform/methanol (2:1, v/v) solvent system to disrupt the cell wall is examined to give highest yield. The use of cellulolytic enzymes (amylase, cellulase, papain, pectinase, hemicellulose, β-glucosidases, β-1,3-glucomannanase and xylanase) to disintegrate the cell wall is comparatively the cleaner and mild method. Combination of various treatments is suggested to provide higher extraction yield of lipids (Patel et al. 2018).

Choice of solvent(s) for efficient lipid extraction depends on several factors including the nature of lipids to be extracted (polar or non-polar), volatility (for easy separation after extraction), absence of toxic/mutagenic/reactive impurities (to avoid reaction with the lipids), ease of formation of an aqueous two-phase system (to remove non-lipids compounds), cost and health and environment safety concerns. The two classical methods, namely, Folch method (published in 1957) and Bligh and Dyer method (published in 1959), are preferred to extract total lipids from solid tissues and biological fluids of plants and animals, respectively (Folch et al. 1957; Bligh and Dyer 1959). Both the methods are based on the use of chloroform/methanol/water solvent system for lipid extraction with few differences in the protocol (Folch et al. 1957; Bligh and Dyer 1959). Owing to the toxicity of chloroform and methanol, Hara and Radin in 1978 proposed n-hexane/isopropanol (3:2 v/v) solvent to extract lipids from rat or mouse brain tissues, whereas Smedes proposed the use of isopropanol/cyclohexane/water (8:10:11 v/v/v) mixture for lipid extraction from marine tissues (Hara and Radin 1978; Smedes 1999). Much faster single-step procedure for total lipid extraction from green microalgae using chloroform/methanol (2:1, v/v) solvent system followed by addition of 0.73% NaCl water was developed by Axelsson and Gentili (Axelsson and Gentili 2014). Acidification of extraction medium helps disrupt ionic interactions of charged, polar lipids and thus increases the recovery of acidic phospholipids from sample tissues (Retra et al. 2008). Soxhlet extraction method using proper binary solvent system could provide higher yield of lipids but continuous heating and exposure to high temperature could lead to lipid oxidation and degradation of heat-labile compounds. Supercritical CO_2 extraction (SCE) of lipids offers selective extraction of metabolites; however, yield can be increased by adding ethanol as co-solvent. High energy demand and cost are associated with microwave-assisted or ultrasonication-assisted lipid extraction methods. These methods are not suitable for large-scale extraction of lipids (Fig. 16.5).

16.4.2 Lipid Detection and Quantification

Qualitative analysis of lipid components is based on their reactions with types of reagents such as acidic ferric chloride. It detects cholesterol and cholesteryl esters as red to violet spots. Examples of lipid types and the reagents used are reviewed by Fuchs et al. (2011). Conventional methods of lipid quantification include gravimetric assays, transesterification and chromatography coupled with spectroscopy. Following lipid extraction and solvent evaporation, weight of total lipids can be determined gravimetrically. Based on the processing method and sample size ranging from 0.5 to 5 gm or up to 100 mg, this method is categorized as micro- and macro-gravimetric method, respectively. It is a cheaper and faster method but does not provide information regarding the classes of lipids. Most commonly employed

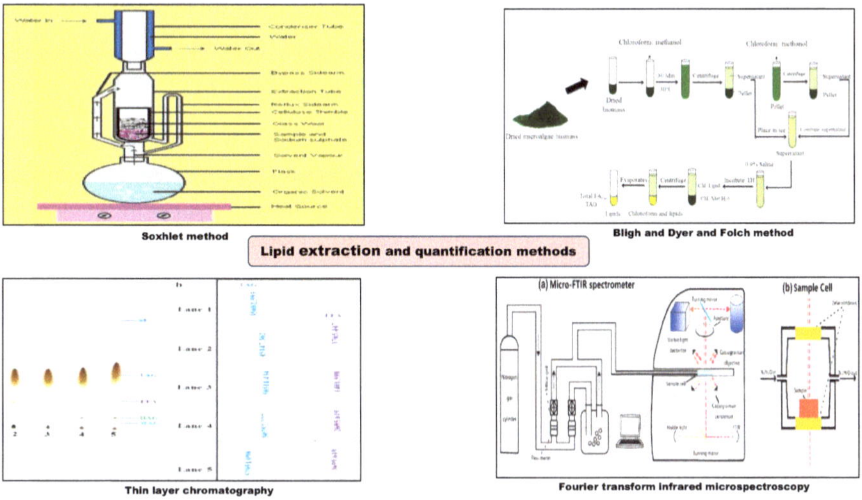

Fig. 16.5 Methods to extract and quantify lipids

chromatographic methods of lipid analysis include TLC, GC and HPLC, used either alone or coupled with MS, MS/MS, FID and TOF.

Commercially available pre-coated chromatoplates made from good-quality absorbent materials, including high-performing silica bonded phases and impregnated layers, greatly improve the reproducibility of lipid separation and characterization. Long-chain triacylglycerols can be separated from short chain (based on mol. wt.) using mixture of petroleum ether and diethyl ether during TLC. Normal-phase TLC containing silicic acid can separate lipids on the basis of degree of saturation. Ease of handling, low cost, quicker analysis of small samples and reproducibility of results are some of the benefits of TLC (Fig. 16.5). Nevertheless, lipid oxidation of particularly unsaturated fatty acids during exposure to atmospheric oxygen is the main disadvantage associated with TLC. Due to excellent sensitivity, molecular specificity and precision, GC/MS and LC/MS are the current gold standard methods for the quantification and analysis of metabolic flux distribution. GC or LC is usually coupled to MS or MS/MS, FID and TOF MS. Various ionization techniques for MS for lipid analysis are fast atom bombardment (FABMS), paper spray ionization, electrospray ionization (ESI) and matrix-assisted laser desorption and ionization. Pyrolysis-GC/MS with existing computational tools yield a clear profile of analyte with a carbon-chain length of C8 to C40. MALDI-TOF MS is a sensitive and faster method for the compositional analysis of intact, non-derivatized polar and non-polar lipids. Ionization of analytes outside the mass spectrophotometer avoids chemical separation of extracts and reduces time for analysis and the need for solvents. HPLC is mostly coupled with ultraviolet (UV)-visible or MS detectors. However, evaporative light scattering detector (ELSD) is identified as more useful detector for lipid quantification than UV-visible and less expensive and easier to use than MS detectors. ELSD allows the simultaneous identification of triacylglycerols as well as neutral lipids. HPLC-ELSD-MS makes possible to quantify all lipid classes while maintains efficacy by reducing material losses associated with purification or derivatization steps required for GC analysis (Jones et al. 2012) (Fig. 16.6a).

Among the spectroscopic methods, Fourier transform infrared (FITR) spectroscopy supported with multivariate analysis is a high-throughput online, fast and non-invasive method for lipid quantification that utilizes small amounts of whole cells for real-time monitoring and screening of cultures (Signori et al. 2016; Laurens

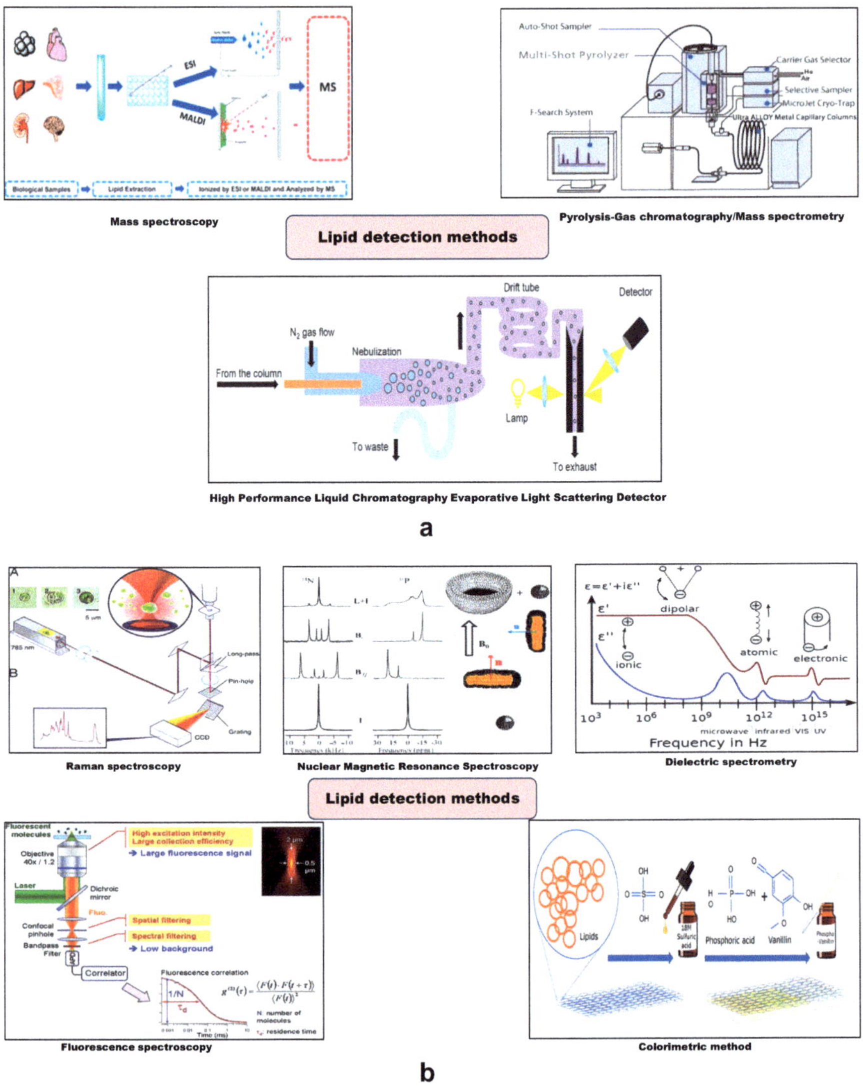

Fig.16.6 (**a**) Methods to detect lipids. (**b**) Methods to detect lipids

et al. 2019). FITR micro-spectroscopy has been used by various biologists to investigate the allocation of carbon (based on a shifting ratio of macromolecules inside the cell) towards lipids in various microalgal species. Nevertheless, signal interference is the main drawback of FTIR to quantify lipids in microbial cells (Fig. 16.5).

NMR spectroscopy is a non-destructive method which requires minimal sample preparation and eliminates the need for chromatographic separation or derivatization of the sample to determine the composition of complex biological samples. Liquid or solid samples can be analysed by liquid- or solid-state NMR. ^{1}HNMR can quantify different lipid classes such as steroids and some pigments (carotenoids and chlorophylls) with non-overlapping peaks.^{13}CNMR can also be efficiently used for the quantification of different lipid components with minimal risk of chemical alterations. Although initial investment in NMR is higher, it requires five times lesser time to analyse compared to GC/MS at the same operating cost. Semi-solid method of NMR called high-resolution magic angle spinning (HR-MAS) when used on intact biological sample spun at low frequencies at a 54.7° angle relative to the magnetic field results in well-resolved spectra with narrow peaks (Andrew 1981).

Time domain (TD) NMR is based on different relaxation times of hydrogen nuclei in various phases of biological molecule. Hydrogen nuclei signal from lipids is separated from the signals derived from other compounds by applying a spin-echo NMR pulse sequence. Thus, lipid can be quantified using appropriate calibration; however, it requires large amount of sample (up to 1 gram of lyophilized biomass). Besides it cannot differentiate between neutral lipids (i.e., TAGs) and polar or membrane lipids. NMR spectroscopic methods are applied to various microorganisms for the efficient detection of different classes of TAGs, free fatty acids, unsaturated and saturated fatty acids (Fig. 16.6b).

Raman spectroscopy is a rapid, non-destructive valuable method for in vivo lipid biomonitoring (degree of lipid unsaturation). It is based on inelastic scattering of monochromatic light produced by a laser in the visible, near infrared or UV range (Wei et al. 2014). Due to overlap of strong Raman signal of fluorescent pigments and carotenoids with lipid peaks, it interferes with lipid detection and quantification. Hence, cosmic noise removal and fluorescence background subtraction is necessary during the post-processing of Raman signal for spectral analysis. Further, photo-damage of specific samples by the use of significantly higher excitation laser power to produce distinguishable signals and a long integration time (100 ms—1 s per pixel) limits its application in living systems (Fig. 16.6b).

Fluorescence spectroscopic method of detection of neutral lipids in biological samples depends upon the emission of fluorescent signals upon interaction of lipophilic fluorescent dye molecules such as BODIPY 505/515 and Nile Red, with lipids at proper excitation and emission wavelengths. Nile Red (9-diethylamino-5H-benzo[α]phenoxazine-5-one) interaction with neutral lipids in shorter emission wavelengths (< 590 nm) and with polar lipids (> 590 nm) helps to differentiate the two types of lipids (Huang et al. 2009). Uptake of the dye depends upon the cell viability and some of the cell treatment approaches that increase its permeability across thick walls including microwaves, ultrasonic treatment, EDTA and mild cell pre-bleaching. BODIPY 505/515

(4,4-difluoro-1,3,5,7-tetramethyl-4-bora-3a,4a-diaza-s-indacene), a highly lipophilic green fluorophore, can stain a variety of cells irrespective of cell wall properties and has higher fluorescent quantum yield, a narrower emission spectrum, higher sensitivity and better reproducibility compared to Nile Red. Nevertheless, non-linear response between intensity and oil content limits the use of fluorescence spectroscopy.

Dielectric spectrometry is an automated, rapid, non-invasive and label-free method to determine the composition of lipids in a cell. This method is based on the differences in permittivity and conductivity of cell components. It measures the dielectric properties of a medium as a function of frequency which is sensitive to cellular lipid content but is not affected by changes in cell concentration (Fig. 16.6b).

Colorimetric method of lipid quantification although is a method of choice due to its fast response and simplified sample handling but the preliminary steps require cell disruption and lipid extraction. The sulfo-phospho-vanillin (SPV) method was developed for the quantification of serum total lipids. Quantification of lipids by SPV method is based on (1) reaction of phosphoric acid with vanillin to form phosphate esters with higher reactivity of the carbonyl group (known as the SPV reagent); (2) reaction of unsaturated lipids with concentrated sulfuric acid (≥95%) at 90–100 °C to form carbonium ion); and (3) reaction of lipid products with SPV reagent (reaction of carbonium ion with the carbonyl group of a phosphovanillin ester) which leads to the formation of a stable pink chromophore compound followed by measuring its absorbance at 530 nm (Knight et al. 1972). Owing to the acid-thermal degradation of lipophilic proteins and pigments, the SPV assay quantifies only lipids (Fig. 16.6b).

Among the many conventional and non-conventional methods available for lipid detection and quantification, the selection of a method is dictated by the suitability of source sample, quantity and stage of sample (dry/wet), desired level of identification, types of fatty acids, their length, level of saturation/unsaturation and finally the accessibility of equipment and resources in the laboratory.

16.5 Identification and Characterization of Nucleic Acids

Ribonucleic acid (RNA) and deoxyribonucleic acid (DNA) are the two classes of nucleic acid which are present essentially in all the living forms of life. These biopolymers are made up of monomeric components which include 5-carbon sugar, a phosphate group and a nitrogenous base. Nucleotides in specific sequence are bonded to form a spiral backbone and are assembled into either chains of bases or base pairs. They encode and store genetic information in the form of nucleotide sequence and play crucial role in directing protein synthesis. Innovative developments in diagnostic tools, molecular tests and transgenics greatly depend upon the successful extraction and amplification of nucleic acids.

Stability of nucleic acids during sample collection, preparation, transport and storage is a pre-requisite for obtaining pure and higher yields of intact nucleic acid following its extraction. The widely used phenol-chloroform extraction method separates the upper (aqueous) phase containing the DNA and the lower (organic) phase containing denatured proteins followed by precipitation and washing of DNA with 70% ethanol and dissolution in Tris-EDTA buffer or nuclease free water. Single-step method of RNA extraction is based on the use of guanidinium isothiocyanate. It separates the upper phase containing the total RNA from the lower phase containing DNA and proteins followed by recovery of RNA by precipitation with isopropanol and treatment with DNAase. Density gradient centrifugation method using cesium chloride (CsCl)/ethidium bromide (EtBr) is employed to separate and obtain the high yield of various DNAs as independent bands. The separation is based on the differences in density between the cesium ion and water and intercalation of EtBr. Owing to the requirement of expensive ultracentrifuge and considerable time, extraction using ultracentrifuge is difficult to perform, and the use of EtBr in large amounts is harmful (Price et al. 2009; Tan and Yiap 2009) (Fig. 16.7).

Solid-phase extraction method utilizes an insoluble positively charged siliceous core particle such as silica matrices, glass particles, diatomaceous earth and anion-exchange carriers rather than liquid phenol (Price et al. 2009). This method is safer with reduced cross contamination and DNA loss and incorporated into many commercial kits. The commercial method is based on the use of spin column operated by centrifugal force that allows DNA to immobilize/adsorb in the presence of a chaotropic salt solution followed by its washing and elution in TE buffer. This technique facilitates rapid extraction and purification of rRNA as well as single-stranded and double-stranded DNA from various types of clinical specimens. Simple and convenient method of DNA extraction known as magnetic bead method such as miniMag (bioMerieux) and MagNA Pure (Roche) requires only small amount of specimen and eliminates the need for repeated centrifugation, vacuum filtration, column separation and use of organic solvents. Negatively charged beads

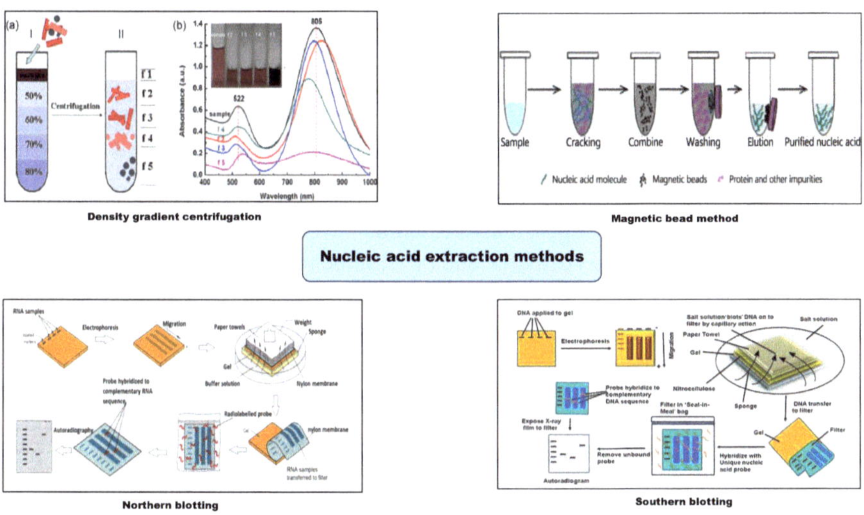

Fig. 16.7 Methods to extract and characterize nucleic acids

selectively bind with proteins and cellular debris and leave isolated DNA in solution (Price et al. 2009; Tan and Yiap 2009) (Fig. 16.7).

Nucleic acid hybridization is based on the base pairing between complementary strands of RNA and DNA, two strands of DNA or two strands of RNA and thus provides a means for detection of DNA or RNA sequences. Specific nucleotide sequence can be detected in total cell DNA/RNA by hybridization with a radiolabelled or fluorescent tagged nucleic acid probe. Southern blotting developed by E.M. Southern is used for the detection of nucleic acid sequences. It is based upon the separation of restriction endonuclease fragments of DNAby gel electrophoresis, transfer to nitrocellulose/nylon membrane and denaturation of one of the strands followed by incubation with radiolabelled probe for its hybridization to complementary sequence. Northern blotting, variant of Southern blotting technique, detects RNA instead of DNA and is frequently used in the studies of gene expression. In situ hybridization utilizes radioactive or fluorescent probes to specific cells/subcellular structures/intact chromosomes and detects chromosomal regions that contain a gene of interest or specific mRNAs in different types of cells within a tissue (Fig. 16.7).

Amplification of DNA (or RNA after reverse transcription) by polymerase chain reaction may provide 100,000 copies for its efficient detection as compared to other methods including blotting. Specificity of amplification is based on the hybridization of oligonucleotide primers to complementary sequences on the template molecule. Complementary or cDNA obtained by the reverse transcription of either total RNA or mRNA using reverse transcriptase is used as a template for real-time quantitative PCR. Quantity of cDNA/template is identified during real-time PCR by using a variety of fluorescent chemistries. Fluorescent signals are generated by either using hydrolysis probes such as TaqMan® probes or a double-stranded DNA binding dye such as SYBR® Green dye (Fig. 16.8a).

16.5.1 Low- and High-Throughput Methods to Sequence Nucleic Acids

Elucidation of nucleic acid sequence either by traditional low-throughput first-generation sequencing technique called Sanger sequencing or by high-throughput sequencing (HTS) makes it possible to examine the cellular genomic and transcriptomic signatures of various diseases and developmental stages. Sanger sequencing/chain termination method determines the nucleotide sequence by the selective incorporation of chain-terminators (dideoxynucleotides (ddNTPs)) by DNA polymerase during in vitro replication. The PCR-amplified DNA products are checked and read by electrophoresing them on denaturing PAGE. Later massively parallel or next-generation or second-generation sequencing (NGS) methods allow the entire genome to be sequenced once. Among the various HTS platforms, Illumina platforms dominate the sequencing industry. Illumina's bridge amplification method is based on the generation of small "clusters" on an Illumina flow cell followed by creation of multiple primer hybridization steps and thus multiple sequencing start

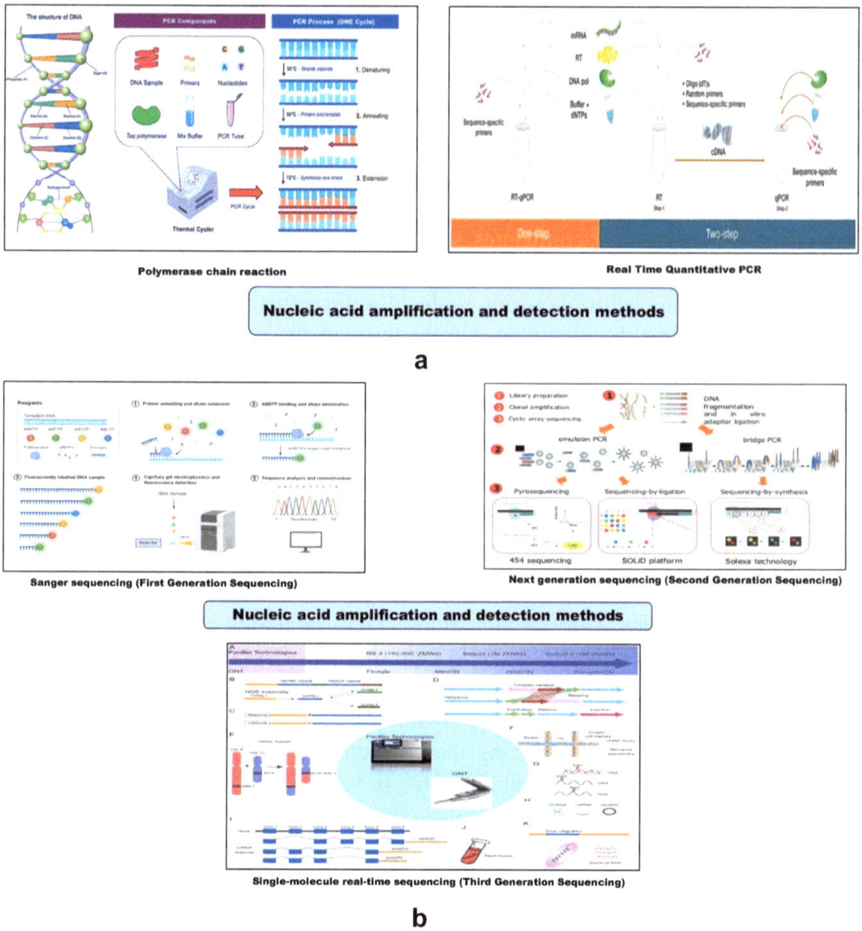

Fig. 16.8 (**a**) Methods to amplify and detect nucleic acids. (**b**) Sequencing methods to amplify and detect nucleic acids

points. This facilitates the paired-end sequencing, i.e., sequencing of both ends of the original template molecule. Sequencing information from paired-end reads help in identifying splice variants in RNA-seq and large structural variants like inversions. Overall, substitution error rate of 0.11% is observed using Illumina platforms (Minoche et al. 2011). PCR amplifies DNA within an emulsified droplet in Ion Torrent and 454 methods. Light and hydrogen ions detect each nucleotide incorporation event during sequencing in 454 (pyrosequencing) and Ion Torrent, respectively. Since the nucleotides used lack a blocking moiety, the entire homopolymer is incorporated during one cycle. This makes sequencing of homopolymers difficult by Ion Torrent and 454 methods (Fig. 16.8b).

The Pacific Biosciences Real-Time Sequencer (PacBio RS) allows single-molecule real-time sequencing (SMRT) also known as third generation or long read sequencing. It uses unmodified polymerase attached to the bottom of zero-mode wave-guides (ZMWs)—small wells along with the freely flowing fluorescently labelled nucleotides in the solution. Construction of well allows the detection of fluorescence occurring only by the bottom of well. The RS uses video imaging of fluorescent nucleotides and records the event when fluorescent label gets detached from the nucleotide upon its incorporation into the DNA strand (Schadt et al. 2010). This method also detects DNA modifications like cytosine methylations. Nanopore sequencers developed by companies such as Oxford Nanopore and NABsys are massively parallel sequencers and do not base on sequencing by synthesis. This method is based on electrophoresing molecules through a pore of a membrane followed by measuring the electrical current through the pore as molecules passes through it (Thompson and Oliver 2012). Characterization of the current of a pore over time helps to determine the molecule that traverses the pore as well as its order. This technology might be useful in sequencing of unmodified miRNA, mRNA and protein molecules (Fig. 16.8b).

16.6 Conclusions

Complex biomolecules or organic macromolecules including carbohydrates, lipids, proteins and nucleic acids (RNA, DNA) are the important constituents of cellular organelles. Owing to their vital role in fundamental biological processes such as cell division, development, morphogenesis, maintenance and metabolic activities of different tissues in the body, deducing their structure and properties would help in better understanding of functional attributes in various cellular and metabolic processes. Developments and advancements in the technology to characterize and identify the biomolecules not only reduce the cost but also the processing time exponentially. Advanced preparative and analytical approaches for separation, identification and quantification of biomolecules address many challenging biological issues related to the identification of potential therapeutic and diagnostic biomarkers for personalized treatment, drug discovery and development and food processing operations/industries to improve nutritional composition. Their invaluable contribution in numerous fields such as medical diagnosis, virology, forensic biology, biotechnology, anthropology and biological systematic may improve the understanding of the biological processes.

Conflict of Interests The authors disclose no potential conflict of interests.

References

Akin O, Temelli F, Koseoğlu S (2012) Membrane applications in functional foods and nutraceuticals. Crit Rev Food Sci Nutr 52(4):347–371

Aktas EG, Acar MH (2019) Polysulfone-based amphiphilic copolymers: effect of hydrophilic content on morphology and performance of ultrafiltration membranes. J Appl Polym Sci 137(4):48306

Andrew ER (1981) Magic angle spinning in solid state n.m.r. spectroscopy. Philos Transac Royal Soc A Math Phys Sci 299:505–520
Axelsson M, Gentili F (2014) A single-step method for rapid extraction of total lipids from green microalgae. PLoS One 9(2):e89643
Balashova N, Giannakakis A, Brown AC, Koufos E, Benz R et al (2018) Generation of a recombinant Aggregatibacter actinomycetemcomitans RTX toxin in Escherichia coli. Gene 672:106–114
Bell DJ et al (1982) In: Fiechter A (ed) Advances in biochemical engineering/biotechnology. Springer Verlag, Berlin/Heidelberg/New York, p 117
Benedek K, Dong S, Karger BL (1984) J Chromatography A 317:227–243. https://doi.org/10.1016/s0021-9673(01)91662-0
Bligh EG, Dyer WJ (1959) A rapid method of total lipid extraction and purification. Can J Biochem Physiol 37(8):911–917
Burgess RR (2018) A brief practical review of size exclusion chromatography: rules of thumb, limitations, and troubleshooting. Protein Expr Purif 150:81–85
Caltabiano AM, Foley JP, Striegel AM (2018) Aqueous size-exclusion chromatography of polyelectrolytes on reversed-phase and hydrophilic interaction chromatography columns. J Chromatogr A 1532:161–174
Chon JH, Zarbis-Papastoitsis G (2011) Advances in the production and downstream processing of antibodies. Nat Biotechnol 28:458–463
Ciric J, Oostland J, de Vries JW, Woortman AJ, Loos K (2012) Size exclusion chromatography with multi detection in combination with matrix-assisted laser desorption ionization-time-of-flight mass spectrometry as a tool for unraveling the mechanism of the enzymatic polymerization of polysaccharides. Anal Chem 84(23):10463–10470
Cohen SA, Benedek K, Tapuhi Y, Ford JC, Karger BL (1985) Conformational effects in the reversed-phase liquid chromatography of ribonuclease A. Anal Biochem 144:275–284
Cordes A, Kula MR (1986) Process design for large-scale purification of formate dehydrogenase from Candida boidinii by affinity partition. J Chromatogr B Biomed Appl 376:375–384
Davies KM et al (2020) The evolution of flavonoid biosynthesis: a bryophyte perspective. Front Plant Sci 11:7
Desfontaine V, Losacco GL, Gagnebin Y, Pezzatti J, Farrell WP (2018) Applicability of supercritical fluid chromatography—mass spectrometry to metabolomics. I—optimization of separation conditions for the simultaneous analysis of hydrophilic and lipophilic substances. J Chromatogr A 1562:96–107
Domingues DS et al (2014) Detection of roasted and ground coffee adulteration by HPLC and by amperometric and by post-column derivatization UV-vis detection. Food Chem 146:353–362
Doyen A, Beaulieu L, Saucier L, Pouliot Y, Bazinet L et al (2011) Impact of ultrafiltration membrane material on peptide separation from a snow crab byproduct hydrolysate by electrodialysis with ultrafiltration membranes. J Agric Food Chem 59(5):1784–1792
Farsad-Naeimi A et al (2020) Sugar consumption, sugar sweetened beverages and attention deficit hyperactivity disorder: a systematic review and meta-analysis. Complement Ther Med 53:102512
Folch J, Lees M, Sloane Stanley GH (1957) A simple method for the isolation and purification of total lipides from animal tissues. J Biol Chem 226(1):497–509
Fuchs B, Süss R, Teuber K, Eibisch M, Schiller J (2011) Lipid analysis by thin-layer chromatography—a review of the current state. J Chromatogr A 1218(19):2754–2774
Gray CJ et al (2019) Advancing solutions to the carbohydrate sequencing challenge. J Am Chem Soc 141(37):14463–14479
Hara A, Radin NS (1978) Lipid extraction of tissues with a low-toxicity solvent. Anal Biochem 90(1):420–426
Hermodson M, Mahoney WC (1983) Separation of peptides by reversed-phase high-performance liquid chromatography. Methods Enzymol 91:352–359
Hofmann J, Hahm HS, Seeberger PH, Pagel K (2015) Identification of carbohydrate anomers using ion mobility-mass spectrometry. Nature 526(7572):241–244

Huang GH, Chen G, Chen F (2009) Rapid screening method for lipid production in alga based on Nile red fluorescence. Biomass Bioenergy 33(10):1386–1392
Huang K, Cui M, Ye F, Li Y, Zhang D (2016) Global profiling of the signaling network of papillary thyroid carcinoma. Life Sci 147:9–14
Hui W et al (2013) Aberrant expression of signaling proteins in essential thrombocythemia. Ann Hematol 92(9):1229–1238
Jane G Muir JG, Rose R, Rosella O, Liels K, Barrett JS, Shepherd SJ, Gibson PR (2009). Measurement of short-chain carbohydrates in common Australian vegetables and fruits by high-performance liquid chromatography (HPLC). J Agric Food Chem. https://doi.org/10.1021/jf802700e
Jawhar NM (2009) Tissue microarray: a rapidly evolving diagnostic and research tool. Annu Saudi Med 29(2):123–127
Jones J, Manning S, Montoya M, Keller K, Poenie M (2012) Extraction of algal lipids and their analysis by HPLC and mass spectrometry. J Am Oil Chem Soc 89:1371–1381
Kallberg K, Johansson HO, Bulow L (2012) Multimodal chromatography: an efficient tool in downstream processing of proteins. Biotechnol J 7(12):1485–1495
Kim MK et al (2016) Differential protein expression in small intestinal neuroendocrine tumors and liver metastases. Pancreas 45(4):528–532
Knight JA, Anderson S, Rawle JM (1972) Chemical basis of the sulfo-phospho-vanillin reaction for estimating total serum lipids. Clin Chem 18(3):199–202
Laurens L et al (2019) Solvent-free spectroscopic method for high-throughput, quantitative screening of fatty acids in yeast biomass. Anal Methods 11(1):58–69
Li X, Wang W, Chen J (2017) Recent progress in mass spectrometry proteomics for biomedical research. Sci China Life Sci 60:1093–1113
Liu B et al (2014) Structural diversity in salmonella O antigens and its genetic basis. FEMS Microbiol Rev 38(1):56–89
Losacco GL et al (2020) Applicability of supercritical fluid chromatography-mass spectrometry to metabolomics. II-assessment of a comprehensive library of metabolites and evaluation of biological matrices. J Chromatogr A 1620:461021
Macklin A, Khan S, Kislinger T (2020) Recent advances in mass spectrometry based clinical proteomics: applications to cancer research. Clin Proteomics 17:17
Manjare SD, Dhingra K (2019) Supercritical fluids in separation and purification: a review. Mater Sci Energy Technol 2(3):463–484
Marichal-Gallardo PA, Alvarez MM (2012) State-of-the-art in downstream processing of monoclonal antibodies: process trends in design and validation. Biotechnol Prog 28(4):899–916
Mariño K, Bones J, Kattla JJ, Rudd PM (2010) A systematic approach to protein glycosylation analysis: a path through the maze. Nat Chem Biol 6(10):713–723
Marrakchi H, Lanéelle M-A, Daffé M (2014) Mycolic acids: structures, biosynthesis, and beyond. Chem Biol. https://doi.org/10.1016/j.chembiol.2013.11.011
Minoche AE, Dohm JC, Himmelbauer H (2011) Evaluation of genomic high-throughput sequencing data generated on Illumina HiSeq and genome analyzer systems. Genome Biol 12(11):R112
Morak M et al (2009) Differential activity-based gel electrophoresis for comparative analysis of lipolytic and esterolytic activities. J Lipid Res 50(7):1281–1292
Nilsson T et al (2010) Mass spectrometry in high-throughput proteomics: ready for the big time. Nat Methods 7(9):681–685
Patel A, Mikes F, Matsakas L (2018) An overview of current pretreatment methods used to improve lipid extraction from oleaginous micro-organisms. Molecules (Basel, Switzerland) 23(7):1562
Pieper K, Winter J, Heinzel-Gutenbrunner M, Margraf-Stiksrud J (2019) Association between a new sugar index and caries experience: results of a cross-sectional field study. Caries Res 53(2):160–167
Price CW, Leslie DC, Landers JP (2009) Nucleic acid extraction techniques and application to the microchip. Lab Chip 9(17):2484–2494
Rajesh S, Jayalakshmi A, Senthilkumar S, Sankar HSH, Mohan DR (2011) Performance evaluation of poly(amide-imide) incorporated cellulose acetate ultrafiltration membranes in

the separation of proteins and its fouling propensity by AFM imaging. Ind Eng Chem Res 50(24):14016–14029

Rakete S, Glomb MA (2013) A novel approach for the quantitation of carbohydrates in mash, wort, and beer with RP-HPLC using 1-naphthylamine for precolumn derivatization. J Agric Food Chem 61(16):3828–3833

Ras M, Lefebvre D, Derlon N, Paul E, Girbal-Neuhauser E (2011) Extracellular polymeric substances diversity of biofilms grown under contrasted environmental conditions. Water Res 45(4):1529–1538

Resende A, Catarino S, Vitor G, de Pinho M (2013) Separation and purification by ultrafiltration of white wine high molecular weight polysaccharides. Ind Eng Chem Res 52:8875–8879

Retra K, Bleijerveld OB, van Gestel RA, Tielens AG, van Hellemond JJ (2008) A simple and universal method for the separation and identification of phospholipid molecular species. Rapid Commun Mass Spectrom 22(12):1853–1862

Rosa PA, Azevedo AM, Sommerfeld S, Bäcker W, Aires-Barros MR (2011) Aqueous two-phase extraction as a platform in the biomanufacturing industry: economical and environmental sustainability. Biotechnol Adv 29(6):559–567

Rudd PM, Elliott T, Cresswell P, Wilson IA, Dwek RA (2001) Glycosylation and the immune system. Science 291(5512):2370–2376

Rühmann B, Schmid J, Sieber V (2014) Fast carbohydrate analysis via liquid chromatography coupled with ultra violet and electrospray ionization ion trap detection in 96-well format. J Chromatogr A 1350:44–50

Ruiz-Matute AI, Hernández-Hernández O, Rodríguez-Sánchez S, Sanz ML, Martínez-Castro I (2011) Derivatization of carbohydrates for GC and GC-MS analyses. J Chromatogr B Anal Technol Biomed Life Sci 879(17–18):1226–1240

Schadt EE, Turner S, Kasarskis A (2010) A window into third-generation sequencing. Hum Mol Genet 19(R2):R227–R240

Sechi S, Oda Y (2003) Quantitative proteomics using mass spectrometry. Curr Opin Chem Biol 7(1):70–77

Signori L et al (2016) Assessing an effective feeding strategy to optimize crude glycerol utilization as sustainable carbon source for lipid accumulation in oleaginous yeasts. Microb Cell Factories 15:75

Smedes F (1999) Determination of total lipid using non-chlorinated solvents. Analyst 124:1711–1718

Sohlenkamp C, Geiger O (2016) Bacterial membrane lipids: diversity in structures and pathways. FEMS Microbiol Rev 40(1):133–159

Song SH, Park KU, Lee JH, Kim EC, Kim JQ, Song J.J (2009) Electrospray ionization-tandem mass spectrometry analysis of the mycolic acid profiles for the identification of common clinical isolates of mycobacterial species. Microbiol Methods. https://doi.org/10.1016/j.mimet.2009.01.023

Tan SC, Yiap BC (2009) DNA, RNA, and protein extraction: the past and the present. J Biomed Biotechnol 2009:574398

Thammasena R, Fu CW, Liu JH, Liu DC (2020) Evaluation of nutrient content, physicochemical and functional properties of desalted duck egg white by ultrafiltration as desalination. Anim Sci J 91(1):e13339

Thompson JF, Oliver JS (2012) Mapping and sequencing DNA using nanopores and nanodetectors. Electrophoresis 33(23):3429–3436

Tornabene TG, Lloyd RE, Holzer G, Oro J (1980) Lipids as a principle for the identification of archaebacteria. Life Sci Space Res 18:109–121

van de Velde B, Guillarme D, Kohler I (2020) Supercritical fluid chromatography—mass spectrometry in metabolomics: past, present, and future perspectives. J Chromatogr B 1161:122444

Vidya B et al (2020) Purification and characterization of β-galactosidase from newly isolated aspergillus terreus (KUBCF1306) and evaluating its efficacy on breast cancer cell line (MCF-7). Bioorg Chem 94:103442

Wei X, Jie D, Cuello JJ, Johnson DJ, Qiu Z et al (2014) Microalgal detection by Raman microspectroscopy. TrAC Trend Anal Chem 53:33–40
Xing DY, Chan SY, Chung TS (2012) Molecular interactions between polybenzimidazole and [EMIM] OAc, and derived ultrafiltration membranes for protein separation. Green Chem 14(5)
Xu G, Amicucci MJ, Cheng Z, Galermo AG, Lebrilla CB (2017) Revisiting monosaccharide analysis—quantitation of a comprehensive set of monosaccharides using dynamic multiple reaction monitoring. Analyst 143(1):200–207
Zhang DY, Ye F, Gao L, Liu X, Zhao X et al (2009) Proteomics, pathway array and signaling network-based medicine in cancer. Cell Div 4:20
Zhou B, Zhang M, Fang ZX, Liu Y (2015) Effects of ultrasound and microwave pretreatments on the ultrafiltration desalination of salted duck egg white protein. Food Bioprod Process 96:306–313

Correction to: Free Radicals and Their Relation to Diseases and Protection Against Them

Saheem Ahmad, Hamda Khan, Zeeshan Rafi,
Uzma Shahab, Jalaluddin Mohammad Ashraf,
Mohammad Kaleem Ahmad, Kirtanjot Kaur,
Ramendra Pati Pandey, Safia Habib, and Moinuddin

Correction to:
Chapter 13 in: R. L. Singh et al. (eds.), *Clinical Applications of Biomolecules in Disease Diagnosis*,
https://doi.org/10.1007/978-981-97-4723-8_13

The original version of this book was inadvertently published with an incorrect affiliation for the authors Hamda Khan, Safia Habib, Moinuddin, and Zeeshan Rafi in Chapter 13. The correct affiliation should be read as:

Hamda Khan, Safia Habib and Moinuddin
Department of Biochemistry, J.N. Medical College, Aligarh Muslim University, Aligarh, India

Zeeshan Rafi
Department of Bioengineering, Integral University, Lucknow, Uttar Pradesh, India

The affiliation of the chapter authors has been updated in the original version with this erratum.

The updated version of this chapter can be found at
https://doi.org/10.1007/978-981-97-4723-8_13

R. L. Singh et al. (eds.), *Clinical Applications of Biomolecules in Disease Diagnosis*, https://doi.org/10.1007/978-981-97-4723-8_17

MIX
Papier aus verantwortungsvollen Quellen
Paper from responsible sources
FSC® C105338

If you have any concerns about our products,
you can contact us on
ProductSafety@springernature.com

In case Publisher is established outside the EU,
the EU authorized representative is:
Springer Nature Customer Service Center GmbH
Europaplatz 3, 69115 Heidelberg, Germany

Printed by Libri Plureos GmbH
in Hamburg, Germany